Electronics texts for engineers and scientists

Editors: H. Ahmed
Reader in Microelectronics
Department of Physics, University of Cambridge
and
P. J. Spreadbury
Lecturer in Engineering, University of Cambridge

Transmission and propagation of electromagnetic waves

Transmission and propagation of electromagnetic waves

SECOND EDITION

K. F. SANDER
Professor of Electronic Engineering
University of Bristol

G. A. L. REED
Lecturer in Electronic Engineering
University of Bristol

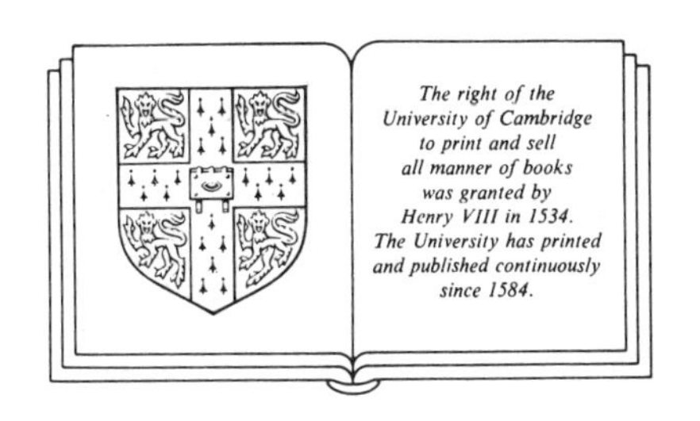

CAMBRIDGE UNIVERSITY PRESS
CAMBRIDGE
LONDON NEW YORK NEW ROCHELLE
MELBOURNE SYDNEY

Published by the Press Syndicate of the University of Cambridge
The Pitt Building, Trumpington Street, Cambridge CB2 1RP
32 East 57th Street, New York, NY 10022, USA
10 Stamford Road, Oakleigh, Melbourne 3166, Australia

First published 1978
Second edition 1986

Printed in Great Britain by J. W. Arrowsmith Ltd., Bristol

British Library cataloguing in publication data

Sander, K. F.
Transmission and propagation of electromagnetic
waves. – 2nd ed. – (Electronics texts for
engineers and scientists)

1. Electromagnetic waves – Transmission
I. Title II. Reed, G. A. L. III. Series
530.1'41 QC665.T7

Library of Congress cataloguing-in-publication data

Sander, K. F. (Kenneth Frederick)
Transmission and propagation of electromagnetic waves.

(Electronic texts for engineers and scientists)
Bibliography: p.
Includes index.
1. Electromagnetic waves – Transmission.
I. Reed, G. A. L. (Geoffrey Alexander Leslie)
II. Title. III. Series.
QC665.T7S26 1986 537 86–2322

2nd edition ISBN 0 521 32409 2 hard covers
ISBN 0 521 31192 6 paperback

1st edition ISBN 0 521 21924 8 hard covers
ISBN 0 521 29312 X paperback

AS

Contents

Preface to first edition

It may be felt, with good reason, that there is hardly room for another book on electromagnetic waves, and therefore some justification for presenting such a book is called for. No excuse is needed for regarding electromagnetic waves as fundamental, and a subject with which engineering undergraduates should be acquainted. Modern communications systems, in the widest sense, depend on the propagation and transmission of electromagnetic waves and without an understanding of the theory it is not possible to appreciate why systems have developed in the way that they have. The object of this book is to provide the knowledge of electromagnetic waves necessary for such an understanding, and subsequently to apply this theory to line, waveguide and radio systems. In a sense therefore later chapters provide examples of the theory contained in the earlier ones.

The theory of plane waves is dealt with in chapter 1, and the theory of transmission lines and waveguides in chapter 3. Chapter 4 deals with radiation from a source. The application of the theory of these chapters is contained in chapters 6, 7 and 8 dealing with line systems, waveguide systems and radio systems respectively. Chapter 5 covers standard transmission line theory, which is essentially about waves in one dimension and applicable to waveguides and plane waves as well as to transmission lines proper. It has been the object of the authors not to present 'rules' without explanation or derivation.

It is assumed that the reader has previously become acquainted with Maxwell's equations, through a treatment such as that in C. W. Oatley's *Electric and magnetic fields* of this series. A summary of vector formulae is provided as an appendix, but an acquaintance with vector methods is presumed. Certain sections are marked with a dagger. This indicates that they may be omitted on a first reading, either because the content is peripheral to the main stream, or that the mathematics required is of a somewhat higher standard than elsewhere. At the end of each chapter a summary of the principal points is given, together with formulae specifically useful in other chapters.

In a book of this nature it is natural that much of the contents is derived from other literature, by selection, condensation and assembly. This applies particularly to the sections closely related to modern practice. The same literature also provides further reading on the topics, so that both acknowledgments for technical material and suggestions for further reading have been placed in a postscript.

It is right and proper however to record here our appreciation to those who have played a part in the forming of this book. It is not feasible to mention them all by name but in particular we wish to thank Dr J. Aitken for his help with the section on microstrip line and Drs R. F. Burbidge and J. A. Jones for their general comments. Finally, with a great sense of debt, we thank Mrs A. Tyler for her unfailing and unflagging efforts to cope with the typing of the manuscript, with its seemingly ceaseless revisions.

Preface to second edition

Apart from the correction of errors the second edition incorporates an additional chapter introducing topics associated with optical fibres. In conformity with the title of the book this material is concerned with propagation and not with optical communication systems. It is once again a pleasure to thank Mrs A. R. Tyler for her care in typing the manuscript.

1

Plane electromagnetic waves

The transmission by electrical means of any sort of information involves propagation of energy by electromagnetic waves, and thus the study of such waves is fundamental to the subject matter of this book. Unfortunately the theory of electromagnetic waves is relatively complicated compared to the theory of simple wave motion in which only a scalar variable such as pressure is involved. Electromagnetic waves concern the way in which vector quantities, the electric field and the magnetic field, vary in space and time. It is thus necessary to set up vector equations relating these quantities. The equations, known as *Maxwell's equations* after the theory's originator, are completely general and although some very general results can be derived, such as showing that the energy in an electromagnetic field travels at a certain speed, it is necessary to consider particular situations. These are chosen on account of mathematical simplicity and for the approximation they provide to real situations in which we are interested. Apart from time there are three space variables: in Cartesian co-ordinates, x, y, z. The simplest case to consider is when variations occur in only one of these variables, and phenomena arising in this situation are the concern of the present chapter. It is evident that solutions under this restriction cannot correspond completely to physically realisable situations, but it is found that these solutions provide an insight into many physical phenomena. The derivation of such solutions is therefore profitable as well as being mathematically the most simple.

The following sections will discuss the existence and characterisation of such waves in free space (or an unbounded dielectric medium) and the effect of conduction in the medium. Attention will then turn to phenomena which occur when, instead of a single medium, two different media are present separated by a plane interface. The laws of reflection and refraction follow.

1.1 Maxwell's equations

The electric and magnetic fields in a time-varying situation are related to each other by Maxwell's equations. (For the derivation and justification of these equations, see Oatley's *Electric and magnetic fields*, chapters 9 and 10.)

The equations are

$$\text{curl}\, \boldsymbol{E} = -\partial \boldsymbol{B}/\partial t \tag{1.1}$$

$$\text{curl}\, \boldsymbol{H} = \partial \boldsymbol{D}/\partial t + \boldsymbol{J} \tag{1.2}$$

$$\text{div}\, \boldsymbol{B} = 0 \tag{1.3}$$

$$\text{div}\, \boldsymbol{D} = \rho \tag{1.4}$$

In these equations, $\boldsymbol{D}$ is the *electric displacement*, $\boldsymbol{E}$ the *electric field strength*, $\boldsymbol{B}$ is the *magnetic flux density* and $\boldsymbol{H}$ the *magnetic field strength*. All are functions of position and time. This functional dependence may be shown explicitly as, for example,

$$\boldsymbol{E}(x, y, z, t) \quad \text{or} \quad \boldsymbol{E}(\boldsymbol{r}, t)$$

The remaining symbols are ρ, the *electric charge density*, and $\boldsymbol{J}$, the *electric current density*. This latter quantity has the significance that the current crossing a vector area $\mathrm{d}\boldsymbol{S}$ (fig. 1.1) is $\boldsymbol{J} \cdot \mathrm{d}\boldsymbol{S}$. Since current is rate of flow of charge, a relation exists between $\boldsymbol{J}$ and ρ, namely the equation of conservation of charge:

$$\text{div}\, \boldsymbol{J} + \frac{\partial \rho}{\partial t} = 0 \tag{1.5}$$

It is in fact implied by (1.2) and (1.4). Taking the divergence of (1.2) we have

$$0 = \frac{\partial}{\partial t}(\text{div}\, \boldsymbol{D}) + \text{div}\, \boldsymbol{J}$$

Substitution for div $\boldsymbol{D}$ from (1.4) then yields (1.5).

The current density $\boldsymbol{J}$ in (1.2) may include two terms; one, $\boldsymbol{J}_s$, associated with the actual source of electromagnetic disturbance and the

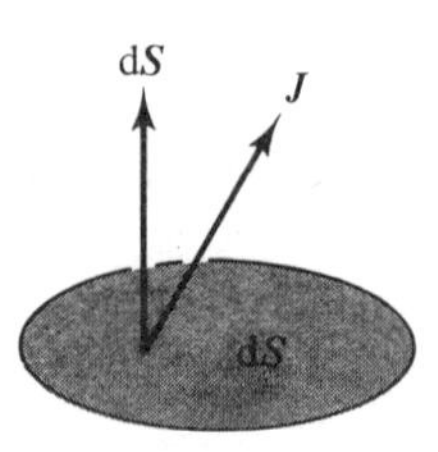

Fig. 1.1. The significance of the current density vector $\boldsymbol{J}$.

other, $\boldsymbol{J}_c$, associated with conduction currents produced in a medium as a result of the electric field. In many problems we shall be concerned with conditions outside the limited regions containing sources and we can take $\boldsymbol{J}_s$ to be zero.

With equations (1.1) to (1.4) have to be associated the constitutive equations for the medium, relating $\boldsymbol{D}$ with $\boldsymbol{E}$, $\boldsymbol{B}$ with $\boldsymbol{H}$, and $\boldsymbol{J}_c$ with $\boldsymbol{E}$. We shall be primarily concerned with linear isotropic media, for which we may write

$$\boldsymbol{D} = \varepsilon \boldsymbol{E} \tag{1.6}$$

$$\boldsymbol{B} = \mu \boldsymbol{H} \tag{1.7}$$

$$\boldsymbol{J}_c = \sigma \boldsymbol{E} \tag{1.8}$$

ϵ, the *permittivity*, μ, the *permeability*, and σ, the *conductivity*, may be functions of position, although we shall almost always consider media for which they are all constants; i.e. linear, isotropic homogeneous media. The simplest instance is free space.

Certain anisotropic media of interest need to be described by equations of the form

$$\left.\begin{aligned} B_x &= \mu_{11}H_x + \mu_{12}H_y + \mu_{13}H_z \\ B_y &= \mu_{21}H_x + \mu_{22}H_y + \mu_{23}H_z \\ B_z &= \mu_{31}H_x + \mu_{32}H_y + \mu_{33}H_z \end{aligned}\right\} \tag{1.9}$$

Similar equations can replace (1.6) and (1.8).

In SI units, used throughout this book, the above quantities have the units designated in table 1.1.

Table 1.1

Quantity	Unit	Symbol
$\boldsymbol{D}$ Electric displacement	coulomb per square metre	C m^{-2}
$\boldsymbol{E}$ Electric field strength	volt per metre	V m^{-1}
$\boldsymbol{B}$ Magnetic flux density	tesla	T
$\boldsymbol{H}$ Magnetic field strength	ampere per metre	A m^{-1}
$\boldsymbol{J}$ Electric current density	ampere per square metre	A m^{-2}
ρ Electric charge density	coulomb per cubic metre	C m^{-3}
σ Conductivity	siemens per metre	S m^{-1}
μ Permeability	henry per metre	H m^{-1}
ε Permittivity	farad per metre	F m^{-1}

1.1.1 Linear isotropic media

A great simplification results when linear isotropic media are considered. Fortunately this covers many situations of practical interest, including free space. The use of (1.6), (1.7) and (1.8) in (1.1) and (1.2) yields

$$\operatorname{curl} \boldsymbol{E} = -\mu \partial \boldsymbol{H}/\partial t \tag{1.10}$$

$$\operatorname{curl} \boldsymbol{H} = \varepsilon\, \partial \boldsymbol{E}/\partial t + \boldsymbol{J}_s + \sigma \boldsymbol{E} \tag{1.11}$$

In these equations, μ, ε, and σ can in principle be functions of position, but analysis is only simple in the homogeneous case.

1.1.2 Sinusoidal variations in time: phasor notation

The fields we shall have most occasion to consider are such that at any point in space each quantity varies sinusoidally in time. This may be compared with steady-state a.c. circuit analysis, where the device of phasors is introduced to facilitate analysis. A similar technique may be applied to field problems. Thus in (1.10) and (1.11), the variables $\boldsymbol{E}$, $\boldsymbol{H}$, $\boldsymbol{J}$ will vary sinusoidally with time, but will be of different amplitudes and phases. We shall have for example

$$\boldsymbol{E}(\boldsymbol{r}, t) = \boldsymbol{E}_0(\boldsymbol{r}) \cos[\omega t + \theta(\boldsymbol{r})] \tag{1.12}$$

in which $\boldsymbol{E}_0$, θ will not involve time. We introduce a complex quantity, $\mathbf{E}$, known as a *phasor*, defined by

$$\mathbf{E}(\boldsymbol{r}) = \boldsymbol{E}_0(\boldsymbol{r}) \exp[\mathrm{j}\theta(\boldsymbol{r})]$$

which enables us to write (1.12) in the form

$$\begin{aligned} \boldsymbol{E}(\boldsymbol{r}, t) &= \operatorname{Re}[\mathbf{E}(\boldsymbol{r})\, \mathrm{e}^{\mathrm{j}\omega t}] \\ &= \tfrac{1}{2}[\mathbf{E}(\boldsymbol{r})\, \mathrm{e}^{\mathrm{j}\omega t} + \mathbf{E}^*(\boldsymbol{r})\, \mathrm{e}^{-\mathrm{j}\omega t}] \end{aligned} \tag{1.13}$$

In a similar manner we may express $\boldsymbol{H}$ and $\boldsymbol{J}$ in terms of their respective phasor equivalents $\mathbf{H}$, $\mathbf{J}$ by the relations

$$\begin{aligned} \boldsymbol{H}(\boldsymbol{r}, t) &= \operatorname{Re}[\mathbf{H}(\boldsymbol{r})\, \mathrm{e}^{\mathrm{j}\omega t}] \\ &= \tfrac{1}{2}[\mathbf{H}(\boldsymbol{r})\, \mathrm{e}^{\mathrm{j}\omega t} + \mathbf{H}^*(\boldsymbol{r})\, \mathrm{e}^{-\mathrm{j}\omega t}] \end{aligned} \tag{1.14}$$

$$\begin{aligned} \boldsymbol{J}(\boldsymbol{r}, t) &= \operatorname{Re}[\mathbf{J}(\boldsymbol{r})\, \mathrm{e}^{\mathrm{j}\omega t}] \\ &= \tfrac{1}{2}[\mathbf{J}(\boldsymbol{r})\, \mathrm{e}^{\mathrm{j}\omega t} + \mathbf{J}^*(\boldsymbol{r})\, \mathrm{e}^{-\mathrm{j}\omega t}] \end{aligned} \tag{1.15}$$

We now substitute (1.13) and (1.14) in (1.10), remembering that since the operation curl does not involve the time variable

$$\text{curl}\,[\mathbf{E}(\boldsymbol{r})\,e^{j\omega t}] = e^{j\omega t}\,\text{curl}\,\mathbf{E}$$

We have

$$\tfrac{1}{2}[e^{j\omega t}\,\text{curl}\,\mathbf{E} + e^{-j\omega t}\,\text{curl}\,\mathbf{E}^*] = -\mu\frac{\partial}{\partial t}[e^{j\omega t}\mathbf{H} + e^{-j\omega t}\mathbf{H}^*]$$

The second term on either side is the complex conjugate of the first since all the operations occurring in performing curl or $\partial/\partial t$ are real. The complete equation is therefore satisfied if

$$e^{j\omega t}\,\text{curl}\,\mathbf{E} = -\mu\frac{\partial}{\partial t}[e^{j\omega t}\mathbf{H}]$$

or

$$\text{curl}\,\mathbf{E} = -j\omega\mu\mathbf{H} \tag{1.16}$$

A similar analysis for (1.11) gives

$$\text{curl}\,\mathbf{H} = j\omega\varepsilon\mathbf{E} + \mathbf{J} \tag{1.17}$$

These equations, (1.16), (1.17), are the phasor equivalents of (1.10), (1.11).

The use of the phasor concept depends on the fact that the variables are related linearly, and the equations are purely real. We note that the requirement of linearity does not prevent μ and ε being functions of position, although they must be independent of the electric and magnetic fields. The constitutive equations (1.6) to (1.8), and (1.9), are linear with real coefficients and do not involve time. Hence the relationships between the phasor equivalents are

$$\mathbf{D} = \varepsilon\mathbf{E}$$

$$\mathbf{B} = \mu\mathbf{H}$$

$$\mathbf{J} = \sigma\mathbf{E}$$

or

$$B_x = \mu_{11}H_x + \mu_{12}H_y + \mu_{13}H_z$$

$$B_y = \mu_{21}H_x + \mu_{22}H_y + \mu_{23}H_z$$

$$B_z = \mu_{31}H_x + \mu_{32}H_y + \mu_{33}H_z$$

In much of this book we shall work almost exclusively with phasor equivalents. They will be distinguished from the real physical fields as above, by the use of italic type-face for the latter, and roman type-face for the phasors. The principal quantities are given in table 1.2.

Table 1.2

Physical variable	Phasor equivalent	Relation
$\boldsymbol{E}$	$\mathbf{E}$	$\boldsymbol{E} = \mathrm{Re}\,(\mathbf{E}\,\mathrm{e}^{\mathrm{j}\omega t})$
$\boldsymbol{H}$	$\mathbf{H}$	$\boldsymbol{H} = \mathrm{Re}\,(\mathbf{H}\,\mathrm{e}^{\mathrm{j}\omega t})$
$\boldsymbol{B}$	$\mathbf{B}$	$\boldsymbol{B} = \mathrm{Re}\,(\mathbf{B}\,\mathrm{e}^{\mathrm{j}\omega t})$
$\boldsymbol{D}$	$\mathbf{D}$	$\boldsymbol{D} = \mathrm{Re}\,(\mathbf{D}\,\mathrm{e}^{\mathrm{j}\omega t})$
$\boldsymbol{J}$	$\mathbf{J}$	$\boldsymbol{J} = \mathrm{Re}\,(\mathbf{J}\,\mathrm{e}^{\mathrm{j}\omega t})$

We may finally write the equations for a linear isotropic homogeneous medium with sinusoidal time variations in the form

$$\mathrm{curl}\,\mathbf{E} = -\mathrm{j}\omega\mu\,\mathbf{H} \tag{1.18}$$

$$\mathrm{curl}\,\mathbf{H} = \mathrm{j}\omega\varepsilon\,\mathbf{E} + \mathbf{J}_\mathrm{s} + \sigma\mathbf{E} \tag{1.19}$$

Of the remaining equations, (1.3) follows from (1.18), and (1.4) provides a method of finding ρ given $\mathbf{J}_\mathrm{s}$. It should be noted that if $\mathbf{J}_\mathrm{s}$ is zero in any region of space then from (1.19) div $\mathbf{E}$ is also zero, and no distributed charge exists.

1.2 Plane waves

We look for simple solutions of (1.18) and (1.19) in a region of space containing no sources, and initially with zero conductivity. It must be remembered that a field of necessity must have sources, and excluding them from one region of space merely places them elsewhere (perhaps at infinity). We shall consider later how fields corresponding to our solutions could be realised in practice.

The equations we start with under the assumptions of this section are therefore

$$\mathrm{curl}\,\mathbf{E} = -\mathrm{j}\omega\mu\,\mathbf{H} \tag{1.20}$$

$$\mathrm{curl}\,\mathbf{H} = \mathrm{j}\omega\varepsilon\,\mathbf{E} \tag{1.21}$$

We look for solutions which are dependent on only one space variable, which we take to be z. Such solutions must be the same across any plane for which z is constant, and hence the term *plane waves* is bestowed.

In Cartesian components with $\boldsymbol{a}_x$. $\boldsymbol{a}_y$, $\boldsymbol{a}_z$ the unit vectors in the x, y, z directions,

$$\text{curl } \mathbf{E} = \left(\frac{\partial E_z}{\partial y} - \frac{\partial E_y}{\partial z}\right) \boldsymbol{a}_x + \left(\frac{\partial E_x}{\partial z} - \frac{\partial E_z}{\partial x}\right) \boldsymbol{a}_y + \left(\frac{\partial E_y}{\partial x} - \frac{\partial E_x}{\partial y}\right) \boldsymbol{a}_z$$

Dependence only on z reduces this to

$$\text{curl } \mathbf{E} = -\frac{\partial E_y}{\partial z} \boldsymbol{a}_x + \frac{\partial E_x}{\partial z} \boldsymbol{a}_y$$

and similarly

$$\text{curl } \mathbf{H} = -\frac{\partial H_y}{\partial z} \boldsymbol{a}_x + \frac{\partial H_x}{\partial z} \boldsymbol{a}_y$$

Substitution in (1.20) and (1.21) yields

$$-\frac{\partial E_y}{\partial z} = -j\omega\mu H_x \tag{1.22}$$

$$\frac{\partial E_x}{\partial z} = -j\omega\mu H_y \tag{1.23}$$

$$0 = -j\omega\mu H_z \tag{1.24}$$

$$-\frac{\partial H_y}{\partial z} = j\omega\varepsilon E_x \tag{1.25}$$

$$\frac{\partial H_x}{\partial z} = j\omega\varepsilon E_y \tag{1.26}$$

$$0 = j\omega\varepsilon E_z \tag{1.27}$$

We deduce from (1.24) and (1.27) that for such solutions E_z, H_z vanish. Further, (1.22) and (1.26) concern only E_y and H_x, and (1.23) and (1.25) concern only E_x and H_y. Consider the latter pair first. Eliminating H_y we have

$$\frac{\partial^2 E_x}{\partial z^2} = -j\omega\mu \frac{\partial H_y}{\partial z} = j\omega\mu \cdot j\omega\varepsilon E_x$$

Therefore

$$\frac{\partial^2 E_x}{\partial z^2} + \omega^2 \mu\varepsilon E_x = 0 \tag{1.28}$$

This equation has two independent solutions

$$E_x = A\, e^{-j\beta z} \tag{1.29}$$

and

$$\mathrm{E}_x = \mathrm{B}\,\mathrm{e}^{\mathrm{j}\beta z} \tag{1.30}$$

where A, B are constants and

$$\beta = \omega\sqrt{(\mu\varepsilon)} \tag{1.31}$$

When E_x is given by (1.29), substitution into (1.23) yields

$$\mathrm{H}_y = \mathrm{A}(\beta/\omega\mu)\,\mathrm{e}^{-\mathrm{j}\beta z} = \mathrm{A}\sqrt{(\varepsilon/\mu)}\,\mathrm{e}^{-\mathrm{j}\beta z} \tag{1.32}$$

and when E_x is given by (1.30)

$$\mathrm{H}_y = -\mathrm{B}\sqrt{(\varepsilon/\mu)}\,\mathrm{e}^{\mathrm{j}\beta z} \tag{1.33}$$

Apart from sign we see that the ratio of E_x to H_y is a constant for the medium. This constant, $\sqrt{(\mu/\varepsilon)}$, has the dimensions of impedance, will be denoted by the symbol ζ, and is termed the *wave impedance*:

$$\zeta = \sqrt{(\mu/\varepsilon)} \tag{1.34}$$

To understand the significance of these equations we must return to the physical fields obtained by applying (1.13). Equations (1.29) and (1.32) yield, assuming A is real constant,

$$E_x(z, t) = A\cos(\omega t - \beta z) \tag{1.35}$$

$$H_y(z, t) = \zeta^{-1} A\cos(\omega t - \beta z) \tag{1.36}$$

At any point with co-ordinate z, the electric and magnetic fields vary sinusoidally in time; they are in phase with each other; and they are mutually perpendicular. The disposition at time $t = 0$ is shown in figure 1.2. It is permissible to regard the term $\cos(\omega t - \beta z)$ as representing wave propagation in the direction of positive z: a constant value of electric field (or magnetic field) is experienced by an observer moving so that $\omega t - \beta z$ is constant. In a time interval of Δt, he has to move so as to increase βz by $\omega\Delta t$ in order to keep $\cos(\omega t - \beta z)$ at the same value, i.e. he moves

$$\Delta z = \omega\Delta t/\beta.$$

This is illustrated in fig. 1.3. The required velocity is

$$v = \omega/\beta \tag{1.37}$$

which is called the *phase velocity*, since it is the velocity at which surfaces of constant phase travel. Using (1.31), in this instance

$$v = 1/\sqrt{(\mu\varepsilon)} \tag{1.38}$$

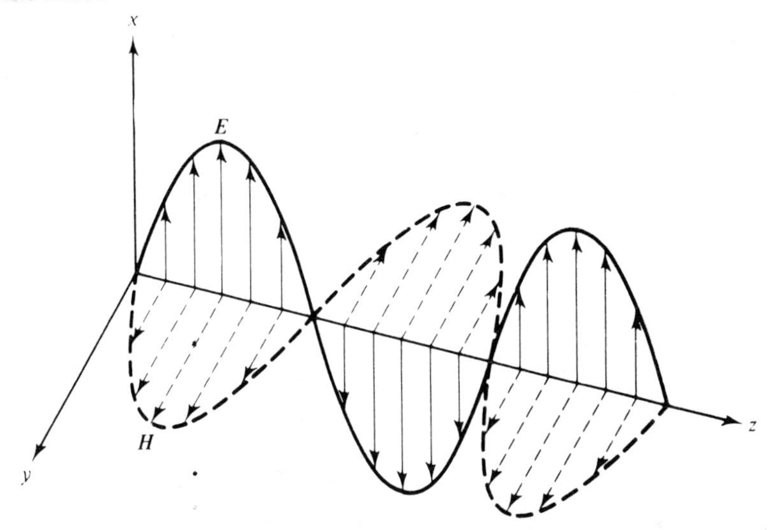

Fig. 1.2. The spatial variation of the electric and magnetic fields defined by (1.35) and (1.36).

Here v is a constant of the medium, although in other cases that we shall consider it will be found that the velocity defined by (1.37) depends on frequency ω.

Using the definition for v, the term $\cos(\omega t - \beta z)$ can be rewritten $\cos \omega(t - z/v)$. A signal of angular frequency ω thus suffers a time delay z/v whilst being propagated a distance z. This delay is the same for all frequencies, so that such propagation is said to be *non-dispersive*. If the phase velocity depends on ω the propagation is *dispersive*.

The periodicity in the z-co-ordinate is given by

$$\lambda = 2\pi v/\omega = v/f \tag{1.39}$$

where f is the frequency equal to $\omega/2\pi$. The quantity λ is called the *wavelength*.

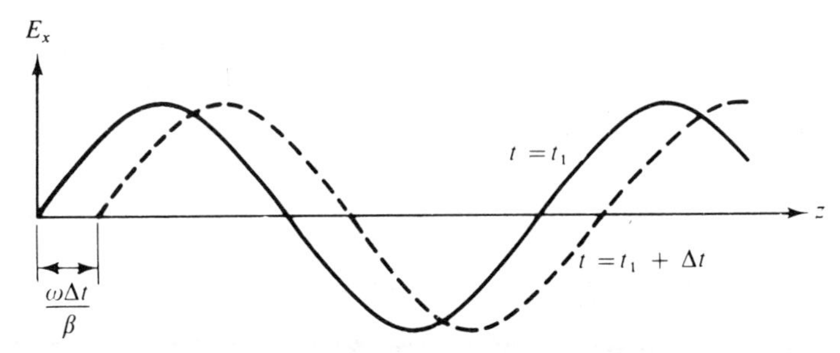

Fig. 1.3. The change of spatial distribution of E_x with variation of time.

The fields described by (1.29) and (1.32) or (1.35) and (1.36) are said to represent a *plane monochromatic wave* in the forward (positive z) direction, and the wave is also said to be *plane-polarised*. 'Monochromatic' merely means 'single frequency'. 'Plane-polarisation' means that the electric field is at all times and places parallel to the same plane, in this case the xz-plane, as well as being perpendicular to the direction of propagation. It is to be noted that E_x, H_y, Oz form a right-handed set of vectors: the vector product $\boldsymbol{E} \times \boldsymbol{H}$ is in the positive z-direction, the direction of propagation.

We can now proceed to the other solutions. Equations (1.30) and (1.33) yield a factor $\cos(\omega t + \beta z)$ in place of $\cos(\omega t - \beta z)$. They evidently refer to a backward wave, in the negative z-direction. E_x, H_y, and the direction of propagation still form a right-handed set, since both H_y and the direction of propagation have been reversed.

Equations (1.22) and (1.26) may be treated in a way similar to the previous analysis. We find

$$\frac{\partial^2 \mathrm{E}_y}{\partial z^2} = \mathrm{j}\omega\mu \frac{\partial \mathrm{H}_x}{\partial z} = -\omega^2 \mu\varepsilon\, \mathrm{E}_y$$

Hence

$$\mathrm{E}_y = \mathrm{A}\, \mathrm{e}^{-\mathrm{j}\beta z} \tag{1.40}$$

or

$$\mathrm{E}_y = \mathrm{B}\, \mathrm{e}^{\mathrm{j}\beta z} \tag{1.41}$$

The corresponding H_x is

$$\mathrm{H}_x = -\mathrm{A}\frac{\beta}{\omega\mu}\, \mathrm{e}^{-\mathrm{j}\beta z} = -\zeta^{-1} \mathrm{A}\, \mathrm{e}^{-\mathrm{j}\beta z} \tag{1.42}$$

or

$$\mathrm{H}_x = \mathrm{B}\frac{\beta}{\omega\mu}\, \mathrm{e}^{\mathrm{j}\beta z} = \zeta^{-1}\, \mathrm{B}\, \mathrm{e}^{\mathrm{j}\beta z} \tag{1.43}$$

A little consideration will show that (1.40) and (1.42) represent the plane wave of (1.29) and (1.32) with the plane of polarisation rotated through $\frac{1}{2}\pi$, so that **E** lies in the yz-plane and **H** in the xz-plane. $\mathbf{E} \times \mathbf{H}$ is still in the direction of propagation. Similarly (1.41) and (1.43) represent a backward wave with a rotated plane of polarisation.

1.2.1 General form for plane-polarised plane waves

The solutions obtained above can all be represented in the same form: if the direction of propagation is the unit vector $\boldsymbol{\nu}$, and **E**, **H** are the

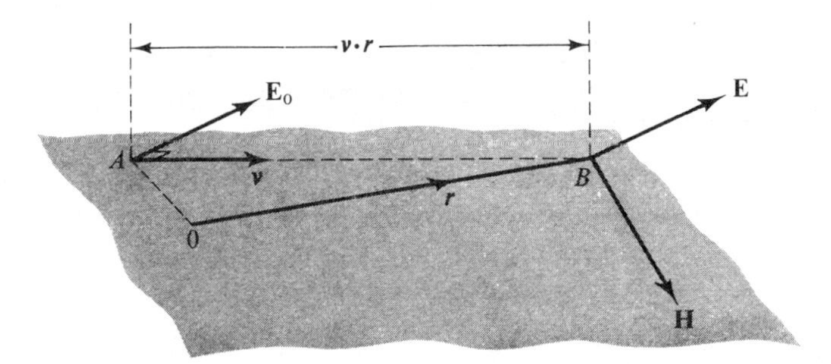

Fig. 1.4. The general plane wave.

electric and magnetic vectors,

$$\mathbf{E} = \mathbf{E}_0 \exp(-j\beta\boldsymbol{\nu}\cdot\boldsymbol{r}) \tag{1.44}$$

$$\mathbf{H} = \zeta^{-1}\boldsymbol{\nu}\times\mathbf{E} \tag{1.45}$$

$$\mathbf{E} = \zeta\mathbf{H}\times\boldsymbol{\nu} \tag{1.46}$$

where

$$\mathbf{E}_0\cdot\boldsymbol{\nu} = 0 \tag{1.47}$$

This is illustrated in fig. 1.4. It will be found that the four cases considered previously correspond respectively to

$$\boldsymbol{\nu} = \boldsymbol{a}_z, \quad \mathbf{E}_0 = \mathrm{E}_0\boldsymbol{a}_x \quad \text{or} \quad \mathrm{E}_0\boldsymbol{a}_y$$

$$\boldsymbol{\nu} = -\boldsymbol{a}_z, \quad \mathbf{E}_0 = \mathrm{E}_0\boldsymbol{a}_x \quad \text{or} \quad \mathrm{E}_0\boldsymbol{a}_y$$

1.2.2 Waves in free space

In free space

$$\mu = \mu_0 = 4\pi\times10^{-7}\ \mathrm{Hm}^{-1}$$

$$\varepsilon = \varepsilon_0 = 8.854\times10^{-12}\ \mathrm{Fm}^{-1}$$

Of these, the value of μ_0 is defined, and ε_0 is experimentally determined from static measurements. Using these values in (1.38) yields a value for velocity extremely close to the accepted value for that of light in free space. This was a first indication, now of course accepted, that light is an electromagnetic radiation. The wavelength of light may be found experimentally directly in terms of the metre, and in the International System the metre is defined to be 1 650 763.73 wavelengths in vacuum of a certain krypton line in the red region of the visible spectrum. Taking the velocity of light to be $2.998\times10^8\ \mathrm{m\,s}^{-1}$, (1.39) gives f as about 4.9×10^{14} Hz. Optical frequencies are therefore very high. By contrast, the experiments of Hertz in 1888, which gave experimental verification of Maxwell's theory, were carried out with radiation for which the wavelenth was about 1.5 m. The frequency was therefore about 2×10^8 Hz, or 200 MHz.

It is common usage to denote the velocity of light in free space with the symbol *c*. This practice will be adhered to as far as possible in this book although occasionally the symbol may be used for the propagation velocity in situations where the medium is nearly always (but not necessarily) free space, as for example in a hollow tube waveguide. The meaning will be clear from the context.

1.2.3 Circular polarisation

The plane-polarised waves already discussed are not the only form plane waves may take. We may superpose the two orthogonally polarised waves

$$\mathrm{E}_x = \mathrm{A}\,\mathrm{e}^{-\mathrm{j}\beta z}, \qquad \mathrm{H}_y = \zeta^{-1}\mathrm{A}\,\mathrm{e}^{-\mathrm{j}\beta z}$$

and

$$\mathrm{E}_y = \mathrm{B}\,\mathrm{e}^{-\mathrm{j}\beta z}, \qquad \mathrm{H}_x = -\zeta^{-1}\mathrm{B}\,\mathrm{e}^{-\mathrm{j}\beta z}$$

The resultants for the vectors are given by

$$\mathbf{E} = \mathrm{E}_x\boldsymbol{a}_x + \mathrm{E}_y\boldsymbol{a}_y = \mathrm{e}^{-\mathrm{j}\beta z}(\mathrm{A}\boldsymbol{a}_x + \mathrm{B}\boldsymbol{a}_y) \tag{1.48}$$

$$\mathbf{H} = \zeta^{-1}\,\mathrm{e}^{-\mathrm{j}\beta z}(\mathrm{A}\boldsymbol{a}_y - \mathrm{B}\boldsymbol{a}_x) \tag{1.49}$$

Now if A and B are both real, the expressions in brackets are constant vectors, and the resultant is a plane-polarised wave with a new plane of polarisation. Suppose however B is imaginary and equal to jA. We find for the instantaneous electric field

$$\begin{aligned} \boldsymbol{E}(z, t) &= \mathrm{Re}\,[\mathrm{e}^{-\mathrm{j}\beta z}(A\boldsymbol{a}_x + \mathrm{j}\,A\boldsymbol{a}_y)\,\mathrm{e}^{\mathrm{j}\omega t}] \\ &= A[\boldsymbol{a}_x \cos(\omega t - \beta z) - \boldsymbol{a}_y \sin(\omega t - \beta z)] \end{aligned} \tag{1.50}$$

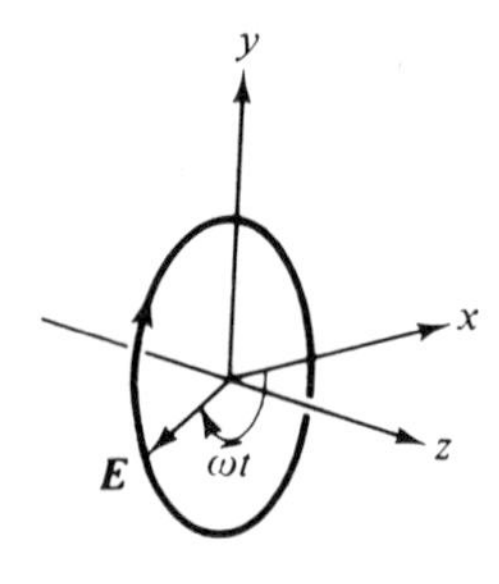

Fig. 1.5. The electric vector for left-handed circular polarisation.

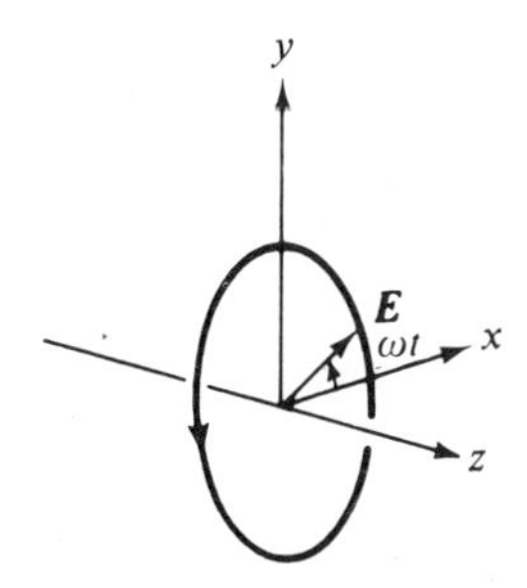

Fig. 1.6. The electric vector for right-handed circular polarisation.

Consider the significance of (1.50) at $z=0$. Figure 1.5 shows the instantaneous direction of $\boldsymbol{E}$, as seen looking into the oncoming wave: the $\boldsymbol{E}$ vector is always of length A and is rotating in a clockwise direction with a frequency the same as that of the radiation. Equation (1.50) is said to describe a plane wave with *left-hand circular polarisation.* By taking $\mathrm{B}=-\mathrm{j}A$ we obtain a wave for which the $\boldsymbol{E}$ vector describes a circle in an anticlockwise direction, and said to be of *right-hand circular polarisation.* this is shown in fig. 1.6. The formulae for $\boldsymbol{H}$ will show that $\boldsymbol{H}$ is at any instant perpendicular to $\boldsymbol{E}$.

In the more general case when B/A is an arbitrary complex number the locus of $\boldsymbol{E}$ will be an ellipse, described either clockwise or anticlockwise. The most general wave may therefore be elliptically polarised with right- or left-handed polarisation.

Equations (1.48) and (1.49) may be interpreted as saying that an arbitrary elliptically polarised wave can be broken down into two plane-polarised waves. It is also possible, and necessary on occasion, to regard a plane-polarised wave as made up of right-handed and left-handed circularly polarised waves. The possibility of decomposition has the great advantage that in the first instance we need only consider plane-polarisation, obtaining other results by superposition.

1.3 Plane waves in a conducting medium

We have assumed above that the medium considered is not a conductor. The formal analysis which we have carried out can be simply made to apply to the more general case. In a sourceless region with finite conductivity, we can replace (1.21) with

$$\mathrm{curl}\,\mathbf{H}=\mathrm{j}\omega\varepsilon\,\mathbf{E}+\sigma\mathbf{E}=\mathrm{j}\omega(\varepsilon+\sigma/\mathrm{j}\omega)\mathbf{E} \qquad (1.51)$$

This is formally the same as (1.21), but with ε replaced by the complex quantity $\varepsilon + \sigma/j\omega$. The solutions given by (1.29) and (1.32) will hold, giving expressions of the form

$$\left.\begin{aligned} E_x &= A\,e^{-\gamma z} \\ H_y &= \zeta'^{-1} A\,e^{-\gamma z} \end{aligned}\right. \tag{1.52}$$

in which

$$\left.\begin{aligned} \gamma &= j\omega[\mu(\varepsilon + \sigma/j\omega)]^{\frac{1}{2}} = \omega\sqrt{(\mu\varepsilon)}[1 + \sigma/j\omega\varepsilon]^{\frac{1}{2}} \\ \zeta' &= \left[\frac{\mu}{\varepsilon + \sigma/j\omega}\right]^{\frac{1}{2}} = \sqrt{\left(\frac{\mu}{\varepsilon}\right)}\,[1 + \sigma/j\omega\varepsilon]^{-\frac{1}{2}} \end{aligned}\right\} \tag{1.53}$$

Before trying to understand the significance of these results let us get some idea of the possible magnitude of the (dimensionless) ratio $\sigma/\omega\varepsilon$. Take first a good conductor. For copper at room temperature

$$\sigma \approx 5.9 \times 10^{7}\ \mathrm{S\,m^{-1}}$$

$$\varepsilon = \varepsilon_0 = 8.864 \times 10^{-12}\ \mathrm{F\,m^{-1}}$$

At 50 Hz

$$\sigma/\omega\varepsilon = 2.1 \times 10^{16}$$

At the highest 'radio' frequency of, say, 100 GHz (10^{11} Hz), the ratio is still of the order of 10^{6}. Hence for copper, and for almost any metallic conductor, we may safely consider $\sigma/\omega\varepsilon \gg 1$.

Next consider a good dielectric, say polyethylene. The losses in such a material are not directly due to ohmic conductivity, but arise through dielectric hysteresis loss, so that σ will be a function of frequency. For dielectrics the loss is specified through the *loss angle* δ, defined by

$$\tan\delta = \sigma/\omega\varepsilon \tag{1.54}$$

By considering a parallel-plate capacitor filled with the dielectric (fig. 1.7(a)), it will be seen that δ is the angle shown in the phasor diagram in

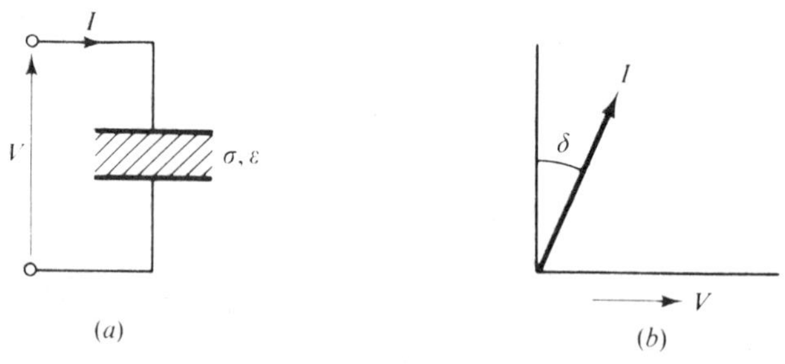

Fig. 1.7. Defining the loss angle δ of a dielectric. (*a*) Circuit. (*b*) Phasor diagram.

fig. 1.7(b). For polyethylene in the frequency range 50 Hz to 1 GHz

$$\varepsilon = 2.3\,\varepsilon_0$$

and

$$2\times 10^{-4} < \tan\delta < 3\times 10^{-4}$$

Hence for such a class of material $\sigma/\omega\varepsilon \ll 1$ over the frequency range of interest.

1.3.1 Relaxation time

The ratio σ/ε may be interpreted as an angular frequency characteristic of the medium: for good conductors, 'useful' frequencies are much smaller than this critical frequency, and for good dielectrics they are much greater. An alternative interpretation of the ratio may be obtained. Let us return to the time-dependent form of Maxwell's equations, (1.1) to (1.4). Taking the divergence of (1.2) with $\boldsymbol{J} = \sigma\boldsymbol{E}$ we have

$$\frac{\partial}{\partial t}\operatorname{div}\boldsymbol{D} + \sigma\operatorname{div}\boldsymbol{E} = 0$$

In a homogeneous medium for which $\boldsymbol{D} = \varepsilon\boldsymbol{E}$, (1.4) enables us to rewrite this result

$$\frac{\partial\rho}{\partial t} + \frac{\sigma}{\varepsilon}\rho = 0$$

whence

$$\rho = \rho_0 \exp(-\sigma t/\varepsilon)$$

The ratio ε/σ is termed here the *relaxation time*. From previously given values, $\varepsilon/\sigma = 1.5\times 10^{-19}$ for copper. Thus any initially distributed volume charge within the conductor would disappear to the surface in a time of the order of 10^{-18} s.

1.3.2 Waves in lowloss dielectric

We consider two situations depending on $\sigma/\omega\varepsilon$ being very large or very small.

We have to evaluate γ and ζ' from (1.53), and then apply the results to interpret (1.52). In so far as ζ' is a multiplier, when $\sigma/\omega\varepsilon \approx 10^{-3}$ we may take $\zeta' \approx \sqrt{(\mu/\varepsilon)}$. However, γ occurs in an exponential, being there multiplied by z. Using the binomial theorem

$$\gamma = \mathrm{j}\omega\sqrt{(\mu\varepsilon)}[1+\sigma/\mathrm{j}\omega\varepsilon]^{\frac{1}{2}} \approx \mathrm{j}\omega\sqrt{(\mu\varepsilon)}(1+\sigma/2\mathrm{j}\omega\varepsilon)$$

Hence

$$\gamma = \tfrac{1}{2}\sigma\sqrt{(\mu/\varepsilon)} + \mathrm{j}\omega\sqrt{(\mu\varepsilon)}$$

and $$\exp(-\gamma z) = \exp\{[-\tfrac{1}{2}\sigma\sqrt{(\mu/\varepsilon)} - j\omega\sqrt{(\mu\varepsilon)}]z\} \tag{1.55}$$

Applying (1.13) to find the instantaneous value of the field components

$$\left.\begin{aligned} E_x(z,t) &= A\exp(-\alpha z)\cos(\omega t - \beta z) \\ H_y(z,t) &= \zeta^{-1}A\exp(-\alpha z)\cos(\omega t - \beta z) \end{aligned}\right\} \tag{1.56}$$

where $$\alpha = \tfrac{1}{2}\sigma\sqrt{(\mu/\varepsilon)}, \quad \beta = \omega\sqrt{(\mu\varepsilon)}, \quad \zeta = \sqrt{(\mu/\varepsilon)} \tag{1.57}$$

We see that the presence of the σ term gives rise to a decrease of amplitude as the wave progresses.

Similarly (1.30) and (1.33) yield

$$E_x(z,t) = B\exp(\alpha z)\cos(\omega t + \beta z)$$
$$H_y(z,t) = -\zeta^{-1}B\exp(\alpha z)\cos(\omega t + \beta z)$$

Since the wave is travelling in the negative z-direction, $\exp(\alpha z)$ represents attenuation in the direction of propagation. The phase velocity is still given by (1.38).

Apart from attenuation, the properties of the plane wave are unchanged when the medium becomes slightly conductive.

The propagation constant

It has been shown that when losses are taken into consideration the dependence on the z-co-ordinate of the components of the electric and magnetic fields is most conveniently expressed through a factor $\exp(-\gamma z)$, as exemplified by (1.52). This factor is characteristic of any form of wave propagation in one dimension, and in general the quantity γ is called the *propagation constant.* In terms of real and imaginary parts

$$\gamma = \alpha + j\beta \tag{1.58}$$

in which α is the *attenuation constant* and β the *phase constant.* This notation is consistent with the usage adopted in the previous discussion.

Measure of attenuation

From (1.56) the amplitude of a forward plane wave is proportional to $\exp(-\alpha z)$. The units of α are *nepers* per unit length. The power loss at any point will be proportional to $|E|^2$, and hence proportional to $\exp(-2\alpha z)$.

It is customary to measure attenuation in decibels. The measure in decibels of power P_2 to power P_1 is calculated from the formula $10\log_{10}(P_2/P_1)$. Hence the measure in decibels of the attenuation introduced by a propagation distance z is

$$10\log_{10}[\exp(2\alpha z)] = 8.686\alpha z \text{ dB} \tag{1.59}$$

The attenuating properties of the medium may be expressed in the form L dB m^{-1} where

$$L = 8.686\alpha \tag{1.60}$$

Example

Find the attenuation constant and phase velocity of plane waves at a frequency of 10 GHz in polyethylene, for which

$$\mu = \mu_0, \quad \varepsilon = 2.3\varepsilon_0, \quad \tan\delta = 2\times10^{-4}$$

Solution. The phase velocity from (1.38) is given by

$$v = \frac{1}{\sqrt{(\mu\varepsilon)}} = \frac{\sqrt{(\varepsilon_0/\varepsilon)}}{\sqrt{(\mu_0\varepsilon_0)}}$$

$$= \frac{3\times10^8}{\sqrt{2.3}}$$

$$= 1.98\times10^8 \text{ m s}^{-1}$$

The attenuation constant is given by

$$\alpha = \tfrac{1}{2}\sigma\sqrt{(\mu/\varepsilon)} = \sigma\omega\sqrt{(\mu\varepsilon)}/2\omega\varepsilon = (\pi f \tan\delta)/v$$

Hence

$$\alpha = \frac{\pi\times10^{10}\times2\times10^{-4}}{1.98\times10^8}$$

$$= 3.17\times10^{-2} \text{ nepers m}^{-1}$$

Or, using (1.60),

$$L = 0.28 \text{ dB m}^{-1}$$

1.3.3 Waves in good conductors: eddy currents

In this situation $\sigma/\omega\varepsilon \gg 1$. We then have

$$\gamma \approx j\omega\sqrt{(\mu\sigma/j\omega)}$$

$$= \sqrt{(\omega\mu\sigma)}\exp(j\pi/4) = (1+j)\sqrt{(\omega\mu\sigma/2)} \tag{1.61}$$

$$\zeta' \approx \sqrt{(j\omega\mu/\sigma)}$$

$$= \sqrt{(\omega\mu/\sigma)}\exp(j\pi/4) = (1+j)\sqrt{(\omega\mu/2\sigma)} \tag{1.62}$$

Equations (1.29) and (1.32) then yield

$$\mathrm{E}_x = \mathrm{A}\exp(-\gamma z) = \mathrm{A}\exp[-(1+\mathrm{j})z/l] \tag{1.63}$$

$$\mathrm{H}_y = \mathrm{A}\sqrt{(\sigma/\omega\mu)}\exp(-\tfrac{1}{4}\mathrm{j}\pi)\exp[-(1+\mathrm{j})z/l] \tag{1.64}$$

where

$$l = \sqrt{(2/\omega\mu\sigma)} \tag{1.65}$$

Applying (1.14) to find the physical fields

$$E_x(z, t) = A\exp(-z/l)\cos[\omega t - (z/l)] \tag{1.66}$$

$$H_y(z, t) = A\sqrt{(\sigma/\omega\mu)}\exp(-z/l)\cos[\omega t - (z/l) - \tfrac{1}{4}\pi] \tag{1.67}$$

It is still possible to interpret (1.66) and (1.67) as describing a wave with wavelength $2\pi l$. The attenuation over this length, however, is $\exp(+2\pi)$, so that the concept of a progressive wave cannot be maintained. The magnetic field lags the electric field by $\pi/4$, in contrast to the previous situation. The quantity l defined in (1.65) has dimension of length. Values of l for copper are shown in table 1.3.

Table 1.3

f (Hz)	l (mm)
50	9
10^6	0.06
10^9	0.002

With small values of l like those occurring above 1 MHz, it is evident that the field inside bulk material is virtually zero. The effect being investigated can only be of interest in a very small layer near the surface of the solid material. l is then regarded as a measure of penetration and is called the *skin depth*. At lower frequencies of interest the skin depth is of the order 1 cm, so that the field can extend throughout relatively thick pieces of material. This case is of interest, for example, in power transformers. Throughout the frequency range, the major consideration is the loss resulting from the flow of eddy currents, be it in power transformers or in the walls of waveguides.

Power dissipated in a thick conductor

The case we shall be primarily interested in is that for which the skin depth is small compared with other dimensions. This means that we can consider an infinite block of material with a plane surface at which the fields are excited, as shown in fig. 1.8.

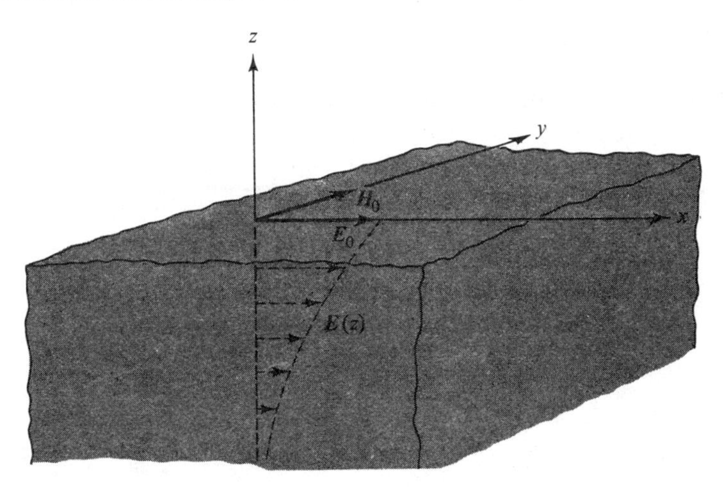

Fig. 1.8. The penetration of a plane wave into a good conductor.

Let the magnetic field at the surface $z=0$ be $H_0 \cos[\omega t - \frac{1}{4}\pi]$. Then (1.66) and (1.67) give

$$\left.\begin{aligned} E_x(z,t) &= \sqrt{(\omega\mu/\sigma)}H_0 \exp(-z/l)\cos[\omega t-(z/l)] \\ H_y(z,t) &= H_0 \exp(-z/l)\cos[\omega t-(z/l)-\tfrac{1}{4}\pi] \end{aligned}\right\} \tag{1.68}$$

The instantaneous rate of loss of energy in a slice parallel to the surface at distance z inside the slab, of thickness $\mathrm{d}z$ and unit area in the xy-plane is

$$\sigma[E_x(z,t)]^2\,\mathrm{d}z = H_0^2\omega\mu \exp(-2z/l)\cos^2[\omega t-(z/l)]\,\mathrm{d}z$$

Since the average of the square of a cosine or a sine over a complete period of its argument is equal to $\frac{1}{2}$, the total power loss can be written as

$$\begin{aligned} P &= \int_0^\infty \tfrac{1}{2}H_0^2\omega\mu \exp(-2z/l)\,\mathrm{d}z \\ &= \tfrac{1}{4}H_0^2\omega\mu l \\ &= \tfrac{1}{2}H_0^2\sqrt{(\omega\mu/2\sigma)} \\ &= \tfrac{1}{2}H_0^2/\sigma l \end{aligned} \tag{1.69}$$

Equation (1.69) can be written in the form

$$P = \tfrac{1}{2}H_0^2 R_s \tag{1.70}$$

where $R_s = 1/\sigma l$ is the *skin resistance*, and is numerically equal to the resistance per square of a sheet of the material of thickness l.

1.3.4 General solution

The extreme cases of $\sigma/\omega\varepsilon$ much less or much greater than unity have been treated since results can be obtained in simple forms, and also correspond to the practically useful situations. If neither approximation is justified it becomes necessary to evaluate, for example,

$$\gamma = \alpha + \mathrm{j}\beta = \mathrm{j}\omega\sqrt{(\mu\varepsilon)}\sqrt{(1+\sigma/\mathrm{j}\omega\varepsilon)}$$

This can be carried through numerically in several ways. It may be noted that an explicit solution can be obtained as follows. We have

$$(\alpha + \mathrm{j}\beta)^2 = -\omega^2\mu\varepsilon(1+\sigma/\mathrm{j}\omega\varepsilon) = -\xi + \mathrm{j}\eta$$

then $$\alpha^2 - \beta^2 = -\xi, \qquad 2\alpha\beta = \eta$$

Elimination of β and solution of the quadratic in α^2 gives

$$\alpha^2 = \tfrac{1}{2}[-\xi + \sqrt{(\xi^2+\eta^2)}]$$

$$\beta^2 = \tfrac{1}{2}[\xi + \sqrt{(\xi^2+\eta^2)}]$$

1.3.5 Perfect conductors

Although even good conductors have some resistivity, it is possible to consider a conductor which is perfect. In many cases this is an adequate approximation, and in others it leads to the first step in finding a solution. A perfect conductor may be characterised by $\sigma \to \infty$, or, since $\boldsymbol{J} = \sigma\boldsymbol{E}$, by $\boldsymbol{E} = 0$ with $\boldsymbol{J}$ indeterminate. We may let $\sigma \to \infty$ in the previous analysis for good conductors, which shows that $l \to 0$. The current will therefore flow in an infinitesimally thin sheet at the surface. Inside the perfect conductor, $\boldsymbol{E}$ and $\boldsymbol{H}$ will be zero.

1.4 Reflection and refraction of plane waves

Reflection and refraction, both familiar effects in optics, are also important in understanding the propagation of radio waves. The two cases are in fact only superficially similar, in that plane waves by definition fill space whereas in the optical context it is pencil beams that are considered. However, pencil beams in optics are in fact a very large number of wavelengths in diameter, and it turns out that if this is so, they may be treated as plane waves.

The phenomena concern the behaviour of waves at surfaces across which there is a discontinuity in properties: for example, the junction of two media of different dielectric constants. The continuous equations, (1.1) to (1.4), cannot be directly applied, so it is first necessary to consider the continuity conditions which must hold across such surfaces.

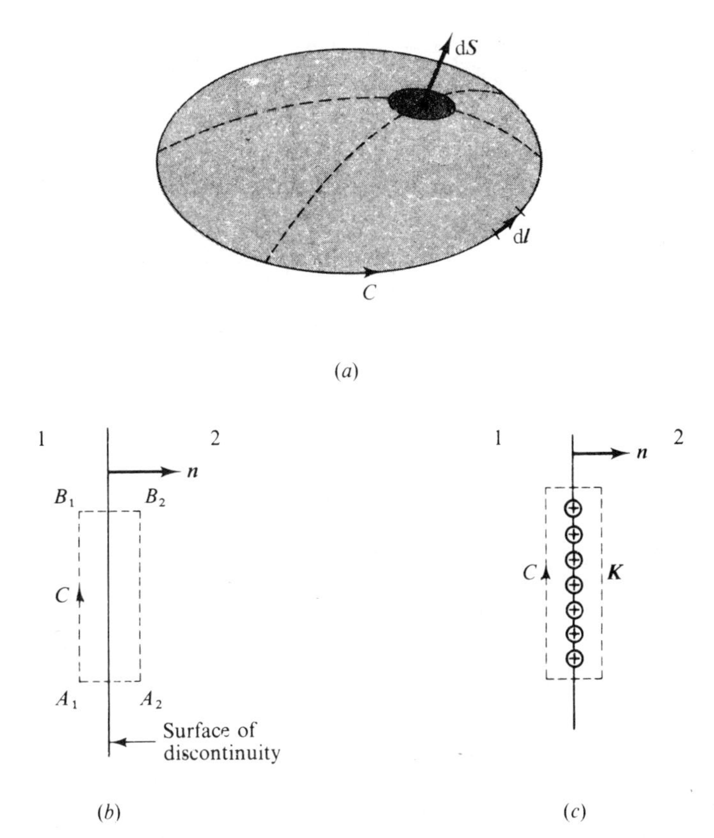

Fig. 1.9 (*a*). Illustrating the integral relation of (1.71). (*b*) Contour crossing surface of discontinuity. (*c*) Contour containing current sheet $\boldsymbol{K}$.

1.4.1 Continuity conditions

Both (1.1) and (1.2) are similar when expressed in integral form. Considering (1.1) we have as an alternative form

$$\oint_C \boldsymbol{E} \cdot \mathrm{d}\boldsymbol{l} = -\int_S \frac{\partial \boldsymbol{B}}{\partial t} \cdot \mathrm{d}\boldsymbol{S} \tag{1.71}$$

where the contour C and the surface S are typified in fig. 1.9(*a*). (It may be noted that the only restriction on S is that it finishes on C.) Consider evaluating this integral relation over the contour and surface of fig. 1.9(*b*), in which the contour is plane and embraces the surface separating two different media. If the contour is made to shrink laterally onto the joining surface, the area becomes vanishingly small, so that, providing $\partial \boldsymbol{B}/\partial t$ is finite on both sides of the discontinuity, the right-hand side of (1.71) tends to zero. The left-hand side takes the form

$$\int_{A_1}^{B_1} \boldsymbol{E} \cdot \mathrm{d}\boldsymbol{l} - \int_{A_2}^{B_2} \boldsymbol{E} \cdot \mathrm{d}\boldsymbol{l}$$

since the other two sides also are indefinitely small. Hence

$$\int_{A_1}^{B_1} \boldsymbol{E} \cdot \mathrm{d}\boldsymbol{l} = \int_{A_2}^{B_2} \boldsymbol{E} \cdot \mathrm{d}\boldsymbol{l}$$

The plane of the contour can be rotated, as long as it contains the normal $\boldsymbol{n}$ from medium 1 into medium 2, so that there must be continuity of the components of $\boldsymbol{E}$ tangential to the surface on both sides. This may be written

$$[\boldsymbol{E}_\mathrm{t}]_2 = [\boldsymbol{E}_\mathrm{t}]_1$$

or

$$[\boldsymbol{n} \times \boldsymbol{E}]_2 = [\boldsymbol{n} \times \boldsymbol{E}]_1 \tag{1.72}$$

A similar argument may be applied to (1.2), provided neither medium is a perfect conductor. This is because a current sheet of infinitesimal thickness can flow in the surface of a perfect conductor, so that the current density is in fact infinite.

This situation is shown in fig. 1.9(c), where a current sheet of density $\boldsymbol{K}$ per unit length is in the surface directed into the paper. Assuming $\partial \boldsymbol{D}/\partial t$ is finite we shall have

$$\oint_C \boldsymbol{H} \cdot \mathrm{d}\boldsymbol{l} = \text{current enclosed by } C$$

The components of $\boldsymbol{H}$ parallel to $\boldsymbol{K}$ on both sides must be continuous; the components perpendicular to $\boldsymbol{K}$ must satisfy

$$[H_\perp]_2 - [H_\perp]_1 = K$$

These conditions may be combined in a vector form

$$[\boldsymbol{n} \times \boldsymbol{H}]_2 - [\boldsymbol{n} \times \boldsymbol{H}]_1 = \boldsymbol{K} \tag{1.73}$$

Two other equations may be obtained. The integral equivalent of (1.3) is

$$\oint \boldsymbol{B} \cdot \mathrm{d}\boldsymbol{S} = 0$$

Applying this to a drum-shaped region as in fig. 1.10(a), shrinking the drum on to the surface of discontinuity gives

$$[\boldsymbol{B} \cdot \boldsymbol{n}]_2 = [\boldsymbol{B} \cdot \boldsymbol{n}]_1 \tag{1.74}$$

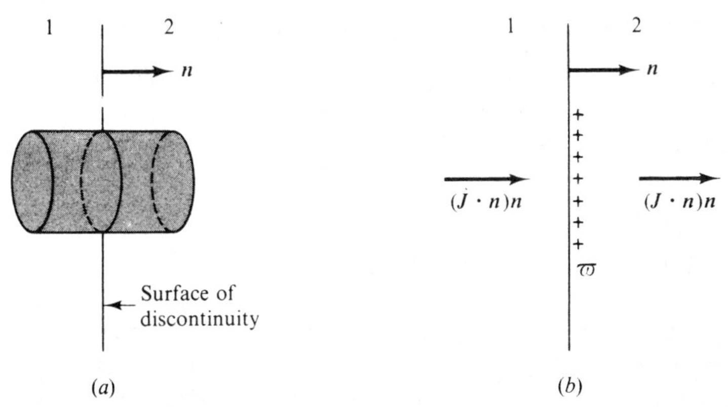

Fig. 1.10 (*a*). Surface of integration containing a surface of discontinuity. (*b*) Charges and currents at the surface between two conducting media.

Equation (1.4) gives

$$\oint \boldsymbol{D} \cdot \mathrm{d}\boldsymbol{S} = \int \rho \mathrm{d}v$$

ρ can be infinite, in that a surface charge can exist on the surface of a conductor, giving rise to an infinite value for the volume charge density. If the surface charge is ϖ per unit area we have

$$[\boldsymbol{D} \cdot \boldsymbol{n}]_2 - [\boldsymbol{D} \cdot \boldsymbol{n}]_1 = \varpi \tag{1.75}$$

A further equation arises if the junction is between two conducting media, since charge and current will be related. This situation is shown in fig. 1.10(*b*). Conservation of charge requires that

$$[\boldsymbol{J} \cdot \boldsymbol{n}]_2 - [\boldsymbol{J} \cdot \boldsymbol{n}]_1 = -\frac{\partial \varpi}{\partial t} \tag{1.76}$$

Combining with (1.75) we have

$$\left[\left(\frac{\partial \boldsymbol{D}}{\partial t} + \boldsymbol{J}\right) \cdot \boldsymbol{n}\right]_2 - \left[\left(\frac{\partial \boldsymbol{D}}{\partial t} + \boldsymbol{J}\right) \cdot \boldsymbol{n}\right]_1 = 0 \tag{1.77}$$

It is worth commenting on the significance of these various equations: (1.72) and (1.73) are the conditions necessary to obtain solutions; (1.74) must be satisfied if (1.72) is satisfied; (1.75) expresses a condition if the surface charge is known, but otherwise is a formula for finding the surface charge density.

1.4.2 Reflection from a perfect conductor at normal incidence

The problem is illustrated in fig. 1.11, in which we consider a plane wave falling normally onto a perfect plane conductor. The medium to the left of the conductor is taken as lossless, with parameters μ, ε. Without loss of generality the incident wave is described by

$$\left.\begin{aligned} E_x &= E_0\, e^{-j\beta z} \\ H_y &= \zeta^{-1} E_0\, e^{-j\beta z} \end{aligned}\right\} \tag{1.78}$$

The mathematical problem is to find a solution of Maxwell's equations which contains (1.78) together with additional terms which ensure that the electric field is zero at $z = 0$, the surface of the conductor. It may be fairly guessed that a reflected wave must exist, for which (from (1.30) and (1.33))

$$E_x = B\, e^{j\beta z}$$

$$H_y = -\zeta^{-1} B\, e^{j\beta z}$$

We must therefore try a complete solution

$$E_x = E_0\, e^{-j\beta z} + B\, e^{j\beta z}$$

$$H_y = \zeta^{-1}(E_0\, e^{-j\beta z} - B\, e^{j\beta z})$$

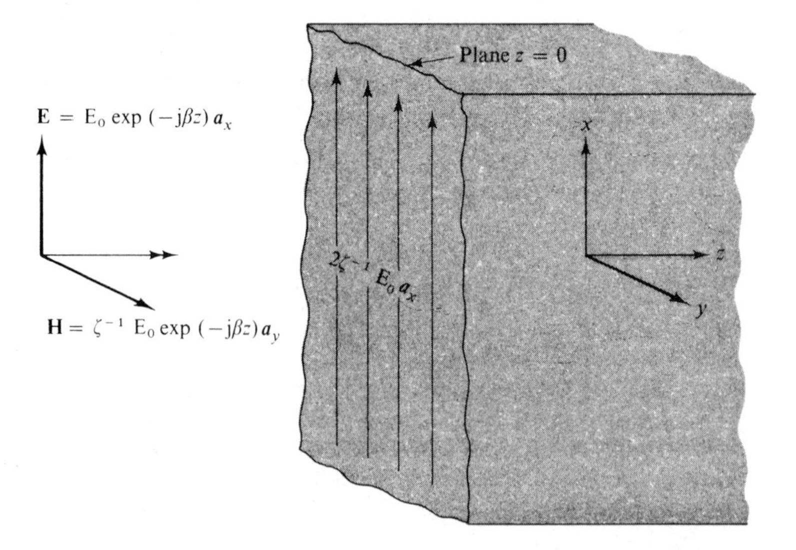

Fig. 1.11. A plane wave incident normally on a perfect conductor.

The single boundary condition is $E_x = 0$ at $z = 0$. Since E_x is a complex amplitude, this condition ensures that the electric field is zero at $z = 0$ for all time. The condition is satisfied if $B = -E_0$, giving finally

$$E_x = E_0(e^{-j\beta z} - e^{j\beta z}) = -2jE_0 \sin \beta z \tag{1.79}$$

$$H_y = \zeta^{-1}E_0(e^{-j\beta z} + e^{j\beta z}) = 2\zeta^{-1}E_0 \cos \beta z \tag{1.80}$$

This solution is in fact the only one, since it can be shown that the conditions imposed lead to uniqueness.

The physical fields corresponding to (1.79) and (1.80) are

$$\begin{aligned} E_x(z, t) &= \mathrm{Re}\,(-2jE_0 \sin\beta z\, e^{j\omega t}) = 2E_0 \sin\beta z \sin \omega t \\ H_y(z, t) &= \mathrm{Re}\,(2\zeta^{-1}E_0 \cos \beta z\, e^{j\omega t}) = 2\zeta^{-1}E_0 \cos \beta z \cos \omega t \end{aligned} \tag{1.81}$$

These describe *standing waves*: the amplitude of each field is a function of position, with periodic maxima and zeros, as contrasted with a progressive wave as described by (1.78) for which the amplitude is independent of position. The concept of standing waves is of general application and will be discussed in a wider context later in this book. In the present context we are solely concerned with derivation of solutions.

Equation (1.80) shows that at $z = 0$ (more precisely, infinitesimally to the left of the plane $z = 0$),

$$H_y = 2\zeta^{-1}E_0$$

Inside the conductor, $H_y = 0$. Hence by (1.73) there must be a current sheet in the surface of the conductor. To give the correct discontinuity in **H**, this current sheet must flow in the positive x-direction and be of magnitude $2\zeta^{-1}E_0$. This is shown in fig. 1.11.

1.4.3 Reflection from a perfect conductor at oblique incidence

Once the incidence is other than normal we have to consider the polarisation of the wave: the normal to the plane surface and the direction of the incident wave define a plane (the *plane of incidence*), and results will depend on the way the field vectors lie with respect to this plane. Invoking the possibility of superposition, we treat the case of the $\boldsymbol{E}$ vector lying (i) perpendicular to the plane of incidence (parallel to the conducting surface), and (ii) parallel to the plane of incidence. These two cases are illustrated in fig. 1.12.

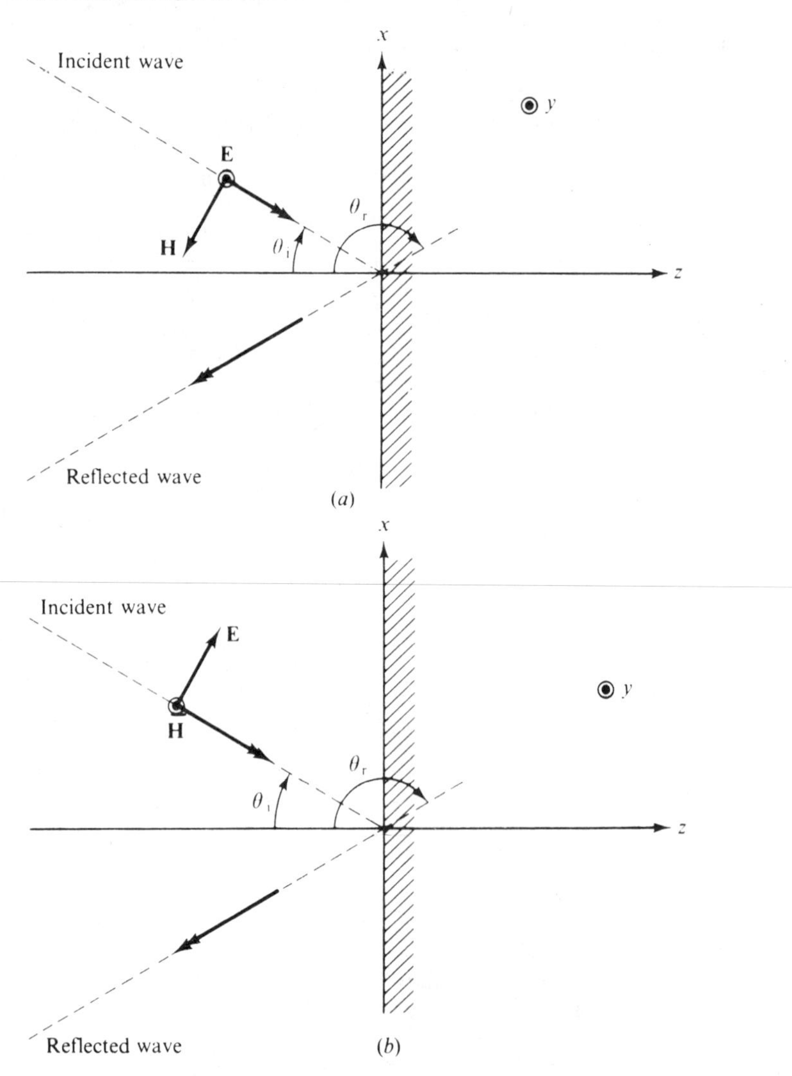

Fig. 1.12. A plane wave incident obliquely on a perfect conductor. Electric vector (*a*) perpendicular to, (*b*) parallel to plane of incidence.

(i) **E** *perpendicular to the plane of incidence*. It is necessary to use the general form of equations (1.44) to (1.47) to obtain suitable expressions for the incident wave: the vector in the direction of propagation is $-\boldsymbol{a}_x \sin\theta_i + \boldsymbol{a}_z \cos\theta_i$ using axes and angles defined in fig. 1.12(*a*). The electric field has direction $\boldsymbol{a}_y$, and the magnetic field direction $-\boldsymbol{a}_x \cos\theta_i - \boldsymbol{a}_z \sin\theta_i$. Thus

$$\mathbf{E} = \mathrm{E}_0 \boldsymbol{a}_y \exp[\mathrm{j}\beta(x \sin\theta_i - z \cos\theta_i)] \qquad (1.82)$$

$$\mathbf{H}=\zeta^{-1}\mathrm{E}_0(-\boldsymbol{a}_x \cos\theta_\mathrm{i}-\boldsymbol{a}_z \sin\theta_\mathrm{i}) \exp[\mathrm{j}\beta(x\sin\theta_\mathrm{i}-z\cos\theta_\mathrm{i})] \quad (1.83)$$

A suitable trial expression for a reflected wave may be obtained by replacing E_0 and θ_i in (1.82) and (1.83) by E_r and θ_r, which are both regarded as unknowns. The total electric field is then given by

$$\mathrm{E}_y=\mathrm{E}_0\exp[\mathrm{j}\beta(x\sin\theta_\mathrm{i}-z\cos\theta_\mathrm{i})]+\mathrm{E}_\mathrm{r}\exp[\mathrm{j}\beta(x\sin\theta_\mathrm{r}-z\cos\theta_\mathrm{r})] \quad (1.84)$$

This expression has to vanish at $z=0$ *for all values of* x. This requires

$$\mathrm{E}_0\exp(\mathrm{j}\beta x\sin\theta_\mathrm{i})=-\mathrm{E}_\mathrm{r}\exp(\mathrm{j}\beta x\sin\theta_\mathrm{r})$$

Hence

$$\mathrm{E}_\mathrm{r}=-\mathrm{E}_0 \quad (1.85)$$

$$\sin\theta_\mathrm{r}=\sin\theta_\mathrm{i} \quad (1.86)$$

The only non-trivial solution of (1.86) is

$$\theta_\mathrm{r}=\pi-\theta_\mathrm{i} \quad (1.87)$$

This is shown in fig. 1.12(*a*) and expresses the well-known optical result that the angle of reflection is equal to the angle of incidence. The expressions for the reflected wave are therefore, using (1.85) and (1.87),

$$\mathbf{E}=-\mathrm{E}_0\boldsymbol{a}_y\exp[\mathrm{j}\beta(x\sin\theta_\mathrm{i}+z\cos\theta_\mathrm{i})] \quad (1.88)$$

$$\mathbf{H}=-\zeta^{-1}\mathrm{E}_0(\boldsymbol{a}_x\cos\theta_\mathrm{i}-\boldsymbol{a}_z\sin\theta_\mathrm{i})\exp[\mathrm{j}\beta(x\sin\theta_\mathrm{i}+z\cos\theta_\mathrm{i})] \quad (1.89)$$

The total fields may be found by combining (1.88) with (1.82) and (1.89) with (1.83).

(ii) **E** *parallel to plane of incidence.* Applying the geometry of fig. 1.12(*b*), the expressions for the incident wave are

$$\mathbf{E}=\mathrm{E}_0(\boldsymbol{a}_x\cos\theta_\mathrm{i}+\boldsymbol{a}_z\sin\theta_\mathrm{i})\exp[\mathrm{j}\beta(x\sin\theta_\mathrm{i}-z\cos\theta_\mathrm{i})] \quad (1.90)$$

$$\mathbf{H}=\zeta^{-1}\mathrm{E}_0\boldsymbol{a}_y\exp[\mathrm{j}\beta(x\sin\theta_\mathrm{i}-z\cos\theta_\mathrm{i})] \quad (1.91)$$

As before we add to these a similar wave obtained by replacing E_0 by E_r and θ_i by θ_r. It is immediately apparent that equating E_x to zero at $z=0$ gives an equation of the same form as (1.84), with solutions given by (1.85) and (1.87). The same law of reflection therefore holds as in the previous case. Expressions for the reflected wave are given by

$$\mathbf{E}=+\mathrm{E}_0(-\boldsymbol{a}_x\cos\theta_\mathrm{i}+\boldsymbol{a}_z\sin\theta_\mathrm{i})\exp[\mathrm{j}\beta(x\sin\theta_\mathrm{i}+z\cos\theta_\mathrm{i})] \quad (1.92)$$

$$\mathbf{H}=+\zeta^{-1}\mathrm{E}_0\boldsymbol{a}_y\exp[\mathrm{j}\beta(x\sin\theta_\mathrm{i}+z\cos\theta_\mathrm{i})] \quad (1.93)$$

It is worthy of comment that the law of reflection results from the necessity to match the phase constants parallel to the interface on both sides. Since the exponential term containing the phase constants is common to any type of wave motion, the same law of reflection applies to all forms of wave motion.

1.5 Refraction at a plane interface between dielectric media

The treatment of refraction may be illustrated by considering a plane wave incident obliquely onto the plane interface between two dielectric media of different electrical parameters.

1.5.1 Electric vector perpendicular to the plane of incidence

The geometry of the situation is shown in fig. 1.13. The incident wave is described by (1.82) and (1.83). We may suppose a reflected wave, expressions for which are obtained by replacing E_0, θ_i by E_r, θ_r as described earlier. The parameters β, ζ now have to be given suffices $\beta_1 = \omega\sqrt{(\mu_1\varepsilon_1)}$, $\zeta_1 = \sqrt{(\mu_1/\varepsilon_1)}$. We have to allow a forward wave in the second medium, expressions for which will be given by (1.82) and (1.83) with E_0 replaced by E_t and θ_i by ϕ, and β, ζ taking values $\omega\sqrt{(\mu_2\varepsilon_2)}$, $\sqrt{(\mu_2/\varepsilon_2)}$, respectively. The components of **E** and **H** parallel to the

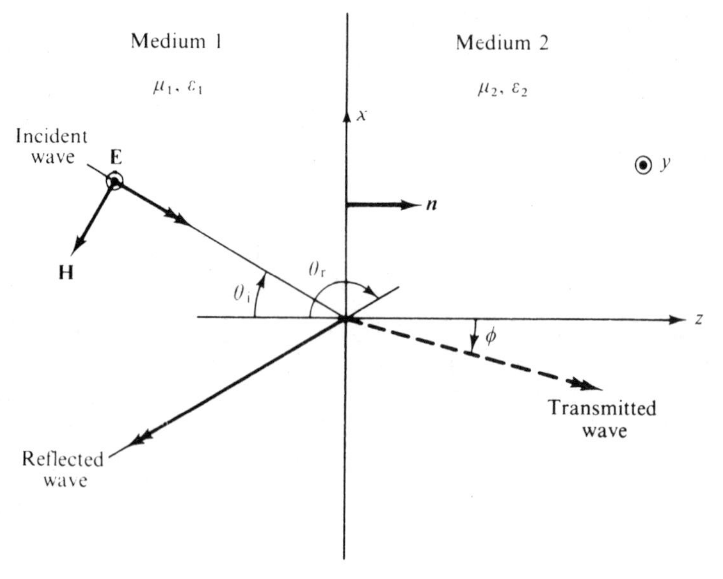

Fig. 1.13. Refraction at the plane interface between dielectric media.

surface $z=0$ have to be continuous. For $z=0$ in medium 1

$$E_y = E_0 \exp(j\beta_1 x \sin\theta_1) + E_r \exp(j\beta_1 x \sin\theta_r) \tag{1.94}$$

$$H_x = -\zeta_1^{-1} \cos\theta_i E_0 \exp(j\beta_1 x \sin\theta_i) - \zeta_1^{-1} \cos\theta_r E_r \exp(j\beta_1 x \sin\theta_r) \tag{1.95}$$

For $z=0$ in medium 2

$$E_y = E_t \exp(j\beta_2 x \sin\phi) \tag{1.96}$$

$$H_x = -\zeta_2^{-1} \cos\phi E_t \exp(j\beta_2 x \sin\phi) \tag{1.97}$$

The two values of E_y given by (1.94) and (1.96) must be identical for all x, and similarly the two values of H_x. Such identity can only be possible if the exponential factors are identical. Hence

$$\beta_1 \sin\theta_i = \beta_1 \sin\theta_r = \beta_2 \sin\phi$$

This implies, as before,

$$\theta_r = \pi - \theta_i$$

and

$$\sqrt{(\mu_1\varepsilon_1)} \sin\theta_i = \sqrt{(\mu_2\varepsilon_2)} \sin\phi$$

or

$$\sin\phi = \sqrt{\left(\frac{\mu_1\varepsilon_1}{\mu_2\varepsilon_2}\right)} \sin\theta_i \tag{1.98}$$

Equation (1.98) is equivalent to Snell's law in optics.

The equality of amplitudes in (1.94) and (1.96) gives

$$E_0 + E_r = E_t$$

and for (1.95) and (1.97)

$$\zeta_1^{-1} \cos\theta_i (E_0 - E_r) = \zeta_2^{-1} \cos\phi\, E_t$$

Hence

$$\frac{E_t}{E_0} = \frac{2\zeta_2 \cos\theta_i}{\zeta_2 \cos\theta_i + \zeta_1 \cos\phi} \tag{1.99}$$

$$\frac{E_r}{E_0} = \frac{\zeta_2 \cos\theta_i - \zeta_1 \cos\phi}{\zeta_2 \cos\theta_i + \zeta_1 \cos\phi} \tag{1.100}$$

$\cos\phi$ in these equations can be eliminated by use of (1.98).

1.5.2 Electric vector parallel to the plane of incidence

The geometry of the situation remains as in fig. 1.13, except that the electric and magnetic fields now have the configuration shown in fig.

1.12(*b*). The expressions for the various waves are of the form given by (1.90) and (1.91). We have for $z = 0$ in medium 1

$$E_x = E_0 \cos\theta_i \exp(j\beta_1 x \sin\theta_i) + E_r \cos\theta_r \exp(j\beta_1 x \sin\theta_r)$$

$$H_y = \zeta_1^{-1}[E_0 \exp(j\beta_1 x \sin\theta_i) + E_r \exp(j\beta_1 x \sin\theta_r)]$$

and for $z = 0$ in medium 2

$$E_x = E_t \cos\phi \exp(j\beta_2 x \sin\phi)$$

$$H_y = \zeta_2^{-1}[E_t \exp(j\beta_2 x \sin\phi)]$$

As before the two expressions for E_x and two for H_y must be identically equal. It follows that $\theta_r = \pi - \theta_i$ and that (1.98) still holds. In addition we have

$$(E_0 - E_r)\cos\theta_i = E_t \cos\phi$$

$$\zeta_1^{-1}(E_0 + E_r) = \zeta_2^{-1} E_t$$

whence

$$\frac{E_t}{E_0} = \frac{2\zeta_2 \cos\theta_i}{\zeta_1 \cos\theta_i + \zeta_2 \cos\phi} \tag{1.101}$$

$$\frac{E_r}{E_0} = \frac{\zeta_1 \cos\theta_i - \zeta_2 \cos\phi}{\zeta_1 \cos\theta_i + \zeta_2 \cos\phi} \tag{1.102}$$

Example

Radio waves are reflected at the surface of the earth. Assuming that the incident wave is plane, and that the earth is plane and of dielectric constant ε_r, obtain expressions for the magnitude of the reflected waves as functions of angles of incidence, for both vertically and horizontally polarised waves.

(i) *Vertical polarisation.* This is taken to mean that the electric field is perpendicular to the earth, and therefore in the plane of incidence. Equation (1.102) therefore applies to the problem. We also have $\zeta_1 = \sqrt{(\mu_0/\varepsilon_0)}$, $\zeta_2 = \sqrt{(\mu_0/\varepsilon_r\varepsilon_0)}$.

Hence

$$\frac{E_r}{E_0} = \frac{\sqrt{(\varepsilon_r)}\cos\theta - \cos\phi}{\sqrt{(\varepsilon_r)}\cos\theta + \cos\phi} = R_v(\theta) \tag{1.103}$$

where $\sin\phi = (1/\sqrt{\varepsilon_r})\sin\theta$

(ii) *Horizontal polarisation.* The electric field is now parallel to the earth and therefore perpendicular to the plane of incidence. We apply (1.100) to find

$$\frac{E_r}{E_0} = \frac{\cos\theta - \sqrt{(\varepsilon_r)}\cos\phi}{\cos\theta + \sqrt{(\varepsilon_r)}\cos\phi} = R_h(\theta) \tag{1.104}$$

$R_v(\theta)$, $R_h(\theta)$ are Fresnel reflection coefficients. They are shown in fig. 1.14 as functions of θ, calculated with $\varepsilon_r = 4, 9, 25$.

It may be noted that the real earth is characterised by a conductivity as well as by a dielectric constant. In accordance with previous analysis this can be allowed for by making ε_r complex. ε_r in the above expressions should be replaced by

$$\varepsilon_r(1 - j\sigma/\omega\varepsilon_r\varepsilon_0)$$

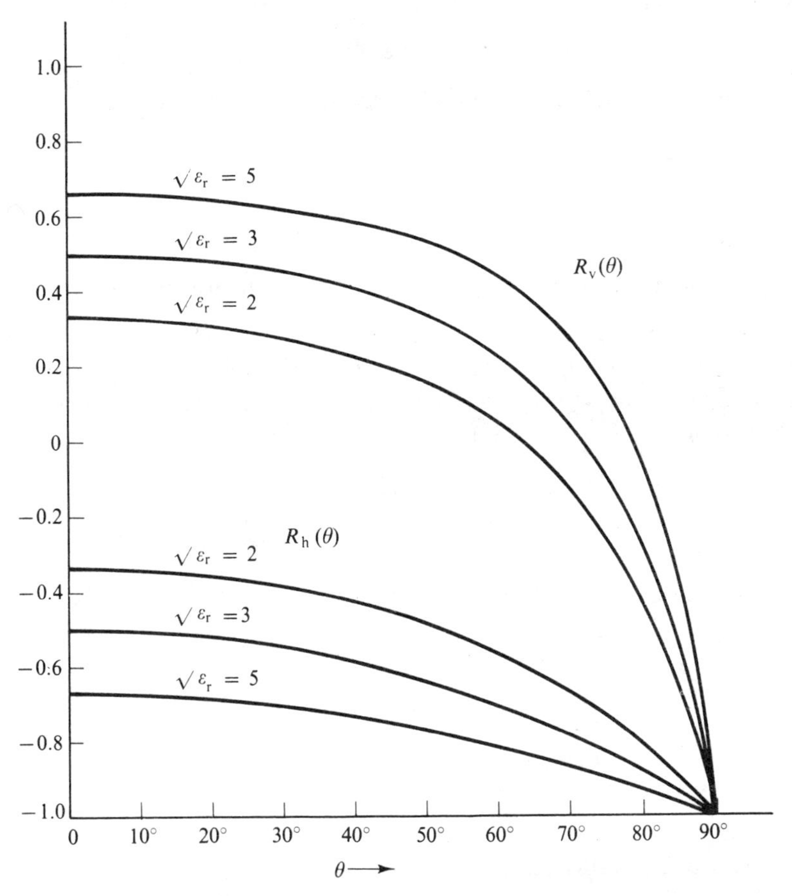

Fig. 1.14. The Fresnel reflection coefficients defined by (1.103) and (1.104).

At high enough frequencies the imaginary part will become negligibly small, so that as far as reflection is concerned the earth will behave like a lossless dielectric.

1.5.3 Optical media

The media normally used in optical systems (e.g. glass) are non-magnetic, so that for two such media $\mu_1 = \mu_2 = \mu_0$. Equation (1.98) can then be compared to Snell's law in the form

$$\sin \phi = (n_1/n_2)\sin \theta_i$$

to give
$$n_1/n_2 = \sqrt{(\varepsilon_1/\varepsilon_2)}$$

Hence the refractive index is given in terms of permittivity by

$$n = \sqrt{(\varepsilon/\varepsilon_0)} = \sqrt{\varepsilon_r}$$

The velocity of light in a medium of refractive index n is given by

$$v = c/n$$

where c is the velocity in vacuo.

1.6 Reflection at the surface of an imperfect conductor

No metallic mirror or reflector is made of perfectly conducting material. Some idea of the efficiency of ordinary conductors as reflectors may be obtained by considering reflection at normal incidence. The geometry of the problem is that of fig. 1.11. We take incident and reflected waves

$$E_x = E_0 \exp(-j\beta_1 z), \quad H_y = \zeta_1^{-1} E_0 \exp(-j\beta_1 z)$$

and
$$E_x = E_r \exp(j\beta_1 z), \quad H_y = -\zeta_1^{-1} E_r \exp(j\beta_1 z)$$

There will be a transmitted 'wave' of the form given by (1.63) and (1.64)

$$E_x = A \exp[-(1+j)z/l]$$

$$H_y = A\sqrt{(\sigma/\omega\mu_2)} \exp(-j\pi/4) \exp[-(1+j)z/l]$$

Equating the total electric field on both sides of $z = 0$ gives

$$E_0 + E_r = A$$

For the magnetic field

$$\zeta_1^{-1}(E_0 - E_r) = A\sqrt{(\sigma/\omega\mu_2)} \exp(-j\pi/4)$$

which may be written

$$\kappa(E_0 - E_r) = A$$

where $$\kappa = \zeta_1^{-1}\sqrt{(\omega\mu_2/\sigma)}\exp(j\pi/4)$$

Hence $$\frac{E_r}{E_0} = \frac{\kappa - 1}{\kappa + 1}$$

The power associated with a plane wave will be shown to be proportional to $|E|^2$. We then have an expression for the reflection coefficient

$$\left|\frac{E_r}{E_0}\right|^2 = \left|\frac{\kappa - 1}{\kappa + 1}\right| = \frac{|\kappa|^2 + 1 - 2\,\mathrm{Re}(\kappa)}{|\kappa|^2 + 1 + 2\,\mathrm{Re}(\kappa)}$$

1.7 Worked examples

1. Explain the concept of wave impedance for plane waves travelling in a lossless medium, and show how this may be extended to cover the case of a conductive medium.

Show that the reflection coefficient for plane waves travelling in vacuo and falling with normal incidence upon a medium having wave impedance ζ is given by

$$(\zeta - \zeta_0)/(\zeta + \zeta_0)$$

Hence or otherwise show that when a plane wave is incident normally upon a medium of permeability μ_0, permittivity ε_0 and high conductivity σ the fraction of power absorbed is approximately

$$2(2\omega\varepsilon_0/\sigma)^{\frac{1}{2}}$$

(Cambridge University; Electrical Sciences Tripos)

Solution. See the relevant parts of §1.2. The required plane-wave solution is $E_x = E_0 \exp(-j\beta z)$, $H_y = \sqrt{(\varepsilon/\mu)}E_x = \zeta^{-1}E_x$. This may be extended to cover the case of a conducting medium by replacing ε with $\varepsilon + \sigma/j\omega$.

Since incidence is normal we have, for free space ($z < 0$),

$$E_x = E_i \exp(-j\beta_0 z) + E_r \exp(j\beta_0 z)$$

$$H_y = \zeta_0^{-1}[E_i \exp(-j\beta_0 z) - E_r \exp(j\beta_0 z)]$$

and for the medium ($z > 0$),

$$E_x = E_t \exp(-j\beta z)$$

$$H_y = \zeta^{-1}E_t \exp(-j\beta z)$$

where $\beta_0 = \omega\sqrt{(\mu_0\varepsilon_0)}$, $\zeta_0 = \sqrt{(\mu_0/\varepsilon_0)}$, $\beta = \omega\sqrt{(\mu\varepsilon)}$, and $\zeta = \sqrt{(\mu/\varepsilon)}$. At $z = 0$, E_x and H_y are both continuous, thus

$$E_i + E_r = E_t$$

$$\zeta_0^{-1}(E_i - E_r) = \zeta^{-1}E_t$$

Hence
$$E_i + E_r = \zeta\zeta_0^{-1}(E_i - E_r)$$

$$E_r/E_i = (\zeta - \zeta_0)/(\zeta + \zeta_0)$$

To find the fraction of power absorbed we calculate $1 - |E_r/E_i|^2$. Now for a highly conducting medium $\zeta = \sqrt{(j\omega\mu/\sigma)}$, $|\zeta| \ll 1$.

Hence
$$\left|\frac{\zeta - \zeta_0}{\zeta + \zeta_0}\right|^2 \approx |-(1 - 2\zeta\zeta_0^{-1})|^2$$

$$= (1 - 2\zeta\zeta_0^{-1})(1 + 2\zeta^*\zeta_0^{-1})$$
$$\approx 1 - 4\zeta_0^{-1}\,\mathrm{Re}\,(\zeta)$$

neglecting powers of $\zeta\zeta_0^{-1}$ other than the first. We therefore find the fraction of power absorbed is given by

$$4\sqrt{\left(\frac{\varepsilon_0}{\mu_0}\right)}\sqrt{\left(\frac{\omega\mu}{2\sigma}\right)}$$

Assuming $\mu = \mu_0$ for a conductor, the answer follows.

2. A plane wave at a frequency of 10 GHz falls normally onto a plane slab of material of thickness 8 mm. The dielectric constant of the material is 2.5. Estimate the proportion of incident power transmitted.

Solution. Take the first air/dielectric interface as $z = -l$ and the second as $z = 0$, with the wave incident at $z = -l$. For $z > 0$ there is a transmitted wave only. In the slab and on the incident side there are both incident and reflected waves. We have therefore

for $z < -l$,
$$E_x = E_1 \exp(-j\beta_1 z) + E_2 \exp(j\beta_1 z)$$
$$H_y = \zeta_1^{-1}[E_1 \exp(-j\beta_1 z) - E_2 \exp(j\beta_1 z)]$$

for $-l < z < 0$,
$$E_x = E_3 \exp(-j\beta_2 z) + E_4 \exp(j\beta_2 z)$$
$$H_y = \zeta_2^{-1}[E_3 \exp(-j\beta_2 z) - E_4 \exp(j\beta_2 z)]$$

and for $z > 0$,
$$E_x = E_5 \exp(-j\beta_1 z)$$
$$H_y = \zeta_1^{-1}E_5 \exp(-j\beta_1 z)$$

where $\beta_1=\omega\surd(\mu_0\varepsilon_0)$, $\zeta_1=\surd(\mu_0/\varepsilon_0)$, $\beta_2=\omega\surd(\mu\varepsilon)$, $\zeta_2=\surd(\mu/\varepsilon)$. At $z=0$, and $z=-l$, E_x and H_y are continuous. Hence

$$E_3+E_4=E_5$$

$$\zeta_2^{-1}(E_3-E_4)=\zeta_1^{-1}E_5$$

$$E_1\exp(j\beta_1 l)+E_2\exp(-j\beta_1 l)=E_3\exp(j\beta_2 l)+E_4\exp(-j\beta_2 l)$$

$$\zeta_1^{-1}[E_1\exp(j\beta_1 l)-E_2\exp(-j\beta_1 l)]=\zeta_2^{-1}[E_3\exp(j\beta_2 l)-E_4\exp(-j\beta_2 l)]$$

From the first two equations

$$E_3=\tfrac{1}{2}E_5(1+\zeta_1^{-1}\zeta_2)$$

$$E_4=\tfrac{1}{2}E_5(1-\zeta_1^{-1}\zeta_2)$$

From the second pair

$$\begin{aligned}E_1\exp(j\beta_1 l)&=\tfrac{1}{2}[E_3\exp(j\beta_2 l)+E_4\exp(-j\beta_2 l)]\\&\quad+\tfrac{1}{2}\zeta_1\zeta_2^{-1}[E_3\exp(j\beta_2 l)-E_4\exp(-j\beta_2 l)]\\&=\tfrac{1}{4}E_5[(1+\zeta_1\zeta_2^{-1})(1+\zeta_1^{-1}\zeta_2)\exp(j\beta_2 l)\\&\quad+(1-\zeta_1\zeta_2^{-1})(1-\zeta_1^{-1}\zeta_2)\exp(-j\beta_2 l)]\\&=E_5[\cos\beta_2 l+\tfrac{1}{2}j(\zeta_1\zeta_2^{-1}+\zeta_1^{-1}\zeta_2)\sin\beta_2 l]\end{aligned}$$

Hence

$$\begin{aligned}|E_1/E_5|^2&=\cos^2\beta_2 l+\tfrac{1}{4}(\zeta_1\zeta_2^{-1}+\zeta_1^{-1}\zeta_2)^2\sin^2\beta_2 l\\&=1+\tfrac{1}{4}(\zeta_1\zeta_2^{-1}-\zeta_1^{-1}\zeta_2)^2\sin^2\beta_2 l\end{aligned}$$

With the situation given

$$\beta_2 l=\omega l\surd(\mu\varepsilon)=2.64\text{ rad}$$

$$\zeta_1/\zeta_2=\surd 2.5$$

Hence

$$|E_5/E_1|^2=0.95$$

1.8 Summary

We have seen that, in a linear medium, real electric and magnetic fields which vary sinusoidally in time may be described by phasor quantities. The instantaneous, real, fields may be derived from the phasor quantities according to the formulae $\text{Re}[\mathbf{E}\exp(j\omega t)]$, $\text{Re}[\mathbf{H}\exp(j\omega t)]$.

A linear isotropic homogeneous medium will sustain plane electromagnetic waves. The wave-like properties are contained in a factor $\exp[-(\alpha+j\beta)z]$, in which α is the attenuation constant in nepers m^{-1}, and β is the phase constant in rad m^{-1}. The attenuation in decibels is given by 8.686α dB m^{-1}. The phase velocity is given by ω/β m s^{-1}.

In a plane wave, $\boldsymbol{E}$ and $\boldsymbol{H}$ are orthogonal to each other, and propagation is in the direction of the vector $\boldsymbol{E}\times\boldsymbol{H}$. The ratio E/H is the wave impedance and equals $\sqrt{(\mu/\varepsilon)}$ for a non-conducting medium.

A slightly conducting medium gives an attenuation constant approximately equal to $\frac{1}{2}\sigma\sqrt{(\mu/\varepsilon)}$.

The field in a good conductor decays exponentially with $\alpha=\beta=\sqrt{(\omega\mu\sigma/2)}=l^{-1}$, where l is the skin depth. The power dissipated is as though the current were flowing in a sheet at the surface of resistance $(\sigma l)^{-1}$ per unit square.

When a region is composed of different media the electric and magnetic fields satisfy continuity conditions across the surface separating the different media: the components of $\boldsymbol{D}$ and $\boldsymbol{B}$ normal to the surface of discontinuity are continuous across the surface (unless surface charges are present); the components of $\boldsymbol{E}$ and $\boldsymbol{H}$ parallel to the surface are continuous across the surface (unless one medium is a perfect conductor).

In a perfect conductor, no time-varying electric or magnetic field can persist. The electric field must be normal to the surface and the magnetic field parallel to it, with a current sheet to ensure that the internal magnetic field is zero.

When a plane wave falls on a plane separating two media reflection and refraction take place according to the following laws: the incident, reflected and refracted waves together with the normal to the plane surface all lie in one plane; the angle of reflection is equal to the angle of incidence; the angle of refraction ϕ is related to the angle of incidence θ by the relation $\sqrt{(\mu_1\varepsilon_1)}\sin\theta=\sqrt{(\mu_2\varepsilon_2)}\sin\phi$. The magnitudes of reflected and transmitted waves may be calculated by applying the continuity conditions.

Formulae

For a plane-polarised plane wave travelling in direction $\boldsymbol{\nu}$,

$$\mathbf{E}=\mathbf{E}_0\exp(-\mathrm{j}\beta\boldsymbol{r}\cdot\boldsymbol{\nu})$$

where $\mathbf{E}_0\cdot\boldsymbol{\nu}=0$, and

$$\mathbf{H}=\zeta^{-1}\boldsymbol{\nu}\times\mathbf{E}$$

$$\mathbf{E}=\zeta\mathbf{H}\times\boldsymbol{\nu}$$

where $\zeta=\sqrt{(\mu/\varepsilon)}$. The phase velocity is given by

$$v_p=\omega/\beta$$

In a good dielectric of loss angle δ, the propagation constant is given by

$$\gamma = \alpha + \mathrm{j}\beta$$

where $\alpha = \frac{1}{2}\sigma\sqrt{(\mu/\varepsilon)} = (\omega \tan \delta)/2v$, $\beta = \omega\sqrt{(\mu\varepsilon)}$. Attenuation equals 8.686α dB m^{-1}.

In a good conductor,

$$\gamma = (1+\mathrm{j})\sqrt{(\tfrac{1}{2}\omega\mu\sigma)}, \qquad \zeta = \sqrt{(\mathrm{j}\omega\mu/\sigma)}$$

$$\text{Skin depth } l = \sqrt{(2/\omega\mu\sigma)}$$

$$\text{Skin resistance } R_s = 1/\sigma l = \sqrt{(\omega\mu/2\sigma)}$$

The Fresnel reflection coefficients are

$$R_v(\theta) = \frac{\sqrt{\varepsilon_r}\cos\theta - \cos\phi}{\sqrt{\varepsilon_r}\cos\theta + \cos\phi}$$

$$R_h(\theta) = \frac{\cos\theta - \sqrt{\varepsilon_r}\cos\phi}{\cos\theta + \sqrt{\varepsilon_r}\cos\phi}$$

where

$$\sin\theta = \sqrt{\varepsilon_r}\sin\phi$$

1.9 Problems

1. Determine the velocity of propagation and attenuation constant (in dB m^{-1}) of plane waves of frequency 5 MHz travelling in media for which:
 (i) $\varepsilon_r = 2.1$, $\tan\delta = 2 \times 10^{-4}$ (polytetrafluoroethylene, PTFE);
 (ii) $\varepsilon_r = 4.0$, $\tan\delta = 6 \times 10^{-2}$ (polyvinylchloride, PVC).
 Find also the attenuation constant for plane waves in water (for which $\varepsilon_r = 80$, $\sigma = 5$ S m^{-1}) at frequencies of 10 kHz, 100 kHz and 1 MHz.

2. Show that the expression

$$\boldsymbol{E} = (E_1\boldsymbol{a}_x + \mathrm{j}E_1\boldsymbol{a}_y)\exp[\mathrm{j}(\omega t - \beta_1 z)]$$

represents the electric field of a circularly polarised wave. Write down corresponding expressions for (*a*) the magnetic field of the above wave, and (*b*) the electric field of the oppositely-sensed circularly polarised wave.

An electromagnetic plane-wave disturbance is launched at $z = 0$ in a certain medium. Discuss the effect on the resulting wave when the propagation constants for the two senses of circularly polarised wave are different, say β_1 and β_2.

(Cambridge University, part question, Electrical Science Tripos.)

3. By considering the expression for the Fresnel reflection coefficients associated with a plane wave incident obliquely on the plane interface between two dielectrics, show that the only situation in which there is no reflected wave

occurs when the electric vector is parallel to the plane of incidence and the angle of incidence is given by $\tan\theta_i = \sqrt{(\varepsilon_1/\varepsilon_2)}$.

4. The reflectivity of a metal is defined as the ratio of reflected power to incident power for a plane wave normally incident onto a plane surface of the metal. At a wavelength of 500 nm the reflectivities for aluminium, silver and steel are found to be 89%, 95% and 56% respectively. Estimate the conductivity of each metal at this wavelength.

5. The conductivity of copper, measured at 24 GHz, is $3.05\times10^7\ \mathrm{S\,m^{-1}}$. Calculate the attenuation coefficient, phase coefficient, phase velocity and wavelength of a uniform plane wave propagating in copper at this frequency.

Show that the skin depth is sufficiently small to justify the concept of surface current density. Calculate the relaxation time and show that the redistribution of charge in the walls of a copper waveguide, operated at 24 GHz, takes place in a time which is very short compared with the periodic time of the guided wave.

$$\varepsilon_0 \approx 1/(36\pi\times10^9)\ \mathrm{F\,m^{-1}}$$

$$\mu_0 = 4\pi\times10^{-7}\ \mathrm{H\,m^{-1}}$$

(Southampton University)

6. The plane surface of a slab of dielectric material of permittivity ε_2 is covered with a plane layer of another dielectric of permittivity ε_1. A plane wave is normally incident from air onto the covered slab. Show that no reflection takes place if the thickness of the layer is one-quarter of the wavelength in the layer and $\varepsilon_1 = \sqrt{(\varepsilon_2\varepsilon_0)}$.

2

Energy flow in the electromagnetic field

Although the theory required to treat electromagnetic transmission systems concerns only a particular class of fields, certain aspects, such as power flow, must be handled by means of general theorems. These results have their parallel in a.c. circuit theory. Power and energy depend on products or powers of electric and magnetic fields, and so the phasor method is not immediately applicable. We therefore take equations in the forms of (1.1) to (1.4), relating instantaneous quantities.

2.1 Poynting's theorem

The derivation of this theorem uses the vector identity

$$\operatorname{div}(\boldsymbol{E}\times\boldsymbol{H})=\boldsymbol{H}\cdot\operatorname{curl}\boldsymbol{E}-\boldsymbol{E}\cdot\operatorname{curl}\boldsymbol{H} \tag{2.1}$$

We have

$$\operatorname{curl}\boldsymbol{E}=-\partial\boldsymbol{B}/\partial t$$

$$\operatorname{curl}\boldsymbol{H}=\partial\boldsymbol{D}/\partial t+\boldsymbol{J}$$

Substitution of these into (2.1) yields

$$\operatorname{div}(\boldsymbol{E}\times\boldsymbol{H})=-\boldsymbol{H}\cdot\partial\boldsymbol{B}/\partial t-\boldsymbol{E}\cdot\partial\boldsymbol{D}/\partial t-\boldsymbol{E}\cdot\boldsymbol{J} \tag{2.2}$$

The right-hand side of this equation may be interpreted as follows. In electrostatics it is shown that a small change in the field produces an increase in internal energy $\boldsymbol{E}\cdot\delta\boldsymbol{D}$ per unit volume. Similarly for a magnetostatic field the increase is $\boldsymbol{H}\cdot\delta\boldsymbol{B}$. If it be assumed that these results hold also for time varying fields, the rates of increase of electric and magnetic internal energies per unit volume will be $\boldsymbol{E}\cdot(\partial\boldsymbol{D}/\partial t)$, $\boldsymbol{H}\cdot(\partial\boldsymbol{B}/\partial t)$ respectively. The last term $\boldsymbol{E}\cdot\boldsymbol{J}$ is the rate per unit volume at which the electric field is doing work. (In the absence of sources this will be heat dissipation.) Hence (2.2) can be interpreted as an energy-balance equation.

We use (2.2) to find the energy-balance equation for a finite volume. Consider a surface S surrounding a volume V, and integrate both sides of (2.2) over the volume. The integral on the left-hand side is a form to which Gauss' theorem may be applied

$$\int \operatorname{div}(\boldsymbol{E}\times\boldsymbol{H})\,\mathrm{d}v = \oint \boldsymbol{E}\times\boldsymbol{H}\cdot\mathrm{d}\boldsymbol{S}$$

in which $\mathrm{d}\boldsymbol{S}$ is the vector area of an element of S, directed outwards from the volume. If we denote by U_e and U_m the energy densities of the electric and magnetic fields respectively, we therefore have

$$\oint \boldsymbol{E}\times\boldsymbol{H}\cdot\mathrm{d}\boldsymbol{S} = -\int_V \left[\frac{\partial U_\mathrm{e}}{\partial t}+\frac{\partial U_\mathrm{m}}{\partial t}+\boldsymbol{E}\cdot\boldsymbol{J}\right]\mathrm{d}v$$

or

$$\oint \boldsymbol{E}\times\boldsymbol{H}\cdot\mathrm{d}\boldsymbol{S}+\int \boldsymbol{E}\cdot\boldsymbol{J}\,\mathrm{d}v = -\frac{\partial}{\partial t}\int (U_\mathrm{e}+U_\mathrm{m})\,\mathrm{d}v \tag{2.3}$$

The right-hand side of (2.3) gives the rate of loss of internal electrical energy within the volume. The left-hand side must therefore represent this loss also: the second term is the rate of conversion to other forms of energy within the volume, so that the first term must represent a power flux from the volume into the region outside. This power flux is given by

$$P = \oint \boldsymbol{E}\times\boldsymbol{H}\cdot\mathrm{d}\boldsymbol{S} \tag{2.4}$$

This can be further interpreted as a local power-flux density given by

$$\boldsymbol{S} = \boldsymbol{E}\times\boldsymbol{H} \tag{2.5}$$

which is called the *Poynting vector*. It is to be realised that the interpretation of $\boldsymbol{S}$ as the local power-flux vector does not follow from (2.4) with mathematical rigour. We can in principle add to $\boldsymbol{E}\times\boldsymbol{H}$ any vector function of position for which the integral over a closed surface vanishes, and still obtain (2.4). Similar objections can be raised to the interpretation of $\boldsymbol{H}\cdot\delta\boldsymbol{B}$, $\boldsymbol{E}\cdot\delta\boldsymbol{D}$ as changes in energy density. However the case in which we are primarily interested is when time variations are sinusoidal. In this case we may show that the time average of the right-hand side of (2.3) is zero, and hence that the time average of $\oint \boldsymbol{E}\times\boldsymbol{H}\cdot\mathrm{d}\boldsymbol{S}$ is indeed the mean energy flow. As will be seen later, for a linear isotropic homogeneous medium $U_\mathrm{e} = \varepsilon E^2/2$. When $E = E_0\cos\omega t$ it is easily shown that $\partial U_\mathrm{e}/\partial t$ is proportional to $\sin 2\omega t$, and hence the time average vanishes. Similar conclusions apply to U_m. We may therefore conclude that, provided we have a closed surface and sinusoidal time variation, the use of (2.4) will give the correct mean power flow.

2.1.1 The field surrounding a wire carrying d.c.

The theory presented in the last section applies to any electromagnetic field. Let us apply it to the case of a long straight wire of circular cross-section carrying a steady current, I. Figure 2.1 shows a length of l of wire of radius a, surrounded by a cylindrical surface. Outside the wire the magnetic field is azimuthal, and at radius r of strength

$$H = I/2\pi r$$

A potential drop equal to the current multiplied by the resistance exists along the wire directed *against* the current. The electric field is therefore along the wire, in the same direction as the current, and of strength

$$E = IR$$

where R is the resistance per unit length.

Forming the product $\boldsymbol{E}\times\boldsymbol{H}$ we see that the Poynting vector is directed *radially inwards* towards the surface of the wire and of magnitude

$$S = EH = I^2R/2\pi r$$

Carrying out the integral in (2.4) the power *outflow* is

$$-(I^2R/2\pi r)2\pi r\, l = -I^2Rl$$

This is precisely the negative of the power dissipated by the resistive drop, and verifies the interpretation given.

2.1.2 Power carried by a plane wave in free space

Consider the plane wave described by

$$E_x = E_0 \cos(\omega t - \beta z)$$

$$H_y = \zeta^{-1}E_0 \cos(\omega t - \beta z)$$

where $\beta = \omega\sqrt{(\mu_0\varepsilon_0)} = \omega/c$, $\zeta = \sqrt{(\mu_0/\varepsilon_0)}$. The constitutive equations are

$$\boldsymbol{D} = \varepsilon_0\boldsymbol{E}, \qquad \boldsymbol{B} = \mu_0\boldsymbol{H}$$

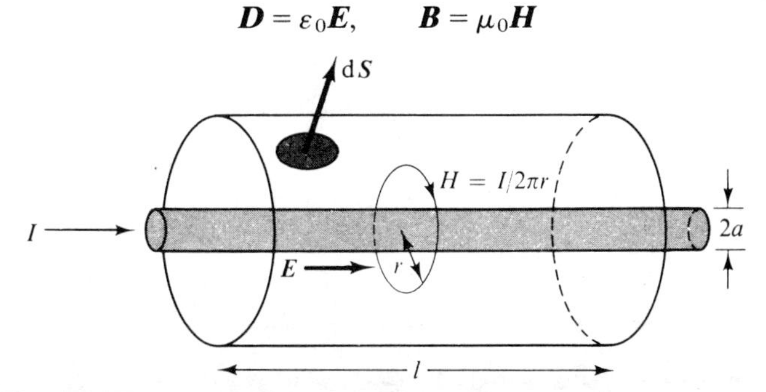

Fig. 2.1. The electric and magnetic fields near a wire carrying d.c.

Hence $\qquad \delta U_e = \varepsilon_0 \boldsymbol{E} \cdot \delta \boldsymbol{E}, \qquad \delta U_m = \mu_0 \boldsymbol{H} \cdot \delta \boldsymbol{H}$

giving $$U_e = \tfrac{1}{2}\varepsilon_0 E^2, \qquad U_m = \tfrac{1}{2}\mu_0 H^2 \tag{2.6}$$

Using the particular forms for E and H, we see for this plane wave

$$U_e = U_m = \tfrac{1}{2}\varepsilon_0 E_0^2 \cos^2(\omega t - \beta z)$$

Averaging over time, we find the mean stored energies are

$$\overline{U_e} = \overline{U_m} = \tfrac{1}{4}\varepsilon_0 E_0^2$$

We also have

$$\overline{\frac{\partial U_e}{\partial t}} = \overline{\frac{\partial U_m}{\partial t}} = \tfrac{1}{2}\varepsilon_0 E_0^2 \, \overline{\sin[2(\omega t - \beta z)]} = 0$$

Since there is no conduction current, (2.5) gives a local power flow

$$\boldsymbol{S} = \boldsymbol{E} \times \boldsymbol{H} = E_0 H_0 \cos^2(\omega t - \beta z)\boldsymbol{a}_z \tag{2.7}$$

Averaging over time, we therefore have a mean power-flow density

$$\overline{S_z} = \tfrac{1}{2}E_0 H_0 = \tfrac{1}{2}\sqrt{(\varepsilon_0/\mu_0)}E_0^2 = \tfrac{1}{2}\sqrt{(\mu_0/\varepsilon_0)}H_0^2 \text{ W m}^{-2} \tag{2.8}$$

We can associate this with a source which would produce a plane wave. Such a source must necessarily be uniform over a plane perpendicular to the direction of propagation, and by symmetry would be expected to radiate away from itself on both sides. Consider the situation shown in fig. 2.2. A current sheet K, of strength $2H_0 \cos \omega t$ A m^{-1} directed in the *negative* x-direction, is situated at $z = 0$. One of the

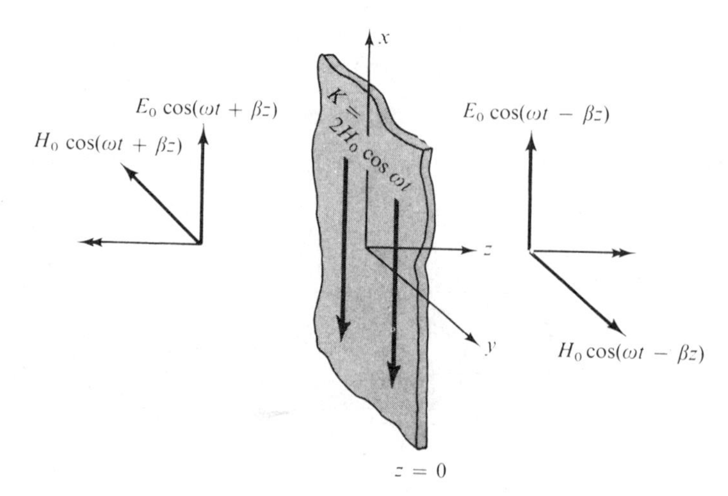

Fig. 2.2. A source of plane waves.

continuity conditions developed in the last chapter requires the tangential electric field to be continuous across the surface $z=0$, which is satisfied. The other requires any discontinuity in the tangential magnetic field to be balanced by a current sheet. The magnetic fields associated with the waves on the two sides are equal and opposite, because of the opposite directions of propagation, so that a discontinuity of $2H_0 \cos \omega t$ has to be accounted for. This is satisfied if the current sheet is as shown. Now the electric field $E_0 \cos \omega t$ is parallel to the current sheet, but in the opposite direction. If the sheet were resistive and carried such a current, the electric field would be parallel and in the *same* direction. Hence in the situation shown the current sheet is absorbing power per unit area of $-2E_0H_0 \cos^2 \omega t$, or in fact is a source *generating* power per unit area $2E_0H_0 \cos^2 \omega t$. This agrees with (2.7) if we allow for power radiated from each side of the sheet. We thus see that the current sheet is a source, the current doing work against the electric field set up by the radiated wave. In a circuit sense, the generation of radiation is, as far as the source is concerned, equivalent to loading the source with a resistance.

2.2 The complex Poynting vector

The complex amplitudes $\mathbf{E}$, $\mathbf{H}$ are related to the instantaneous quantities by the equations

$$\boldsymbol{E}(\boldsymbol{r}, \mathrm{t}) = \mathrm{Re}\,[\mathbf{E}(\boldsymbol{r}) \exp(\mathrm{j}\omega t)]$$

$$= \tfrac{1}{2}[\mathbf{E} \exp(\mathrm{j}\omega t) + \mathbf{E}^* \exp(-\mathrm{j}\omega t)]$$

$$\boldsymbol{H}(\boldsymbol{r}, t) = \tfrac{1}{2}[\mathbf{H} \exp(\mathrm{j}\omega t) + \mathbf{H}^* \exp(-\mathrm{j}\omega t)]$$

Substituting in (2.5)

$$\boldsymbol{E} \times \boldsymbol{H} = \tfrac{1}{4}[\mathbf{E} \exp(\mathrm{j}\omega t) + \mathbf{E}^* \exp(-\mathrm{j}\omega t)] \times [\mathbf{H} \exp(\mathrm{j}\omega t) + \mathbf{H}^* \exp(-\mathrm{j}\omega t)]$$

$$= \tfrac{1}{4}[\mathbf{E} \times \mathbf{H}^* + \mathbf{E}^* \times \mathbf{H}] + \tfrac{1}{4}[\mathbf{E} \times \mathbf{H} \exp(2\mathrm{j}\omega t) + \mathbf{E}^* \times \mathbf{H}^* \exp(-2\mathrm{j}\omega t)]$$

The first term is a constant and may be rewritten

$$\tfrac{1}{2}\,\mathrm{Re}\,[\mathbf{E} \times \mathbf{H}^*] \tag{2.9}$$

The second term is also real, but has a sinusoidal variation in time with angular frequency 2ω, so that its time average is zero. Hence the time average of $\boldsymbol{E} \times \boldsymbol{H}$, which we have shown above to be connected with the mean power flow, is equal to the expression in (2.9). This result leads us to define a *complex Poynting vector*

$$\mathbf{S} = \mathbf{E} \times \mathbf{H}^* \tag{2.10}$$

The mean rate at which energy flows out of the volume is given by the integral of the time average of the Poynting vector over the surface and hence by the expression

$$\bar{P} = \tfrac{1}{2}\,\mathrm{Re}\left[\oint_S \mathbf{S} \cdot \mathrm{d}\boldsymbol{S}\right] \tag{2.11}$$

†2.2.1 Relation to stored energy

The imaginary part of the complex vector may be interpreted by an analysis similar to that presented in the last section. Let us consider the case of a linear isotropic homogeneous medium. The relevant equations are

$$\operatorname{curl} \mathbf{E} = -\mathrm{j}\omega\mu\mathbf{H}$$

$$\operatorname{curl} \mathbf{H} = \mathrm{j}\omega\varepsilon\mathbf{E} + \mathbf{J} \tag{2.12}$$

By analogy with (2.1) we have

$$\operatorname{div}(\mathbf{E} \times \mathbf{H}^*) = \mathbf{H}^* \cdot \operatorname{curl} \mathbf{E} - \mathbf{E} \cdot \operatorname{curl} \mathbf{H}^* \tag{2.13}$$

Since the coefficients in (2.12) are real, we have

$$\operatorname{curl} \mathbf{H}^* = -\mathrm{j}\omega\varepsilon\mathbf{E}^* + \mathbf{J}^*$$

On substitution in (2.13) we find

$$\begin{aligned}\operatorname{div}(\mathbf{E} \times \mathbf{H}^*) &= \mathbf{H}^* \cdot (-\mathrm{j}\omega\mu\mathbf{H}) - \mathbf{E} \cdot (-\mathrm{j}\omega\varepsilon\mathbf{E}^* + \mathbf{J}^*)\\ &= -\mathrm{j}\omega\mu\mathbf{H} \cdot \mathbf{H}^* + \mathrm{j}\omega\varepsilon\mathbf{E} \cdot \mathbf{E}^* - \mathbf{E} \cdot \mathbf{J}^* \end{aligned} \tag{2.14}$$

Now $\mu\mathbf{H} \cdot \mathbf{H}^*$ can be shown to be proportional to the mean of the stored magnetic energy density U_m. From (2.6), $U_\mathrm{m} = \frac{1}{2}\mu\,\boldsymbol{H} \cdot \boldsymbol{H}$. Substituting for $\boldsymbol{H}$ in terms of $\mathbf{H}$, this expression becomes

$$U_\mathrm{m} = \tfrac{1}{8}\mu[2\mathbf{H} \cdot \mathbf{H}^* + \mathbf{H} \cdot \mathbf{H} \exp(2\mathrm{j}\omega t) + \mathbf{H} \cdot \mathbf{H}^* \exp(-2\mathrm{j}\omega t)]$$

Taking the time average of both sides we have

$$\overline{U_\mathrm{m}} = \tfrac{1}{4}\mu\mathbf{H} \cdot \mathbf{H}^*$$

Similarly

$$\overline{U_\mathrm{e}} = \tfrac{1}{4}\varepsilon\mathbf{E} \cdot \mathbf{E}^*$$

Hence (2.14) may be rewritten

$$\operatorname{div}(\mathbf{E} \times \mathbf{H}^*) = 4\mathrm{j}\omega(\overline{U_\mathrm{e}} - \overline{U_\mathrm{m}}) - \mathbf{E} \cdot \mathbf{J}^*$$

Integrating as before over a volume V

$$\tfrac{1}{2}\oint_S \mathbf{E} \times \mathbf{H}^* \cdot \mathrm{d}\boldsymbol{S} = 2\mathrm{j}\omega \int_V (\overline{U_\mathrm{e}} - \overline{U_\mathrm{m}})\,\mathrm{d}v - \tfrac{1}{2}\int_V \mathbf{E} \cdot \mathbf{J}^*\,\mathrm{d}v$$

Taking real parts of each side we have the relation previously deduced, that the real part of the complex Poynting vector is related to the mean power flow. Taking the imaginary parts of each side gives the further result that the imaginary part of the integral of the complex Poynting vector is related to the difference between stored electric and magnetic energies.

†2.3 Radiation pressure and momentum

Corresponding to the power carried by an electromagnetic wave, it is also possible to attribute to the wave a momentum. We shall consider only fields uniform in a plane transverse to the direction of propagation, although the result may be proved more generally.

Any force on a body is by action of the wave on currents and charges induced within the body. The force per unit volume is given by the expression

$$\boldsymbol{F} = \rho \boldsymbol{E} + \boldsymbol{J} \times \boldsymbol{B} \tag{2.15}$$

ρ and $\boldsymbol{J}$ can be expressed in terms of the fields by (1.2) and (1.4). Let us restrict ourselves to the case of plane-polarised plane waves travelling in the positive and negative z-directions, when the only non-zero quantities involved will be E_x, H_y, J_x, and only variation in the z-co-ordinate is permitted. Equations (1.10) and (1.11) reduce in this case (compare § 1.2) to

$$\frac{\partial E_x}{\partial z} = -\frac{\partial B_y}{\partial t}, \qquad -\frac{\partial H_y}{\partial z} = \frac{\partial D_x}{\partial t} + J_x \tag{2.16}$$

The only component of $\boldsymbol{F}$ is F_z. Since div $\boldsymbol{E}$ is zero, ρ is zero and F_z is given from (2.15) by the expression

$$F_z = J_x B_y$$

Substituting from (2.16) we have

$$\begin{aligned} F_z &= -B_y\left(\frac{\partial H_y}{\partial z} + \frac{\partial D_x}{\partial t}\right) \\ &= -B_y\frac{\partial H_y}{\partial z} - \frac{\partial}{\partial t}(D_x B_y) + D_x\frac{\partial B_y}{\partial t} \\ &= -B_y\frac{\partial H_y}{\partial z} - D_x\frac{\partial E_x}{\partial z} - \frac{\partial}{\partial t}(D_x B_y) \\ &= -\frac{1}{2}\frac{\partial}{\partial z}(\varepsilon_0 E_x^2 + \mu_0 H_y^2) - \mu_0\varepsilon_0\frac{\partial}{\partial t}(E_x H_y) \end{aligned}$$

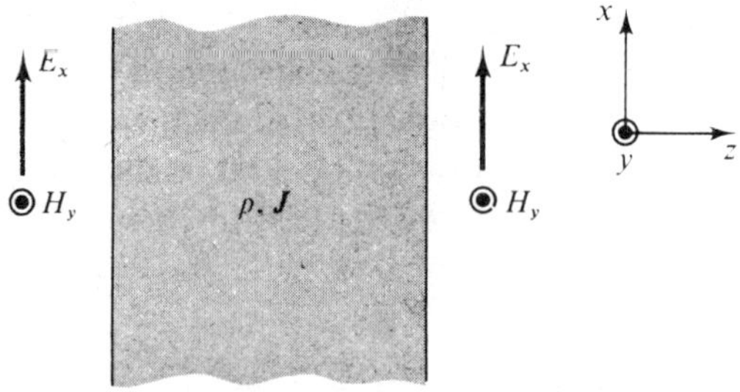

Fig. 2.3. Fields for calculating the pressure exerted by a plane wave.

The values of μ_0 and ε_0 have been used since we may regard other electrical properties as arising from induced currents. Finally, since $E_x H_y$ is the Poynting vector for the case being considered, we have

$$F_z = -\frac{1}{2}\frac{\partial}{\partial z}(\varepsilon_0 E_x^2 + \mu_0 H_y^2) - \mu_0 \varepsilon_0 \frac{\partial \boldsymbol{S}}{\partial t}$$

Let us suppose that the currents are confined to a region $z_1 < z < z_2$, as shown in fig. 2.3, and find the total force per unit area on this region. This total force, $\mathscr{F}_z$, will be given by

$$\mathscr{F}_z = \int_{z_1}^{z_2} F_z \, \mathrm{d}z = [-\tfrac{1}{2}(\varepsilon_0 E_x^2 + \mu_0 H_y^2)]_{z_1}^{z_2} - \mu_0 \varepsilon_0 \int_{z_1}^{z_2} \frac{\partial \boldsymbol{S}}{\partial t} \mathrm{d}z$$

Now at z_1 and z_2 the continuity conditions require E_x and H_y to be continuous. Hence we may write

$$\mathscr{F}_z = [\tfrac{1}{2}(\varepsilon_0 E_x^2 + \mu_0 H_y^2)]_{z_1} - [\tfrac{1}{2}(\varepsilon_0 E_x^2 + \mu_0 H_y^2)]_{z_2} - \mu_0 \varepsilon_0 \int_{z_1}^{z_2} \frac{\partial \boldsymbol{S}}{\partial t} \mathrm{d}z \tag{2.17}$$

where the fields E_x, H_y in the first two terms refer to the values *just outside* the region on which $\mathscr{F}_z$ is acting. The expressions in the square brackets are equal to the total stored energy density of the external fields, so an alternative form for (2.17) is

$$\mathscr{F}_z = U(z_1) - U(z_2) - \mu_0 \varepsilon_0 \int_{z_1}^{z_2} \frac{\partial \boldsymbol{S}}{\partial t} \mathrm{d}z \tag{2.18}$$

We may regard the term $U(z_1)-U(z_2)$ as being an action from the outside of our region, and therefore equal to a force $\mathscr{F}_e$. The resultant force on the region we have calculated to be $\mathscr{F}_z$, so that (2.18) can be written in the form

$$\mathscr{F}_e = \mathscr{F}_z + \frac{\mathrm{d}p}{\mathrm{d}t}$$

in which

$$p = \mu_0\varepsilon_0 \int_{z_1}^{z_2} \boldsymbol{S}\, \mathrm{d}z$$

This equation is to be compared to Newton's law of motion which states that force is equal to rate of change of momentum. This comparison allows us to interpret p as a momentum and therefore we can attribute to the wave a momentum (per unit area) equal to

$$\mu_0\varepsilon_0\boldsymbol{S} = \boldsymbol{S}/c^2 \tag{2.19}$$

†2.3.1 Force on a totally absorbing slab

Consider the situation of a plane wave falling normally onto a non-reflecting totally absorbing slab. We will evaluate the mean force exerted on the slab. We note that when time variations are sinusoidal the $(E_xH_y)/\partial t$ is zero, whatever the relative phases of the components. Hence, since the average of $\partial\boldsymbol{S}/\partial t$ is zero and $U(z_2)=0$, the mean pressure on the slab is, from (2.18) given by

$$\overline{\mathscr{F}_z} = \overline{U(z_1)}$$

For the plane wave, using (2.6),

$$U = U_e + U_m = \varepsilon_0 E^2$$

$$\overline{U} = \tfrac{1}{2}\varepsilon_0 E_0^2$$

Therefore

$$\overline{\mathscr{F}_z} = \tfrac{1}{2}\varepsilon_0 E_0^2 \tag{2.20}$$

Consider now the momentum of the incident wave, given by (2.19),

$$\begin{aligned} p &= \mu_0\varepsilon_0 E_x H_y \\ &= \mu_0\varepsilon_0\sqrt{(\varepsilon_0/\mu_0)}\, E_0^2 \cos^2(\omega t - \beta z) \\ &= \frac{1}{c}\varepsilon_0 E_0^2 \cos^2(\omega t - \beta z) \end{aligned}$$

The mean momentum per unit volume is therefore

$$\bar{p} = \tfrac{1}{2}\varepsilon_0 E_0^2/c$$

This momentum is totally absorbed, and we would expect the mechanical force to be given by the rate of change of momentum. The momentum per unit area carried by the wave and lost in the slab in one second is that contained in a cylinder of unit cross-sectional area and length equal to c. Thus the momentum per unit area lost each second is equal to

$$c\bar{p} = \tfrac{1}{2}\varepsilon_0 E_0^2$$

This is equal to the force per unit area and in agreement with (2.20).

The idea developed here for a uniform wave can be also developed in a general way, giving the general expression for momentum density as $\boldsymbol{D} \times \boldsymbol{B}$.

We may obtain some idea of the magnitude involved by considering a plane wave carrying 1 kW m^{-2}. From (2.8) we see that

$$\tfrac{1}{2}\surd(\varepsilon_0/\mu_0)E_0^2 = 10^3$$

Hence
$$\tfrac{1}{2}\varepsilon_0 E_0^2 = 10^3 \times \surd(\mu_0\varepsilon_0) = \tfrac{1}{3} \times 10^{-5}$$

The pressure is therefore equal to $\frac{1}{3} \times 10^{-5}$ N m^{-2}

2.4 Energy flow from sources in a finite region

In communication systems one is interested in the energy flow produced by a radiating source at distances very great compared with source dimensions or wavelength. The rate of emission of energy from a source is finite, so that at a large distance from the source region the power density must be inversely proportional to the square of the distance; to find out what implication this result has for $\boldsymbol{E}$, $\boldsymbol{H}$ and $\boldsymbol{S}$ we must turn to Maxwell's equations for an approximate solution valid for large r. The co-ordinate system appropriate to the situation is spherical polar, as shown in fig. 2.4.

2.4.1 The distant field

In a source-free region of free space the relevant equations, for sinusoidal time variation, are

$$\text{curl } \mathbf{E} = -\mathrm{j}\omega\mu\,\mathbf{H}$$

$$\text{curl } \mathbf{H} = \mathrm{j}\omega\varepsilon\,\mathbf{E}$$

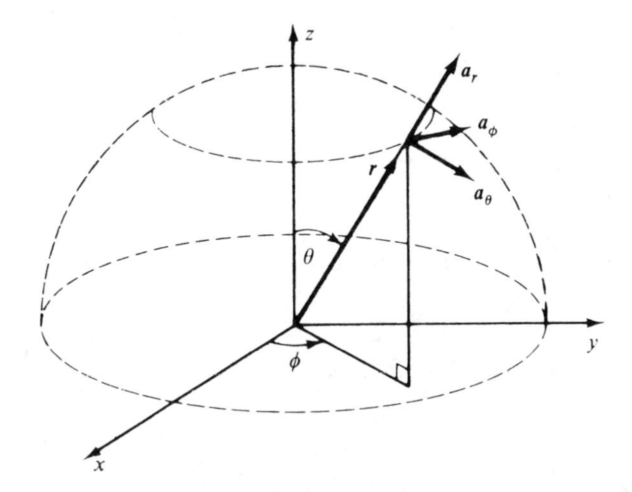

Fig. 2.4. Spherical polar co-ordinates.

In spherical polar co-ordinates

$$\operatorname{curl} \boldsymbol{F} = \frac{1}{r \sin\theta}\left[\frac{\partial}{\partial\theta}(\sin\theta F_\phi) - \frac{\partial F_\theta}{\partial\phi}\right]\boldsymbol{a}_r$$

$$+\frac{1}{r}\left[\frac{1}{\sin\theta}\frac{\partial F_r}{\partial\phi} - \frac{\partial}{\partial r}(rF_\phi)\right]\boldsymbol{a}_\theta + \frac{1}{r}\left[\frac{\partial}{\partial r}(rF_\theta) - \frac{\partial F_r}{\partial\theta}\right]\boldsymbol{a}_\phi \qquad (2.21)$$

where $\boldsymbol{a}_r$, $\boldsymbol{a}_\theta$, $\boldsymbol{a}_\phi$ are unit vectors in the radial, polar and azimuthal directions respectively as shown in fig. 2.4. and F_r, F_θ, F_ϕ are the components of $\boldsymbol{F}$ according to $\boldsymbol{F} = F_r\boldsymbol{a}_r + F_\theta\boldsymbol{a}_\theta + F_\phi\boldsymbol{a}_\phi$.

It has been stated above that the radial power flow must decrease as r^{-2}. This power flow is dependent on the radial component of the Poynting vector $\mathbf{E}\times\mathbf{H}^*$,

$$S_r = E_\theta H_\phi^* - E_\phi H_\theta^*$$

We may guess that all the field components involved (transverse to the radius vector) decrease proportional to r^{-1}. Consider the magnitudes of the three terms in (2.21). If F_θ and F_ϕ decrease proportional to r^{-1}, the radial term in curl $\boldsymbol{F}$ must decrease at least as fast as r^{-2}. We therefore deduce that if E_θ, E_ϕ are proportional to r^{-1} then H_r is proportional to r^{-2}, and similarly if H_θ, H_ϕ are proportional to r^{-1} then E_r is proportional to r^{-2}. Therefore, neglecting terms involving E_r, H_r, the θ- and ϕ-components of curl $\mathbf{E} = -j\omega\mu\mathbf{H}$ give

$$-\frac{1}{r}\frac{\partial}{\partial r}(r\mathrm{E}_\phi) \approx -j\omega\mu\mathrm{H}_\theta \tag{2.22}$$

$$\frac{1}{r}\frac{\partial}{\partial r}(r\mathrm{E}_\theta) \approx -j\omega\mu\mathrm{H}_\phi \tag{2.23}$$

Similarly curl $\mathbf{H} = j\omega\varepsilon\mathbf{E}$ yields

$$-\frac{1}{r}\frac{\partial}{\partial r}(r\mathrm{H}_\phi) \approx j\omega\varepsilon\mathrm{E}_\theta \tag{2.24}$$

$$\frac{1}{r}\frac{\partial}{\partial r}(r\mathrm{H}_\theta) \approx j\omega\varepsilon\mathrm{E}_\phi \tag{2.25}$$

These four equations can be compared with (1.22) to (1.27), and may be treated in precisely the same way. Equations (2.23) and (2.24) involve E_θ and H_ϕ, and may be written

$$\frac{\partial}{\partial r}(r\mathrm{E}_\theta) = -j\omega\mu(r\mathrm{H}_\phi)$$

$$\frac{\partial}{\partial r}(r\mathrm{H}_\phi) = -j\omega\varepsilon(r\mathrm{E}_\theta)$$

Hence

$$\frac{\partial^2}{\partial r^2}(r\mathrm{E}_\theta) + \omega^2\mu\varepsilon(r\mathrm{E}_\theta) = 0 \tag{2.26}$$

and

$$r\mathrm{E}_\theta = \mathrm{A}\,\mathrm{e}^{\pm jkr}$$

where

$$k = \omega\sqrt{(\mu\varepsilon)} = \omega/v$$

These solutions will correspond to a physical $E_\theta(\boldsymbol{r}, t)$ given by

$$E_\theta(\boldsymbol{r}, t) = r^{-1}A\cos(\omega t \pm kr)$$

By the same discussion as in chapter 1, these represent waves propagating in a radial direction, outwards for the negative sign and inwards for the positive sign. We are only interested in outward waves, since sources are restricted to a finite volume. The 'constant' A may be in fact a function of the co-ordinates θ and ϕ, which do not appear in (2.26). The required solution is therefore

$$\mathrm{E}_\theta = r^{-1}\,\mathrm{e}^{-jkr}\mathrm{Q}(\theta, \phi)$$

where $\mathrm{Q}(\theta, \phi)$ is determined by the disposition of the sources. Substitution into (2.23) gives

$$\mathrm{H}_\phi = r^{-1}\sqrt{(\varepsilon/\mu)}\,e^{-jkr}\mathrm{Q}(\theta, \phi) \tag{2.27}$$

Similarly (2.22) and (2.25) yield the expressions

$$E_\phi = r^{-1}\,e^{-jkr}Q'(\theta, \phi)$$

$$H_\theta = -r^{-1}\sqrt{(\varepsilon/\mu)}\,e^{-jkr}Q'(\theta, \phi)$$

These equations may be interpreted to give the following properties of distant waves.

(i) They are transverse, since the radial component of both **E** and **H** decrease more rapidly with distance than do the transverse components.

(ii) **E**, **H** and the direction of propagation $\boldsymbol{a}_r$ form a right-handed set of vectors.

(iii) The ratio E/H is equal to the wave impedance $\zeta = \sqrt{(\mu/\varepsilon)}$.

The results may be summed up in formulae similar to those for plane waves, equations (1.44) to (1.47).

$$\mathbf{E} = r^{-1}\,e^{-jkr}\mathbf{Q}(\theta, \phi) \tag{2.28}$$

$$\mathbf{H} = \zeta^{-1}\,\boldsymbol{a}_r \times \mathbf{E} \tag{2.29}$$

$$\mathbf{E} = \zeta\mathbf{H} \times \boldsymbol{a}_r \tag{2.30}$$

where

$$\mathbf{Q} \cdot \boldsymbol{a}_r = 0 \tag{2.31}$$

At large distances the decay proportional to r^{-1} is so slow over a few wavelengths that we have virtually a plane wave, which may be polarised in the same way as plane waves (see chapter 1). These distant spherical waves are often said to be *quasi-plane.*

2.4.2 Power flow at a great distance

We shall frequently require to know the distribution of power at a great distance from the radiating source. The direction and density of power at a given place may be determined from the Poynting vector, which takes on a simple form for quasi-plane waves.

We have

$$\mathbf{S} = \mathbf{E} \times \mathbf{H}^* = \mathbf{E} \times \zeta^{-1}(\boldsymbol{a}_r \times \mathbf{E}^*)$$
$$= \zeta^{-1}(\mathbf{E} \cdot \mathbf{E}^*)\boldsymbol{a}_r$$

in which (2.31) has been used.

An alternative expression is

$$\mathbf{S} = \zeta(\mathbf{H} \cdot \mathbf{H}^*)\boldsymbol{a}_r$$

The power flow from a distant source is thus (as expected) radial, and the mean density is

$$P = \tfrac{1}{2}\zeta^{-1}\mathbf{E} \cdot \mathbf{E}^* = \tfrac{1}{2}\zeta\mathbf{H} \cdot \mathbf{H}^* \text{ W m}^{-2} \tag{2.32}$$

2.5 Worked example

A long straight circular wire carries a current of such high frequency that the current is distributed as though the surface of the wire were plane. If $\overline{W_m}$ is the average stored magnetic energy and P the power dissipated per unit length show that

$$\tfrac{1}{2}\oint \mathbf{E}\times\mathbf{H}^*\cdot \mathrm{d}\boldsymbol{S} = -2\mathrm{j}\omega\overline{W_m} - P$$

where the integral is taken over the surface of a unit length of the wire.

An impedance per unit length, $Z_s = R_s + \mathrm{j}X_s$, is defined as the voltage drop along a unit length of the surface divided by the total current. Show that, if I is the peak current,

$$P + 2\mathrm{j}\omega\overline{W_m} = \tfrac{1}{2}|\mathrm{I}|^2 Z_s$$

and

$$R_s = X_s = \sqrt{(\omega\mu/2\sigma)}/2\pi a$$

Solution. As far as the field inside the wire is concerned the problem is virtually planar, so that the results of § 1.3.3 may be applied. We then have to concern ourselves with a width equal to the circumference of the wire, $2\pi a$. Equations (1.63) and (1.64) give

$$\mathrm{E}_x = \mathrm{H}_0\sqrt{(\mathrm{j}\omega\mu/\sigma)}\,\mathrm{e}^{-\gamma z}$$

$$\mathrm{H}_y = \mathrm{H}_0\,\mathrm{e}^{-\gamma z}$$

where

$$\gamma = (1+\mathrm{j})\sqrt{(\omega\mu\sigma/2)}$$

We reckon z from the surface of the wire inwards, and because of the small skin depth we may consider that z can increase indefinitely. The current in the wire must be related to the surface magnetic field by

$$\mathrm{I} = 2\pi a\mathrm{H}_0$$

We have, bearing in mind that d$\boldsymbol{S}$ is outward

$$\tfrac{1}{2}\oint \mathbf{E}\times\mathbf{H}^*\cdot \mathrm{d}\boldsymbol{S} = -\tfrac{1}{2}\sqrt{(\mathrm{j}\omega\mu/\sigma)}\,2\pi a|\mathrm{H}_0|^2$$

$$= -\pi a(1+\mathrm{j})\sqrt{(\omega\mu/2\sigma)}|\mathrm{H}_0|^2$$

The power dissipated per unit area has been shown to be $\tfrac{1}{2}|\mathrm{H}_0|^2\sqrt{(\omega\mu/2\sigma)}$. Hence

$$P = \pi a|\mathrm{H}_0|^2\sqrt{(\omega\mu/2\sigma)}$$

The mean density of stored magnetic energy is $\tfrac{1}{4}\mu\mathbf{H}\cdot\mathbf{H}^*$. Hence

$$\overline{W_m} = \tfrac{1}{4}\mu\,2\pi a\int_0^\infty |\mathrm{H}_0|^2 \exp\left[-2z\sqrt{(\omega\mu\sigma/2)}\right]\mathrm{d}z$$

$$= \tfrac{1}{2}\pi a|\mathrm{H}_0|^2\sqrt{(\mu/2\omega\sigma)}$$

Comparing the two sides of the given equation we find that it is verified. The impedance defined is equal to

$$Z_s = E/\mathrm{I} = \sqrt{(\mathrm{j}\omega\mu/\sigma)}/2\pi a$$

$$\tfrac{1}{2}Z_s|\mathrm{I}|^2 = \tfrac{1}{2}\cdot 2\pi a\sqrt{(\mathrm{j}\omega\mu/\sigma)}|\mathrm{H}_0|^2$$

$$= P + 2\mathrm{j}\omega\,\overline{W_\mathrm{m}}$$

2.6 Summary

We have shown that at any point in an electromagnetic field the Poynting vector $\boldsymbol{S} = \boldsymbol{E}\times\boldsymbol{H}$ may be interpreted in terms of power flow: the power flowing in a direction $\boldsymbol{n}$ is $\boldsymbol{E}\times\boldsymbol{H}\cdot\boldsymbol{n}$ W m^2.

A complex Poynting vector may be defined in terms of phasor quantities by $\mathbf{S} = \mathbf{E}\times\mathbf{H}^*$. The time-average of the power flow along $\boldsymbol{n}$ is equal to $\tfrac{1}{2}\,\mathrm{Re}\,(\mathbf{S}\cdot\boldsymbol{n})$. The imaginary part of $\mathbf{S}$ may be interpreted in terms of mean electric and magnetic energy densities.

A plane wave exerts a force on a body on which it falls. At normal incidence when the body is opaque to the wave, the mean radiation pressure is equal to the mean energy density in the field immediately adjacent to the surface on which the wave is incident. A plane wave has the attribute of momentum with a density equal to $\boldsymbol{S}/c^2$ per unit volume.

At a great distance from a distribution of sources around the origin the wave produced is quasi-plane: **E**, **H** and the direction of propagation are related as in a plane wave. Each field component transverse to the radius vector decays as r^{-1}; the radial components decay as r^{-2}. The transverse components may vary in any way with direction from the origin. The power flow is radial.

For a progressive plane wave or quasi-plane wave the mean power flow is given in terms of the phasor fields by $\tfrac{1}{2}\zeta\mathbf{H}\cdot\mathbf{H}^*$ or $\tfrac{1}{2}\zeta^{-1}\mathbf{E}\cdot\mathbf{E}^*$.

Formulae

Poynting vector	$\boldsymbol{S} = \boldsymbol{E}\times\boldsymbol{H}$
Power flow	$= \oint \boldsymbol{E}\times\boldsymbol{H}\cdot \mathrm{d}\boldsymbol{S}$
Complex Poynting vector	$\mathbf{S} = \mathbf{E}\times\mathbf{H}^*$
Average power flow	$= \tfrac{1}{2}\,\mathrm{Re}\oint \mathbf{E}\times\mathbf{H}^*\cdot \mathrm{d}\boldsymbol{S}$

Fields at a great distance from the source

$$\mathbf{E} = r^{-1}\,\mathrm{e}^{-jkr}\mathbf{Q}(\theta, \phi)$$

$$\mathbf{H} = \zeta^{-1}\boldsymbol{a}_r \times \mathbf{E}$$

$$\mathbf{E} = \zeta\mathbf{H} \times \boldsymbol{a}_r$$

where $\zeta = \sqrt{(\mu/\varepsilon)}, \qquad \mathbf{Q} \cdot \boldsymbol{a}_r = 0$

Average density of radial power flow

$$P = \tfrac{1}{2}\zeta^{-1}\mathbf{E} \cdot \mathbf{E}^* = \tfrac{1}{2}\zeta\mathbf{H} \cdot \mathbf{H}^*$$

2.7 Problems

1. A coaxial line supplies a load with direct current I_0 at voltage V_0. At any point along the line the electric field is radial and given by

$$E = V_0/r \ln (b/a)$$

in which a, b are the inner and outer radii respectively. The magnetic field is circumferential and given by

$$H = I_0/2\pi r$$

Show that the power flow is correctly predicted by Poynting's theorem.

2. Discuss the significance of each term in the following equation:

$$\oint \boldsymbol{E} \times \boldsymbol{H} \cdot \mathrm{d}\boldsymbol{S} = -\frac{\partial}{\partial t}\int (\tfrac{1}{2}\mu H^2 + \tfrac{1}{2}\varepsilon E^2)\,\mathrm{d}v - \int \sigma E^2\,\mathrm{d}v$$

The total electric and magnetic fields resulting from an incident plane wave falling on a load are given by

$$\mathrm{E}_x = \mathrm{E}_\mathrm{i}\,\mathrm{e}^{-j\beta z} + \mathrm{E}_\mathrm{r}\,\mathrm{e}^{j\beta z}$$

$$\mathrm{H}_y = \zeta^{-1}(\mathrm{E}_\mathrm{i}\,\mathrm{e}^{-j\beta z} - \mathrm{E}_\mathrm{r}\,\mathrm{e}^{jbz})$$

By consideration of the complex Poynting vector show that the total power absorbed by the load per unit area is given by

$$\tfrac{1}{2}\zeta^{-1}(|\mathrm{E}_\mathrm{i}|^2 - |\mathrm{E}_\mathrm{r}|^2)$$

3. It was shown in chapter 1, (1.69) that a plane wave at normal incidence on a block of conducting material produced a mean power loss of $\frac{1}{2}H_0^2\sqrt{(\omega\mu/2\sigma)}$, in which H_0 is the magnetic field at the surface. Show that this result can be obtained from Poynting's theorem applied at the surface.

4. A plane wave falls at normal incidence onto a plane surface of a perfect conductor. Determine the resultant current flow and hence calculate the mean pressure exerted by the radiation in terms of the incident power density.

5. Estimate the radiation pressure produced by sunlight normally incident on a perfect reflector, given that the energy is in the form of plane waves and is 2 kW.m^{-2}.

3

Guided waves

In chapter 1 the plane-wave solutions of Maxwell's equations were derived, and the principal properties investigated. Communications systems use free-space electromagnetic waves, but they also use waves confined by cables or waveguides. It is the purpose of this chapter to investigate the existence and properties of waves guided by conductors, either in the form of two-conductor systems or of hollow tubes.

The simplest possible configurations are those for which the system of conductors is uniform in one dimension: a linear axis exists and the cross-section of the system in a plane perpendicular to that axis is the same wherever the plane is taken along the axis. This means for example that a two-wire system is restricted to straight and parallel wires. In the first instance, the conductors are assumed to be perfect: as discussed earlier this implies that the component of the electric field tangential to the surface vanishes. Any system of conductors must be surrounded by a medium, which will be assumed linear, isotropic and homogeneous. Application to real systems requires the lifting, to some degree, of all these restrictions: no system of conductors can be perfectly straight and parallel, the conductor cross-section cannot be exactly constant, the conductor will not be perfect, and a dielectric will not be lossless.

The assumption of uniformity along an axis, taken as the z-axis, enables us to look for solutions dependent on the z-co-ordinate through the factor $\exp(-\gamma z)$. The application of boundary conditions will be in the xy-plane. For simplicity we shall work in Cartesian co-ordinates, although on occasion it will be necessary to use cylindrical polars.

3.1 Non-uniform plane waves

In chapter 1 it was found that solutions existed for which there was no variation in the plane perpendicular to the direction of propagation. For these waves the electric and magnetic fields were transverse to the direction of propagation. For this reason such waves are referred to as *uniform transverse electric and magnetic waves*: *uniform TEM waves* for short. We first look for *TEM* solutions which are non-uniform, in that the fields are dependent on the x and y co-ordinates, and investigate the type of structure which will support such waves.

Maxwell's equations in Cartesian co-ordinates for a linear isotropic homogeneous medium (equations (1.20) and (1.21)) are

$$\frac{\partial E_z}{\partial y}-\frac{\partial E_y}{\partial z}=-j\omega\mu H_x \tag{3.1}$$

$$\frac{\partial E_x}{\partial z}-\frac{\partial E_z}{\partial x}=-j\omega\mu H_y \tag{3.2}$$

$$\frac{\partial E_y}{\partial x}-\frac{\partial E_x}{\partial y}=-j\omega\mu H_z \tag{3.3}$$

$$\frac{\partial H_z}{\partial y}-\frac{\partial H_y}{\partial z}=j\omega\varepsilon E_x \tag{3.4}$$

$$\frac{\partial H_x}{\partial z}-\frac{\partial H_z}{\partial x}=j\omega\varepsilon E_y \tag{3.5}$$

$$\frac{\partial H_y}{\partial x}-\frac{\partial H_x}{\partial y}=j\omega\varepsilon E_z \tag{3.6}$$

We are investigating the existence of *TEM* solutions in general. We therefore place $E_z = H_z = 0$ in the equations, obtaining

$$-\frac{\partial E_y}{\partial z}=-j\omega\mu H_x \tag{3.7}$$

$$\frac{\partial E_x}{\partial z}=-j\omega\mu H_y \tag{3.8}$$

$$\frac{\partial E_y}{\partial x}-\frac{\partial E_x}{\partial y}=0 \tag{3.9}$$

$$-\frac{\partial H_y}{\partial z}=j\omega\varepsilon E_x \tag{3.10}$$

$$\frac{\partial H_x}{\partial z}=j\omega\varepsilon E_y \tag{3.11}$$

$$\frac{\partial H_y}{\partial x}-\frac{\partial H_x}{\partial y}=0 \tag{3.12}$$

Equations (3.7), (3.8), (3.10) and (3.11) are identical to (1.22), (1.23), (1.25) and (1.26) from which our plane-wave theory developed. We have however in addition (3.9) and (3.12) to satisfy.

Equations (3.7) and (3.8) enable us to express H_x and H_y in terms of E_x, E_y in the form

$$\left.\begin{aligned} H_x &= \frac{1}{j\omega\mu}\frac{\partial E_y}{\partial z} \\ H_y &= -\frac{1}{j\omega\mu}\frac{\partial E_x}{\partial z} \end{aligned}\right\} \tag{3.13}$$

Substitution into (3.12) shows that

$$\frac{\partial}{\partial x}\left(\frac{\partial E_x}{\partial z}\right)+\frac{\partial}{\partial y}\left(\frac{\partial E_y}{\partial z}\right)=0$$

or

$$\frac{\partial}{\partial z}\left(\frac{\partial E_x}{\partial x}+\frac{\partial E_y}{\partial y}\right)=0$$

Since E_x and E_y are not independent of z we must satisfy

$$\frac{\partial E_x}{\partial x}+\frac{\partial E_y}{\partial y}=0 \tag{3.14}$$

Equation (3.9) can only be satisfied if it is possible to express E_x and E_y as the gradient of a third function V in the form

$$\left.\begin{aligned} E_x &= -\frac{\partial V}{\partial x} \\ E_y &= -\frac{\partial V}{\partial y} \end{aligned}\right\} \tag{3.15}$$

Substitution into (3.14) then shows that V must satisfy the equation

$$\frac{\partial^2 V}{\partial x^2}+\frac{\partial^2 V}{\partial y^2}=0 \tag{3.16}$$

We obtain a further condition by substituting (3.13) into (3.10) and (3.11), when we find that both E_x and E_y satisfy equations of the form

$$\frac{\partial^2 E_x}{\partial z^2}+\omega^2\mu\varepsilon E_x=0$$

Use of (3.15) implies that V also satisfies the equation

$$\frac{\partial^2 V}{\partial z^2}+\omega^2\mu\varepsilon V=0 \tag{3.17}$$

This equation has solutions

$$V(x, y, z)=\psi(x, y)\exp(\pm j\beta z) \tag{3.18}$$

where $$\beta = \omega\sqrt{(\mu\varepsilon)} \tag{3.19}$$

Equation (3.16) shows that $\psi(x, y)$ must satisfy

$$\frac{\partial^2\psi}{\partial x^2} + \frac{\partial^2\psi}{\partial y^2} = 0 \tag{3.20}$$

Equations (3.15) show that **E** is the gradient vector of the function $V = \psi \exp(\pm j\beta z)$ as far as the x and y co-ordinates are concerned, therefore normal to any surface on which ψ is a constant. Hence the function $\psi(x, y)$ is determined from (3.20) subject to ψ being constant over the conducting surfaces.

3.1.1 Transmission lines: the principal wave

The existence of non-uniform *TEM* modes depends on the existence of a solution to (3.20). This is Laplace's equation in two dimensions, and is the equation obeyed by the electrostatic potential function. Such functions have many particular properties, and among them is that a potential function (in a region not containing any sources) attains its extreme values on the boundary of the region: it has no true maxima or minima. This implies that, in the case of two-dimensional fields, if the potential is constant on a closed contour it must be constant throughout the region inside. A constant value for $\psi(x, y)$ leads to zero values of all the field components in a *TEM* wave. The non-uniform *TEM* wave cannot therefore exist inside a hollow tube, but only where there are two distinct conductors. On the perimeter of each of these ψ will be constant, but the values may be different. As long as a non-trivial solution for $\psi(x, y)$ can be found, the *TEM* wave will exist at any frequency and (3.19) shows that it will travel with the velocity of plane waves associated with the medium between the conductors. This wave is termed the *principal wave*. As we shall see later, other waves are possible, but the principal wave alone has the property of existing at any frequency.

3.1.2 The parallel-plate transmission line

The simplest example of a two-wire structure is akin to the parallel-plate capacitor in electrostatics. Consider two long parallel conducting strips separated by a dielectric, as shown in fig. 3.1(a). If, as in the parallel-plate capacitor, we ignore fringing fields, we can consider the strips to be parts of infinite parallel plates which are shown in section in fig. 3.1(b).

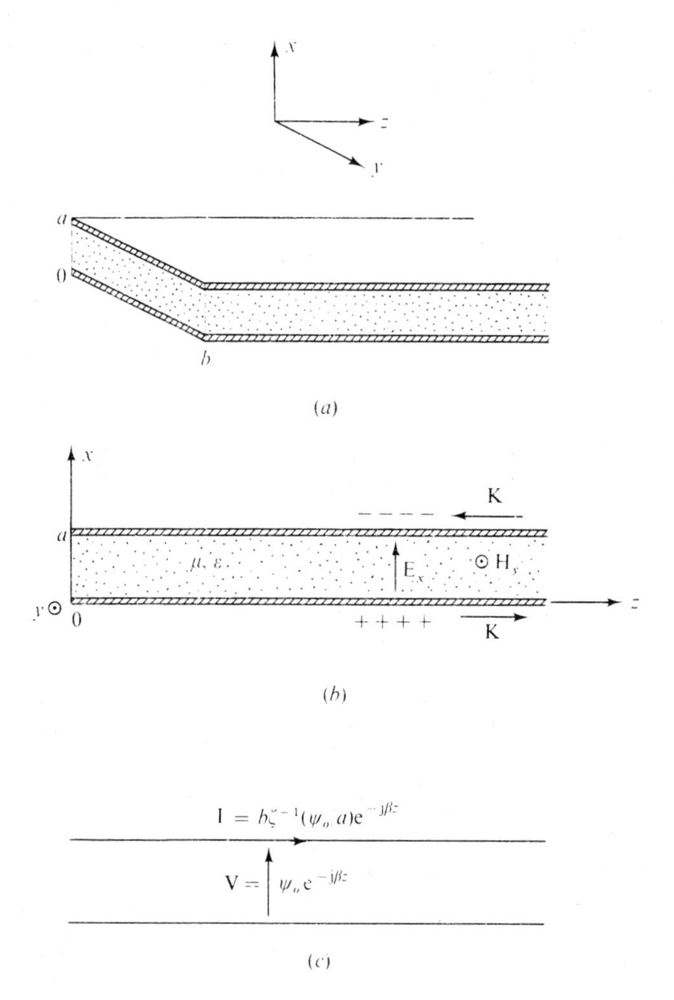

Fig. 3.1 (*a*) Strips. (*b*) Parallel plates. (*c*) Equivalent circuit of parallel-plate transmission line.

We will further make the field quantities independent of the y-co-ordinate, in the direction of the width of the plates.

The equation for ψ (equation (3.20)) then reduces to

$$\frac{\partial^2 \psi}{\partial x^2} = 0$$

The appropriate solution is

$$\psi = \psi_0 x/a$$

which is zero on $x = 0$ and a constant ψ_0 on $x = a$. The forward wave will then be described by

$$\mathrm{V} = (\psi_0 x/a)\,\mathrm{e}^{-\mathrm{j}\beta z}$$

Use of (3.13), (3.15), (3.18), (3.19) yields $\mathrm{E}_y = \mathrm{H}_x = 0$ and

$$\mathrm{E}_x = -(\psi_0/a)\,\mathrm{e}^{-\mathrm{j}\beta z} \tag{3.21}$$

$$\mathrm{H}_y = (\beta/\omega\mu)\mathrm{E}_x = \zeta^{-1}\mathrm{E}_x \tag{3.22}$$

These two field quantities are identical to those in a plane wave, and our solution describes a plane wave not in free space but in a region bounded by conducting planes at $x = 0, a$. In this sense the solution is trivial, but it will be shown later that other types of wave can be carried by the two parallel plates which are not so easily predictable.

The electric field intensity must be accompanied by charges on the plates, of density $+\varepsilon\mathrm{E}_x$ at $x = 0$ and $-\varepsilon\mathrm{E}_x$ at $x = a$. A current must flow longitudinally, of such magnitude as will annul the tangential magnetic field at the conducting surfaces. This current will be of magnitude H_y at $x = 0$ flowing in the positive z-direction, and H_y at $x = a$ flowing in the negative z-direction. Equation (3.21) tells us that we can properly talk about potential difference between the two plates, as long as we restrict ourselves to any one plane perpendicular to the z-axis. If we apply the solution to the transmission line shown in fig. 3.1(a) we may draw the circuit equivalent shown in fig. 3.1(c). The potential difference, since E_x is constant for given z, is $\psi_0 \exp(-\mathrm{j}\beta z)$, and the current in a strip of width b is $b\zeta^{-1}(\psi_0/a)\exp(-\mathrm{j}\beta z)$ in the direction shown. The current goes up one plate and returns down the other as in a conventional electric circuit. It is appropriate therefore to regard this parallel-plate structure when supporting the given fields as a transmission line.

3.1.3 The coaxial transmission line

The Cartesian co-ordinates used above will lead to a solution in closed form only when the boundaries are planes. We can write (3.18), (3.15) and (3.20) in a more general form. We have, for a forward wave,

$$\mathrm{V} = \psi\,\mathrm{e}^{-\mathrm{j}\beta z} \tag{3.23}$$

$$\mathbf{E} = -\mathrm{e}^{-\mathrm{j}\beta z}\boldsymbol{\nabla}_{\mathrm{t}}\psi \tag{3.24}$$

$$\nabla_{\mathrm{t}}^2\psi = 0 \tag{3.25}$$

where $\boldsymbol{\nabla}_{\mathrm{t}}$ denotes the gradient operator in the two co-ordinates perpendicular to the z-direction. Equation (3.13) then yields

$$\mathbf{H} = (\beta/\omega\mu)\boldsymbol{a}_z \times \mathbf{E} \tag{3.26}$$

We note that the electric and magnetic field lines are orthogonal.

For the coaxial geometry of fig. 3.2 it is appropriate to work in cylindrical polars. We can avoid directly solving (3.25) by recognising that ψ is the electrostatic potential function which takes on constant values on the circles $\rho = a$, $\rho = b$. Electrostatic theory gives the appropriate function as

$$\psi = \psi_0 \frac{\ln(b/\rho)}{\ln(b/a)} \tag{3.27}$$

This is zero on the outer conductor at $\rho = b$ and ψ_0 on the inner conductor at $\rho = a$. Hence $\mathbf{E}$ is purely radial, given by

$$\mathrm{E}_\rho = -\frac{\partial\psi}{\partial\rho}\mathrm{e}^{-\mathrm{j}\beta z} = \frac{\psi_0}{\rho \ln(b/a)}\mathrm{e}^{-\mathrm{j}\beta z} \tag{3.28}$$

The magnetic field is then given by (3.26) to be purely azimuthal about the axis and of value

$$\mathrm{H}_\phi = \frac{\psi_0}{\ln(b/a)}\frac{1}{\zeta\rho}\mathrm{e}^{-\mathrm{j}\beta z} \tag{3.29}$$

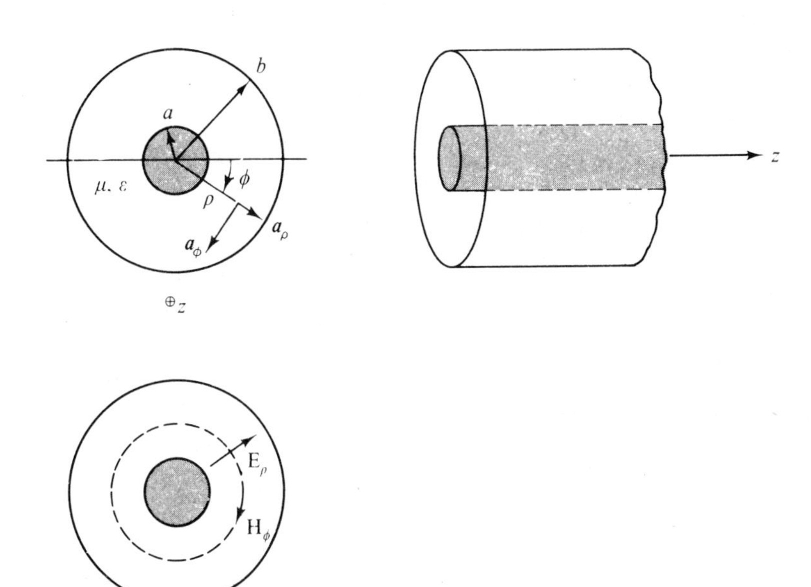

Fig. 3.2. Coaxial geometry.

This magnetic field must be supported by a current in the inner conductor flowing in the positive z-direction, of magnitude $2\pi\rho H_\phi$, giving

$$I = \psi_0 \frac{2\pi}{\ln(b/a)} \frac{1}{\zeta} e^{-j\beta z} \tag{3.30}$$

The potential difference between the conductors is

$$V_{ab} = \psi(a) e^{-j\beta z} = \psi_0 e^{-j\beta z} \tag{3.31}$$

The ratio of potential difference to current for a progressive wave is called the *characteristic impedance* of the transmission line and in a coaxial line is given by the ratio of (3.31) to (3.30),

$$Z_0 = \frac{1}{2\pi} \zeta \ln(b/a) = \frac{1}{2\pi} \sqrt{\left(\frac{\mu}{\varepsilon}\right)} \ln(b/a) \tag{3.32}$$

3.1.4 The general transmission line

The detailed analysis has been carried out above for two structures used as transmission lines. Any pair of conductors will act as a transmission line, but it is not always easy to find a solution to (3.25). This however only prevents calculation of the characteristic impedance. The velocity of propagation down the line is independent of the conductor cross-section: it depends only on the medium and is the same as the velocity of plane waves in the unbounded medium.

3.2 Waveguide modes

The *TEM* solutions derived have been shown to apply to transmission-line structures. Further generalisation must be made to allow non-zero components of electric and magnetic fields in the axial direction. We need in fact to consider only two cases: transverse electric, with $E_z = 0$, $H_z \neq 0$, or transverse magnetic, with $H_z = 0$, $E_z \neq 0$. Since Maxwell's equations are linear, two separate sets of solutions can be added together to give the most general solution. It will be shown that *TE* and *TM* waves can exist inside hollow conducting tubes, and so are usually associated with waveguides.

3.2.1 Transverse electric modes

In equations (3.1) to (3.6) we now place E_z equal to zero, and assume that the dependence on z is through the function $\exp(-\gamma z)$. Equations

(3.1) and (3.2) become

$$E_y = -j\omega\mu H_x/\gamma \tag{3.33}$$

$$E_x = j\omega\mu H_y/\gamma \tag{3.34}$$

These tell us that, as with *TEM* waves, in a transverse plane the electric and magnetic fields are mutually orthogonal and that their magnitudes are in the ratio

$$E_t/H_t = j\omega\mu/\gamma \tag{3.35}$$

Equations (3.5) and (3.4) become

$$j\omega\varepsilon E_y + \gamma H_x = -\frac{\partial H_z}{\partial x}$$

$$j\omega\varepsilon E_x - \gamma H_y = \frac{\partial H_z}{\partial y}$$

Substitution for E_x and E_y from (3.33) and (3.34) then gives

$$H_x = -\frac{\gamma}{\gamma^2 + \omega^2\mu\varepsilon}\frac{\partial H_z}{\partial x} \tag{3.36}$$

$$H_y = -\frac{\gamma}{\gamma^2 + \omega^2\mu\varepsilon}\frac{\partial H_z}{\partial y} \tag{3.37}$$

These expressions satisfy the relation imposed by (3.6), and show that in the transverse plane $\mathbf{H}_t$ can be expressed as the gradient of a function. Substitution of these various expressions in (3.3) yields an equation in H_z only:

$$\frac{\partial^2 H_z}{\partial x^2} + \frac{\partial^2 H_z}{\partial y^2} + (\gamma^2 + \omega^2\mu\varepsilon)H_z = 0 \tag{3.38}$$

or

$$\nabla_t^2 H_z + (\gamma^2 + \omega^2\mu\varepsilon)H_z = 0 \tag{3.39}$$

It is necessary to solve this equation subject to any boundary conditions applicable. The normal component of **H** at the surface of a perfect conductor must vanish, as must also the tangential electric field. Within the framework of the systems being investigated, the axial magnetic field is not restricted so the mutual orthogonality of $\mathbf{E}_t$ and $\mathbf{H}_t$ in a transverse plane assures the one condition being satisfied if the other is satisfied. Equations (3.36) and (3.37) show that $\mathbf{H}_t$ is everywhere normal to surfaces on which H_z is constant, and hence the condition that the component of $\mathbf{H}_t$ normal to a contour vanish is that

$$\boldsymbol{n} \cdot \boldsymbol{\nabla}_t H_z = \frac{\partial H_z}{\partial n} = 0 \tag{3.40}$$

where $\boldsymbol{n}$ denotes the normal to the contour.

It is known that solutions of (3.38) or (3.39) subject to the normal derivative being zero round a closed contour only exist for certain values of the expression $\gamma^2+\omega^2\mu\varepsilon$ which appears as a coefficient in the equation. This implies that for given frequency certain discrete values of γ alone are possible, each giving rise to a particular field configuration. Each such configuration is termed a *mode.*

The parallel-plate system

It is instructive to compare the solutions obtained for the parallel-plate system considered earlier in association with a strip transmission line. With the same simplification, assuming no dependence on the y-coordinate, (3.38) becomes

$$\frac{\partial^2 H_z}{\partial x^2}+\kappa^2 H_z=0$$

where

$$\kappa^2=\gamma^2+\omega^2\mu\varepsilon$$

The solution of this equation is

$$H_z=(A\cos\kappa x+B\sin\kappa x)\exp(-\gamma z)$$

The boundary condition (3.40) requires $\partial H_z/\partial x$ to vanish at $x=0, a$. The condition at $x=0$ implies that $B=0$, and

$$H_z=A\cos\kappa x\exp(-\gamma z)$$

The further condition at $x=a$ can only be satisfied if $A\sin\kappa a$ vanishes. An unrestricted value of κ requires A to be zero, and no solutions exist. For those values of κ for which $\sin\kappa a$ is zero, however, there will be solutions for H_z. Hence we consider

$$\kappa a=m\pi \qquad (m=1, 2, 3, \ldots)$$

The various components are then given by the following equations

$$\gamma_m^2+\omega^2\mu\varepsilon=(m\pi/a)^2$$

$$H_z=A\cos(m\pi x/a)\exp(-\gamma_m z)$$

$$H_x=A(\gamma_m a/m\pi)\sin(m\pi x/a)\exp(-\gamma_m z)$$

$$H_y=0$$

$$E_x=0$$

$$E_y=-j\omega\mu H_x/\gamma_m$$

For each value of m this set of equations gives expressions for the field components, the forms of which differ from one value of m to the next and are characteristic of the different modes. The most important difference to be noted is that γ takes on a different value for each m at a given frequency, and further that γ^2 may be positive or negative, implying either real or purely imaginary values for γ_m. A purely imaginary value for γ_m is interpreted as giving a progressive wave through the factor $\exp(-j\beta_m z)$. A real value must correspond to exponential attenuation. For each value of m a frequency exists, given in this instance by $\omega_m = [m\pi/a\sqrt{(\mu\varepsilon)}]$ such that if $\omega > \omega_m$, γ is imaginary, and if $\omega < \omega_m$, γ is real. This frequency is known as the *cut-off frequency* of the mode. This is shown diagrammatically in fig. 3.3. It will be recalled that the phase velocity of a wave was defined in (1.37) by

$$v_p = \omega/\beta$$

The ordinate in fig. 3.3 is normalised by introducing the phase velocity of a plane wave in the unbounded medium, $v = 1/\sqrt{(\mu\varepsilon)}$. For a particular value of m and a particular frequency, a point such as P may be located on the diagram. The slope of the line OP is then equal to the phase velocity at the frequency and mode in question. As the frequency increases the phase velocity tends towards that in the unbounded medium, but near cut-off it is very large. The *TE* modes are thus dispersive, as defined in chapter 1.

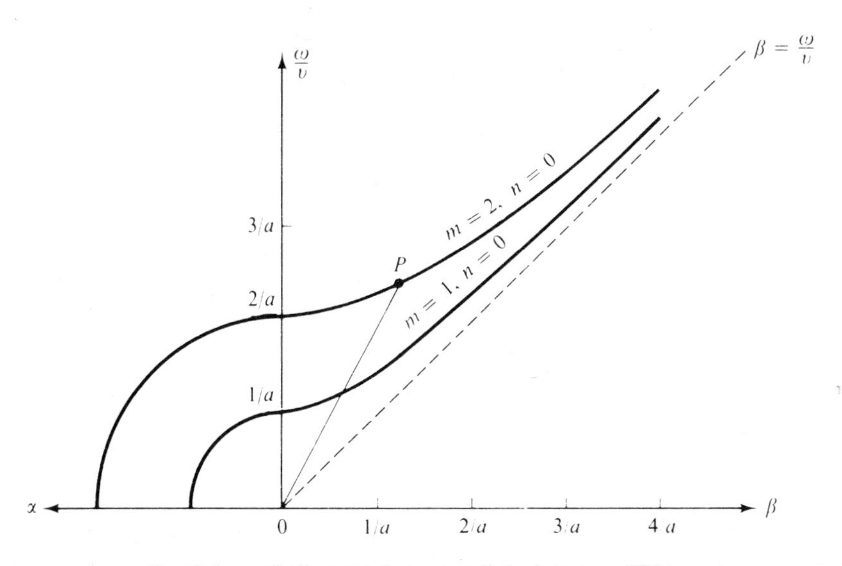

Fig. 3.3. ω-β diagram for a parallel-plate waveguide.

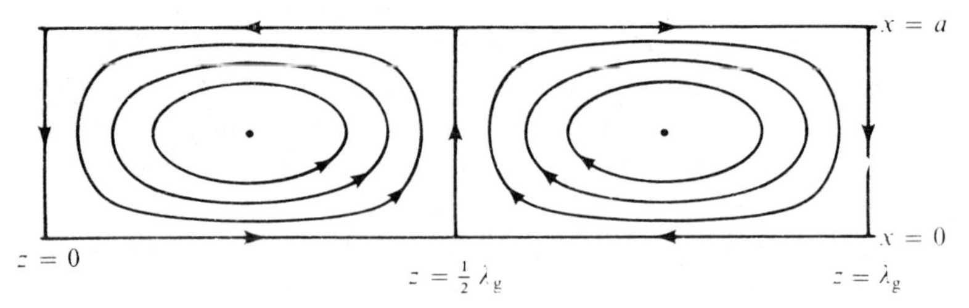

Fig. 3.4. Magnetic lines of force in a parallel-plate waveguide.

In fig. 3.4 are shown the magnetic field lines in the xz-plane for a propagating mode. The periodicity in the axial direction is given by

$$\lambda_g = 2\pi/\beta = v_p/f$$

This is the *guide wavelength*, to be carefully distinguished from λ, the wavelength of a plane wave, given by

$$\lambda = v/f$$

We note that $\lambda_g > \lambda$ since $v_p > v$.

We should also investigate the wall currents supporting the modes. These are to be found by considering the components of magnetic field tangential to the walls. The relevant transverse component is H_y, which is identically zero, and there are therefore no currents flowing in the direction of the wave. The axial component H_z requires a current sheet in the y-direction: in the wall at $x = 0$ the current density will be $A \exp(-\gamma z)$ in the positive y-direction; in the wall at $x = a$ the current density will be $A(-1)^m$ in the negative y-direction. This is shown in fig. 3.5. The pattern of current flow is thus quite different to that in the

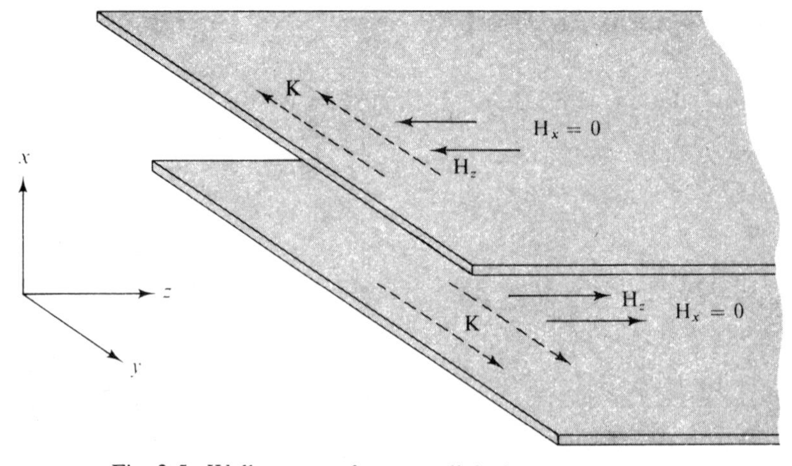

Fig. 3.5. Wall currents for a parallel-plate waveguide.

same system supporting the *TEM* transmission-line mode of propagation.

It may be observed that walls can be inserted at planes perpendicular to the y-axis without altering the field pattern. Such walls would then carry currents, both axial and transverse.

3.2.2 Transverse magnetic waves

The analysis for the situation in which $H_z = 0$, $E_z \neq 0$ is carried through from equations (3.1) to (3.6) in a fashion precisely analogous to that for *TE* waves (§3.2.1). It is found that

$$E_x = \gamma H_y / j\omega\varepsilon \tag{3.41}$$

$$E_y = -\gamma H_x / j\omega\varepsilon \tag{3.42}$$

leading again to the conclusion that the electric and magnetic field vectors are mutually orthogonal in a transverse plane, and that

$$E_t / H_t = \gamma / j\omega\varepsilon$$

E_x and E_y are expressed in terms of E_z in the form of a gradient function

$$E_x = -\frac{\gamma}{\gamma^2 + \omega^2\mu\varepsilon} \frac{\partial E_z}{\partial x} \tag{3.43}$$

$$E_y = -\frac{\gamma}{\gamma^2 + \omega^2\mu\varepsilon} \frac{\partial E_z}{\partial y} \tag{3.44}$$

The longitudinal electric field E_z must satisfy the equation

$$\frac{\partial^2 E_z}{\partial x^2} + \frac{\partial^2 E_z}{\partial y^2} + (\gamma^2 + \omega^2\mu\varepsilon) E_z = 0 \tag{3.45}$$

or

$$\nabla_t^2 E_z + (\gamma^2 + \omega^2\mu\varepsilon) E_z = 0 \tag{3.46}$$

The boundary condition is that the tangential electric field shall vanish on the conductor. This obviously requires E_z to be zero on the surfaces. The fact that $\mathbf{E}_t$ is the gradient vector of E_z means that $\mathbf{E}_t$ is normal to a surface E_z constant, and hence there can be no tangential component at the conductor surface. The normal magnetic field also vanishes at the surface because of the orthogonality between $\mathbf{E}_t$ and $\mathbf{H}_t$. The sole condition to be imposed is therefore that $E_z = 0$ on conductor surfaces.

3.3 General formulation for *TE* and *TM* waves

The theory above has been phrased in Cartesian co-ordinates, but the results can be re-written in a more general vector form, as was done for the transmission line in equations (3.23) to (3.26).

3.3.1 *TE* waves

These are characterised by the longitudinal magnetic field H_z which satisfies (3.39), repeated here,

$$\nabla_t^2 H_z + (\gamma^2 + \omega^2 \mu\varepsilon) H_z = 0 \tag{3.47}$$

and is subject to the condition that the derivative of H_z in the direction of the normal to the guide wall vanishes. Solutions to this equation determine each mode and the form of H_z for that mode. In terms of H_z the transverse magnetic field is given by the vector form of (3.36) and (3.37)

$$\mathbf{H}_t = -\frac{\gamma}{\gamma^2 + \omega^2 \mu\varepsilon} \boldsymbol{\nabla}_t H_z \tag{3.48}$$

Finally, the electric field is, from (3.33) and (3.34), given by

$$\mathbf{E}_t = (j\omega\mu/\gamma)\mathbf{H}_t \times \mathbf{a}_z \tag{3.49}$$

Following earlier usage $j\omega\mu/\gamma$ is termed the wave impedance for the mode.

3.3.2 *TM* waves

These are characterised by E_z satisfying (3.46), repeated here,

$$\nabla_t^2 E_z + (\gamma^2 + \omega^2 \mu\varepsilon) E_z = 0 \tag{3.50}$$

subject to the condition that E_z vanishes at the guide walls. Then from (3.43) and (3.44) we have

$$\mathbf{E}_t = -\frac{\gamma}{\gamma^2 + \omega^2 \mu\varepsilon} \boldsymbol{\nabla}_t E_z.$$

Finally from (3.41) and (3.42)

$$\mathbf{H}_t = (j\omega\varepsilon/\gamma)\boldsymbol{a}_z \times \mathbf{E}_t$$

The ratio $\gamma/j\omega\varepsilon$ is the wave impedance for the *TM* wave.
This general formulation can be applied to waveguides of any cross-section. The most common waveguides are rectangular or circular and we shall investigate these two types in greater detail.

3.4 Rectangular waveguide: *TE* modes

The form of the waveguide is shown in fig. 3.6, together with the Cartesian co-ordinate system appropriate to the rectangular geometry.

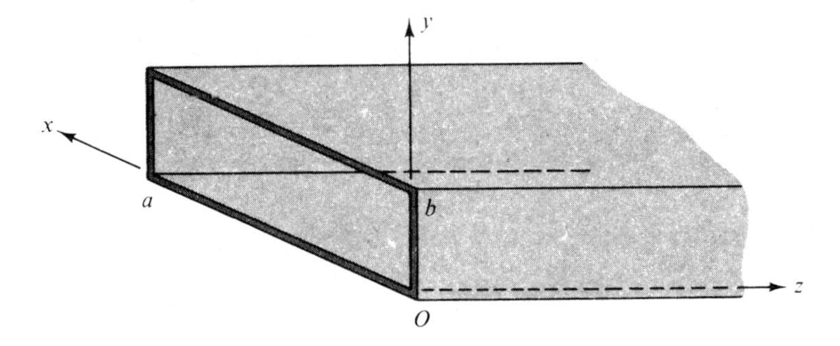

Fig. 3.6. Rectangular waveguide.

We are able to use equations (3.33) to (3.38) directly. We first obtain solutions of (3.38) for H_z which satisfy the boundary conditions of (3.40). For the present geometry these conditions are

$$\frac{\partial H_z}{\partial x}=0 \quad \text{on } x=0,\ x=a$$

$$\frac{\partial H_z}{\partial_y}=0 \quad \text{on } y=0,\ y=b$$

The form of (3.38) suggests solutions of the form

$$H_z=(A\cos px+B\sin px)(C\cos qy+D\sin qy)\,e^{-\gamma z} \tag{3.51}$$

Substitution into (3.38) shows that p, q and γ must satisfy the equation

$$-(p^2+q^2)+\gamma^2+\omega^2\mu\varepsilon=0 \tag{3.52}$$

Differentiating (3.51) we have

$$\frac{\partial H_z}{\partial x}=p(-A\sin px+B\cos px)(C\cos qy+D\sin qy)\,e^{-\gamma z}$$

$$\frac{\partial H_z}{\partial y}=q(A\cos px+B\sin px)(-C\sin qy+D\cos qy)\,e^{-\gamma z}$$

The conditions $\partial H_z/\partial x=0$ on $x=0$ and $\partial H_z/\partial y=0$ on $y=0$ can only be satisfied for all x, y, z if $B=D=0$. We thus have

$$\frac{\partial H_z}{\partial x}=-ACp\sin px\cos qy\;e^{-\gamma z}$$

$$\frac{\partial H_z}{\partial y}=-ACq\cos px\sin qy\;e^{-\gamma z}$$

At $x = a$, $\partial H_z/\partial x = 0$, which is only possible if

$$\sin pa = 0$$

Similarly at $y = b$, $\partial H_z/\partial y = 0$, requiring

$$\sin qb = 0$$

We therefore conclude that p and q can only take discrete values

$$\left.\begin{aligned} p &= m\pi/a \\ q &= n\pi/b \end{aligned}\right\} \quad m, n = 0, 1, 2, \ldots$$

Further $$H_z = A \cos(m\pi x/a)\cos(n\pi y/b)\,e^{-\gamma z}$$

in which A is an arbitrary constant.

There is thus a set of modes described by the values of m and n, each mode being conveniently designated by the symbol TE_{mn}. The propagation constant for the TE_{mn} mode is, from (3.52), given by

$$\gamma^2 = -[\omega^2\mu\varepsilon - (m\pi/a)^2 - (n\pi/b)^2] \tag{3.53}$$

As discussed earlier, for a progressive wave γ must be imaginary, so for the TE_{mn} mode to propagate it is necessary that

$$\omega^2\mu\varepsilon > (m\pi/a)^2 + (n\pi/b)^2 = \omega_{mn}^2\mu\varepsilon$$

For each mode a cutoff frequency exists below which propagation is not possible. The cut-off frequency is given by

$$f_{mn} = \omega_{mn}/2\pi = \tfrac{1}{2}v[(m/a)^2 + (n/b)^2]^{\frac{1}{2}} \tag{3.54}$$

where $v = 1/\sqrt{(\mu\varepsilon)}$ is the velocity of plane waves in the unbounded medium.

In terms of f_{mn} we may write (3.53) in the form

$$\gamma^2 = -(2\pi/v)^2(f^2 - f_{mn}^2)$$

Hence $$\left.\begin{aligned} \gamma &= j\beta = j(2\pi/v)(f^2 - f_{mn}^2)^{\frac{1}{2}} \quad \text{for } f > f_{mn} \\ \gamma &= \alpha = (2\pi/v)(f_{mn}^2 - f^2)^{\frac{1}{2}} \quad \text{for } f < f_{mn} \end{aligned}\right\} \tag{3.55}$$

For frequencies greater than f_{mn}, the TE_{mn} mode is propagated with phase constant β, or a phase velocity

$$v_{mn} = \omega/\beta = vf/(f^2 - f_{mn}^2)^{\frac{1}{2}} > v \tag{3.56}$$

For frequencies less than f_{mn} the fields are attenuated, with attenuation coefficient α nepers per metre given by (3.55).

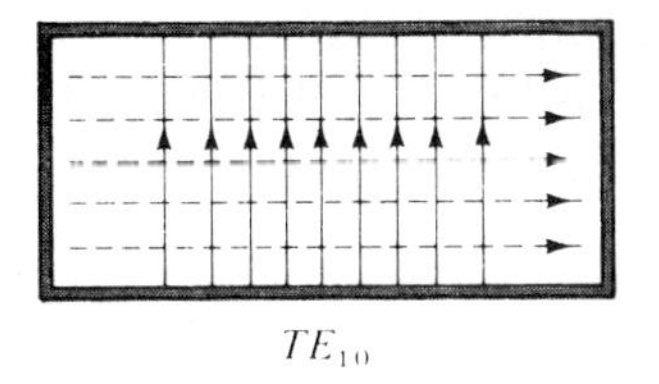

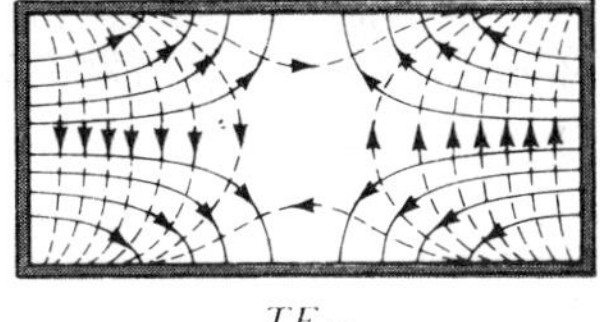

Fig. 3.7. Transverse field patterns of *TE* modes in rectangular waveguide. Solid lines show electric field, pecked lines magnetic field.

Considering only a propagating wave, (3.36) and (3.37) give

$$H_x = jA\beta \frac{m\pi}{a}\left[\left(\frac{m\pi}{a}\right)^2 + \left(\frac{n\pi}{b}\right)^2\right]^{-1} \sin\frac{m\pi x}{a}\cos\frac{n\pi y}{b}\exp(-j\beta z)$$

$$H_y = jA\beta \frac{n\pi}{b}\left[\left(\frac{m\pi}{a}\right)^2 + \left(\frac{n\pi}{b}\right)^2\right]^{-1} \cos\frac{m\pi x}{a}\sin\frac{n\pi y}{b}\exp(-j\beta z)$$

Finally from (3.33) and (3.34)

$$E_x = \omega\mu H_y/\beta$$

$$E_y = -\omega\mu H_x/\beta$$

Some modal patterns are shown in fig. 3.7. It will be observed that the analysis previously done for the parallel-plate system refers to the TE_{m0} modes in the rectangular guide.

3.4.1 The dominant mode

If we assume that a and b are unequal, and in particular $a > b$, (3.54) shows that the lowest cut-off frequency corresponds to the case $m = 1, n = 0$, when

$$f_{10} = v/2a \tag{3.57}$$

The next lowest will be either f_{20} or f_{11} depending on the relative magnitudes of a and b. There is thus a range of frequencies for which only one mode can propagate. This is called the *dominant mode*.

It is instructive to express the cut-off condition in terms of wavelength. Since v is the velocity of a *TEM* wave in the medium, the wavelength of a *TEM* wave of frequency f_{10} is given by $\lambda = v/f_{10}$. Equation (3.57) then shows that

$$\lambda = 2a$$

This cut-off frequency is therefore that for which the wide dimension is one half-wavelength.

When a source of any sort is used to generate waves in a waveguide it will in general produce all modes in varying proportions. However if only propagation in the lowest mode is possible at the frequency used, all higher modes will be attenuated. For many purposes the single mode is desirable. In cases where the frequency is such that propagation is possible at several modes special precautions are needed, since departures from ideal guides will lead to one mode generating another, and *mode coupling* takes place.

3.4.2 Degeneracy

A square waveguide, for which $a = b$, exhibits the phenomenon known as *degeneracy*: the mode TE_{mn} differs only from TE_{nm} by rotation through a right angle. The cut-off frequencies are identical, and no dominant mode exists. This means that a square guide is liable to mode coupling. Any symmetry in cross-section, as in a circular guide for example, gives rise to degeneracy.

3.4.3 Standard sizes for rectangular waveguides

The frequency range over which the dominant mode is usable is limited because, apart from the necessity to keep below the cut-off frequency of the next mode, the phase velocity varies rapidly as the lower cut-off is approached. A range of standard sizes has evolved, a selection of which is detailed in table 3.1.

3.4.4 Guide wavelength

The axial periodicity of any field component is given by $\exp(-j\beta z)$. The guide wavelength is therefore given by

$$\lambda_g = 2\pi/\beta = v_{mn}/f$$

Since $v_{mn} > v$ by (3.56), λ_g is always greater than the wavelength of a *TEM* wave in the same medium.

Table 3.1. *Some standard rectangular guides*

Type	Inside dimensions (in)	(mm)	TE_{10} cut-off (GHz)	Range (GHz)
WG 10	2.840 × 1.340	72.14 × 34.04	2.080	2.60–3.95
WG 12	1.872 × 0.872	47.55 × 22.15	3.155	3.95–5.85
WG 14	1.372 × 0.622	34.85 × 15.80	4.285	5.85–8.20
WG 16	0.900 × 0.400	22.86 × 10.16	6.56	8.20–12.4
WG 18	0·622 × 0.311	15.80 × 7.90	9.49	12.4–18.0
WG 22	0.280 × 0.140	7.11 × 3.56	21.10	26.5–40.0

Example

WG 16 has internal dimensions 22.9 mm × 10.2 mm and is air-filled. Find the five lowest cut-off frequencies. It is recommended for use in the dominant mode for frequencies between 8.20 and 12.40 GHz. Find the phase velocity and guide wavelength at these extreme frequencies in terms of free-space *TEM* values.

Solution. In (3.54) we have to insert

$$a = 0.0229 \text{ m}$$

$$b = 0.0102 \text{ m}$$

$$v = c = 3 \times 10^8 \text{ m s}^{-1}$$

We find

$$f_{10} = 6.55 \text{ GHz}$$

$$f_{20} = 13.1 \text{ GHz}$$

$$f_{01} = 14.7 \text{ GHz}$$

$$f_{11} = 16.0 \text{ GHz}$$

$$f_{21} = 19.7 \text{ GHz}$$

From (3.56),

$$v_{10}/c = f/(f^2 - f_{10}^2)^{\frac{1}{2}}$$

We have

$$\lambda_g = v_{10}/f = c/(f^2 - f_{10}^2)^{\frac{1}{2}}$$

Hence

$$\lambda_g/\lambda = f/(f^2 - f_{10}^2)^{\frac{1}{2}}$$

At 12.40 GHz

$$v_{10}/c = \lambda_g/\lambda = 1.178$$

At 8.20 GHz

$$v_{10}/c = \lambda_g/\lambda = 1.662$$

Example
Find the attenuation coefficient in dB m^{-1} for the lowest *TE* mode in WG 16 at 6 GHz.

Solution. From (3.55)

$$\alpha = 2\pi(f_{10}^2 - f^2)^{1/2}/c$$

For the TE_{10} mode, $f_{10} = 6.55$ GHz.

Hence $\alpha = 55$ nepers per metre

$$= 478 \text{ dB m}^{-1}$$

This is more meaningful if expressed as the attenuation for a length equal to the width of the wider side, 22.9 mm. This is found to be 10.9 dB.

3.5 Rectangular waveguide: *TM* modes

The longitudinal electric field satisfies (3.45). E_z is therefore of the same form as (3.51), coupled with (3.52). The boundary condition is simpler; E_z is to vanish on $x = 0$, a, and $y = 0$, b. This requires firstly that

$$E_z = A \sin px \sin qx\, e^{-\gamma z}$$

and secondly that

$$p = m\pi/a,\ q = n\pi/b$$

The required solution is therefore

$$E_z = A \sin(m\pi x/a) \sin(n\pi y/b)$$

The propagation constant is given by the same relation as for the *TE* modes, (3.53). It is to be noticed however that whereas a TE_{10} mode was possible, the lowest possible *TM* mode is TM_{11} since $E_z = 0$ for either m or n zero. The cut-off frequencies for the modes are precisely those given by (3.54). For frequencies greater than cut-off in the TM_{mn} mode we may write

$$E_x = -jA\beta \frac{m\pi}{a}\left[\left(\frac{m\pi}{a}\right)^2 + \left(\frac{n\pi}{b}\right)^2\right]^{-1} \cos\frac{m\pi x}{a} \sin\frac{n\pi y}{b} \exp(-j\beta z)$$

$$E_y = -jA\beta \frac{n\pi}{b}\left[\left(\frac{m\pi}{a}\right)^2 + \left(\frac{n\pi}{b}\right)^2\right]^{-1} \sin\frac{m\pi x}{a} \cos\frac{n\pi y}{b} \exp(-j\beta z)$$

where β is given by (3.55).

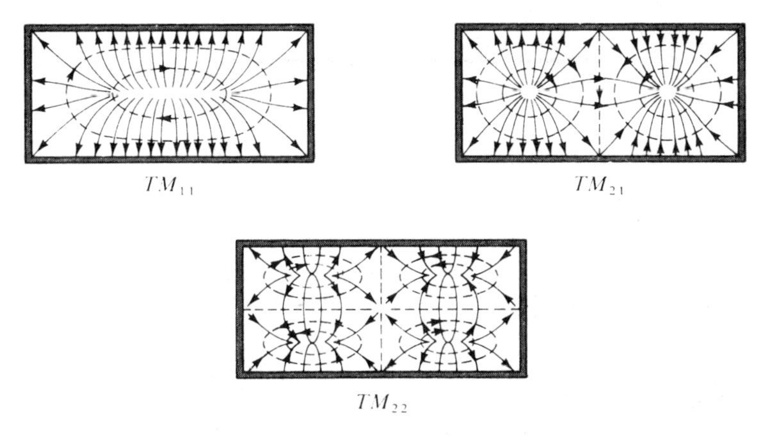

Fig. 3.8. Transverse field patterns of *TM* modes in rectangular waveguide.

The components of magnetic field are then given by

$$H_x = -\omega\varepsilon E_y/\beta$$

$$H_y = \omega\varepsilon E_x/\beta$$

Some modal patterns are shown in fig. 3.8.

Since the cut-off frequencies for the *TM* and *TE* series of modes are the same, in general more than one mode is able to propagate at any one frequency. However the TE_{10} is a true dominant mode, since the TM_{10} mode does not exist.

3.6 Wall currents

The system of wall currents will differ in each mode although the procedure for determining them is the same. It will be illustrated for the important case of the dominant mode TE_{10}. If $\boldsymbol{n}$ is the unit vector into the waveguide wall from the interior, the density of the current sheet in the wall which is consistent with zero magnetic field inside the wall is, from (1.73), given by

$$\mathbf{K} = -\boldsymbol{n} \times \mathbf{H}$$

This is illustrated in fig. 3.9.

The non-zero magnetic field components for the TE_{10} mode are given by

$$H_z = A \cos(\pi x/a) \exp(-j\beta z) \tag{3.58}$$

$$H_x = jA(\beta a/\pi) \sin(\pi x/a) \exp(-j\beta z) \tag{3.59}$$

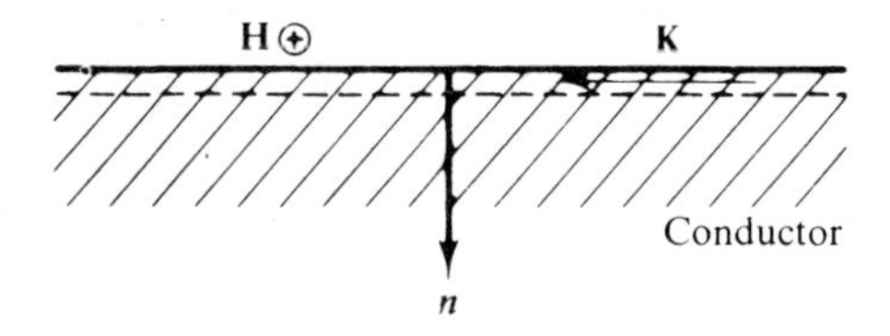

Fig. 3.9. Determination of current associated with a magnetic field.

The longitudinal component gives rise to current flow transverse to the axis, whereas the transverse magnetic field gives an axial flow. This is illustrated in fig. 3.10 and the results given in table 3.2.

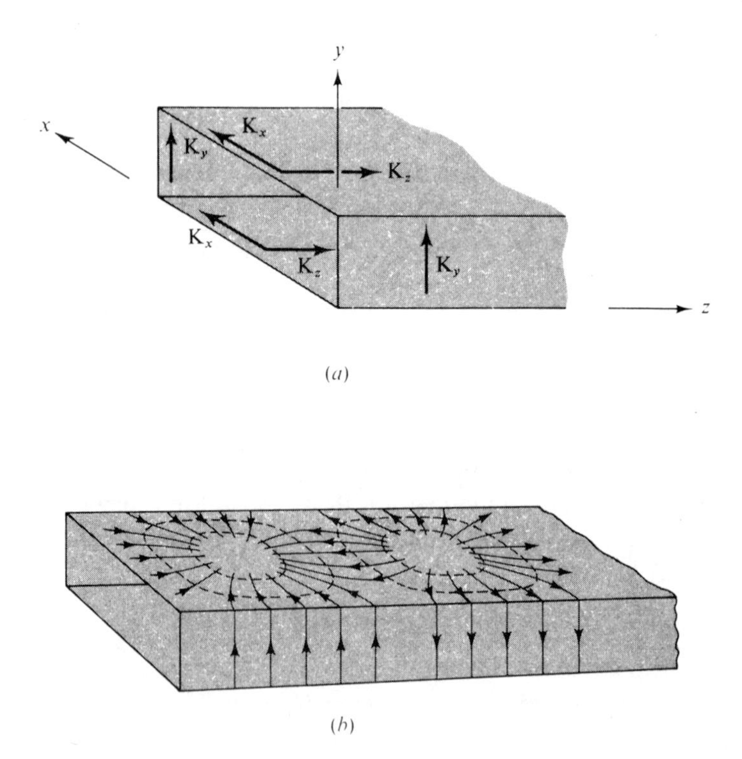

Fig. 3.10. Wall currents in rectangular waveguide. (*a*) Relation to magnetic field. (*b*) TE_{10} mode.

Table 3.2

Wall	$\boldsymbol{n}$	**H**	**K**
$x=0$	$-\boldsymbol{a}_x$	$\mathrm{A}\boldsymbol{a}_z$	$-\mathrm{A}\boldsymbol{a}_y$
$x=a$	$\boldsymbol{a}_x$	$-\mathrm{A}\boldsymbol{a}_z$	$\mathrm{A}\boldsymbol{a}_y$
$y=0$	$-\boldsymbol{a}_y$	$\mathrm{A}\left[\mathrm{j}\beta\frac{a}{\pi}\sin\frac{\pi x}{a}\boldsymbol{a}_x+\cos\frac{\pi x}{a}\boldsymbol{a}_z\right]$	$\mathrm{A}\left[\cos\frac{\pi x}{a}\boldsymbol{a}_x-\mathrm{j}\beta\frac{a}{\pi}\sin\frac{\pi x}{a}\boldsymbol{a}_z\right]$
$y=b$	$\boldsymbol{a}_y$	$\mathrm{A}\left[\mathrm{j}\beta\frac{a}{\pi}\sin\frac{\pi x}{a}\boldsymbol{a}_x+\cos\frac{\pi x}{a}\boldsymbol{a}_z\right]$	$\mathrm{A}\left[-\cos\frac{\pi x}{a}\boldsymbol{a}_x+\mathrm{j}\beta\frac{a}{\pi}\sin\frac{\pi x}{a}\boldsymbol{a}_z\right]$

In this table the common factor $\exp(-\mathrm{j}\beta z)$ has been suppressed.

3.7 Power flow

The power carried along the guide can be evaluated with the aid of the Poynting vector. Consider the dominant mode. The magnetic field components are given in (3.58) and (3.59). The only electric field component is

$$\mathrm{E}_y = -\omega\mu \mathrm{H}_x/\beta$$

The Poynting vector $\mathbf{E}\times\mathbf{H}^*$ has two components, an axial one given by $-\mathrm{E}_y\mathrm{H}_x^*$, and a transverse one in the x-direction of $\mathrm{E}_y\mathrm{H}_z^*$. E_y and H_x are in phase with each other, whereas E_y and H_z are in quadrature, so that the time-averaged power flow is axial. The transverse flow is oscillating in space also, since no power flows into the perfectly conducting walls. The mean axial power flow is given by

$$P=\tfrac{1}{2}\,\mathrm{Re}\int_0^a\int_0^b -\mathrm{E}_y\mathrm{H}_x^*\,\mathrm{d}x\,\mathrm{d}y$$

$$=\frac{1}{2}\frac{\omega\mu}{\beta}\int_0^a\int_0^b |\mathrm{H}_x|^2\,\mathrm{d}x\,\mathrm{d}y$$

$$=\frac{1}{2}\frac{\omega\mu}{\beta}|\mathrm{A}|^2\beta^2\frac{a^2}{\pi^2}\int_0^a\int_0^b \sin^2\frac{\pi x}{a}\,\mathrm{d}x\,\mathrm{d}y$$

$$=\frac{1}{2}|\mathrm{A}|^2\,\omega\mu\beta\frac{a^2}{\pi^2}\cdot\frac{ab}{2}$$

Substitution for β and rearrangement reduces the expression to

$$P=|\mathrm{A}|^2\surd(\mu/\varepsilon)\,a^3bf\,(f^2-f_{10}^2)^{\frac{1}{2}}/v^2$$

Alternatively

$$P=|\mathrm{A}|^2\surd(\mu/\varepsilon)\,a^3b/\lambda\lambda_g$$

3.7.1 Power flow with two modes present

Let us suppose we have two modes denoted by m_1, n_1 and m_2, n_2. The total transverse fields may be written in the form

$$\mathbf{E}_t^{(1)} + \mathbf{E}_t^{(2)}, \ \mathbf{H}_t^{(1)} + \mathbf{H}_t^{(2)}$$

where the superscripts (1) and (2) denote the two modes. The axial Poynting vector will thus take the form

$$\mathbf{E}_t^{(1)} \times \mathbf{H}_t^{(1)*} + \mathbf{E}_t^{(2)} \times \mathbf{H}_t^{(2)*} + \mathbf{E}_t^{(1)} \times \mathbf{H}_t^{(2)*} + \mathbf{E}_t^{(2)} \times \mathbf{H}_t^{(1)*}$$

Integration over the cross-section will involve the four terms separately. The first one will give the power of mode 1 by itself, and the second the power of mode 2 by itself. The third term will be proportional to

$$\int_0^a \int_0^b \sin \frac{m_1 \pi x}{a} \cos \frac{n_1 \pi y}{b} \sin \frac{m_2 \pi x}{a} \cos \frac{n_2 \pi y}{b} \, dx \, dy$$

If $m_1 \neq m_2$ and $n_1 \neq n_2$ as was supposed, this integral vanishes by the well-known properties used in Fourier series. Similarly the last term is zero. Hence when two modes are present they behave independently as far as power flow is concerned. This result can easily be extended to any number of modes, and also to the case of one *TE* mode and one *TM* mode.

3.8 Circular waveguide

To investigate the modes possible in a conducting tube of circular cross-section we have to put (3.47) or (3.50) in polar co-ordinates, as shown in fig. 3.11. The appropriate form is

$$\frac{1}{\rho} \frac{\partial}{\partial \rho} \left(\rho \frac{\partial H_z}{\partial \rho} \right) + \frac{1}{\rho^2} \frac{\partial^2 H_z}{\partial \phi^2} + \kappa^2 H_z = 0$$

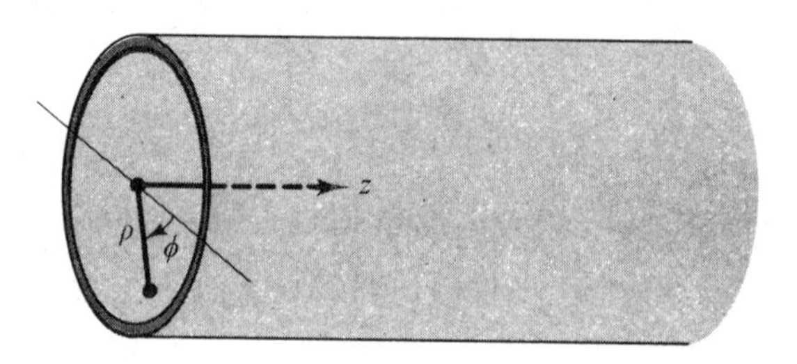

Fig. 3.11. Co-ordinates for circular waveguide.

where
$$\kappa^2 = \gamma^2 + \omega^2 \mu\varepsilon$$

This equation can be solved in terms of Bessel functions, and before proceeding with derivation of modal patterns the nature of the solution will be briefly discussed.

3.8.1 Bessel functions

We are concerned with the solution of the equation

$$\frac{1}{\rho}\frac{\partial}{\partial \rho}\left(\rho \frac{\partial \psi}{\partial \rho}\right) + \frac{1}{\rho^2}\frac{\partial^2 \psi}{\partial \phi^2} + \kappa^2 \psi = 0$$

The variable ψ may represent either E_z or H_z. A solution to this equation may be shown to be

$$\psi = [AJ_m(\kappa\rho) + BY_m(\kappa\rho)][C \cos m\phi + D \sin m\phi] \tag{3.60}$$

where $J_m(u)$, $Y_m(u)$ are Bessel functions of the first and second kind respectively. They satisfy the equation

$$\frac{1}{u}\frac{\mathrm{d}}{\mathrm{d}u}\left(u \frac{\mathrm{d}w}{\mathrm{d}u}\right) + \left(1 - \frac{m^2}{u^2}\right) w = 0$$

in which $w(u)$ is any linear combination of $J_m(u)$ and $Y_m(u)$. The function $Y_m(u)$ has a singularity at $u = 0$, but $J_m(u)$ is well behaved for all values of u.

In the context of the hollow circular tube we must make $B = 0$, since there can be no singularity of field components away from the conducting surface. The constant m in (3.60) must be an integer, since in the present geometry an increase in angle of 2π must lead to the same value of a field component. We note that if we were considering a coaxial system then it would no longer be right to make $B = 0$, for the axis would not then be in a physically relevant region.

The form of the second bracket in (3.60) merely expresses the arbitrariness of the origin of ϕ in a symmetrical geometry, and may thus be replaced by cos $m\phi$. We have therefore to consider a solution

$$\psi = J_m(\kappa\rho) \cos m\phi \tag{3.61}$$

The form of $J_m(u)$ for a few values of m is illustrated in fig. 3.12. We have to put in the additional restriction on ψ that either ψ (in the case of *TM* modes) or $\partial\psi/\partial\rho$ (in the case of *TE* modes) must be zero at the wall of the waveguide, $\rho = a$. Either of these conditions implies a restriction on possible values of κ. It may be shown that the oscillatory nature

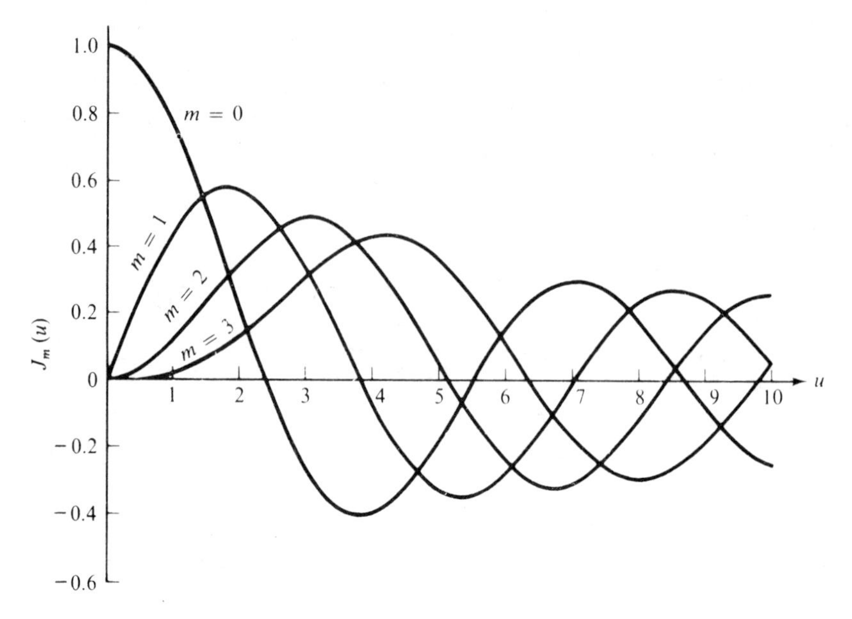

Fig. 3.12. Bessel functions of the first kind, $J_m(u)$.

exhibited in Fig. 3.12 carries on indefinitely, and $J_m(u)$ is zero at a sequence of points

$$J_m(u)=0, \quad u=t_{mn} \qquad (m=0, 1, 2, \ldots, n=1, 2, \ldots)$$

It is obvious that $J_m(u)$ has then a sequence of maxima and minima at each of which the derivative $J'_m(u)$ vanishes. This gives rise to a second sequence

$$J'_m(u)=0, \quad u=s_{mn} \qquad (m=0, 1, 2, \ldots, n=1, 2, \ldots)$$

Some values are given in table 3.3. (The root s_{01} is taken as the first non-zero value of u for which $J'_0(u)$ vanishes.) We now see that, in the

Table 3.3

	t_{mn}				s_{mn}		
$n \backslash m$	0	1	2	$n \backslash m$	0	1	2
1	2.405	3.832	5.136	1	3.832	1.841	3.054
2	5.520	7.016	8.417	2	7.016	5.331	6.706
3	8.654	10.173	11.620	3	10.173	8.536	9.969
	$J_m(t_{mn})=0$				$J'_m(s_{mn})=0$		

case of a *TM* mode for which ψ in (3.61) represents E_z (which must vanish at $\rho = a$), κ must satisfy

$$J_m(\kappa a) = 0$$

or

$$\kappa a = t_{mn} \tag{3.62}$$

In the case of a *TE* mode when ψ is to be identified with H_z, for which $\partial H_z/\partial \rho = 0$ at $\rho = a$, we have

$$J'_m(\kappa a) = 0$$

$$\kappa a = s_{mn} \tag{3.63}$$

In either case we have a sequence of discrete values of κ, to be compared with the result for rectangular guide.

3.8.2 *TE* modes

We have

$$H_z = AJ_m(\kappa\rho)\cos m\phi \exp(-\gamma z) \tag{3.64}$$

As we have seen, the condition that the radial derivative of H_z must vanish at the surface of the guide requires κ to take only the discrete values given by (3.63). The propagation constant is given therefore by the equation

$$\gamma^2 = -\omega^2\mu\varepsilon + (s_{mn}/a)^2 \tag{3.65}$$

As with the rectangular guide, for any given m and n, ω must exceed a certain value before γ becomes imaginary and a progressive wave possible. For propagation in the TE_{mn} mode, ω must be greater than ω_{mn}, where

$$f_{mn} = \omega_{mn}/2\pi = s_{mn}/[2\pi a\sqrt{(\mu\varepsilon)}] = s_{mn}v/2\pi a \tag{3.66}$$

When $\omega > \omega_{mn}$, $\quad \gamma = j\beta = j[\mu\varepsilon(\omega^2 - \omega_{mn}^2)]^{\frac{1}{2}}$

When $\omega < \omega_{mn}$, $\quad \gamma = \alpha = [\mu\varepsilon(\omega_{mn}^2 - \omega^2)]^{\frac{1}{2}}$ (3.67)

Inspection of table 3.3 shows that the lowest cut-off frequency corresponds to TE_{11}, for which

$$f_{11} = 1.841v/2\pi a = 0.293v/a$$

The wavelength of a uniform *TEM* wave of this frequency is given by

$$\lambda = v/f_{11} = 3.413a$$

Roughly speaking, the wavelength corresponding to cut-off frequency is about one-half of the circumference.

Expressions for the components of the electric and magnetic fields in the TE_{mn} mode may be derived from (3.48) and (3.49). We find

$$\left.\begin{aligned} H_\rho &= -A(j\beta/\kappa)J'_m(\kappa\rho)\cos m\phi \\ H_\phi &= A(j\beta/\kappa)(m/\kappa\rho)J_m(\kappa\rho)\sin m\phi \\ E_\rho &= \omega\mu H_\phi/\beta \\ E_\phi &= -\omega\mu H_\rho/\beta \end{aligned}\right\} \quad (3.68)$$

where κ is given by (3.63) and β is calculated from (3.67) for the particular mode. It should be noted that the notation $J'_m(\kappa\rho)$ has the significance

$$\left[\frac{d}{du}J_m(u)\right]_{u=\kappa\rho}$$

3.8.3 *TM* modes

In this case

$$E_z = AJ_m(\kappa\rho)\cos m\phi \quad (3.69)$$

At $\rho = a$, E_z must vanish, and hence the possible values of κ are given by (3.62), giving for the propagation constant

$$\gamma^2 = -\omega^2\mu\varepsilon + (t_{mn}/a)^2$$

We thus have a series of cut-off frequencies defined by

$$f_{mn} = \omega_{mn}/2\pi = t_{mn}v/2\pi a \quad (3.70)$$

The components may be found by applying (3.48) and (3.49), as before:

$$\left.\begin{aligned} E_\rho &= -A(j\beta/\kappa)J'_m(\kappa\rho)\cos m\phi \\ E_\phi &= A(j\beta/\kappa)(m/\kappa\rho)J_m(\kappa\rho)\sin m\phi \\ H_\rho &= -\omega\varepsilon E_\phi/\beta \\ H_\phi &= \omega\varepsilon E_\rho/\beta \end{aligned}\right\} \quad (3.71)$$

Some field configurations for both *TE* and *TM* modes are shown in fig. 3.13. It will be noted from table 3.3 that the *TM* mode lowest in cut-off frequency is TM_{01} and that the cut-off frequency is higher than that for TE_{11}. The latter is thus a true dominant mode, although it must be realised that the mode pattern shown in fig. 3.13 can be oriented in any

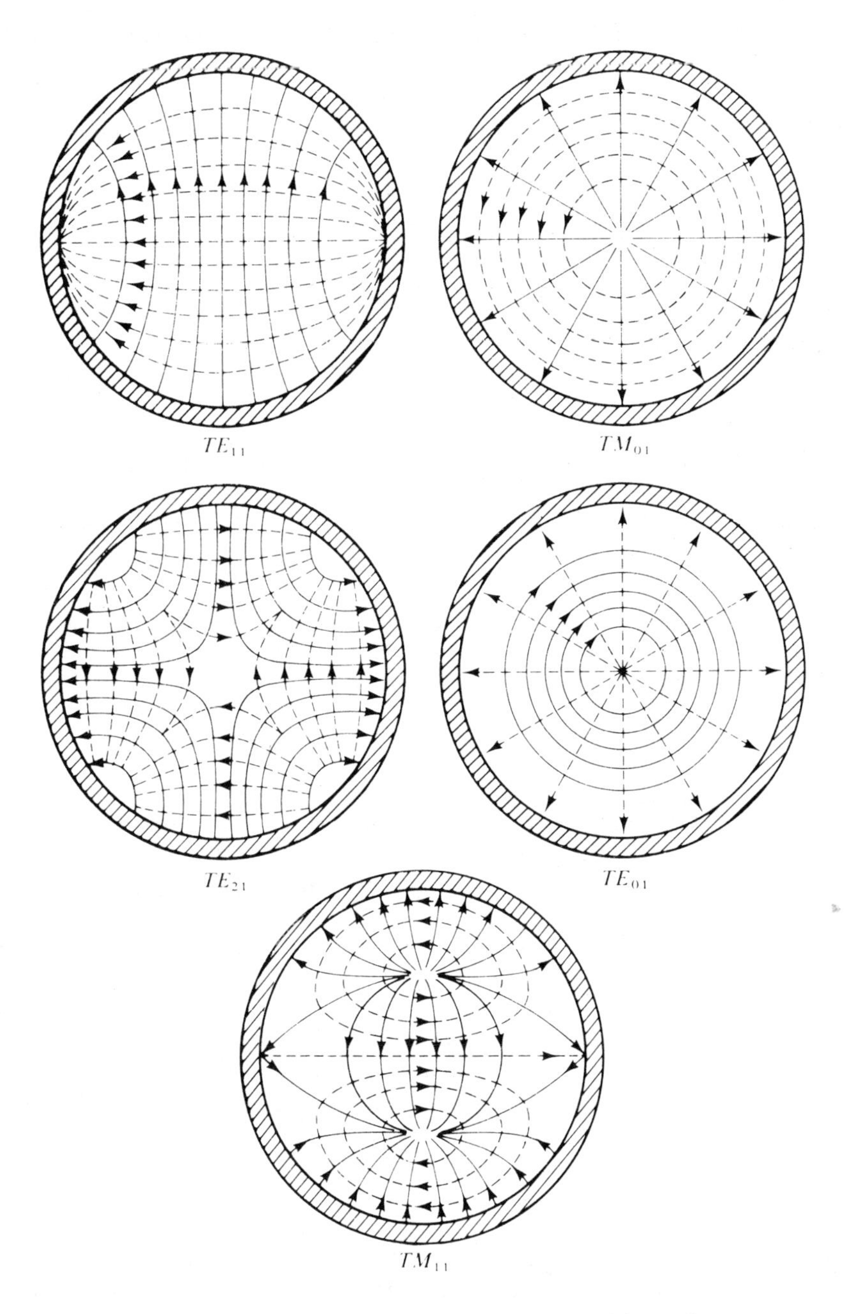

Fig. 3.13. Transverse fields for circular waveguide modes.

direction across the guide so that a real degeneracy exists. It will also be noted from table 3.3 that TE_{01} and TM_{11} have the same cut-off frequency. In table 3.4 the first few modes are placed in order of increasing cut-off frequency. In this table the product of guide radius, in cm, and cut-off frequency, in GHz, is given for an air-filled guide.

Table 3.4

Mode	af_{mn} (cm GHz)
TE_{11}	8.79
TM_{01}	11.48
TE_{21}	14.58
TE_{01}, TM_{11}	18.30
TE_{12}	25.45
TM_{02}	26.36

3.8.4 Power flow

The power associated with a mode can be evaluated as before using the Poynting vector. The axial component is $E_\rho H_\phi^* - E_\phi H_\rho^*$. For the *TE* mode we have

$$\mathsf{S} = \frac{\omega\mu}{\beta}(|H_\rho|^2 + |H_\phi|^2)$$

$$= \frac{\omega\mu}{\beta}\frac{|A|^2\beta^2}{\kappa^4}\left\{[\kappa J'_m(\kappa\rho)]^2 \cos^2 m\phi + \left[\frac{m}{\rho}J_m(\kappa\rho)\right]^2 \sin^2 m\phi\right\}$$

The total average power is obtained by integrating S over the cross-section of the guide:

$$P = \tfrac{1}{2}\,\mathrm{Re}\int_0^a \int_0^{2\pi} \mathsf{S}\rho \, d\phi \, d\rho$$

From the nature of the expression for S it is evident that evaluation of the integral involves considerable algebraic manipulation. It is shown in § 10.1 that the power flow in the TE_{mn} mode can be expressed in the form

$$P = |A|^2 \omega\mu\beta a^4 A_{mn} \tag{3.72}$$

in which A_{mn} depends only on the mode indices. The numerical results are given in table 3.5. A similar analysis for the *TM* modes yields

$$P = |A|^2 (\beta^3/\omega\varepsilon) a^4 B_{mn} \tag{3.73}$$

The coefficients B_{mn} are given in table 3.6.

Table 3.5. *The coefficients A_{mn}, equation (3.72)*

$n \backslash m$	$A_{mn} \times 10^3$		
	0	1	2
1	17.4	55.3	11.4
2	2.87	3.19	1.56
3	0.349	0.794	0.492

Table 3.6. *The coefficients B_{mn}, equation (3.73)*

$n \backslash m$	$B_{mn} \times 10^3$		
	0	1	2
1	73.2	17.4	6.87
2	5.97	2.87	1.63
3	1.55	0.946	0.628

3.8.5 Wall currents

The wall currents follow from a knowledge of the tangential magnetic fields. With a *TE* mode there will always be a circulating current because of the axial magnetic field, which is never zero over the whole circumference. There may also be an axial current, depending on the value of the circumferential magnetic field: for *TM* modes (3.71) shows that H_ϕ is never zero on the circumference except at isolated points; for TE_{0n} modes, however, the magnetic field is purely radial so that there is no axial current.

3.8.6 Piston attenuators

At a frequency below the cut-off of a particular mode, that mode will be exponentially attenuated. We have from (3.65)

$$\alpha^2 = (s_{mn}/a)^2 - (\omega/v)^2$$

or

$$\alpha^2 = (t_{mn}/a)^2 - (\omega/v)^2$$

If a field is excited at a frequency well below the cut-off of the dominant mode, the field in the guide will, within a few radii from the source, become virtually pure TE_{11} with an attenuation constant in an

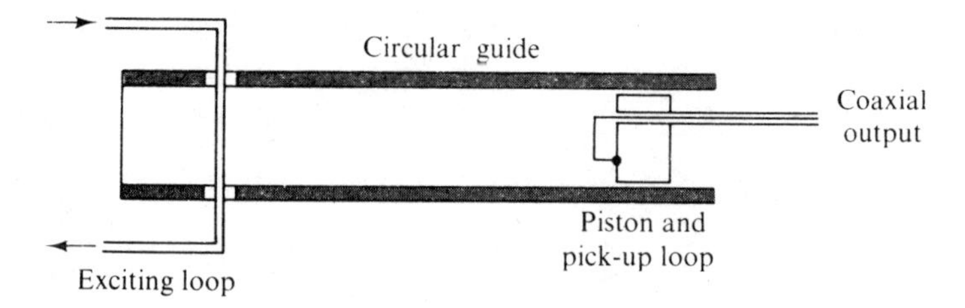

Fig. 3.14. Schematic diagram of a piston attenuator.

air-filled guide of

$$\alpha = 1.841/a \text{ nepers m}^{-1} = 15.99/a \text{ dB m}^{-1}$$

A precision-made circular guide excited at one end and containing a sliding piston mounting a pick-up loop thus forms an attenuator which can be absolutely calculated in terms of the tube diameter. This is shown schematically in fig. 3.14. Such a device is known as a piston attenuator. It should be noted that a correction may be needed as the frequency is increased.

3.8.7 Higher modes in coaxial line

The investigation of non-uniform *TEM* waves included consideration of the coaxial structure only in so far as it could support such a wave. Coaxial lines are virtually always used in such a way, but it is relevant to note that waveguide modes are also possible. The analysis follows through on the same lines as for the circular guide. In the case of *TM* modes the form of E_z is given by (3.60). It is required that E_z vanish on the inner and outer conductors, $\rho = a, b$. Hence

$$AJ_m(\kappa b) + BY_m(\kappa b) = 0$$

and

$$AJ_m(\kappa a) + BY_m(\kappa a) = 0$$

This is only possible if

$$-A/B = Y_m(\kappa b)/J_m(\kappa b) = Y_m(\kappa a)/J_m(\kappa a)$$

The possible values for κ are to be found from the equation

$$J_m(\kappa b)Y_m(\kappa a) - Y_m(\kappa b)J_m(\kappa a) = 0$$

This equation can be shown to yield discrete values for κa, for a given ratio of b/a. Waveguide *TM* modes thus exist. Similarly it is found that *TE* modes exist. The chief importance of this result is that in order to ensure principal-wave (*TEM*) transmission a coaxial line must be of such diameter that at the highest relevant frequency all waveguide modes are below cut-off. Thus at discontinuities in the line, for example at junctions, the waveguide modes excited will be attenuated away from the discontinuity.

3.9 'Voltage' and 'current' in a waveguide

It is only with a *TEM* wave on transmission line that it is possible to associate a voltage and current, in the circuit interpretation, with the field at a given point on the line. With a single mode in a waveguide, however, the variables are still limited to two, being measures of the electric field and of the magnetic field. At a given point along the guide, the pattern of the electric and magnetic fields is determined by the mode in question, and it is only the amplitude which is undetermined. With a progressive wave we have seen that the ratio E_t/H_t is a constant, the wave impedance for the mode. In the presence of a backward progressive wave as well as a forward one, the ratio of E_t/H_t will vary, and the amplitudes will have to be specified independently.

It is evident that in any normal sense a unique voltage does not exist in a waveguide. There are not two terminals to define it. However, we may arbitrarily specify a 'voltage' proportional to the electric field strength of a forward progressive wave, and associate with it a 'current' to give the correct power flow. We shall illustrate this for the dominant mode in a rectangular waveguide.

†3.9.1 Equivalent circuit for dominant-mode transmission in a rectangular guide

The relevant formulae for the TE_{10} mode are

$$\left.\begin{aligned} H_z &= A\cos(\pi x/a)\exp(-j\beta z) \\ H_x &= j(A\beta a/\pi)\sin(\pi x/a)\exp(-j\beta z) \\ E_y &= -\omega\mu H_x/\beta = -jA\omega\mu(a/\pi)\sin(\pi x/a)\exp(-j\beta z) \\ \beta &= [\mu\varepsilon(\omega^2-\omega_{10}^2)]^{\frac{1}{2}} \\ \omega_{10} &= \pi/[a\sqrt{(\mu\varepsilon)}] \\ P &= |A|^2\omega\mu\beta a^3 b/4\pi^2 \end{aligned}\right\} \qquad (3.74)$$

The electric field strength can be written in terms of the maximum amplitude E_m, occurring at $x = a/2$, in the form

$$E_y = E_m \sin(\pi x/a) \exp(-j\beta z)$$

in which
$$E_m = -jA\omega\mu a/\pi$$

We will, quite arbitrarily, define a voltage at the co-ordinate z by the (dimensionally correct) equation

$$V(z) = aE_m \exp(-j\beta z)$$

We can then express E_y and H_x in terms of $V(z)$ as follows:

$$E_y = [V(z)/a] \sin(\pi x/a)$$

$$H_x = -[V(z)\beta/\omega\mu a] \sin(\pi x/a) \tag{3.75}$$

The constant A can be related to V by the equation

$$|A| = (\pi/a^2\omega\mu)|V(z)|$$

and hence the power flow at z is given by

$$P = (1/4\pi^2)(\pi/a^2\omega\mu)^2 \times \omega\mu\beta a^3 b|V|^2$$

$$= \tfrac{1}{4}(\beta/\omega\mu)(b/a)|V|^2 \tag{3.76}$$

If we treat $|V|$ as peak 'voltage', we would expect the power flow in a forward wave on a transmission line to be written in the form

$$P = \tfrac{1}{2}|V|^2/Z_0 \tag{3.77}$$

Identifying (3.76) and (3.77) gives an expression for Z_0:

$$Z_0 = 2\omega\mu a/\beta b$$

The associated current for a forward wave on the transmission line should then be given by

$$I(z) = V(z)/Z_0 = V(z)\beta b/2\omega\mu a$$

Comparing with 3.75 we see that H_x can be written in the form

$$H_x = -(2/b)I(z) \sin(\pi x/a)$$

Since H_x has dimensions A m^{-1}, this equation is dimensionally correct. With these definitions we can conduct analysis of transmission along a waveguide operating in the dominant mode as though it were a transmission line of which the characteristic impedance and phase velocity are given by

$$Z_0 = \frac{2\omega\mu a}{\beta b} = \frac{2a}{b}\sqrt{\left(\frac{\mu}{\varepsilon}\right)} \frac{\omega}{(\omega^2 - \omega_{10}^2)^{\frac{1}{2}}}$$

$$v_p = \frac{\omega}{\beta} = \frac{\omega}{[\mu\varepsilon(\omega^2 - \omega_{10}^2)]^{\frac{1}{2}}}$$

It should be noted that Z_0 can be changed by any constant multiplier, provided that associated changes are made in equations relating electric and magnetic field strengths to the defined voltage and current. This shows the arbitrariness of the definitions.

3.10 Losses in transmission lines and waveguides

The foregoing analysis has assumed perfect conductors and lossless dielectric. In a real system neither assumption is true.

3.10.1 Dielectric losses

Dielectric losses are of greater significance in transmission lines than in waveguides since the latter are self-supporting mechanically, whereas transmission lines frequently need to incorporate dielectric support. The effect of lossy dielectric in the *TEM* wave on transmission lines can be dealt with simply: as in chapter 1 on the same topic, the finite conductivity can be allowed for by taking the permittivity to be complex. A comparison of the two situations will show that the attenuation constant for the principal wave is precisely that for the uniform *TEM* wave. The loss will also modify slightly the characteristic impedance, but at high frequencies this modification is usually unimportant.

In waveguides the introduction of the complex permittivity is not so straightforward: the whole analysis has been conducted on the assumption that γ^2 is real, positive or negative. It is simplest to assume, as will be done to allow for wall losses, that slight losses do not alter the character of the field. This means that the modes and cut-off frequencies are calculated assuming no losses. Then an attenuation coefficient is estimated by replacing the permittivity ε in the formula $\gamma = \mathrm{j}\beta = \mathrm{j}[\mu\varepsilon(\omega^2 - \omega_{mn}^2)]^{\frac{1}{2}}$ by the complex permittivity $\varepsilon + \sigma/\mathrm{j}\omega$. Assuming that the losses are small,

$$(\varepsilon + \sigma/\mathrm{j}\omega)^{\frac{1}{2}} \approx (1 + \sigma/2\mathrm{j}\omega\varepsilon)\sqrt{\varepsilon}$$

Carrying out this substitution we find

$$\alpha = \frac{\sigma}{2}\left[\frac{\mu}{\varepsilon}\left(1 - \frac{\omega_{mn}^2}{\omega^2}\right)\right]^{\frac{1}{2}}$$

3.10.2 Wall losses

The theory presented earlier shows that there will be currents flowing in the walls of the waveguides. These are ideally current sheets but because of the finite conductivity will in practice occupy thin layers near the surface. The case of a current sheet excited by a magnetic field constant over the plane surface of a block of good conductor was analysed in chapter 1 where it was shown that the skin depth in copper at frequencies of the order of 1 GHz and higher was only a few micrometres, very much less than the wavelength and, it might be added, comparable with surface finish. It might be suspected, and it is confirmed by analysis, that as long as the magnetic field at the surface varies only slowly the results in chapter 1 apply locally. Since the skin depth is so small there is no difficulty in meeting the conditions of slow variation. The result relevant to the present situation is that contained in (1.70), enabling the power lost over an area dS of wall surface to be expressed in the form

$$-\mathrm{d}P = \tfrac{1}{2}\mathbf{H} \cdot \mathbf{H}^* R_s \,\mathrm{d}S$$

where $\mathbf{H}$ is the magnetic field at the wall and $R_s = \sqrt{(\omega\mu/2\sigma)}$.

This treatment assumes that the fields are perturbed to a negligible extent from the lossless case. It must be pointed out, however, that the presence of wall resistivity does require a completely different exact analysis: there is no question of the component of $\mathbf{E}$ tangential to the wall vanishing exactly, and the field is not confined to the interior of the guide. These are great complications which make the perturbation technique the only useful one.

The process of estimating the attenuation coefficient is straightforward: the total power lost in a length dz of guide is found from the formula

$$\tfrac{1}{2}R_s \int_C \mathbf{H} \cdot \mathbf{H}^* \,\mathrm{d}l \times \mathrm{d}z$$

where dl is a small element of C, the contour of the cross-section. The rate of *gain* of power along the guide is then given by

$$\frac{\mathrm{d}P}{\mathrm{d}z} = -\tfrac{1}{2}R_s \int_C \mathbf{H} \cdot \mathbf{H}^* \,\mathrm{d}l$$

Now the attenuation coefficient, α, is related to power by

$$P = P_0 \, e^{-2\alpha z}$$

which gives

$$\alpha = -\frac{1}{2P}\frac{dP}{dz}$$

$$= \frac{R_s}{4P}\int_C \mathbf{H}\cdot\mathbf{H}^* \, dl \qquad (3.78)$$

This process will be illustrated by some important examples.

3.10.3 Coaxial transmission line

Equations (3.29) and (3.30) show that for a forward wave we have

$$H_\phi = \frac{I_0}{2\pi\rho}\exp(-j\beta z)$$

There are two walls to be considered, the inner and the outer as shown in fig. 3.15. On the outer wall,

$$|H_\phi| = \frac{I_0}{2\pi b}$$

Therefore

$$\int |H_\phi|^2 \, dl = \left(\frac{I_0}{2\pi b}\right)^2 \cdot 2\pi b$$

Similarly over the inner wall,

$$\int |H_\phi|^2 \, dl = \left(\frac{I_0}{2\pi a}\right)^2 \cdot 2\pi a$$

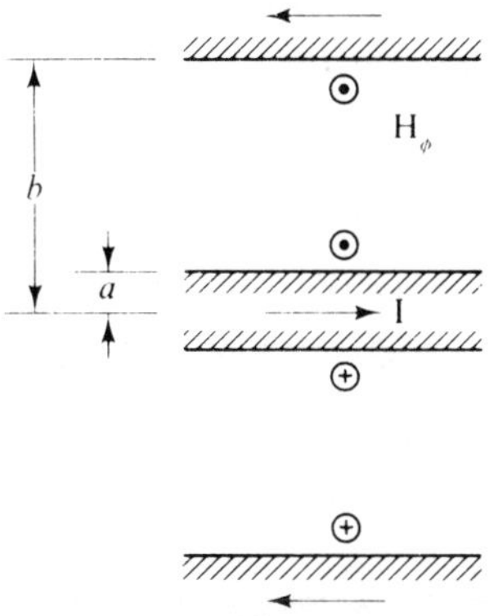

Fig. 3.15. Magnetic fields in a coaxial line.

Over the whole perimeter we can therefore write

$$\oint \mathbf{H}\cdot\mathbf{H}^*\,\mathrm{d}l = \frac{I_0^2}{2\pi}\left(\frac{1}{a}+\frac{1}{b}\right)$$

To find the power flow we can use the expression

$$P = \tfrac{1}{2}Z_0 I_0^2$$

where Z_0 is given by (3.32)

$$Z_0 = \frac{1}{2\pi}\sqrt{\left(\frac{\mu}{\varepsilon}\right)}\,\ln\frac{b}{a}$$

Substituting these various expressions in (3.78) then yields

$$\begin{aligned}\alpha &= \frac{1}{4\pi}\frac{R_s}{Z_0}\left(\frac{1}{a}+\frac{1}{b}\right)\\ &= \frac{1}{2}R_s\sqrt{\left(\frac{\varepsilon}{\mu}\right)}\left(\frac{1}{a}+\frac{1}{b}\right)\left(\ln\frac{b}{a}\right)^{-1}\end{aligned}$$

where $R_s = \sqrt{(\omega\mu/2\sigma)}$ (in which all the parameters refer to the wall material). By adding to this expression the attenuation coefficient arising from the dielectric loss, derived from (1.57) and (1.54) we have an expression for the total attenuation in a coaxial line:

$$\begin{aligned}\alpha &= \frac{1}{2}\omega\sqrt{(\mu\varepsilon)}\tan\delta + \frac{1}{2}R_s\sqrt{\left(\frac{\varepsilon}{\mu}\right)}\left(\frac{1}{a}+\frac{1}{b}\right)\left[\ln\frac{b}{a}\right]^{-1}\\ &= \frac{\pi f}{v}\tan\delta + \frac{1}{2}R_s\sqrt{\left(\frac{\varepsilon}{\mu}\right)}\left(\frac{1}{a}+\frac{1}{b}\right)\left[\ln\frac{b}{a}\right]^{-1}\end{aligned} \qquad (3.79)$$

This expression is shown graphically in fig. 3.16 for the case of a particular coaxial cable filled with polyethylene dielectric. It must be remembered that R_s is proportional to $\sqrt{f}$.

3.10.4 Dominant mode in rectangular waveguide

The magnetic field tangential to the walls was given in table 3.2. Using the values in that table, the required expression to substitute in (3.78) is given by

$$\begin{aligned}\int_C \mathbf{H}\cdot\mathbf{H}^*\,\mathrm{d}l &= 2\int_0^b |A|^2\,\mathrm{d}y + 2\int_0^a |A|^2\left[\left(\frac{\beta a}{\pi}\sin\frac{\pi x}{a}\right)^2 + \left(\cos\frac{\pi x}{a}\right)^2\right]\mathrm{d}x\\ &= |A|^2\left[2b + a + \left(\frac{\beta a}{\pi}\right)^2 a\right]\end{aligned}$$

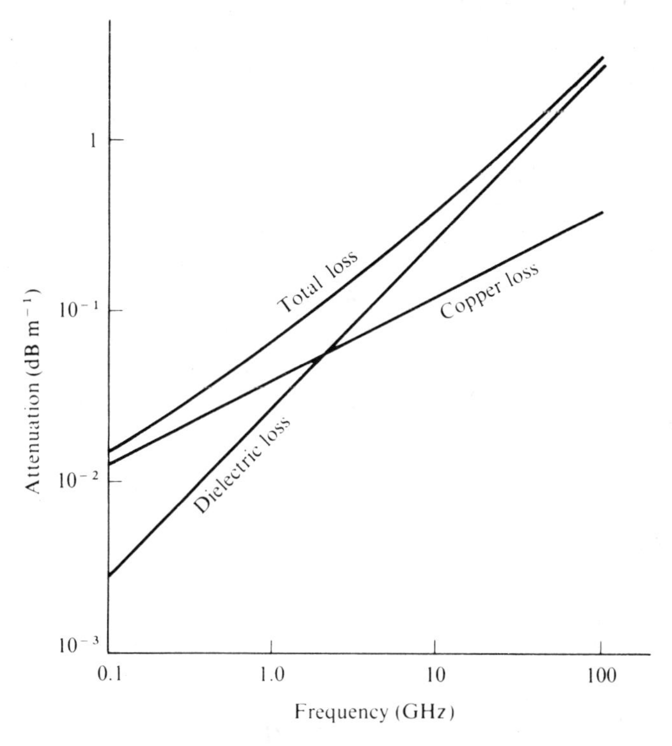

Fig. 3.16. Attenuation coefficient for dielectric filled coaxial line. $b = 12.7$ mm, $b/a = 3.59$; dielectric: $\varepsilon_r = 2.3\varepsilon_0$, $\mu = \mu_0$, $\tan\delta = 2\times 10^{-4}$; copper wall $\sigma = 5.9\times 10^7\,\mathrm{S m^{-1}}$

The expression for P is given by

$$P = |A|^2 \omega\mu\beta a^3 b/4\pi^2$$

Hence

$$\alpha = R_s \frac{\pi^2}{\omega\mu\beta a^3 b}\left[2b + a + \left(\frac{\beta a}{\pi}\right)^2 a\right]$$

To find an expression giving α as an explicit function of frequency we have to use

$$\beta^2 = \omega^2\mu\varepsilon - (\pi/a)^2$$

This gives

$$(\beta a/\pi)^2 + 1 = \omega^2\mu\varepsilon(a/\pi)^2 = (\omega/\omega_{10})^2$$

Substituting and reducing finally yields for α the expression

$$\alpha = \sqrt{\left(\frac{\varepsilon}{\mu}\right)}\frac{R_s}{b}\frac{[1 + (2b/a)(\omega_{10}/\omega)^2]}{[1 - (\omega_{10}/\omega)^2]^{\frac{1}{2}}} \tag{3.80}$$

This expression is shown graphically for WG 16 in fig. 3.17.

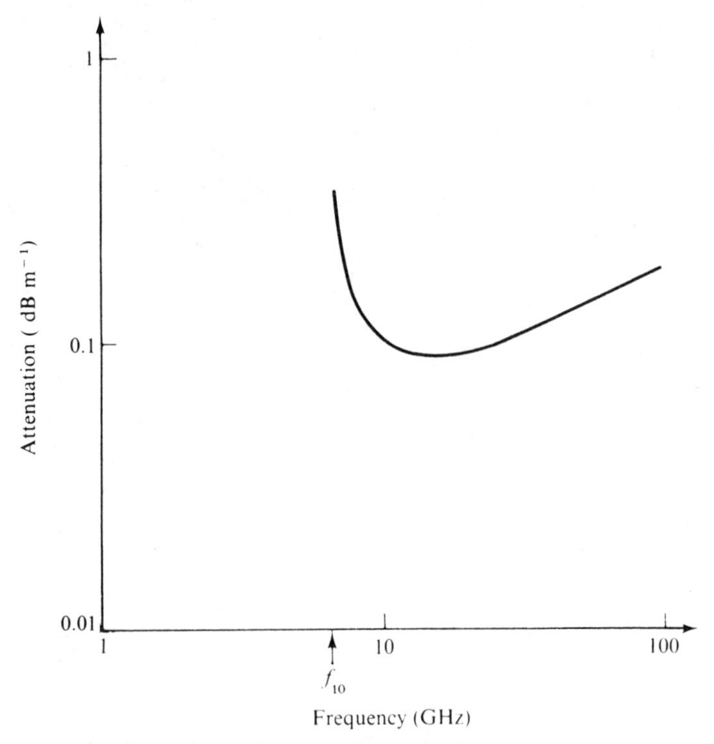

Fig. 3.17. Attenuation coefficient for WG 16 rectangular waveguide. Copper wall $\sigma = 5.9 \times 10^7\ \mathrm{Sm}^{-1}$

3.10.5 TE_{0n} modes in circular waveguide

For TE_{0n} modes the magnetic field tangential to the walls is H_z, which, from (3.64), is given by

$$H_z = AJ_0(\kappa\rho)\exp(-j\beta z)$$

Hence

$$\int_C \mathbf{H}\cdot\mathbf{H}^*\,dl = |A|^2 2\pi a[J_0(\kappa a)]^2$$

The power flow is given by the expression in (3.72)

$$P = \omega\mu\beta a^4 |A|^2 A_{0n}$$

Hence for the TE_{0n} mode

$$\alpha = \frac{R_s}{4\omega\mu\beta a^4 A_{0n}} \cdot 2\pi a[J_0(s_{0n})]^2$$

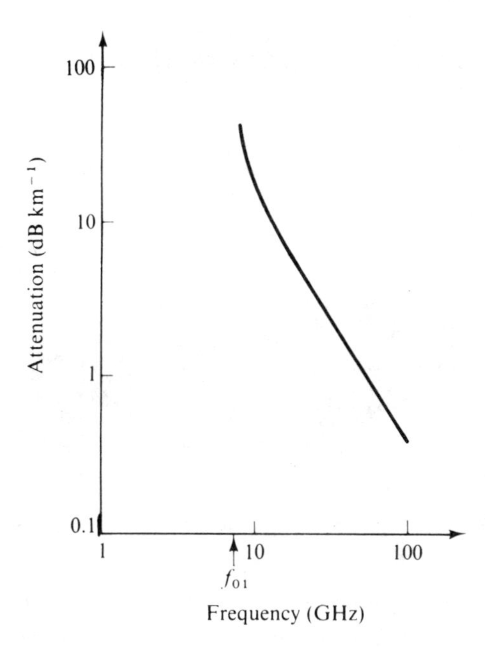

Fig. 3.18. Attenuation coefficient for TE_{01} mode in circular waveguide 50 mm internal diameter. Copper wall $\sigma = 5.9 \times 10^7$ Sm^{-1}

This expression can be simplified by use of the result in § 10.1 where it is shown that

$$A_{0n} = \frac{\pi}{2} \frac{1}{s_{0n}^2} [J_0(s_{0n})]^2$$

Substitution of this result and use of (3.66) to relate s_{0n} to ω_{0n} finally gives

$$\alpha = \frac{R_s}{a} \sqrt{\left(\frac{\varepsilon}{\mu}\right)} \left(\frac{\omega_{0n}}{\omega}\right)^2 \left[1 - \left(\frac{\omega_{0n}}{\omega}\right)^2\right]^{-\frac{1}{2}} \tag{3.81}$$

This expression shows the unique property of these modes: that the attenuation decreases with increase of frequency. The TE_{01} mode is used in the 'long-haul' waveguide discussed in chapter 7. In fig. 3.18 is shown a graph of attenuation versus frequency calculated from (3.81) for a copper guide of 50 mm internal diameter.

†3.11 A resonant cavity

As an application of the waveguide theory some of the properties of waveguide resonators will be illustrated by analysis of a TE_{10} mode

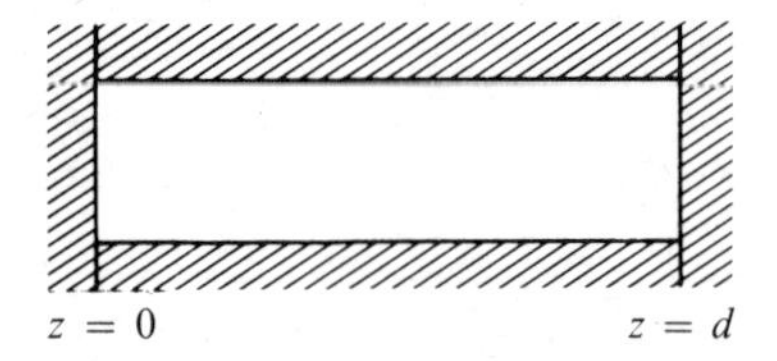

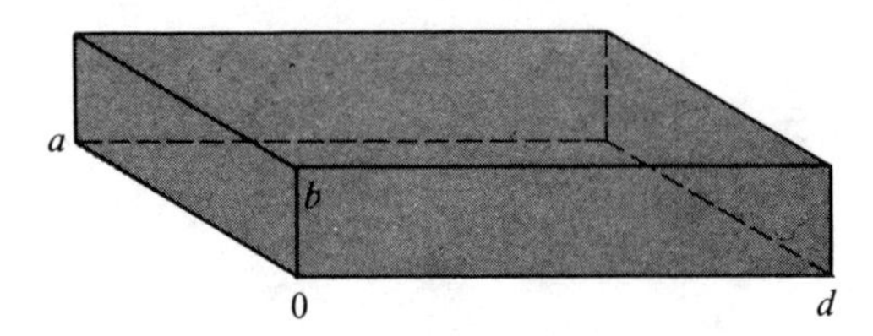

Fig. 3.19. A resonator in rectangular waveguide.

cavity in rectangular guide. We consider a length d of rectangular waveguide of cross-section $a \times b$ which is closed at the two ends by conducting plates, as shown in fig. 3.19. If we assume that the frequencies of interest are such that only the dominant mode propagates the fields inside the cavity can only be backward and forward waves in that mode. Hence we have, following (3.74),

$$H_z = \cos\frac{\pi x}{a}(A\,e^{-j\beta z} + B\,e^{j\beta z})$$

$$H_x = \frac{ja\beta}{\pi}\sin\frac{\pi x}{a}(A\,e^{-j\beta z} - B\,e^{j\beta z})$$

$$E_y = -\frac{ja\omega\mu}{\pi}\sin\frac{\pi x}{a}(A\,e^{-j\beta z} + B\,e^{j\beta z})$$

We must have $E_y = 0$ at $z = 0, d$. The first condition can be satisfied by making $B = -A$, giving

$$\begin{aligned} H_z &= -2Aj\cos\frac{\pi x}{a}\sin\beta z \\ H_x &= \frac{2Aja\beta}{\pi}\sin\frac{\pi x}{a}\cos\beta z \\ E_y &= -\frac{2Aa\omega\mu}{\pi}\sin\frac{\pi x}{a}\sin\beta z \end{aligned} \tag{3.82}$$

It is therefore necessary that $\sin \beta d = 0$, or

$$d = l\pi/\beta = \tfrac{1}{2}l\lambda_g \qquad (l = 1, 2, \ldots) \tag{3.83}$$

Thus a sequence of frequencies is defined, and under the conditions imposed by the assumption of perfect conductors a field can exist only at these frequencies. The ω-β diagram for the dominant mode is the same as in fig. 3.3. β varies smoothly from zero upwards, so that solutions are possible for any value of d. (At higher frequencies, of course, higher modes may propagate and other sequences of possible frequencies occur.) These possible frequencies are termed *resonant frequencies.*

With a real cavity, losses occur in the walls and also in the connection used to couple the cavity to source and load. These losses produce a range about each resonant frequency over which the cavity can be excited, as is true for resonant circuits. The concept of the quality factor, or Q, of a resonant circuit is extended to the cavity, using the definition

$$Q = \omega \frac{\text{stored energy}}{\text{energy lost per second}} \tag{3.84}$$

We therefore have to calculate the stored energy and the energy loss.

†3.11.1 Energy losses

The energy loss in the walls between z and $z + \delta z$ is given by

$$P_w = \delta z \ \tfrac{1}{2}R_s \int_C \mathbf{H} \cdot \mathbf{H}^* \, dl$$

The calculation follows the same lines as that for the attenuation of a progressive wave. Following the results in table 3.2 we find

$$P_w = \delta z \ \tfrac{1}{2}R_s \Big\{ 2\int_0^b 4|A|^2 \sin^2 \beta z \, dy$$

$$+ 2\int_0^a 4|A|^2[(a\beta/\pi)^2 \sin^2(\pi x/a)\cos^2 \beta z + \cos^2(\pi x/a)\sin^2 \beta z]\, dx\Big\}$$

$$= 2\delta z R_s |A|^2\{2b \sin^2 \beta z + a[(a\beta/\pi)^2 \cos^2 \beta z + \sin^2 \beta z]\}$$

We then integrate with respect to z between $z = 0$ and $z = d$, remembering that $\beta d = l\pi$, giving

$$P_w = 2|A|^2 R_s\{2b + a[1 + (a\beta/\pi)^2]\}d/2$$

To this must be added the loss in the end walls, arising because of currents associated with the tangential component H_x. These give a contribution

$$P_e = 2\times\tfrac{1}{2}R_s\int_0^a\int_0^b 4|A|^2(a\beta/\pi)^2\sin^2(\pi x/a)\,dx\,dy$$

$$= 4|A|^2R_s(a\beta/\pi)^2 ab/2$$

†3.11.2 Stored energy

Energy is stored in the electric and magnetic fields with densities $\tfrac{1}{2}\varepsilon E^2$ and $\tfrac{1}{2}\mu H^2$ respectively, in which E and H are instantaneous values. The expressions for the instantaneous fields represented by (3.82) are

$$H_z(x,t) = +2\,A\cos\frac{\pi x}{a}\sin\beta z\sin\omega t$$

$$H_x(x,t) = -2A\,\frac{a\beta}{\pi}\sin\frac{\pi x}{a}\cos\beta z\sin\omega t$$

$$E_y(x,t) = -2A\,\frac{a\omega\mu}{\pi}\sin\frac{\pi x}{a}\sin\beta z\cos\omega t$$

Hence we have for the stored electric energy

$$W_e = \frac{1}{2}\varepsilon\left(\frac{2A\omega\mu a}{\pi}\right)^2\cos^2\omega t\int_0^d\int_0^a\int_0^b \sin^2\frac{\pi x}{a}\sin^2\beta z\,dx\,dy\,dz$$

$$= \frac{1}{2}\varepsilon\left(\frac{2A\omega\mu a}{\pi}\right)^2\cos^2\omega t\,\frac{abd}{4} \tag{3.85}$$

For the magnetic energy we have two terms:

$$W_m = \tfrac{1}{2}\mu 4A^2\sin^2\omega t\int_0^d\int_0^a\int_0^b\left[\left(\frac{a\beta}{\pi}\right)^2\sin^2\frac{\pi x}{a}\cos^2\beta z + \cos^2\frac{\pi x}{a}\sin^2\beta z\right]dx\,dy\,dz$$

$$= 2\mu A^2\sin^2\omega t\left[\left(\frac{a\beta}{\pi}\right)^2 + 1\right]\frac{abd}{4}$$

Now from the expression for β for the TE_{10} mode (3.74),

$$\left(\frac{a\beta}{\pi}\right)^2 + 1 = \omega^2\mu\varepsilon\left(\frac{a}{\pi}\right)^2$$

The stored magnetic energy is therefore given by the expression

$$W_m = \frac{1}{2}\varepsilon\left(\frac{2A\omega\mu a}{\pi}\right)^2 \sin^2 \omega t \frac{abd}{4} \tag{3.86}$$

Comparing (3.85) and (3.86) we see that, as in a resonant circuit, the electric and magnetic stored energies are in phase quadrature with each other, but with the same peak value. The total stored energy is

$$W = \frac{1}{2}\varepsilon\left(\frac{2A\omega\mu a}{\pi}\right)^2 \frac{abd}{4}$$

These various values may now be substituted in (3.84). We find

$$\frac{1}{Q} = \frac{1}{\omega}\frac{P_e + P_w}{W}$$

$$= \frac{R_s\pi^2}{\omega} \times \frac{4ab(a\beta/\pi)^2 + 2d\{2b + a[1 + (a\beta/\pi)^2]\}}{(\omega\mu a)^2 \varepsilon abd}$$

in which d is given by (3.83) and β is related to ω. The simplest form for this complicated expression is obtained by using the variable

$$\lambda = v/f = 1/[f\surd(\mu\varepsilon)] = 2\pi/[\omega\surd(\mu\varepsilon)]$$

$$\beta^2 = \omega^2\mu\varepsilon - (\pi/a)^2 = (2\pi/\lambda)^2[1 - (\lambda/2a)^2]$$

The substitution finally yields the result

$$\frac{1}{Q} = \frac{R_s}{\pi}\sqrt{\left(\frac{\varepsilon}{\mu}\right)}\left\{\frac{\lambda}{b}\left[1 + \frac{2b}{a}\left(\frac{\lambda}{2a}\right)^2\right] + \frac{4}{l}\left[1 - \left(\frac{\lambda}{2a}\right)^2\right]^{3/2}\right\} \tag{3.87}$$

Example

A cavity is to be made from WG 16 of the shortest length which will resonate at 10 GHz. Determine the length of the cavity and estimate the Q-factor.

Solution. It has been shown in an earlier example that, for the dominant mode in WG 16, $f_{10} = 6.55$ GHz. It was also shown that $\lambda_g = c(f^2 - f_{10}^2)^{-1/2}$. Hence, at 10 GHz, $\lambda_g = 0.0397$ m.

The shortest cavity is of length $\frac{1}{2}\lambda_g = 1.99$ cm. We need to know the value of R_s for copper at 10 GHz. At room temperature we take $\sigma = 6 \times 10^7$ Sm^{-1}, $\mu = \mu_0$. Now

$$R_s = (\omega\mu/2\sigma)^{\frac{1}{2}} = (\pi f\mu_0/\sigma)^{\frac{1}{2}}$$

At 1 GHz, $R_s = 8.11 \times 10^{-3}$ Ω. At frequency f_G GHz, $R_s = 8.11 \times 10^{-3}\surd f_G$ Ω. At 10 GHz, $R_s = 25.6 \times 10^{-3}$ Ω. We can now insert values in (3.87), remembering that λ is the free-space wavelength 0.03 m, and that $l = 1$. We find

$$Q = 7.3 \times 10^3$$

3.12 Worked example

Starting from Maxwell's equations, derive the free-space wave equation

$$\frac{\partial^2 E_x}{\partial y^2} + \frac{\partial^2 E_x}{\partial z^2} - \frac{1}{c^2}\frac{\partial^2 E_x}{\partial t^2} = 0$$

with $c^2 = 1/\mu\varepsilon$ for the case where $E_z = E_y = 0$. Show how div $\boldsymbol{D} = 0$ is satisfied.

Discuss the use of a dominant evanescent waveguide mode to make a high-precision attenuator. Sketch the electric and magnetic fields of such an evanescent mode in a rectangular guide.

It is required to make an attenuator for measuring attenuations up to 100 dB with an accuracy of 1 dB. The internal dimensions of the guide are 50 mm × 25 mm. What is the attenuation, in decibels per metre, for such a guide operating at very low frequencies? Assuming that the coupling components to the fields are perfect, what is the maximum frequency at which this attenuator can be used without recalibration?

Maxwell's equations in free space are

$$\text{div}\,\boldsymbol{D} = 0,\ \text{div}\,\boldsymbol{B} = 0,\ \text{curl}\,\boldsymbol{H} = \partial\boldsymbol{D}/\partial t, \text{ and curl}\,\boldsymbol{E} = -\partial\boldsymbol{B}/\partial t.$$

(Cambridge University)

Solution. Since $E_z = E_y = 0$,

$$\text{curl}\,\boldsymbol{E} = \left(0, \frac{\partial E_x}{\partial z}, -\frac{\partial E_x}{\partial y}\right)$$

Hence

$$H_x = 0, \frac{\partial H_y}{\partial t} = -\frac{1}{\mu}\frac{\partial E_x}{\partial z}, \frac{\partial H_z}{\partial t} = \frac{1}{\mu}\frac{\partial E_x}{\partial y}$$

Putting $H_x = 0$ in curl $\boldsymbol{H}$ we have

$$\frac{\partial H_z}{\partial y} - \frac{\partial H_y}{\partial z} = \frac{\varepsilon\,\partial E_x}{\partial t},\ -\frac{\partial H_z}{\partial x} = 0,\ \frac{\partial H_y}{\partial x} = 0$$

Since H_y, H_z cannot be zero or constant in the context of the problem they must be independent of x, and hence E_x is independent of x. This ensures div $\boldsymbol{D} = \text{div}\,(\varepsilon\boldsymbol{E}) \equiv 0$. Substituting for H_y, H_z we have

$$\varepsilon\frac{\partial^2 E_x}{\partial t^2} = \frac{1}{\mu}\left(\frac{\partial^2 E_x}{\partial y^2} + \frac{\partial^2 E_x}{\partial z^2}\right)$$

Hence
$$\frac{\partial^2 E_x}{\partial y^2}+\frac{\partial^2 E_x}{\partial z^2}-\mu\varepsilon\frac{\partial^2 E_x}{\partial t^2}=0$$

which it was required to derive.

We use a time variation of $\exp(j\omega t)$ and assume a z-variation of $\exp(-\gamma z)$. The equation then becomes

$$\frac{\partial^2 E_x}{\partial y^2}+\sigma^2 E_x=0$$

where
$$\sigma^2=\gamma^2+\frac{\omega^2}{c^2}$$

We apply this to the rectangular tube $x=0, a$; $y=0, b$. We find

$$E_x \propto \sin(n\pi y/b)$$

if E_x is to vanish at $y=0$, b as required by the boundary conditions. Hence

$$\gamma^2+(\omega/c)^2=(n\pi/b)^2$$

To make an attenuator we must operate at a frequency for which no mode is propagating, and therefore the field configuration is predominantly that of the dominant mode. This will require $a<b$ and $n=1$. For attenuation, $\gamma=\alpha$ is real. Hence

$$\alpha^2=(\pi/b)^2-(\omega/c)^2$$

At low frequencies $\alpha=\pi/b$. When $b=50\text{ mm}=0.05\text{ m}$, $\alpha=20\pi$ nepers $\text{m}^{-1}=546\text{ dB m}^{-1}$.

As the frequency is increased, α will decrease. It will be 1% in error when

$$(\omega/c)^2=0.02(\pi/b)^2$$

or
$$f=(c/2b)\surd(0.02)=424\text{ MHz}$$

Note that in this example $a<b$ and **E** is x-oriented whereas in the text $a>b$ and **E** is y-oriented.

3.13 Summary

Systems of two or more perfect conductors embedded in a homogeneous dielectric medium and which have an axis of uniformity will support a principal wave in which the electric and magnetic fields are orthogonal and are transverse to the axis of uniformity. Such a wave is propagated with the velocity of a uniform plane wave in the unbounded dielectric medium. In any given plane perpendicular to the axis a unique voltage and current may be defined.

Inside a hollow tube no principal wave exists: it is necessary to have an axial component of either ***E*** or ***H***. Modes are designated transverse magnetic (*TM*) or transverse electric (*TE*). Each mode has its own field pattern and is characterised by a particular cut-off frequency. Propagation of a mode occurs only for frequencies exceeding the cut-off frequency of that mode. Below the cut-off frequency the mode is evanescent. The phase constant has the form $\beta = 2\pi(f^2 - f_c^2)^{\frac{1}{2}}/c$ in which f_c is the cut-off frequency. The phase velocity ω/β varies with frequency, and the wavelength within the tube is different from the wavelength in an unbounded medium.

If the modes are such that there is a range above the lowest cut-off frequency in which only one mode can propagate, that mode is said to be dominant.

Modes are nominally independent. The characteristics of a single mode can be translated into an equivalent transmission line.

A particular mode is associated with its own distribution of current in the waveguide wall. Attenuation may be estimated from these currents by calculating the power loss they would incur in a wall of finite conductivity.

Closed sections of waveguide form cavities which resonate at certain frequencies. A value of Q-factor may be calculated for any given resonance.

Formulae

Formulae for the cases of particular interest in the succeeding chapters are given below.

Coaxial line (*fig.* 3.2)

$$\mathrm{E}_\rho = \rho^{-1}\mathrm{A} \exp(-\mathrm{j}\beta z)$$

$$\mathrm{H}_\phi = \sqrt{(\varepsilon/\mu)}\, \mathrm{E}_\rho$$

$$\beta = \omega\sqrt{(\mu\varepsilon)} = \omega/v$$

$$\mathrm{V} = \int_a^b \mathrm{E}_\rho \, \mathrm{d}\rho = \mathrm{A} \ln(b/a)$$

$$Z_0 = \frac{1}{2\pi}\sqrt{(\mu/\varepsilon)} \ln(b/a)$$

$$P=\tfrac{1}{2}|A|^2 2\pi\sqrt{(\varepsilon/\mu)}\ln(b/a)$$

$$\alpha=\frac{\pi f}{v}\tan\delta+\frac{1}{2}R_s\sqrt{\left(\frac{\varepsilon}{\mu}\right)}\left(\frac{1}{a}+\frac{1}{b}\right)[\ln(b/a)]^{-1}$$

$R_s=\sqrt{(\pi f\mu/\sigma)}$, referring to the wall

Dominant mode in rectangular waveguide, TE_{10} (fig. 3.6)

$$H_z=A\cos(\pi x/a)\exp(-j\beta z)$$

$$H_x=jA\frac{\beta a}{\pi}\sin(\pi x/a)\exp(-j\beta z)$$

$$E_y=-\frac{\omega\mu}{\beta}H_x=-jA\frac{\omega\mu a}{\pi}\sin(\pi x/a)\exp(-j\beta z)$$

$$\beta=[\mu\varepsilon(\omega^2-\omega_{10}^2)]^{\frac{1}{2}}=2\pi f[1-(f_{10}/f)^2]^{\frac{1}{2}}/c$$

$$v_p=c/[1-(f_{10}/f)^2]^{\frac{1}{2}}$$

$$\lambda_g=\lambda/[1-(f_{10}/f)^2]^{\frac{1}{2}}\text{ where }\lambda=c/f$$

$$f_{10}=c/2a$$

$$P=|A|^2\omega\mu\beta a^3 b/4\pi^2$$

$$\alpha=\sqrt{\left(\frac{\varepsilon}{\mu}\right)}\frac{R_s}{b}\frac{1+(2b/a)(f_{10}/f)^2}{[1-(f_{10}/f)^2]^{\frac{1}{2}}}$$

TE_{01} in circular waveguide (fig. 3.11)

$$H_z=AJ_0(s_{01}\rho/a)\exp(-j\beta z)$$

$$H_\rho=jA\frac{\beta a}{s_{01}}J_1(s_{01}\rho/a)\exp(-j\beta z)$$

$$E_\phi\doteq-\frac{\omega\mu}{\beta}H_\rho=-A\frac{j\omega\mu a}{s_{01}}J_1(s_{01}\rho/a)\exp(-j\beta z)$$

$$\beta=[\mu\varepsilon(\omega^2-\omega_{01}^2)]^{\frac{1}{2}}=2\pi f[1-(f_{01}/f)^2]^{\frac{1}{2}}/c$$

$$v_p=c/[1-(f_{01}/f)^2]^{\frac{1}{2}}$$

$$f_{01}=3.832c/2\pi a=0.609c/a$$

$$P=17.4\omega\mu\beta a^4|A|^2\times10^{-3}$$

$$\alpha=\frac{R_s}{a}\sqrt{\left(\frac{\varepsilon}{\mu}\right)}\left(\frac{f_{01}}{f}\right)^2\left[1-\left(\frac{f_{01}}{f}\right)^2\right]^{-\frac{1}{2}}$$

3.14 Problems

1. A coaxial transmission line has an inner of diameter 2 mm. Determine the size of a tube to be used as a concentric outer which will make the characteristic impedance of the line equal to 100 Ω, assuming air-spacing. Find the maximum electric field within the coaxial line when it propagates a forward wave transmitting 10 W of R.F. power.

2. A transmission line is constructed from strips of copper foil 10 cm in width, two such strips being separated by a film 0.5 mm thick of dielectric of relative permittivity 2.5. Estimate the phase velocity and the characteristic impedance.

3. By expanding Maxwell's equations into components, show that the dispersion relation for *TEM* waves is

$$\gamma^2 + \omega^2/c^2 = 0$$

Use this result to calculate the lowest resonant frequency of an air-filled coaxial cavity of length 0.1 m, resonating in the *TEM* mode.

What would be the effect of filling the cavity with a dielectric of relative permittivity 6?

Explain how you would calculate the Q of such a cavity.

(Southampton University.)

4. A hollow rectangular, perfectly conducting pipe is described by the planes $x = 0$, $x = a$, $y = 0$, $y = b$. The z-axis is taken along the pipe. Starting from Maxwell's equations, show that field configurations exist for which only E_x, H_y, H_z are different from zero, and derive expressions for these three components in a forward travelling wave. Show that a minimum frequency exists below which any field of this form is attenuated along the pipe.

A particular waveguide has inside dimensions 1 cm × 2 cm. Determine the two cut-off frequencies possible with the mode described above. For the orientation giving the lowest cut-off frequency determine the maximum electric field strength at the walls when the guide is carrying a power of 10 kW at a frequency of 12 GHz.

(Bristol University.)

5. It is customary to operate waveguide of cut-off frequency f_0 at a nominal frequency of $1.5f_0$. Calculate the dimensions of (i) a rectangular waveguide whose sides are in the ratio 2 : 1 and which operates in the dominant mode, and (ii) a circular waveguide operating in the TE_{11} mode, both suitable for working at 6 GHz.

Suggest a design for a transition from rectangular to circular waveguide.

6. A microwave cavity is formed from a rectangular waveguide of 2.5 cm × 1 cm cross-section by short-circuiting a length equal to the width of the guide. It resonates in a TE_{011} mode. The resonator is coupled to a microwave circuit either by means of a 1 mm long electric probe or by a magnetic loop of 1 mm diameter.

Sketch the resonator and the field distributions. Determine and clearly indicate in the sketch the optimum positions and orientations of the probes, for maximum coupling.

(Southampton University; part question.)

7. Determine the modes which can propagate at 3 GHz in a circular waveguide of diameter 8 cm. Determine the values of the components of electric field strength at the centre of the waveguide when it carries a power of 2 MW.

8. Discuss the causes of loss in transmission lines and waveguides, with particular reference to the way losses change as frequency is varied.

The TE_{01} mode in a circular waveguide is perfectly symmetrical around the guide. The magnetic field at the walls is purely longitudinal, and, for a travelling wave in an air-filled guide, its amplitude H_0 is related to the power flow P by the equation

$$P = \frac{\pi}{2}\sqrt{\left(\frac{\mu_0}{\varepsilon_0}\right)}\,\frac{f\sqrt{(f^2 - f_0^2)}}{f_0^2}\,(aH_0)^2$$

in which a is the guide radius in metres and $f_0(= 0.183/a$ GHz) is the cut-off frequency for the mode.

A particular guide is 6 cm in diameter. The wall material is such that at 1 GHz a magnetic field at the wall of peak value 1 A m^{-1} gives rise to a loss of 4 mW m^{-2}. Estimate the attenuation coefficient, in dB km^{-1}, at a frequency of 8 GHz.

4

Radiation

We have been concerned in the last chapter with waves guided by a structure in linear form. Also important in communications is the propagation between transmitting and receiving antennae. We have considered in chapter 1 some properties of plane waves, and seen in chapter 2 how radiation a long way off from a source approximates to such plane waves. It is now necessary to consider the properties of antennae, both in transmitting and in receiving.

An antenna consists of suitably disposed conductors which will radiate electromagnetic energy when connected to a source of appropriate frequency. When the antenna is being excited, a distribution of current will exist over all the conduction surfaces which is determined solely by the conductor geometry and the frequency. The problem of determining the field radiated by a prescribed current distribution is in essence a purely mathematical one: we shall in due course derive a formula which enables the electric and magnetic fields to be computed. Unfortunately the matter is seldom so simple: it is first necessary to find the current distribution, which is mathematically far more difficult. However, it is frequently possible to make a good estimate of the current distribution, and further to show that the finer structure of the distribution is not of prime importance. This enables us to make good estimates of the radiated field in many cases, and so find out something about the properties of an antenna in transmission. In some cases it may be easier to estimate the local field produced by an antenna and evaluate the distant field from this. To find out directly about properties of the same antenna in reception is difficult, but fortunately a reciprocity theorem exists which can be applied to relate the two situations.

We first consider the problem of determining the field radiated from a given current distribution. In magnetism we would evaluate the magnetic field of a distribution of currents with the aid of the Biot–Savart law, which in essence provides a formula for the field produced by an element of current $i\,\mathrm{d}l$. The magnetic field from such an element at a point whose radius vector with respect to the element is $\boldsymbol{r} = r\boldsymbol{a}$† is given by

$$\boldsymbol{H} = \frac{i}{4\pi r^2} \mathrm{d}\boldsymbol{l} \times \boldsymbol{a} \tag{4.1}$$

We use this formula for frequencies which in some sense are 'low' with results in agreement with experiment, but it is in disagreement with prediction from Maxwell's equations at high frequency. It was shown in chapter 2 that the field a long way from a distribution was proportional to r^{-1}, rather than r^{-2} as predicted by (4.1). We therefore first obtained a replacement for (4.1) which reconciles this apparent disagreement.

4.1 The short oscillating electric dipole

The element we choose to consider is shown diagrammatically in fig. 4.1: a current filament of length $\mathrm{d}l$ carries a current $i = I_0 \cos \omega t$, and in order to conserve charge we must suppose point charges at either end, also alternating. For charge to be conserved we must have i equal to $\mathrm{d}q/\mathrm{d}t$, and hence the charges are $\pm(I_0/\omega)\sin \omega t$. We will for convenience take the current in the direction of the z-axis, when (4.1) can be written in phasor form

$$\mathrm{H}_\phi = \frac{\mathrm{I}\,\mathrm{d}l}{4\pi r^2} \sin \theta \tag{4.2}$$

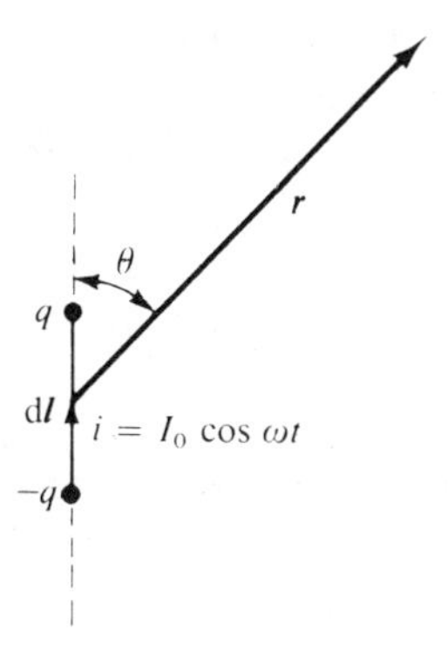

Fig. 4.1. The short oscillating electric dipole.

† The notation $\boldsymbol{a}$ will be used for the unit vector in the direction of r, distinguished by a numerical suffix where necessary.

We will try to reconcile the two aspects of the problem by modifying (4.2). Considering (2.27), a likely form to try is

$$H_\phi = \frac{e^{-jkr}}{r}\left(a + \frac{b}{r}\right)\sin\theta \tag{4.3}$$

where $k = \omega\sqrt{(\mu\varepsilon)}$, and a, b are constants.

This form gives a wave of the correct type for large values of r, and reduces to (4.2) for small r as $\omega \to 0$. Following (2.21), Maxwell's equations for free space in spherical polar co-ordinates are

$$\frac{1}{r\sin\theta}\left[\frac{\partial}{\partial\theta}(\sin\theta E_\phi) - \frac{\partial E_\theta}{\partial\phi}\right] = -j\omega\mu H_r \tag{4.4}$$

$$\frac{1}{r}\left[\frac{1}{\sin\theta}\frac{\partial E_r}{\partial\phi} - \frac{\partial}{\partial r}(rE_\phi)\right] = -j\omega\mu H_\theta \tag{4.5}$$

$$\frac{1}{r}\left[\frac{\partial}{\partial r}(rE_\theta) - \frac{\partial E_r}{\partial\theta}\right] = -j\omega\mu H_\phi \tag{4.6}$$

$$\frac{1}{r\sin\theta}\left[\frac{\partial}{\partial\theta}(\sin\theta H_\phi) - \frac{\partial H_\theta}{\partial\phi}\right] = j\omega\varepsilon E_r \tag{4.7}$$

$$\frac{1}{r}\left[\frac{1}{\sin\theta}\frac{\partial H_r}{\partial\phi} - \frac{\partial}{\partial r}(rH_\phi)\right] = j\omega\varepsilon E_\theta \tag{4.8}$$

$$\frac{1}{r}\left[\frac{\partial}{\partial r}(rH_\theta) - \frac{\partial H_r}{\partial\theta}\right] = j\omega\varepsilon E_\phi \tag{4.9}$$

In these equations **J** has been taken to be zero, since current is restricted to a very small region near the origin.

Assuming H_ϕ is given by (4.3) and $H_r = H_\theta = 0$, we find from equations (4.7) to (4.9) that

$$j\omega\varepsilon E_r = \frac{e^{-jkr}}{r^2}\left(a + \frac{b}{r}\right)2\cos\theta$$

$$j\omega\varepsilon E_\theta = \frac{e^{-jkr}}{r}\left(jka + \frac{jkb}{r} + \frac{b}{r^2}\right)\sin\theta$$

$$j\omega\varepsilon E_\phi = 0$$

We have obtained these expressions for the components of the electric field by assuming an expression for H_ϕ. We now have to check for consistency by using the electric field components in equations (4.4)–(4.6) and identifying the resulting expression for the magnetic field components with the assumed form of (4.3). Since E_ϕ is zero, and there is no ϕ-dependence, (4.4) and (4.5) are satisfied with the assumed H_ϕ. Substituting the calculated E_θ and E_r in (4.6) yields

$$H_\phi = \frac{1}{k^2 r}\left\{\frac{\partial}{\partial r}\left[e^{-jkr}\left(jka + \frac{jkb}{r} + \frac{b}{r^2}\right)\sin\theta\right] + \frac{e^{-jkr}}{r^2}\left(a + \frac{b}{r}\right)2\sin\theta\right\}$$

$$= \frac{e^{-jkr}}{r}\sin\theta\left[a + \frac{b}{r} + \frac{2}{k^2 r^2}(a - jkb)\right]$$

The right-hand side must be identical to (4.3). This is only possible if $a = jkb$. Finally, (4.3) should approach (4.2) as k becomes small. This requires $b = I\,dl/4\pi$. We therefore have self-consistent field components given by

$$H_\phi = \frac{I\,dl}{4\pi}\frac{e^{-jkr}}{r}\left(jk + \frac{1}{r}\right)\sin\theta \tag{4.10}$$

$$E_r = \frac{1}{j\omega\varepsilon}\frac{I\,dl}{4\pi}\frac{e^{-jkr}}{r^2}\left(jk + \frac{1}{r}\right)2\cos\theta \tag{4.11}$$

$$E_\theta = \sqrt{\left(\frac{\mu}{\varepsilon}\right)}\frac{I\,dl}{4\pi}\frac{e^{-jkr}}{r}\left(jk + \frac{1}{r} + \frac{1}{jkr^2}\right)\sin\theta \tag{4.12}$$

These expressions give the field components for a short oscillating electric, or Hertzian, dipole of strength $I\,dl$. We note that, as well as giving (4.2) as $r \to 0$, they yield the expected results as $r \to \infty$:

$$E_r \propto r^{-2}, \qquad E_\theta \propto r^{-1}, \qquad E_\theta/H_\phi = \sqrt{(\mu/\varepsilon)}$$

The discussion leading to the derivation of the exact formulae (4.10) to (4.12) has used the limiting cases of r small and r large. It is instructive to look at the formulae to make more precise the norm by which r is judged large or small. Examination of the formulae shows that the appropriate dimensionless variable is kr, which is equal to $2\pi r/\lambda$. Thus 'r large' means many wavelengths, and 'r small' means much less than a wavelength. For example, the Biot–Savart law will apply at distances much greater than the length dl but much less than a wavelength. Likewise the adjective 'short' applied to the dipole means much less than a wavelength. In contrast to the transmission-line situation where the system is regarded as very long, antennae are structures essentially limited in size, and we shall see that the dimensions relative to the wavelength play a very important rôle.

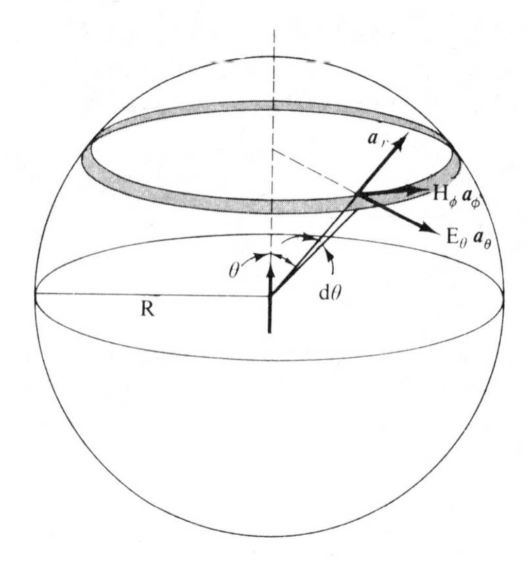

Fig. 4.2. Evaluation of power flowing from a dipole.

4.1.1 Power radiated by a short dipole

We have taken the short dipole as an element in its own right, without considering any physical model or how one could be supplied with an electrical input. We can work out the radiated power and hence the power input without such a model by using Poynting's theorem. Consider the fields on a very large sphere, centred at the dipole, of radius R, as shown in fig. 4.2. Applying the results expressed by (2.11), the mean power crossing the whole sphere will be given by

$$\tfrac{1}{2}\,\mathrm{Re}\oint \mathbf{E}\times\mathbf{H}^*\cdot \boldsymbol{a}\,\mathrm{d}S \tag{4.13}$$

The radial component of $\mathbf{E}\times\mathbf{H}^*$ is in the present case $E_\theta H_\phi^*$. When $kr \gg 1$, from equations (4.10) to (4.12),

$$\sqrt{\left(\frac{\varepsilon}{\mu}\right)}\,E_\theta = H_\phi = \frac{I\,\delta l}{4\pi}\frac{jk}{r}\,e^{-jkr}\sin\theta \tag{4.14}$$

Hence the power density at radius r in direction θ is given by

$$\frac{1}{2}E_\theta H_\phi^* = \frac{1}{2}\sqrt{\left(\frac{\mu}{\varepsilon}\right)}\,|H_\phi|^2 = \frac{1}{2}\sqrt{\left(\frac{\mu}{\varepsilon}\right)}\left|\frac{I\,\delta l}{4\pi r}k\right|^2\sin^2\theta$$

This result should be compared with that obtained using the general result of (2.32). Substituting in (4.13) we have

$$P=\frac{1}{2}\int_0^{2\pi}\int_0^{\pi} \sqrt{\left(\frac{\mu}{\varepsilon}\right)} \left|\frac{\mathrm{I}\,\delta lk}{4\pi R}\right|^2 \sin^2\theta\, R^2\, \mathrm{d}\theta \sin\theta\, \mathrm{d}\phi$$

$$=\frac{1}{2}\sqrt{\left(\frac{\mu}{\varepsilon}\right)}\left(\frac{k\delta l}{4\pi}\right)^2|\mathrm{I}|^2\int_0^{\pi} 2\pi \sin^3\theta\, \mathrm{d}\theta$$

Now
$$\int_0^{\pi} \sin^3\theta\, \mathrm{d}\theta=\int_0^{\pi} (1-\cos^2\theta)\frac{\mathrm{d}}{\mathrm{d}\theta}(-\cos\theta)\,\mathrm{d}\theta$$

$$=\int_{-1}^{+1}(1-u^2)\,\mathrm{d}u$$

$$=\frac{4}{3}$$

Therefore

$$P=\frac{1}{2}\sqrt{\left(\frac{\mu}{\varepsilon}\right)}\frac{2\pi}{3}\left(\frac{\delta l}{\lambda}\right)^2|\mathrm{I}|^2$$

If we consider that the current in the dipole is provided by a current generator, the power radiated is that which would be dissipated in a resistor of value R_a, given by

$$R_a=\sqrt{\left(\frac{\mu}{\varepsilon}\right)}\frac{2\pi}{3}\left(\frac{\delta l}{\lambda}\right)^2 \tag{4.15}$$

This is called the *radiation resistance* of the element. In circuit terms we are imagining the element to be driven by a generator such that a current I flows. The generator will see an equivalent circuit as in fig. 4.3, of which R_a has been calculated but not the reactance X_a.

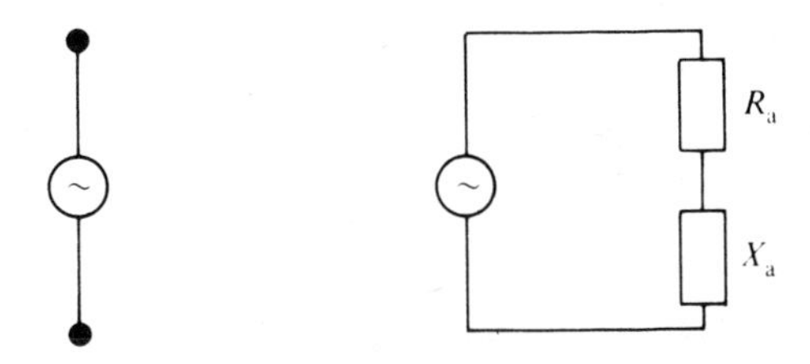

Fig. 4.3. Equivalent circuit for a dipole.

4.2 A linear antenna

The way in which elements of this sort can be combined to give a prescribed current distribution will be illustrated firstly by a linear current filament. Consider the situation shown in fig. 4.4. A current filament lies along the z-axis from $z=-l$ to $z=+l$, carrying current $\mathrm{I}(z)$. If I is not constant a charge distribution exists along the filament, and also at the ends, but this enters into our calculation only implicitly. We can regard a short length $d\zeta$ at S, $z=\zeta$, to be a short dipole, the field components for which are given by equations (4.10) to (4.12). In these equations r_{12} is the distance from the point S to a distant point P. If P is very distant compared to the length of the filament, r_{12} is substantially constant for all points of the filament. Taking O to be a representative point we may write

$$r_{12} = SP \approx OP = r_1 \tag{4.16}$$

The angle θ is likewise virtually the same wherever S lies on the filament. We may not however replace r_{12} by OP in the term $\exp(-\mathrm{j}kr_{12})$, since this is periodic in λ. We have to obtain a better approximation to r_{12} which is good to a length less than λ for distant P. A perpendicular dropped from S to OP defines S'. We have

$$\begin{aligned} SP \approx S'P &= OP - OS\cos\theta \\ &= r_1 - \zeta\cos\theta \end{aligned} \tag{4.17}$$

The error in this equation decreases indefinitely as r_1 becomes larger, Putting these approximations into (4.14), the element $\mathrm{I}(\zeta)\,d\zeta$ at S makes a contribution to H_ϕ at P given by

$$\delta\mathrm{H}_\phi = \frac{\mathrm{I}(\zeta)\,\delta\zeta}{4\pi r_1}\exp[-\mathrm{j}k(r_1 - \zeta\cos\theta)]\mathrm{j}k\sin\theta$$

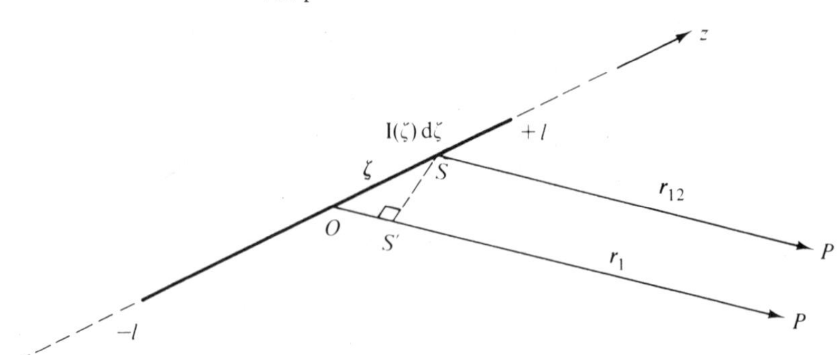

Fig. 4.4. A linear distribution of current.

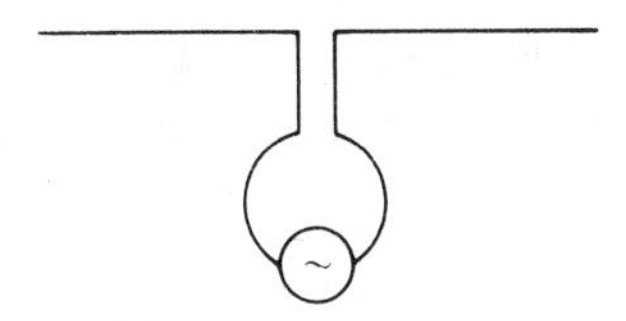

Fig. 4.5. A centre-fed dipole.

For each element, H_ϕ is in the same direction so that the resultant becomes

$$H_\phi = \frac{jk}{4\pi r_1} \exp(-jkr_1) \sin\theta \int_{-l}^{l} I(\zeta) \exp[jk\zeta\cos\theta]\, d\zeta \qquad (4.18)$$

We note that this is of the form predicted by (2.29).

Before proceeding further it is necessary to have an explicit form for $I(\zeta)$. Consider the case of a centre-fed dipole, as shown in fig. 4.5: the current must fall to zero at the ends, and provided it is not too long we would expect the current to decrease monotonically towards the ends. A possible form would be, for example,

$$I = I_0 \cos(\pi z/2l) \qquad (4.19)$$

It must be emphasised that, in the real case when the dipole of fig. 4.5 might be two equal rods fed at the centre, the actual distribution would not be precisely that given by (4.19). The true distribution must yield an electric field which is everywhere normal to the conductor. This condition is in fact sufficient to determine the fields, but the procedure is far from straightforward.

When the expression in (4.19) is substituted in (4.18) it becomes necessary to evaluate the integral

$$\int_{-l}^{l} \cos\frac{\pi\zeta}{2l} \exp[jk\zeta\cos\theta]\, d\zeta$$

Writing the cosine in the form $\frac{1}{2}\{\exp[j(\pi\zeta/2l)] + \exp[-j(\pi\zeta/2l)]\}$ the integration can be performed, yielding finally

$$H_\phi = I_0 \frac{jk}{4\pi r_1} \exp(-jkr_1) \sin\theta \frac{\pi/l}{(\pi/2l)^2 - k^2\cos^2\theta} \cos(kl\cos\theta) \qquad (4.20)$$

If $kl \ll 1$, i.e. when the length is sufficiently small, the expression reduces to

$$H_\phi = I_0 \frac{2l}{\pi} \frac{jk}{4\pi r_1} \exp(-jkr_1) \sin\theta \qquad (4.21)$$

This is, as it should be, the same as (4.14) except that for I dl we have $2I_0(l/\pi)$. This is also to be expected, since for a number of small elements we would have to replace I δl by $\int_{-l}^{l} I(\xi)\,d\zeta$.

4.2.1 The half-wave dipole

A simplification to (4.20) occurs if the frequency is such that $\pi/2l = k$, or $2l = \frac{1}{2}\lambda$. In this case we refer to a *half-wave* dipole. Equation (4.20) becomes

$$H_\phi = \frac{jI_0}{2\pi r_1}\exp(-jkr_1)\frac{\cos(\frac{1}{2}\pi\cos\theta)}{\sin\theta} \tag{4.22}$$

The dependence of $|H_\phi|$ on θ is illustrated in fig. 4.6. On the same plot is shown $|H_\phi|$ for the short dipole as given by (4.21). In both cases the maximum of $|H_\phi|$ is used as a normalising factor.

Equation (4.22) represents the distant field of a real half-wave dipole tolerably well. It may be asked why this is so, and for example what would happen if the dipole were made longer than one half-wavelength. Since we have not derived the distribution represented by (4.19), but only invented it, some justification is needed. It will be noticed that the

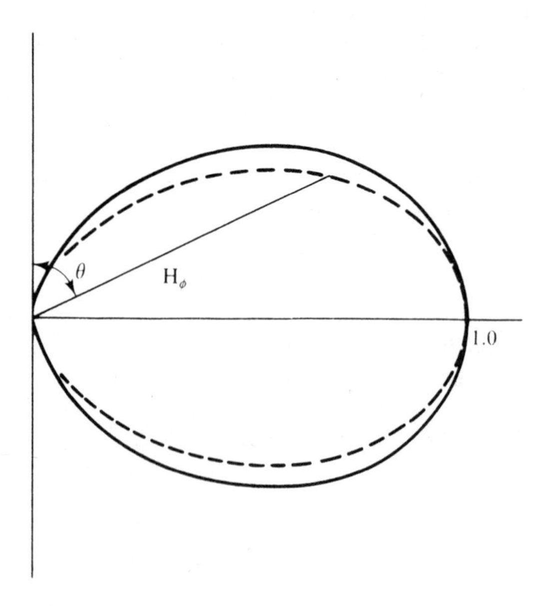

Fig. 4.6. Polar plot of $|H_\phi|$ for a short dipole (full line) and half-wave dipole (pecked line).

integrand in (4.18) is $I(\zeta)$ multiplied by a function periodic in λ. The mathematical implication of this fact is that variations of $I(\zeta)$ on a scale much smaller than λ will only alter the result to a minor extent. Small deviations from the form of (4.19) will not have much effect, as may be shown by using e.g. $I_0(1-|z/l|)$ and similar expressions. Although an analogy between a finite single wire in space and a transmission line may seem far-fetched, both theory and experiment show that the analogy is in some respects close. It was shown in chapter 3 that a waveguide cavity resonated when its length was an integral multiple of $\frac{1}{2}\lambda$. This is true of any wave motion, and if the dipole is regarded in this way it would appear likely that it would in some sense resonate at frequencies for which the length is an integral number of half-wavelengths. From a purely practical standpoint the impedance presented at the centre of a dipole can be measured as a function of frequency, with results of the type shown in fig. 4.7. It is seen that the reactance passes through zero at a frequency close to the expected value. Complete agreement could not possibly be expected: the gap in the middle and the rod diameter are important parameters, and have not been taken into account.

At higher frequencies for which the wavelength is $4l/n$, the appropriate form for current would be

$$I = I_0 \cos(n\pi z/2l) \tag{4.23}$$

In such a case it would be proper to think of an attenuated wave on the dipole, making the amplitudes of maxima and minima decrease as the ends were approached. Such attenuation would be the result of radiated power and also, in a real antenna, loss due to conductor resistance.

As with the short dipole it is possible to estimate a value for radiation resistance by working out the total radiated power.

Radiation resistance for a half-wave dipole

We have, as in (4.13) and referring to fig. 4.2,

$$P = \frac{1}{2}\mathrm{Re}\oint \mathbf{E}\times\mathbf{H}^* \cdot \boldsymbol{a}\,\mathrm{d}S$$

$$= \frac{1}{2}\sqrt{\left(\frac{\mu}{\varepsilon}\right)}\int_0^{2\pi}\int_0^{\pi}|H_\phi|^2 R^2\,\mathrm{d}\theta\sin\theta\,\mathrm{d}\phi$$

$$= \frac{1}{2}\sqrt{\left(\frac{\mu}{\varepsilon}\right)}\,I_0^2\frac{1}{2\pi}\int_0^{\pi}\frac{[\cos(\frac{1}{2}\pi\cos\theta)]^2}{\sin\theta}\,\mathrm{d}\theta$$

The integral can be seen by sketching the curve to be of the order of

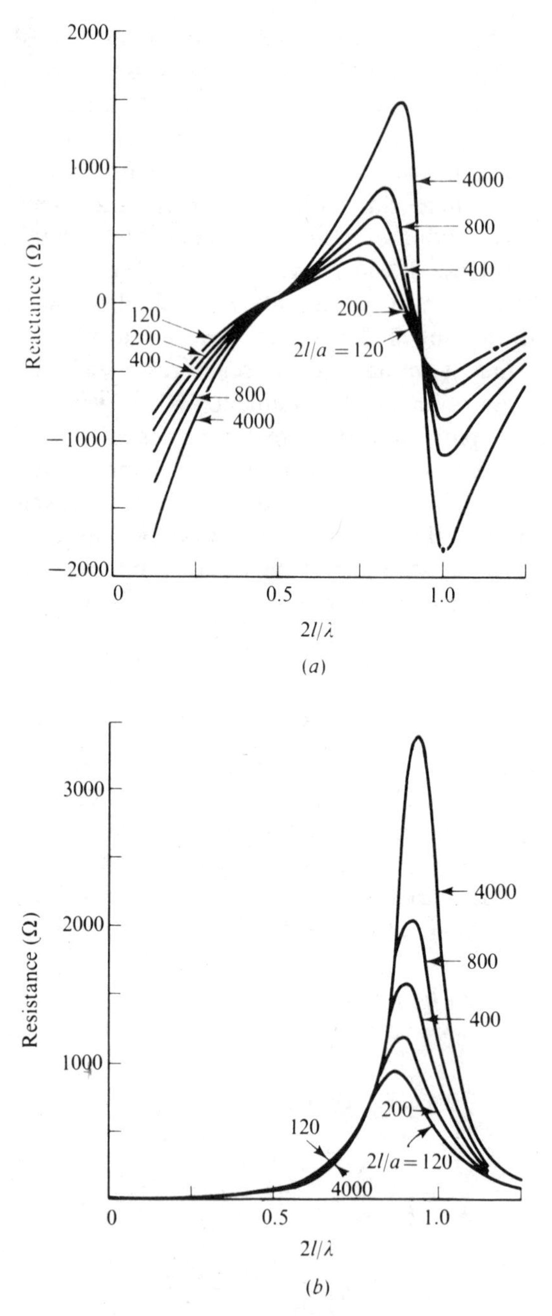

Fig. 4.7. Input resistance and reactance of a half-wave dipole as a function of length and radius of conductors.

unity. The evaluation in terms of tabulated functions is given in § 10.2, where the value is shown to be 1.219. Equating P to $\frac{1}{2}R_a I_0^2$ we have

$$R_a = \frac{1}{2\pi}\sqrt{\left(\frac{\mu_0}{\varepsilon_0}\right)} \times 1.219$$

$$= 73\ \Omega \tag{4.24}$$

It is instructive to compare this result with that for a short dipole. The radiation resistance for the short dipole, the field of which was given in (4.14), was derived in (4.15). To find the radiation resistance when the field is given by (4.21), we replace δl by $2l/\pi$, giving

$$R_a = \sqrt{\left(\frac{\mu}{\varepsilon}\right)} \frac{2\pi}{3}\left(\frac{2l}{\pi\lambda}\right)^2 \tag{4.25}$$

Consider $2l = \lambda/10$ as being short. For free space $\sqrt{(\mu/\varepsilon)} = 377\ \Omega$, giving $R_a = 0.8\ \Omega$. Thus to obtain a total radiated power of 100 W, an r.m.s. current of 11 A would be necessary, whereas from (4.24) only a little over 1 A would be needed for the half-wave dipole. The diagrams in fig. 4.6 show that the patterns of radiation are not much different, so we may deduce that the halfwave dipole is a much more efficient radiator. This is driven home in a practical sense when it is desired to build systems giving high-power radiation at low frequencies of the order of 30 kHz. At such a frequency the wavelength is 10 km, so that any practical radiator is inevitably short in the electrical sense.

4.3 An arbitrary current distribution: the distant field

In order for us to be able to calculate the fields arising from any current distribution we must express the results for the electric dipole in a general form. We are most usually concerned with the distant field, so that E_θ and H_ϕ are given by (4.14). Remembering that $\mathrm{d}\boldsymbol{l}$ was taken in the z-direction we can write vectorially

$$\mathbf{H} = \frac{I}{4\pi}\frac{\mathrm{j}k}{r}\exp(-\mathrm{j}kr)\,\mathrm{d}\boldsymbol{l} \times \boldsymbol{a} \tag{4.26}$$

In the distant field $\mathbf{E}$ is predominantly transverse, so that

$$\mathbf{E} = \sqrt{(\mu/\varepsilon)}\,\mathbf{H} \times \boldsymbol{a} \tag{4.27}$$

We have to perform one further generalisation, which is to remove the dipole from the origin of co-ordinates, as shown in fig. 4.8. In terms of

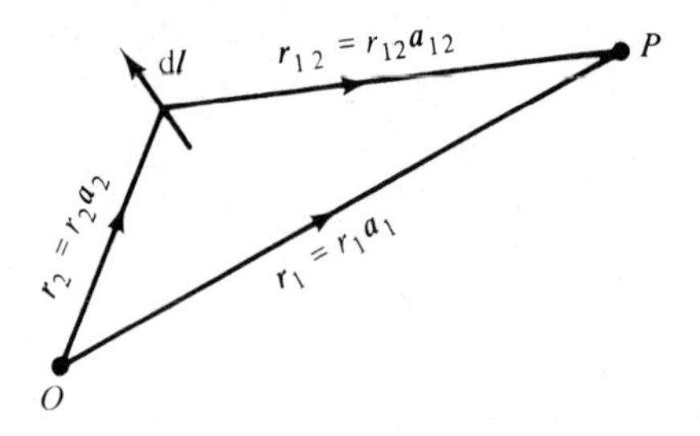

Fig. 4.8. Geometry for a dipole of arbitrary position.

the variables shown (4.26) takes the form

$$\mathbf{H}(\boldsymbol{r}_1)=\frac{\mathrm{I}}{4\pi}\frac{\mathrm{j}k}{r_{12}}\exp\left(-\mathrm{j}kr_{12}\right)\mathrm{d}\boldsymbol{l}\times\boldsymbol{a}_{12} \tag{4.28}$$

The same sort of argument as was used in deriving (4.17) for the linear antenna can now be employed. The unit vector $\boldsymbol{a}_{12}$ and the term r_{12}^{-1} may be replaced by $\boldsymbol{a}_1$ and r_1^{-1} respectively. The construction in fig. 4.4 is repeated in fig. 4.9, when it may be seen that to the required order of accuracy

$$r_{12}=r_1-r_2\cos\gamma=r_1-r_2\boldsymbol{a}_1\cdot\boldsymbol{a}_2 \tag{4.29}$$

Substituting in (4.28) gives the required form as

$$\mathbf{H}(\boldsymbol{r}_1)=\frac{\mathrm{I}}{4\pi}\frac{\mathrm{j}k}{r_1}\exp\left(-\mathrm{j}kr_1\right)\exp\left(\mathrm{j}kr_2\boldsymbol{a}_1\cdot\boldsymbol{a}_2\right)\mathrm{d}\boldsymbol{l}\times\boldsymbol{a}_1 \tag{4.30}$$

$$\mathbf{E}(\boldsymbol{r}_1)=\sqrt{\left(\frac{\mu}{\varepsilon}\right)}\,\mathbf{H}(\boldsymbol{r}_1)\times\boldsymbol{a}_1 \tag{4.31}$$

We can now consider a current distribution $\mathbf{J}(\boldsymbol{r}_2)$. We replace I d$\boldsymbol{l}$ in (4.20) by $\mathbf{J}(\boldsymbol{r}_2)\,\mathrm{d}v$, d$v$ being a volume element at $\boldsymbol{r}_2$. The total field is

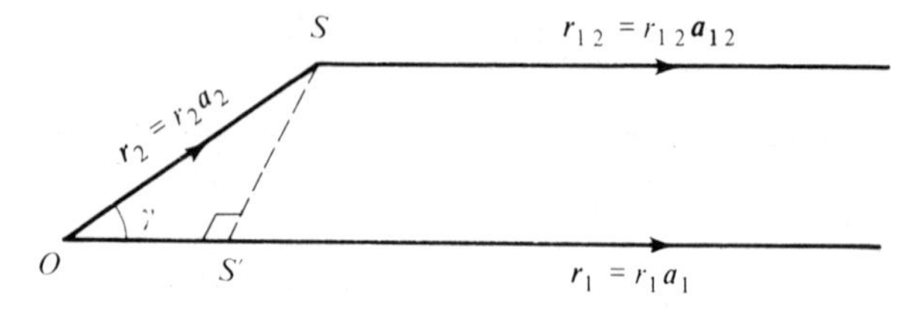

Fig. 4.9. Calculation of path length difference.

obtained by integrating over the entire current distribution. We have

$$\mathbf{H}(\boldsymbol{r}_1)=\frac{1}{4\pi}\frac{\mathrm{j}k}{r_1}\exp(-\mathrm{j}kr_1)\int \exp(\mathrm{j}kr_2\boldsymbol{a}_1\cdot\boldsymbol{a}_2)\mathbf{J}(\boldsymbol{r}_2)\times\boldsymbol{a}_1\,\mathrm{d}v$$

$$=\frac{1}{4\pi}\frac{jk}{r_1}\exp(-jkr_1)\mathbf{K}\times\boldsymbol{a}_1 \tag{4.32}$$

where

$$\mathbf{K}=\int \exp(\mathrm{j}kr_2\boldsymbol{a}_1\cdot\boldsymbol{a}_2)\mathbf{J}(\boldsymbol{r}_2)\,\mathrm{d}v \tag{4.33}$$

It will be noted that $\mathbf{K}$ is a function of the *direction* of $\boldsymbol{r}_1$, and not its magnitude. This is in accord with (2.28) and (2.29). With the same notation, (4.31) may be written

$$\mathbf{E}(\boldsymbol{r}_1)=\frac{\zeta}{4\pi}\frac{\mathrm{j}k}{r_1}\exp(-\mathrm{j}kr_1)(\mathbf{K}\times\boldsymbol{a}_1)\times\boldsymbol{a}_1$$

$$=-\frac{\zeta}{4\pi}\frac{\mathrm{j}k}{r_1}\exp(-\mathrm{j}kr_1)[\mathbf{K}-\boldsymbol{a}_1(\boldsymbol{a}_1\cdot\mathbf{K})] \tag{4.34}$$

The quantity in square brackets is the component of $\mathbf{K}$ transverse to $\boldsymbol{a}_1$. The problem of determining the distant field from a given current distribution therefore depends only on evaluating the quantity $\mathbf{K}$ defined by (4.33). This will be illustrated by the small circular loop antenna.

4.4 The small loop antenna

We consider a circular loop carrying a current I which is of the same phase at every point of the loop. This is only possible if the loop is electrically small, i.e. its diameter is small compared with the wavelength. The geometry is shown in fig. 4.10. We have

$$\boldsymbol{r}_2=b\boldsymbol{a}_x\cos\psi+b\boldsymbol{a}_y\sin\psi$$

$$\mathrm{d}\boldsymbol{l}=b\,\mathrm{d}\psi(-\boldsymbol{a}_x\sin\psi+\boldsymbol{a}_y\cos\psi)$$

$$\boldsymbol{r}_1=r_1(\boldsymbol{a}_x\sin\theta\cos\phi+\boldsymbol{a}_y\sin\theta\sin\phi+\boldsymbol{a}_z\cos\theta$$

Hence

$$r_2\boldsymbol{a}_1\cdot\boldsymbol{a}_2=b\sin\theta\cos(\psi-\phi)$$

Reverting from $\mathbf{J}\,\mathrm{d}v$ to $\mathrm{I}\,\mathrm{d}\boldsymbol{l}$ in (4.33) we have

$$\mathbf{K}(\theta,\phi)=\mathrm{I}\int_0^{2\pi}\exp[\mathrm{j}kb\sin\theta\cos(\psi-\phi)]b(-\boldsymbol{a}_x\sin\psi+\boldsymbol{a}_y\cos\psi)\,\mathrm{d}\psi$$

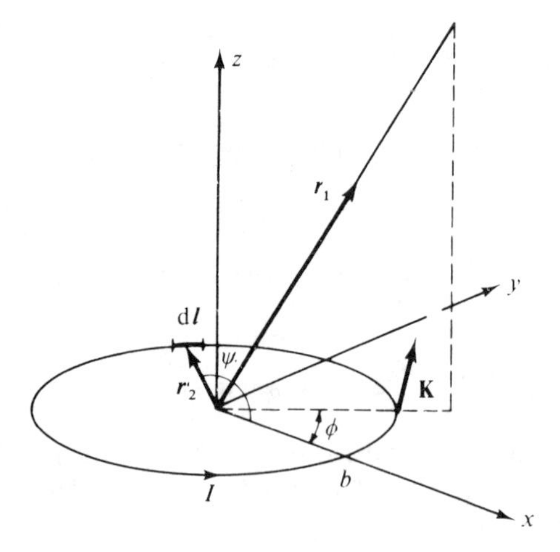

Fig. 4.10. Small circular loop antenna.

Since $kb \ll 1$, we may profitably expand the exponential as a power series giving

$$\mathbf{K}(\theta, \phi) = \mathrm{I}b \int_0^{2\pi} [1 + \mathrm{j}kb \sin\theta \cos(\psi - \phi) + \ldots](-\boldsymbol{a}_x \sin\psi + \boldsymbol{a}_y \cos\psi)\,\mathrm{d}\psi$$

$$\approx \mathrm{j}kb^2\mathrm{I}\pi(-\boldsymbol{a}_x \sin\phi + \boldsymbol{a}_y \cos\phi) \sin\theta$$

We can interpret this in spherical polar co-ordinates by saying that **K** has only a 'ϕ' component K_ϕ given by

$$\mathrm{K}_\phi = \mathrm{j}kb^2\pi\mathrm{I} \sin\theta. \tag{4.35}$$

E and **H** are then obtained from (4.32) and (4.34). The results are

$$\mathrm{E}_\phi = \zeta\left(\frac{ka}{2}\right)^2 \mathrm{I} \sin\theta \frac{\exp(-\mathrm{j}kr_1)}{r_1}$$
$$\mathrm{H}_\theta = -\left(\frac{ka}{2}\right)^2 \mathrm{I} \sin\theta \frac{\exp(-\mathrm{j}kr_1)}{r_1} \tag{4.36}$$

Comparing the results with (4.14) we see that the angular distribution is the same as for the short electric dipole. The loop is often regarded as a small magnetic dipole. Comparing (4.14) with (4.36), the radiation resistance will be obtained from (4.15) if δl is replaced by $4\pi(\frac{1}{2}ka)^2/k$. This yields

$$R_a = \frac{\pi^5}{6} \sqrt{\left(\frac{\mu}{\varepsilon}\right)} \left(\frac{2b}{\lambda}\right)^4 \tag{4.37}$$

For $2b = \lambda/10$ we find

$$R_a = 1.9\,\Omega.$$

4.5 An antenna in reception

In order to develop ideas on this problem it is convenient to consider a particular model. Let us take a wire antenna fed across a small gap somewhere along its length, as indicated in fig. 4.11(a). Here such an antenna is pictured transmitting, being fed by a source at the gap, which we assume to be of negligible magnitude. If we were to solve the transmitting problem correctly we would find a solution of Maxwell's equations for which the component of electric field parallel to the wire vanished on the surface, except across the gap where a finite voltage drop would be developed. Once the external equations had been solved

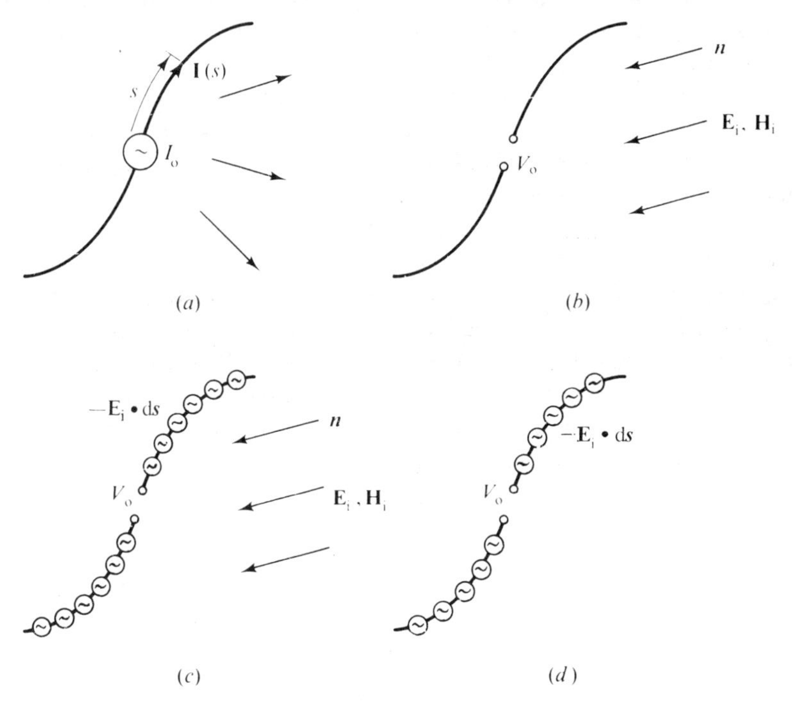

Fig. 4.11. Antenna in reception. (a) Current distribution in transmission. (b) Open-circuit between terminals in the field of an incident wave. (c) Induced currents replaced by generators. (d) The scattered field produced by the generators only.

the magnetic field next to the wire would be known and the currents in the wire could be calculated. A current distribution, denoted by $\mathbf{I}(s)$, would be found, and would be characteristic of the antenna. If it were known, the external field could be expressed, using (4.34), in the form

$$\mathbf{E}_a = -\zeta\frac{jk}{4\pi}\frac{\exp(-jkr_1)}{r_1}\int[\mathbf{I}-\boldsymbol{a}_1(\mathbf{I}\cdot\boldsymbol{a}_1)]\exp(jkr_2\boldsymbol{a}_1\cdot\boldsymbol{a}_2)\,dl \qquad (4.38)$$

We have in the preceding sections used this expression assuming a plausible form for $\mathbf{I}(s)$.

Now in reception we have the situation shown in fig. 4.11(*b*): the antenna is in a plane-wave field originating from a distant source. If the gap is left open a current distribution will be set up in the conductors, producing a scattered field. This scattered field when added to the incident field must result in the component of electric field parallel to the wire being zero on the conducting surface. The voltage appearing across the gap will depend almost entirely on the scattered field: the contribution from the incident field will be very small if the gap is very small. If a load is connected across the gap, current will flow, and to the incident and scattered fields will be added a field which is the same as in transmission. The terminals of the gap will, in a circuit sense, be characterised by the open-circuit voltage and an impedance equal to the radiation impedance earlier defined, as shown in fig. 4.12. To calculate the open-circuit voltage appears to need a complete solution. Fortunately this can be avoided by use of the reciprocal properties of Maxwell's equations. We shall show that a relationship exists between the field in transmission and the scattered field which will enable us to relate the open-circuit voltage in reception to the current distribution in transmission, and hence to the radiation pattern. We first derive a source distribution for the scattered field.

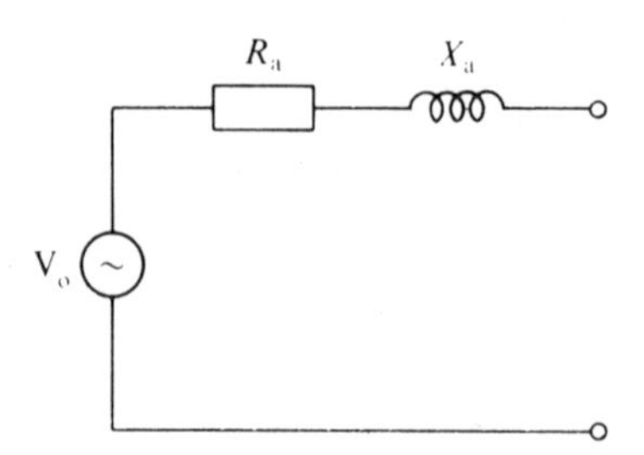

Fig. 4.12. Equivalent circuit of antenna in reception.

4.5.1 The scattered field

The total field in the situation of fig. 4.11(*b*) consists of the incident and scattered fields. The currents flowing are such as to produce an electric field which will precisely cancel the component of $\mathbf{E}_i$ parallel to the wire, thus satisfying the boundary conditions. Let us provide this component by generators, as in fig. 4.11(*c*), instead of by a current. Nothing external will have altered, and in particular the voltage across the gap will be the same. If we now remove the incident field leaving the generators as in fig. 4.11(*d*), the field produced will be the scattered field.

This scattered field is produced by sources over the antenna, and therefore will, like the transmitted field, be quasi-plane at great distances.

4.5.2 The reciprocity theorem

We consider the two separate fields produced by two distinct source distributions, unspecified except that they are limited to a finite region of space, and so produce quasi-plane fields at a distance. Let us denote the two fields by $\mathbf{E}_1$, $\mathbf{H}_1$ and $\mathbf{E}_2$, $\mathbf{H}_2$ produced by source distributions $\mathbf{J}_1$, $\mathbf{J}_2$. We shall show that

$$\int \mathbf{E}_1 \cdot \mathbf{J}_2 \, \mathrm{d}v = \int \mathbf{E}_2 \cdot \mathbf{J}_1 \, \mathrm{d}v$$

in which the integration is taken over a volume embracing all sources of the fields. The fields will satisfy Maxwell's equations and we may write

$$\text{curl}\, \mathbf{E}_1 = -\mathrm{j}\omega\mu \mathbf{H}_1, \qquad \text{curl}\, \mathbf{H}_1 = \mathrm{j}\omega\epsilon \mathbf{E}_1 + \mathbf{J}_1$$

$$\text{curl}\, \mathbf{E}_2 = -\mathrm{j}\omega\mu \mathbf{H}_2, \qquad \text{curl}\, \mathbf{H}_2 = \mathrm{j}\omega\varepsilon \mathbf{E}_2 + \mathbf{J}_2$$

As in the proof of Poynting's theorem (chapter 2), we use the vector identity of (2.1). We may write

$$\begin{aligned} \text{div}\,(\mathbf{E}_1 \times \mathbf{H}_2) &= \mathbf{H}_2 \cdot \text{curl}\, \mathbf{E}_1 - \mathbf{E}_1 \cdot \text{curl}\, \mathbf{H}_2 \\ &= -\mathrm{j}\omega\mu \mathbf{H}_1 \cdot \mathbf{H}_2 - \mathrm{j}\omega\varepsilon \mathbf{E}_1 \cdot \mathbf{E}_2 - \mathbf{E}_1 \cdot \mathbf{J}_2 \end{aligned}$$

and similarly

$$\text{div}\,(\mathbf{E}_2 \times \mathbf{H}_1) = -\mathrm{j}\omega\mu \mathbf{H}_1 \cdot \mathbf{H}_2 - \mathrm{j}\omega\varepsilon \mathbf{E}_1 \cdot \mathbf{E}_2 - \mathbf{E}_2 \cdot \mathbf{J}_1$$

Hence, subtracting, we find

$$\text{div}\,(\mathbf{E}_1 \times \mathbf{H}_2) - \text{div}\,(\mathbf{E}_2 \times \mathbf{H}_1) = \mathbf{E}_2 \cdot \mathbf{J}_1 - \mathbf{E}_1 \cdot \mathbf{J}_2 \qquad (4.39)$$

Integrating over a volume enclosing the sources we have

$$\int (\mathbf{E}_2 \cdot \mathbf{J}_1 - \mathbf{E}_1 \cdot \mathbf{J}_2)\,\mathrm{d}v = \int \operatorname{div}(\mathbf{E}_1 \times \mathbf{H}_2 - \mathbf{E}_2 \times \mathbf{H}_1)\,\mathrm{d}v$$

$$= \oint (\mathbf{E}_1 \times \mathbf{H}_2 - \mathbf{E}_2 \times \mathbf{H}_1) \cdot \mathrm{d}\boldsymbol{S}$$

in which the surface of integration bounds the volume enclosing the sources. It may conveniently be taken as a large sphere of radius R. Since all sources are enclosed, both fields are quasi-plane over this sphere, $\mathbf{E}_1$ and $\mathbf{E}_2$ are perpendicular to $\boldsymbol{R}$. Denoting the unit radial vector by $\boldsymbol{a}$, the magnetic fields are given by

$$\mathbf{H}_1 = \sqrt{(\varepsilon/\mu)}\,\boldsymbol{a} \times \mathbf{E}_1, \qquad \mathbf{H}_2 = \sqrt{(\varepsilon/\mu)}\,\boldsymbol{a} \times \mathbf{E}_2$$

Substituting for $\mathbf{H}_1$ and $\mathbf{H}_2$ we find

$$\mathbf{E}_1 \times \mathbf{H}_2 = \mathbf{E}_2 \times \mathbf{H}_1 = \sqrt{(\varepsilon/\mu)}(\mathbf{E}_1 \cdot \mathbf{E}_2)\boldsymbol{a}$$

The right-hand side of (4.39) therefore vanishes, so that we have

$$\int \mathbf{E}_2 \cdot \mathbf{J}_1\,\mathrm{d}v = \int \mathbf{E}_1 \cdot \mathbf{J}_2\,\mathrm{d}v$$

4.5.3 The open-circuit voltage

We can now apply this result to the source distributions of figs. 4.11(a) and 4.11(d). Let $\mathbf{E}_1$, $\mathbf{J}_1$ refer to the transmitted field and $\mathbf{E}_2$, $\mathbf{J}_2$ to the scattered field. For both fields we interpret $\mathbf{J}\,\mathrm{d}v$ as $\mathrm{I}\,\mathrm{d}\boldsymbol{l}$. We have to consider both the wire and the gap. For the transmitting case $\mathbf{E}_1 \cdot \mathrm{d}\boldsymbol{l}$ is zero over the wire, by the boundary conditions, and since for the scattered field the aerial is on open-circuit, across the gap $\mathrm{I}_2\,\mathrm{d}\boldsymbol{l}$ is zero. Hence

$$\int \mathbf{E}_2 \cdot \mathbf{J}_1\,\mathrm{d}v = \int \mathrm{I}_1 \mathbf{E}_2 \cdot \mathrm{d}\boldsymbol{l} = 0$$

Over the wire $\mathbf{E}_2 \cdot \mathrm{d}\boldsymbol{l} = -\mathbf{E}_\mathrm{i} \cdot \mathrm{d}\boldsymbol{l}$, and across the gap $-\int \mathbf{E}_2 \cdot \mathrm{d}\boldsymbol{l}$ is equal to the open-circuit voltage, V_0. Hence, if the driving current in transmission is I_0,

$$-\int \mathbf{E}_\mathrm{i} \cdot \mathrm{d}\boldsymbol{l}\,\mathrm{I}_1 - \mathrm{V}_0 \mathrm{I}_0 = 0$$

or

$$\mathrm{V}_0 = -\frac{1}{\mathrm{I}_0} \int \mathrm{I}_1 \mathbf{E}_\mathrm{i} \cdot \mathrm{d}\boldsymbol{l} \tag{4.40}$$

This is true for any incident field. Now consider a plane-wave incident field for which

$$\mathbf{E}_\mathrm{i} = \boldsymbol{E}_0 \exp(-\mathrm{j}k\boldsymbol{r} \cdot \boldsymbol{n})$$

where $\boldsymbol{E}_0 \cdot \boldsymbol{n} = 0$. Substituting in (4.10) we have

$$V_0 = -\frac{1}{I_0}\boldsymbol{E}_0 \cdot \int \exp(-jk\boldsymbol{r}_2 \cdot \boldsymbol{n})\mathbf{I}_1 \, dl$$

Since $\mathbf{E}_0$ is perpendicular to $\boldsymbol{n}$, it is the component of $\mathbf{I}_1$ transverse to $\boldsymbol{n}$ that is relevant, and we may write

$$V_0 = -\frac{1}{I_0}\boldsymbol{E}_0 \cdot \int \exp(-jk\boldsymbol{r}_2 \cdot \boldsymbol{n})[\mathbf{I}_1 - \boldsymbol{n}(\mathbf{I}_1 \cdot \boldsymbol{n})] \, dl \qquad (4.41)$$

The integral in this equation can be identified with that in (4.38) provided that $\boldsymbol{a}_1 = -\boldsymbol{n}$. There is thus a relation between the open-circuit voltage in reception and the transmitted field. Let us define a dimensionless quantity $\mathbf{e}_a$ by writing (4.38) in the form

$$\mathbf{E}_a = j\zeta \mathbf{e}_a I_0 \frac{\exp(-jkr_1)}{r_1} \qquad (4.42)$$

where

$$\mathbf{e}_a = -\frac{k}{4\pi}\frac{1}{I_0}\int [\mathbf{I} - \boldsymbol{a}_1(\mathbf{I} \cdot \boldsymbol{a}_1)] \exp(jkr_2\boldsymbol{a}_1 \cdot \boldsymbol{a}_2) \, dl \qquad (4.43)$$

$\mathbf{e}_a$ is characteristic of the antenna and is a function of direction only.

The power flow at distance r_1 from the radiator is radial, as shown by § 2.4.2 and its density in the distant field is given by (2.32) as

$$\frac{1}{2\zeta r_1^2}\mathbf{E}_a \cdot \mathbf{E}_a^*$$

The power per unit solid angle is therefore

$$\frac{1}{2\zeta}\mathbf{E}_a \cdot \mathbf{E}_a^* = \tfrac{1}{2}\zeta I_0^2 \mathbf{e}_a \cdot \mathbf{e}_a^* \qquad (4.44)$$

This gives a physical significance to the magnitude of $\mathbf{e}_a$. Its direction characterises the polarisation of the radiated wave. In terms of $\mathbf{e}_a$, (4.41) and (4.43) give

$$V_0 = +\frac{4\pi}{k}\boldsymbol{E}_0 \cdot \mathbf{e}_a = +2\lambda \boldsymbol{E}_0 \cdot \mathbf{e}_a \qquad (4.45)$$

This enables us to find the open-circuit voltage across the antenna gap in terms of $\mathbf{e}_a$ and the incident field. The sign in the expression has occurred because a particular phase relationship between transmitted and received waves has been assumed. This is not normally the situation, and (4.45) gives magnitude only.

Example

A plane wave of frequency 100 MHz has a peak value of electric field strength of $5\ \mu\text{V m}^{-1}$. Find the maximum value of the open-circuit voltage induced in a half-wave dipole situated in the wave. Find also the maximum power which can be fed to a load, and compare it with the power density in the incident wave.

Solution. From (4.22)

$$\mathrm{E}_\theta = \zeta \mathrm{H}_\phi = \zeta \frac{\mathrm{j}I_0}{2\pi} \frac{\exp(-\mathrm{j}kr_1)}{r_1} \frac{\cos(\frac{1}{2}\pi \cos\theta)}{\sin\theta}$$

Hence from (4.42)

$$\mathrm{e}_\mathrm{a} = \frac{1}{2\pi} \frac{\cos(\frac{1}{2}\pi \cos\theta)}{\sin\theta}$$

The maximum value occurs for $\theta = \frac{1}{2}\pi$, when $\mathrm{e}_\mathrm{a} = 1/2\pi$. The corresponding value of the open-circuit voltage is, from (4.45),

$$\mathrm{V}_0 = 2\lambda E_0 \frac{1}{2\pi}$$

$$= \frac{3}{\pi} \times 5\ \mu\text{V}$$

$$= 4.8\ \mu\text{V}$$

The available power is that delivered into a matched load of resistance R_a, and is therefore given by

$$\frac{1}{2} \frac{|\mathrm{V}_0|^2}{4R_\mathrm{a}} = \frac{1}{8} \frac{(4.8)^2}{73}\ \text{pW}$$

$$= 0.039\ \text{pW}$$

The power carried by the incident wave has the value

$$\frac{1}{2\zeta} |E_0|^2\ \text{W m}^{-2} = 0.033\ \text{pW m}^{-2}$$

At optimum power transfer the antenna has therefore an effective area of $1.2\ \text{m}^2$.

4.6 Aperture theory

It has been stated that on occasion it may be convenient to think not in terms of a current distribution but rather in terms of a local field when determining radiating properties of an antenna. The simplest example is for 'mirror' structures such as the parabolic dish used at microwave frequencies for which the dish dimensions are many wavelengths. This type of problem is akin to the situation in optics, when the size of the structure is several orders of magnitude larger than the wavelength. We shall show that the field produced by a source can be calculated from the knowledge of the electric and magnetic fields over a closed surface containing the entire source distribution, as shown in fig. 4.13. It will be shown that an element of area $\boldsymbol{n}\mathrm{d}S$ on S behaves as a combination of electric and magnetic sources.

The result of the calculation given in the next section is that the electric field at a distant point of radius vector $\boldsymbol{r}_1$ can be regarded as the sum of fields arising from each element of the area of S. It will be shown that the contribution made by the element $\mathrm{d}S$, of outward normal $\boldsymbol{n}$, is expressed by the formula

$$\mathrm{d}\mathbf{E} = \frac{\mathrm{j}k}{4\pi r_{12}} \exp(-\mathrm{j}kr_{12})\{(\boldsymbol{n}\times\mathbf{E}_S)\times\boldsymbol{a}_{12} + \surd(\mu/\varepsilon)[(\boldsymbol{n}\times\mathbf{H}_S)\times\boldsymbol{a}_{12}]\times\boldsymbol{a}_{12}\}\,\mathrm{d}S \tag{4.46}$$

Although this formula looks complicated, it has a simple interpretation. Since we are considering only one element on S, the quantities $\boldsymbol{n}\times\mathbf{E}_S$, $\boldsymbol{n}\times\mathbf{H}_S$ are to be regarded as constants. From (4.27) and (4.28) the fields due to a Hertzian dipole of strength $\mathbf{p}_\mathrm{e} = \mathrm{I}\,\mathrm{d}\boldsymbol{l}$ are given by

$$\begin{aligned}\mathbf{H} &= \frac{\mathrm{j}k}{4\pi r_{12}} \exp(-\mathrm{j}kr_{12})\,\mathbf{p}_\mathrm{e}\times\boldsymbol{a}_{12}\\ \mathbf{E} &= \sqrt{\left(\frac{\mu}{\varepsilon}\right)}\,\mathbf{H}\times\boldsymbol{a}_{12} = \frac{\mathrm{j}k}{4\pi r_{12}} \exp(-\mathrm{j}kr_{12})\sqrt{\left(\frac{\mu}{\varepsilon}\right)}\,(\mathbf{p}_\mathrm{e}\times\boldsymbol{a}_{12})\times\boldsymbol{a}_{12}\end{aligned} \tag{4.47}$$

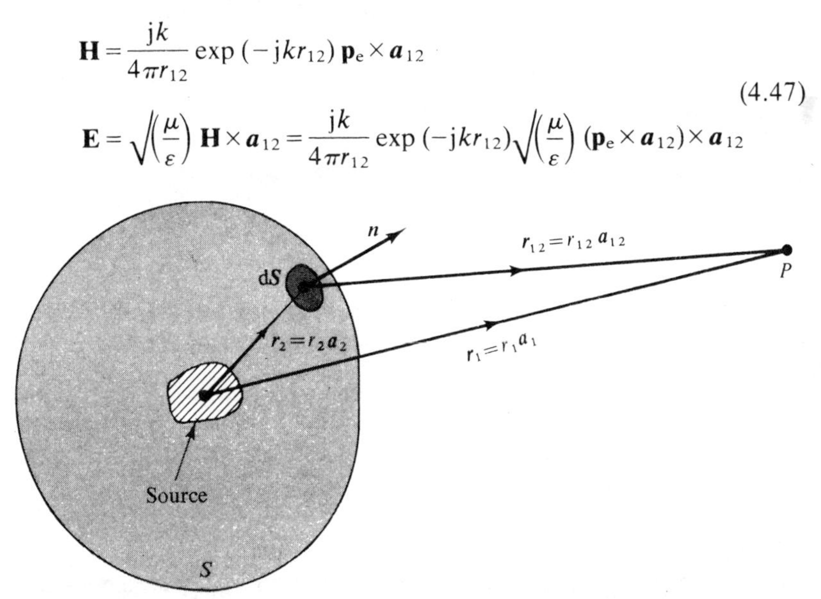

Fig. 4.13. An equivalent surface distribution.

We can compare the second term in (4.46) with the expression for electric field in (4.47) and see that the element of area acts as Hertzian dipole of strength $\boldsymbol{n} \times \mathbf{H}_S \, \mathrm{d}S$. The first term in (4.46) can be associated with fields given by the equations

$$\mathbf{E} = -\frac{\mathrm{j}k}{4\pi r_{12}} \exp(-\mathrm{j}kr_{12})\, \mathbf{p}_\mathrm{m} \times \boldsymbol{a}_{12}$$
$$\mathbf{H} = \sqrt{\left(\frac{\varepsilon}{\mu}\right)} \boldsymbol{a}_{12} \times \mathbf{E} \qquad (4.48)$$

In this expression $\mathbf{p}_\mathrm{m}$ represents a magnetic dipole, analogous to $\mathbf{p}_\mathrm{e}$ in (4.47), so that we can interpret the first term of (4.46) as arising from an equivalent magnetic source of strength $\mathbf{E}_S \times \boldsymbol{n} \, \mathrm{d}S$.

We shall now derive (4.46) before applying it to particular problems.

†4.6.1 Equivalent source distribution

In order to obtain an expression involving the field at P arising from the fields over S we make use of the reciprocity relationship (4.39) integrated over the region lying between S (fig. 4.13) and a sphere of very large radius. The two sets of fields we consider are (i) the situation as shown in fig. 4.13, and (ii) the same geometry with the source within S removed and an electric dipole of strength $\mathbf{p}$ at P. With these fields the same arguments as used in § 4.5.2 show that the integral over the very large sphere vanishes. Further, the only source in the region is the dipole at P. Hence

$$\int \operatorname{div}(\mathbf{E}_1 \times \mathbf{H}_2 - \mathbf{E}_2 \times \mathbf{H}_1)\, \mathrm{d}v = \oint_S (\mathbf{E}_1 \times \mathbf{H}_2 - \mathbf{E}_2 \times \mathbf{H}_1) \cdot (-\boldsymbol{n}\, \mathrm{d}S)$$
$$= -\int \mathbf{E}_1 \cdot \mathbf{J}_2 \, \mathrm{d}v$$

The right-hand side reduces to $-\mathbf{E} \cdot \mathbf{p}$, since $\mathbf{p} = I\, \mathrm{d}\boldsymbol{l} = \mathbf{J}_2 \, \mathrm{d}v$. We can substitute $\mathbf{E}_S$ and $\mathbf{H}_S$ for $\mathbf{E}_1$ and $\mathbf{H}_1$ together with, using (4.47),

$$\mathbf{E}_2 = -\sqrt{\left(\frac{\mu}{\varepsilon}\right)} \frac{\mathrm{j}k}{4\pi r_{12}} \exp(-\mathrm{j}kr_{12})\, \mathbf{p}^\mathrm{T}$$

$$\mathbf{H}_2 = -\frac{\mathrm{j}k}{4\pi r_{12}} \exp(-\mathrm{j}kr_{12})\, \mathbf{p} \times \boldsymbol{a}_{12}$$

in which $\mathbf{p}^\mathrm{T} = \boldsymbol{a}_{12} \times (\mathbf{p} \times \boldsymbol{a}_{12})$, the component of $\mathbf{p}$ transverse to $\boldsymbol{a}_{12}$. In this last equation, a sign change has occurred since $\boldsymbol{r}_{12}$ is directed towards the dipole and not away from it. We can then write for the contribution to $\mathbf{E} \cdot \mathbf{p}$ from $\boldsymbol{n}\, \mathrm{d}S$

$$\mathbf{p} \cdot \delta\mathbf{E} - \frac{\mathrm{j}k}{4\pi r_{12}} \exp(-\mathrm{j}kr_{12})[-\mathbf{E}_S \times (\mathbf{p} \times \boldsymbol{a}_{12}) + \sqrt{\left(\frac{\mu}{\varepsilon}\right)}\, \mathbf{p}^{\mathrm{T}} \times \mathbf{H}_S] \cdot \boldsymbol{n}\, \mathrm{d}S$$

This result enables us to find the component of $\delta\mathbf{E}$ in any direction by suitable choice of $\mathbf{p}$. We note that $\delta\mathbf{E}$ is perpendicular to $\boldsymbol{a}_{12}$, since when $\mathbf{p} = \boldsymbol{a}_{12}$ both $\mathbf{E}_2$ and $\mathbf{H}_2$ vanish. The second term can be simply rearranged using the properties of the triple scalar product in the form

$$\mathbf{p}^{\mathrm{T}} \times \mathbf{H}_S \cdot \boldsymbol{n} = \mathbf{p}^{\mathrm{T}} \cdot \mathbf{H}_S \times \boldsymbol{n}$$

Since $\mathbf{p}^{\mathrm{T}}$ is perpendicular to $\boldsymbol{r}_{12}$, only the same component of $\mathbf{H}_S \times \boldsymbol{n}$ is relevant, so that we can write this term as

$$\mathbf{p}^{\mathrm{T}} \cdot (\mathbf{H}_S \times \boldsymbol{n})^{\mathrm{T}} = \mathbf{p} \cdot (\mathbf{H}_S \times \boldsymbol{n})^{\mathrm{T}}$$

The first term is slightly more complicated, but using the same properties we have

$$\boldsymbol{n} \cdot \mathbf{E}_S \times (\mathbf{p} \times \boldsymbol{a}_{12}) = -\boldsymbol{n} \times \mathbf{E}_S \cdot \boldsymbol{a}_{12} \times \mathbf{p} = -(\boldsymbol{n} \times \mathbf{E}_S) \times \boldsymbol{a}_{12} \cdot \mathbf{p}$$

We thus have finally

$$\mathbf{p} \cdot \delta\mathbf{E} = \frac{\mathrm{j}k}{4\pi r_{12}} \exp(-\mathrm{j}kr_{12})[(\boldsymbol{n} \times \mathbf{E}_S) \times \boldsymbol{a}_{12} + \sqrt{\left(\frac{\mu}{\varepsilon}\right)}\, (\mathbf{H}_S \times \boldsymbol{n})^{\mathrm{T}}] \cdot \mathbf{p}\, \mathrm{d}S$$

Since $\mathbf{p}$ is arbitrary we can deduce

$$\delta\mathbf{E} = \frac{\mathrm{j}k}{4\pi r_{12}} \exp(-\mathrm{j}kr_{12})\left[(\boldsymbol{n} \times \mathbf{E}_S) \times \boldsymbol{a}_{12} + \sqrt{\left(\frac{\mu}{\varepsilon}\right)}\, (\mathbf{H}_S \times \boldsymbol{n})^{\mathrm{T}}\right] \mathrm{d}S$$

The vector $(\mathbf{H}_S \times \boldsymbol{n})^{\mathrm{T}}$ is equal to $[(\boldsymbol{n} \times \mathbf{H}_S) \times \boldsymbol{a}_{12}] \times \boldsymbol{a}_{12}$; by comparison with (4.47) this is the result stated in (4.46).

The concept of replacing the sources internal to S by equivalent dipoles is not a simple one, for the equivalent sources depend not only on the strengths of the electric and magnetic fields at a point on the surface but also on the direction chosen for $\mathrm{d}\boldsymbol{S}$. The equivalence is only true when integration over the whole surface S is carried out.

4.6.2 Application to antennae

When we apply (4.46) to the antenna problem we can make a further simplification by assuming that $\boldsymbol{r}_{12}$ is large compared to the linear dimensions of S. We may then use the approximation of (4.29), enabling us to write, on rearranging the order of the vector products,

$$\mathbf{E}(\boldsymbol{r}_1)=\frac{\mathrm{j}k}{4\pi}\frac{\exp(-\mathrm{j}kr_1)}{r_1}\boldsymbol{a}_1\times\oint_S\left[\sqrt{\left(\frac{\mu}{\varepsilon}\right)}\boldsymbol{a}_1\times(n\times\mathbf{H}_S)-\boldsymbol{n}\times\mathbf{E}_S\right]\times\exp(\mathrm{j}k\boldsymbol{r}_2\cdot\boldsymbol{a}_1)\,\mathrm{d}S \tag{4.49}$$

As it should be, $\mathbf{E}(\boldsymbol{r}_1)$ is perpendicular to $\boldsymbol{a}_1$. It must be noted however that there is a hidden complication: the integral requires that both $\mathbf{E}_S$ and $\mathbf{H}_S$ be specified over S, both of which are produced by the same source, and therefore cannot be independently specified. Some assumption is usually necessary to estimate either quantity. Although the integral is over a closed surface, the formula is commonly applied to situations where the field is significantly different from zero only over a limited area, the aperture, and is assumed to be identically zero over the remaining area of S. This is also an approximation. If the aperture is some way from the source feeding it, it is often adequate to assume that the field in the aperture is quasi-plane. It is usually possible to find a surface of constant phase over which **E** and **H** are perpendicular to each other and to the normal to the surface. The technique will be illustrated by an example.

4.6.3 The uniformly illuminated aperture

As an example of the use of (4.49) we will consider the case of a plane rectangular aperture over which the field is uniform. It will be assumed that both electric and magnetic vectors are in the plane of the aperture and are mutually perpendicular: the geometry is shown in fig. 4.14. In

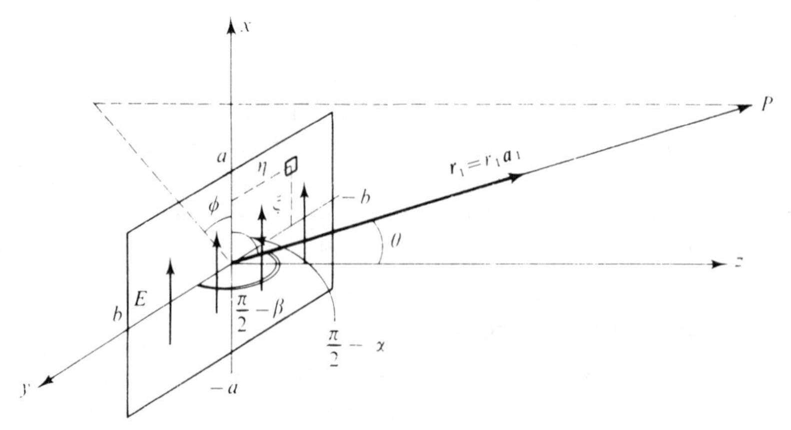

Fig. 4.14. A rectangular aperture.

terms of the Cartesian axes shown

$$\left.\begin{aligned}
&\mathbf{E}_S = \mathrm{E}_0 \boldsymbol{a}_x \qquad (-a<x<a)\\
&\mathbf{H}_S = \zeta^{-1}\mathrm{E}_0 \boldsymbol{a}_y \qquad (-b<y<b)\\
&\boldsymbol{n} = \boldsymbol{a}_z\\
&\boldsymbol{a}_1 = \boldsymbol{a}_x \cos\phi \sin\theta + \boldsymbol{a}_y \sin\phi \sin\theta + \boldsymbol{a}_z \cos\theta\\
&\boldsymbol{r}_2 = \xi \boldsymbol{a}_x + \eta \boldsymbol{a}_y
\end{aligned}\right\} \tag{4.50}$$

where ξ, η define a point in the aperture.

Apart from the exponential function, the integrand in (4.49) is, from (4.50), constant as far as the variables of integration are concerned. We thus have

$$\mathbf{E}(\boldsymbol{r}_1) = \frac{\mathrm{j}k}{4\pi}\frac{\exp(-\mathrm{j}kr_1)}{r_1}\boldsymbol{a}_1 \times [\zeta \boldsymbol{a}_1 \times (\boldsymbol{n} \times \mathbf{H}_S) - \boldsymbol{n} \times \mathbf{E}_S]$$

$$\times \int_{-a}^{a}\int_{-b}^{b} \exp(\mathrm{j}k\boldsymbol{r}_2 \cdot \boldsymbol{a}_1)\,\mathrm{d}\xi\,\mathrm{d}\eta \tag{4.51}$$

The argument of the exponential is most conveniently written in the form

$$\mathrm{j}k\boldsymbol{r}_2 \cdot \boldsymbol{a}_1 = \mathrm{j}k(\xi \sin\alpha + \eta \sin\beta)$$

where α, β are the complements of the angles between $\boldsymbol{a}_1$ and the x, y axes respectively, as shown in fig. 4.14. These angles can be expressed in terms of θ, ϕ if necessary. We then have

$$\int_{-a}^{a}\int_{-b}^{b} \exp(\mathrm{j}k\boldsymbol{r}_2 \cdot \boldsymbol{a}_1)\,\mathrm{d}\xi\,\mathrm{d}\eta$$

$$= \int_{-a}^{a} \exp(\mathrm{j}k\xi \sin\alpha)\,\mathrm{d}\xi \int_{-b}^{b} \exp(\mathrm{j}k\eta \sin\beta)\,\mathrm{d}\eta$$

$$= 4ab\frac{\sin(ka \sin\alpha)}{ka \sin\alpha}\cdot\frac{\sin(kb \sin\beta)}{kb \sin\beta} \tag{4.52}$$

The theory of this section is applied only to cases where the linear aperture dimensions are much greater than the wavelength. In fig. 4.15 is shown the term $[\sin(ka \sin\alpha)]/(ka \sin\alpha)$ for $ka = 100$, corresponding to about 30 wavelengths across the aperture. It will be seen that significant radiation occurs only in directions for which α is small; that is, the radiation is highly directed along the normal to the aperture. The term outside the integral in (4.51) is only a slowly varying function of θ

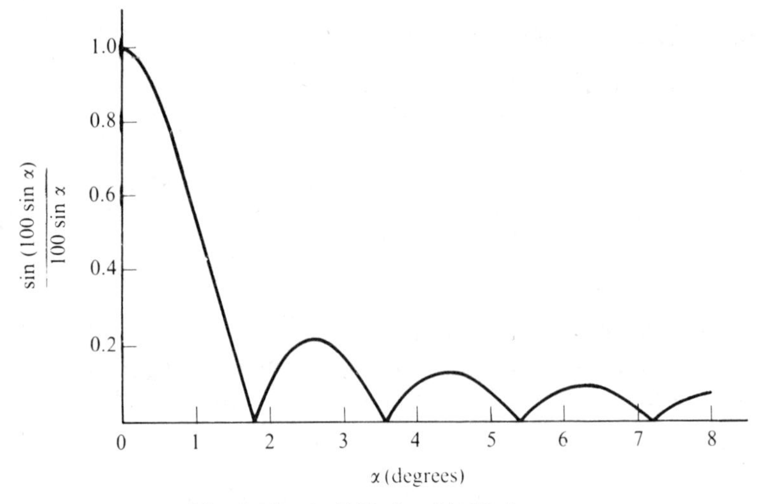

Fig. 4.15. sin (100 sin α)/100 sin α.

and ϕ, so that if both ka and kb are large we may evaluate it along the z-axis with small error. We find, using (4.50) and (4.52)

$$\mathbf{E}(\boldsymbol{r}_1) \approx \mathrm{E}_0 \frac{\mathrm{j}k}{4\pi} \frac{\exp(-\mathrm{j}kr_1)}{r_1} 2\boldsymbol{a}_x \times 4ab \frac{\sin(ka \sin\alpha)}{ka \sin\alpha} \times \frac{\sin(kb \sin b)}{kb \sin\beta} \tag{4.53}$$

Since the aperture field was polarised with the electric field in the x-direction, it is to be expected that the distant field will be similarly polarised.

†4.6.4 A Gaussian beam

The uniform illumination assumed across the aperture of the last section could be produced in the optical context by placing the aperture transverse to a wider parallel beam of coherent light, from a laser for example. Although the predicted pattern would be observed, in order to do so it would be found necessary to go further away from the aperture as the aperture increased in size. It would also be found that over a long section the beam would remain nearly parallel. It is natural to ask if (4.49) can be used to explain this fact. The only approximation made in deriving (4.46) was in assuming that the point of observation was many wavelengths from the surface of integration. This is easily satisfied. In deriving (4.49) two further approximations were made: r_{12} was replaced by r_1 in the denominator, which is satisfactory if r_1 is greater than a few aperture dimensions, and, more sensitive, r_{12} was replaced by $r_1 - r_2 \boldsymbol{a}_1 \cdot \boldsymbol{a}_2$ in the exponential term. We have not investigated quantitatively the accuracy of this latter approximation.

We will consider an aperture in the plane $z = 0$, as indicated in fig. 4.16. We shall take not uniform illumination but, more realistically in the laser context, a Gaussian distribution. Let us first investigate the crucial exponential term. In terms of the co-ordinates in the figure

$$r_{12}^2 = (x - \xi)^2 + (y - \eta)^2 + z^2$$
$$= r^2 \left[1 - \frac{2(x\xi + y\eta)}{r^2} + \frac{\xi^2 + \eta^2}{r^2} \right]$$

For the distant field we retained only the first two terms. For a nearly parallel beam the second and third terms may be of similar magnitude, so we must retain both. To the first approximation therefore

$$r_{12} = r - \frac{x\xi + y\eta}{r} + \frac{1}{2} \frac{\xi^2 + \eta^2}{r} + \ldots \tag{4.54}$$

In the previous context 'distant' means that the third term can be legitimately neglected in forming $\exp(-\mathrm{j}kr_{12})$. This implies

$$ka^2/r \ll 2\pi$$

where a represents the linear dimension. In terms of wavelength we have

$$a^2/\lambda r \approx a^2/\lambda z \ll 1 \tag{4.55}$$

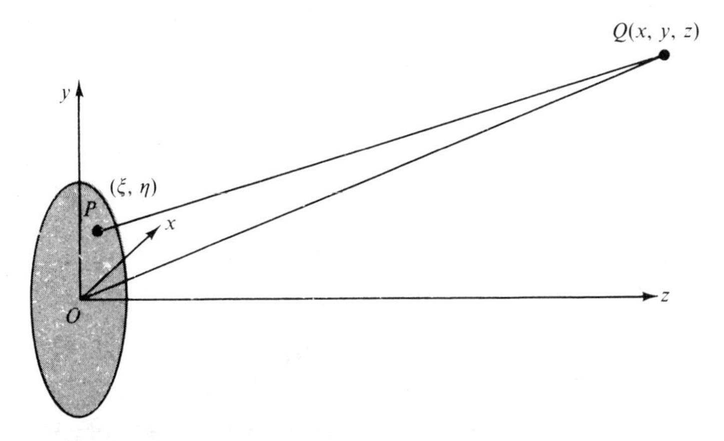

Fig. 4.16. A circular aperture.

This ratio is known as the *Fresnel number* for the aperture, and must be much less than unity for the previously defined 'distant' field calculations to be valid.

Because we are dealing with a nearly parallel beam, simplifications can be made in the integrand of (4.46). We assume that in the aperture **E** is in the x-direction and **H** in the y-direction, related by $\zeta H_y = E_x$. This is permissible provided $\mathbf{E}_x$ changes only slowly on a wavelength scale. The angle between $\boldsymbol{r}_1$ and the normal (the z-axis) will be small, and as before we find

$$\mathbf{E} = \frac{\mathrm{j}k}{4\pi}\frac{1}{r}2\boldsymbol{a}_x \int E_x \exp(-\mathrm{j}kr_{12})\,\mathrm{d}S \tag{4.56}$$

We will take the distribution across the aperture to be given by

$$E_x = E_0 \exp[-(\rho/\rho_0)^2] \tag{4.57}$$

in which ρ is the radius $(\xi^2+\eta^2)^{\frac{1}{2}}$ and ρ_0 is a constant. The spot *intensity* will be proportional to E_x^2. and will be e^{-2} of its centre value for $\rho = \rho_0$. Since the spot intensity falls off very rapidly with radius, the aperture may conveniently be defined by (4.57) and the limits of integration be made infinite. Equation (4.56) thus becomes, substituting from (4.54),

$$E_x = \frac{\mathrm{j}k}{4\pi}\frac{2E_0}{r}\exp(-\mathrm{j}kr)\int_{-\infty}^{\infty}\int_{-\infty}^{\infty}\exp\left[-\frac{(\xi^2+\eta^2)}{\rho_0^2}+\frac{\mathrm{j}k(x\xi+y\eta)}{r} - \frac{\mathrm{j}k(\xi^2+\eta^2)}{2r}\right]\mathrm{d}\xi\,\mathrm{d}\eta \tag{4.58}$$

It is shown in § 10.3 that this expression becomes

$$E_x = \frac{\mathrm{j}k}{4\pi}\frac{2E_0}{r}\exp(-\mathrm{j}kr)\frac{\pi}{\beta}\exp\left(-\frac{k^2\rho^2}{4\beta r^2}\right) \tag{4.59}$$

where

$$\beta = \frac{1}{\rho_0^2} + \frac{\mathrm{j}k}{2r}$$

The expression contains phase information as well as amplitude. Taking the modulus of both sides of (4.59) to obtain amplitude we find

$$|E_x| = \frac{k}{4\pi}\frac{2E_0}{r}\frac{\pi}{|\beta|}\exp\left[-\frac{k^2\rho^2}{4r^2}\mathrm{Re}\left(\frac{1}{\beta}\right)\right]$$

$$= E_0\left[1+\left(\frac{2r}{k\rho_0^2}\right)^2\right]^{-\frac{1}{2}}\exp\left\{-\left(\frac{\rho}{\rho_0}\right)^2\left[1+\left(\frac{2r}{k\rho_0^2}\right)^2\right]^{-1}\right\} \tag{4.60}$$

The radial distribution is determined by the exponential term, and we see that the spot is Gaussian. We may assess the distribution of intensity in the beam by finding the radial extension for which the intensity is e^{-2} of the centre value. From (4.60) this will occur when E_x is e^{-1} of its maximum value, or

$$(\rho/\rho_0)^2 = 1 + (2r/k\rho_0^2)^2 \tag{4.61}$$

The beam will remain substantially parallel providing $(2r/k\rho_0)^2 < 1$. It will have increased its radius by a factor of $\sqrt{2}$ when

$$z \approx r = \tfrac{1}{2}k\rho_0^2 = \pi\rho_0^2/\lambda$$

Equation (4.55) shows that this corresponds to a Fresnel number near unity. For larger values of r the beam diameter is, from (4.61), approximately given by

$$\rho/\rho_0 = 2r/k\rho_0^2$$

To give orders of magnitude let us consider a beam for which $\rho_0 =$ 0.5 mm, $\lambda = 0.6$ μm. We find that for $\rho = \rho_0\sqrt{2}$, $z = 1.3$ m and for $r \approx z =$ 100 m, $\rho = 2.9$ mm.

4.7 Antenna arrays

We have seen that the half-wave dipole forms a convenient radiating element in the sense that it has a suitable radiation resistance. It is not, however, very directional. Directionality can be obtained by the use of arrays of such dipoles, which effectively form a large aperture. This will be illustrated by examples.

4.7.1 An array of two elements

Consider the array of two parallel colinear dipoles shown in fig. 4.17. We will suppose that a current I_1 is maintained in the dipole at O and I_2 in the dipole at A, and that each dipole of itself will radiate the field described by a single dipole in free space. We will use spherical polar co-ordinates with the polar axis Oz parallel to the dipoles, as shown in the figure. The line of centre of the dipoles is taken on the axis Oy. It is convenient to introduce as a separate variable the angle, χ between the radius vector and the axis Oy. This can be expressed in terms of the angles θ, ϕ if necessary.

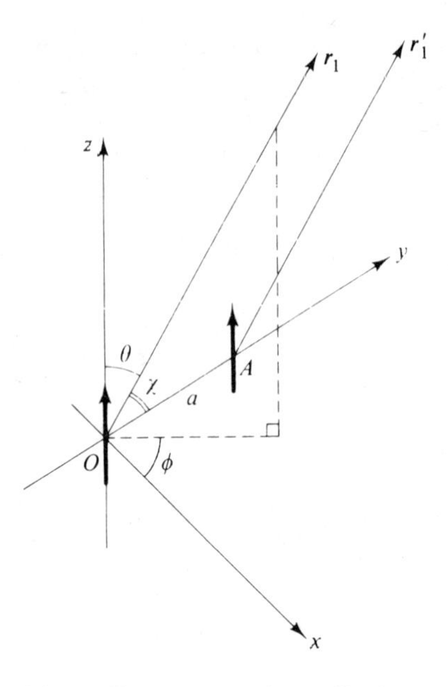

Fig. 4.17. An array of two dipoles.

At a distant point P of radius vector $\boldsymbol{r}$, the magnetic field of the dipole at the origin will be given by (4.22)

$$\mathrm{H}_\phi = \frac{\mathrm{j I_0}}{2\pi r_1} \exp(-\mathrm{j}kr_1) \frac{\cos(\frac{1}{2}\pi \cos\theta)}{\sin\theta}$$

The assumption that the radiation pattern of one dipole is not influenced by its neighbour is not exactly true, but is not far wrong. The radiation resistance will be influenced considerably, and furthermore the two dipoles in the example being considered must be regarded as constituting a two-port network. This however is not relevant to our problem where we consider current feeds. From the parallel disposition of the two dipoles, at a distant point the magnetic field will still be azimuthal and its magnitude will be the sum of the two constituent fields. Further, for a distant point, $\boldsymbol{r}_1$ and $\boldsymbol{r}_1'$ will be virtually parallel so that we have

$$\mathrm{H}_\phi = \frac{\mathrm{j}}{2\pi}\left[\frac{\mathrm{I}_1}{r_1}\exp(-\mathrm{j}kr_1) + \frac{\mathrm{I}_2}{r_1'}\exp(-\mathrm{j}kr_1')\right]\frac{\cos(\frac{1}{2}\pi\cos\theta)}{\sin\theta} \tag{4.62}$$

As in § 4.3 we may, for the distant point, replace r_1' in the denominator by r_1. In the exponential term we use (4.29) in the form

$$r_1' = r_1 - a\cos\chi$$

Hence we have finally from (4.62)

$$H_\phi = \frac{j}{2\pi r_1} \exp(-jkr_1)[I_1 + I_2 \exp(jka \cos\chi)]F(\theta) \qquad (4.63)$$

in which $F(\theta) = \cos(\frac{1}{2}\pi \cos\theta)/\sin\theta$ (4.64)

The resultant radiation pattern is determined by the function

$$[I_1 + I_2 \exp(jka \cos\chi)]F(\theta) \qquad (4.65)$$

The term $F(\theta)$ is characteristic of the identical elements of the array, the half-wave dipole pattern of (4.64). The array contributes the *array factor*

$$M = M_0[I_1 + I_2 \exp(jka \cos\chi)] \qquad (4.66)$$

in which M_0 is a constant chosen so that the maximum value of $|M|$ is unity. This factor depends only on the geometrical disposition of the radiating elements and their strengths. The element itself only appears in the term $F(\theta)$ and (4.63) would be valid for another type of radiator if $F(\theta)$ were changed appropriately.

Even though only two elements are involved in this array a number of different possibilities can be demonstrated.

Broadside array

Let us firstly consider the case when I_1 and I_2 are equal in magnitude and phase. Equation (4.66) gives

$$M = M_0 I_1[1 + \exp(jka \cos\chi)]$$

or

$$|M| = |\cos(\tfrac{1}{2}ka \cos\chi)|$$

To make matters precise let us choose $a = \frac{1}{2}\lambda$, for which $ka = \pi$, when

$$|M| = |\cos(\tfrac{1}{2}\pi \cos\chi)|$$

A section of this array factor is shown in fig. 4.18: the solid pattern can be generated by rotation about the axis Oy. It is to be expected that the maximum exists in a direction perpendicular to Oy, since the path lengths from the two dipoles will then be equal and the two fields in phase.

The overall radiation pattern involves $F(\theta)$ from (4.65) as well as (4.66), and is not rotationally symmetrical about any axis. In fig. 4.19(*a*) and (*b*) are shown the pattern in the plane $z = 0$ and the plane $x = 0$ respectively. In the former case $\theta = \frac{1}{2}\pi$, $\chi = \frac{1}{2}\pi - \phi$, for which

$$|H_\phi| \propto |M|F(\theta) = |\cos(\tfrac{1}{2}\pi \sin\phi)|$$

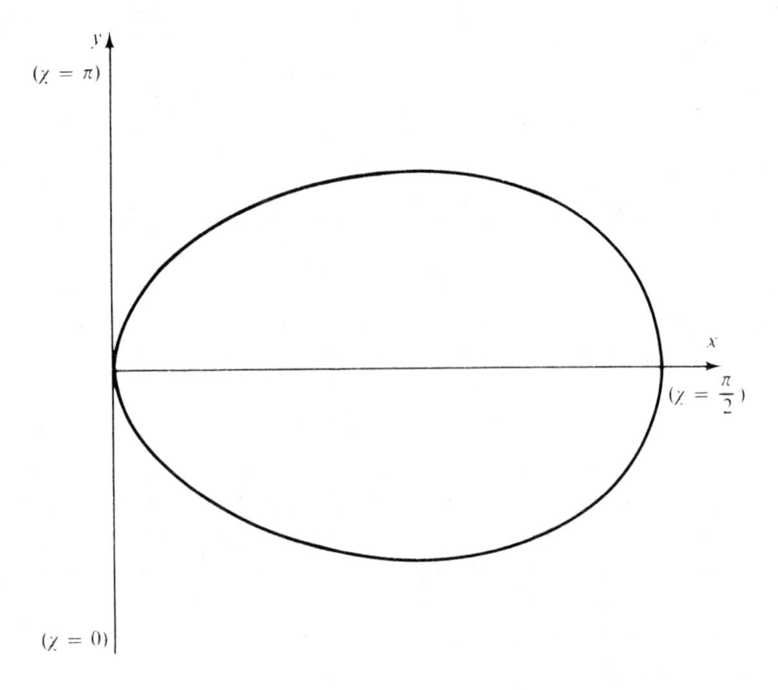

Fig. 4.18. Array factor for broadside array of two dipoles.

In the latter case $\phi = \pm\frac{1}{2}\pi$, $\chi = \frac{1}{2}\pi - \theta$, giving

$$|\mathrm{H}_\phi| \propto |\mathrm{M}|F(\theta) = \cos(\tfrac{1}{2}\pi\cos\theta)\cos(\tfrac{1}{2}\pi\sin\theta)/\sin\theta$$

End-fire array

Instead of taking $I_1 = I_2$, consider $I_2 = -I_1$. We now have from (4.63), keeping $ka = \pi$,

$$|\mathrm{H}_\phi| \propto |\sin(\tfrac{1}{2}\pi\cos\chi)|F(\theta)$$

This is shown in figs. 4.20(*a*) and (*b*) for the same planes as before, $z = 0$ and $x = 0$. It can be seen that the beam, although not completely symmetrical in shape about the y-axis, is generally directed along that axis.

If instead of choosing I_2 in- or anti-phase with I_1 we let $I_2 = I_2\exp(j\psi)$ we have from (4.63), keeping $ka = \pi$,

$$|\mathrm{H}_\phi| \propto \cos[\tfrac{1}{2}(\pi\cos\chi + \psi)]F(\theta)$$

In the plane $z = 0$ ($\theta = \frac{1}{2}\pi$, $\chi = \frac{1}{2}\pi - \phi$) this will have a maximum in the

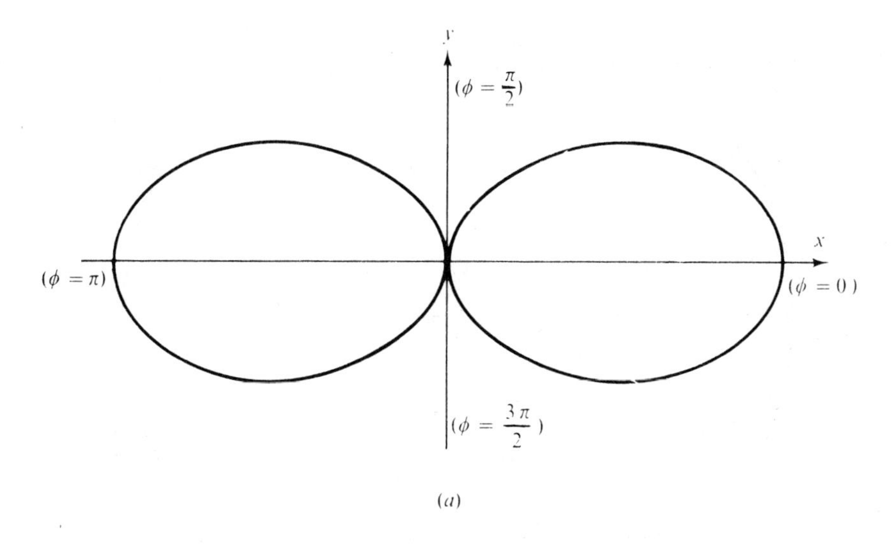

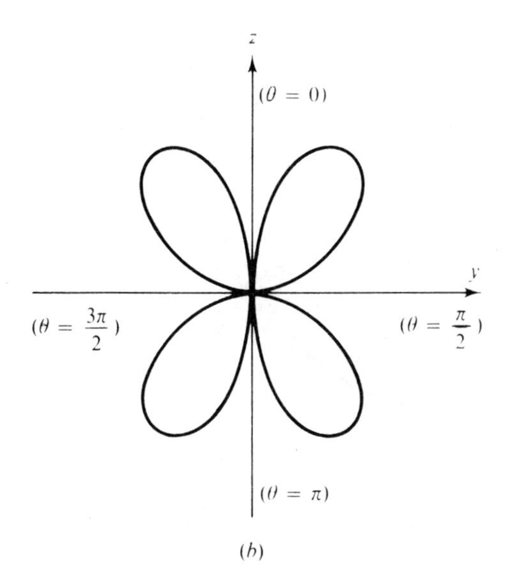

Fig. 4.19. Radiation pattern for array of two dipoles. (*a*)Plane $z = 0$. (*b*) Plane $x = 0$.

direction for which

$$\pi \sin \phi + \psi = 0$$

or

$$\phi = -\sin^{-1}(\psi/\pi)$$

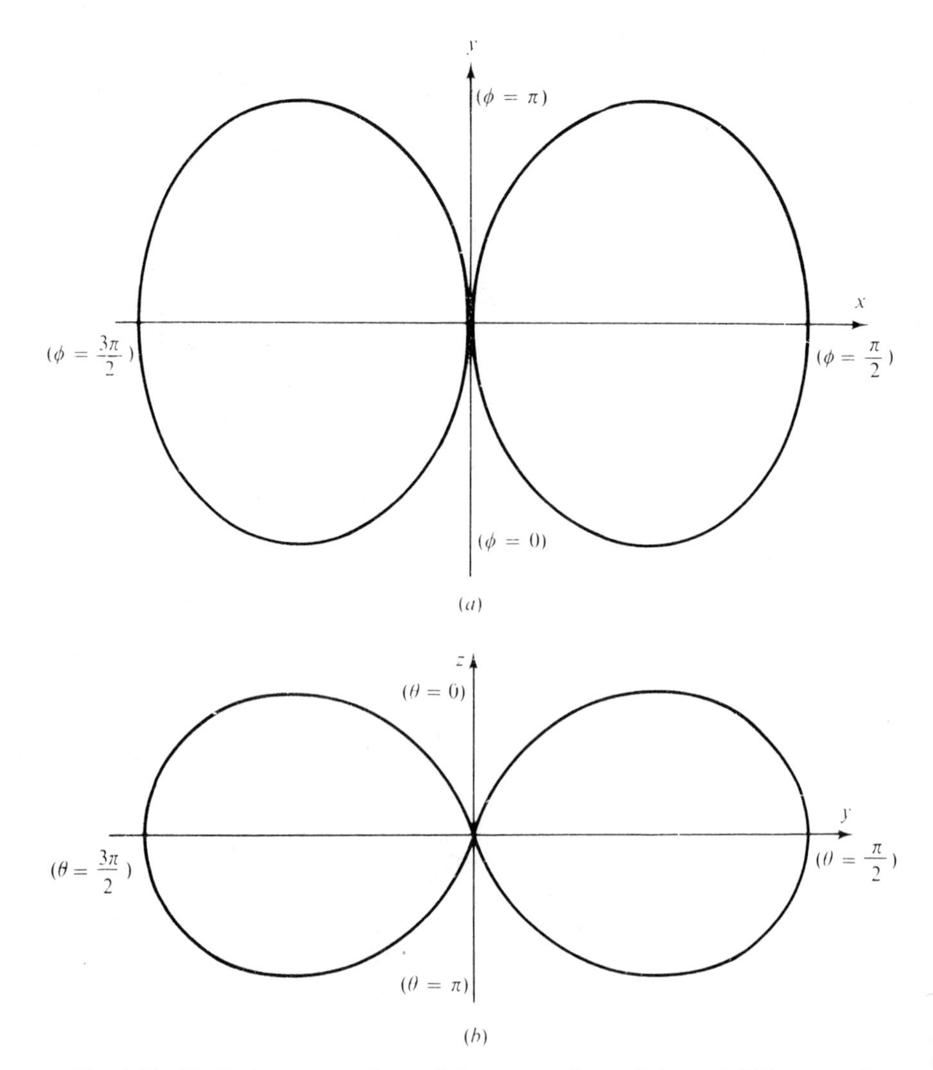

Fig. 4.20. Radiation pattern for end-fire array of two dipoles. (*a*) Plane $z = 0$. (*b*) Plane $x = 0$.

Thus a maximum occurs in directions intermediate between those of the broadside and end-fire arrays.

4.7.2 General arrays

The principle outlined above, the combination of individual fields allowing for path difference, can be applied to any array of elements. An array factor can be determined considering the extension of the two element case to cover a general array of N elements.

An array of N equispaced elements

The geometry is shown in fig. 4.21. By a simple extension of (4.66) the array factor is given by

$$\mathrm{M} = M_0\{\mathrm{I}_1 + \mathrm{I}_2 \exp(\mathrm{j}ka\cos\chi) + \mathrm{I}_3 \exp(2\mathrm{j}ka\cos\chi) + \ldots + \mathrm{I}_s \exp[(s-1)\mathrm{j}ka\cos\chi] + \ldots + \mathrm{I}_N \exp[(N-1)\mathrm{j}ka\cos\chi]\} \tag{4.67}$$

(i) *Co-phased currents of equal magnitude.* In this case we have

$$\mathrm{M} = M_0\mathrm{I}_1 \sum_{s=0}^{N-1} \exp(s\mathrm{j}ka\cos\chi)$$

This is a geometric series which can be summed, giving

$$\mathrm{M} = M_0\mathrm{I}_1 \frac{1-\exp(N\mathrm{j}ka\cos\chi)}{1-\exp(\mathrm{j}ka\cos\chi)}$$

$$= \mathrm{M}_0\mathrm{I}_1 \exp\{\tfrac{1}{2}(N-1)\,\mathrm{j}ka\cos\chi\}\frac{\sin(\frac{1}{2}Nka\cos\chi)}{\sin(\frac{1}{2}ka\cos\chi)}$$

Hence
$$|\mathrm{M}| = \frac{1}{N}\left|\frac{\sin(\frac{1}{2}Nka\cos\chi)}{\sin(\frac{1}{2}ka\cos\chi)}\right| \tag{4.68}$$

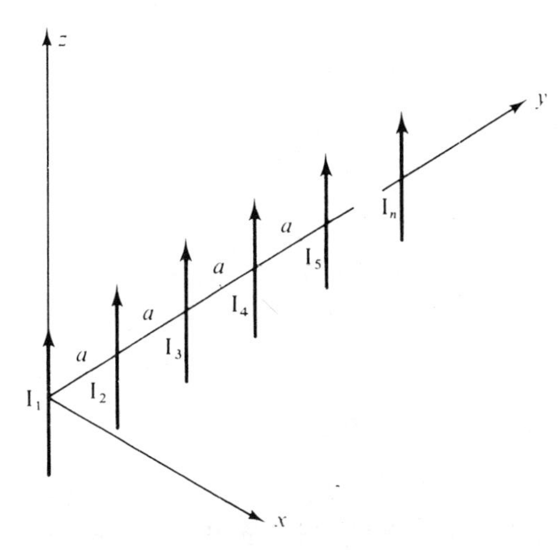

Fig. 4.21. An array of N elements.

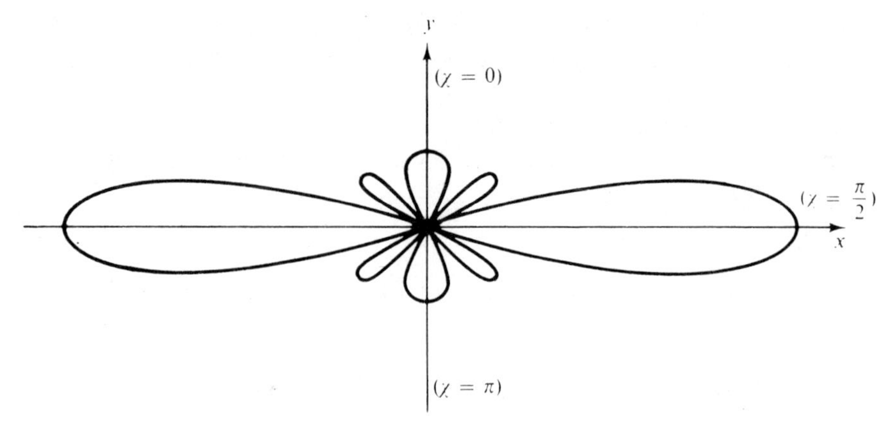

Fig. 4.22. Array factor for uniform array of five elements.

This equation may be compared with (4.53) for a uniformly illuminated aperture, when it will be seen that the uniformity results in both cases in the form of expression $\sin Nx/\sin x$. Equation (4.68) shows that M has a maximum for $\chi = \frac{1}{2}\pi$, broadside, and the null directions are given by the values of χ for which $\frac{1}{2}Nka \cos \chi$ becomes an integral multiple of π. If we take $a = \frac{1}{2}\lambda$ as before, this becomes

$$\chi = \cos^{-1}(2s/N) \tag{4.69}$$

The integer s can take on positive or negative values between unity and the greatest integer in $\frac{1}{2}N$. Subsidiary maxima occur at values of χ given approximately by

$$\tfrac{1}{2}Nka \cos \chi = \tfrac{1}{2}(2s+1)\pi$$

They decrease in magnitude as the inverse of $\sin[(2s+1)\pi/2N]$. These are the *side lobes* of the main pattern.

The case $N = 5$ is illustrated in fig. 4.22.

A convenient indication of the beam width is provided by the separation of the null direction either side of the maximum. These occur for $\chi = \frac{1}{2}\pi \pm \delta$, where we find δ from (4.69) to be given by

$$\delta = \sin^{-1}(2/N) \approx 2/N \qquad (N \gg 1)$$

The separation is thus $4/N$ radians.

(ii) *Binomial loading.* The presence of side lobes is often undesirable. They arise from the uniform loading and the sudden discontinuity at the ends of the array. Consider the case when the currents are given by the formula

$$I_s = I_1\binom{N-1}{s} \qquad (s-0, 1, \ldots, N-1)$$

The binomial coefficient $\binom{N-1}{s}$ is the coefficient of the term x^s in the expansion of $(1+x)^{N-1}$, and may be expressed in the form

$$\binom{N-1}{s} = \frac{(N-1)(N-2)\ldots(N-s)}{s!}$$

From (4.67) we then have

$$\begin{aligned} M &= M_0 I_1 \sum_{s=0}^{N-1} \binom{N-1}{s} \exp(sjka\cos\chi) \\ &= M_0 I_1 [1+\exp(jka\cos\chi)]^{N-1} \\ &= M_0 I_1 \exp[\tfrac{1}{2}(N-1)jka\cos\chi][2\cos(\tfrac{1}{2}ka\cos\chi)]^{N-1} \end{aligned}$$

Hence, when $ka = \pi$,

$$|M| = [\cos(\tfrac{1}{2}\pi\cos\chi)]^{N-1} \tag{4.70}$$

Since $-1<\cos\chi<1$, the expression of (4.70) has no zeros except for $\chi=0$, π, and thus the array has no side lobes. The case $N=5$ is illustrated in fig. 4.23. The penalty which has been paid is that the beam is much wider, as comparison with fig. 4.22 will show. In practice some taper of current magnitudes away from the centre is desirable, a suitable compromise between beam width and side lobe magnitude being accepted.

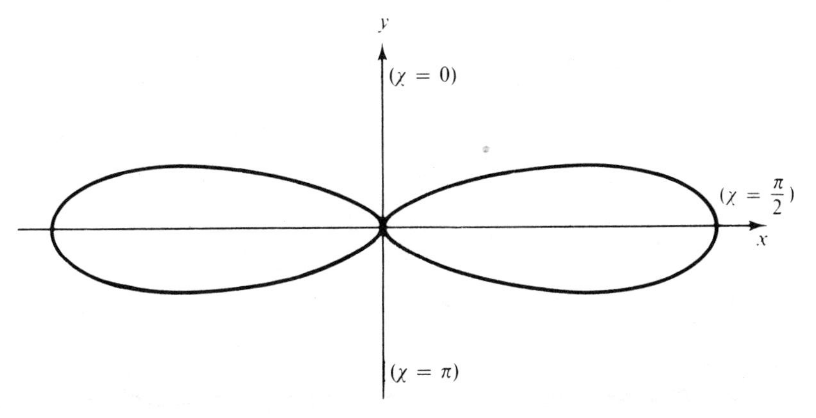

Fig. 4.23. Array factor for binomial loading of five-element array.

(iii) *Progressively phased currents.* The extension to N elements is to have currents given by

$$\mathrm{I}_s = \mathrm{I}_0 \exp(\mathrm{j}s\psi) \qquad (s = 0, 1, \ldots, N-1) \tag{4.71}$$

Substituting in (4.67)

$$\mathrm{M} = M_0 \mathrm{I}_0 \sum_{s=0}^{N-1} \exp\left[s\mathrm{j}(ka \cos\chi + \psi)\right]$$

Hence

$$|\mathrm{M}| = \frac{1}{N}\left|\frac{\sin[\frac{1}{2}N(ka \cos\chi + \psi)]}{\sin(\frac{1}{2}ka \cos\chi + \psi)}\right|$$

Consider the particular case $\psi = -ka$, $ka = \pi$. Equation (4.71) then becomes

$$\mathrm{M} = \frac{1}{N}\left|\frac{\sin[\frac{1}{2}\pi N(1 - \cos\chi)]}{\sin[\frac{1}{2}\pi(1 - \cos\chi)]}\right|$$

This is of the form $\sin Nx/\sin x$, and has a maximum for $\cos\chi = 1$, $\chi = 0$. It is therefore end-fire. The case for $N = 5$ is shown in fig. 4.24. The currents have been chosen as though a wave were progressing in the positive y-direction, the phase retarding by π between each element. It might therefore be thought that the beam should be directed along $\chi = 0$, whereas it is symmetrically directed along $\chi = 0$, π. This is because in the steady state a phase lag of π is indistinguishable from a phase lead of π.

In the above three situations the array factor has been evaluated. The complete radiation pattern is obtained by multiplying by $F(\theta)$. The effect of this on the broadside array will be to give, in the xz-plane, the maximum of the array factor, the radiation pattern of a dipole. The end-fire array will be only slightly modified, since its directions effects are mainly about a line for which $\theta = \frac{1}{2}\pi$.

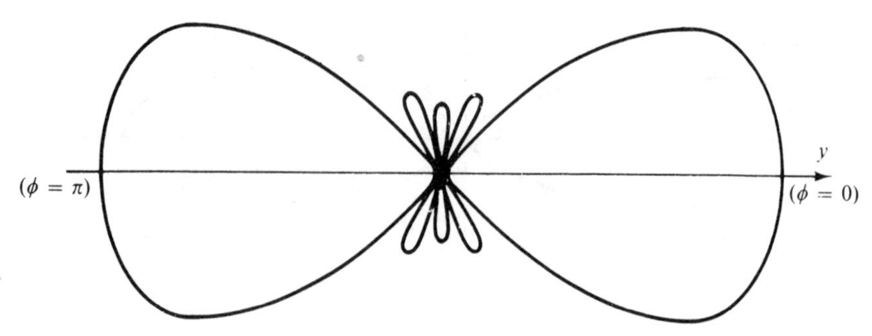

Fig. 4.24. Array factor for five-element end-fire array.

4.7.3 Parasitic elements

In the arrays previously considered each element has been driven from a source. It is possible to have elements which are excited by the field of a primary radiator so that the resultant field is produced by interference between the fields from primary and secondary radiators. As an example consider again the two-element array of fig. 4.17. If we assume that the current in the radiator at A is in phase with the field at A produced by the dipole at O, the array factor (4.66) becomes

$$\mathrm{M} = \mathrm{I}_1 + \mathrm{I}_1 \exp(-\mathrm{j}ka) \cdot \exp(\mathrm{j}ka \cos\chi)$$

Hence
$$|\mathrm{M}| \propto \cos[\tfrac{1}{2}ka(1-\cos\chi)]$$

If we now make $ka = \frac{1}{2}\pi$, or $a = \frac{1}{4}\lambda$,

$$|\mathrm{M}| \propto \cos[\tfrac{1}{4}\pi(1-\cos\chi)]$$

This factor is unity when $\chi = 0$, in the forward direction, and zero when $\chi = \pi$, in the backward direction.

This two-element array forms the basis of the common 'H' configuration used for VHF reception. The front element, or director, will be cut to an experimentally determined length for which optimum performance is obtained. Adjusting the length has the effect of altering the resonant frequency of the director, and hence the phase of the induced current.

Further parasitic elements may be used, resulting in the Yagi-type array shown in fig. 4.25.

4.7.4 Frequency behavior of arrays

It is to be noted that a change of frequency will alter the radiation pattern of an array. With a driven array the alteration in wavelength alters the electrical path lengths and hence alters the radiation pattern.

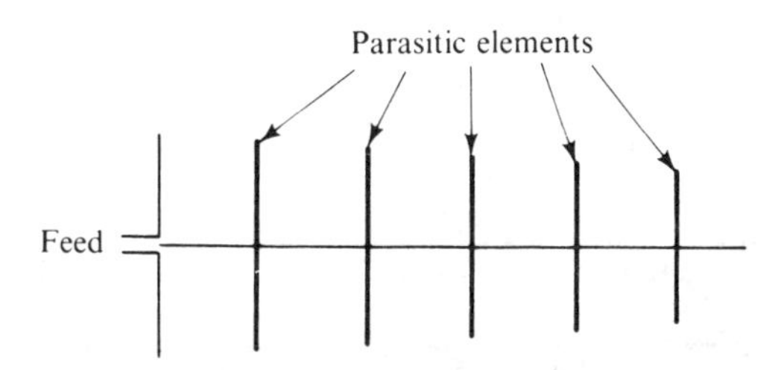

Fig. 4.25. Yagi-type array.

The change in the radiation pattern of the element itself is usually of less significance.

Arrays using parasitic elements are more sensitive to frequency change since not only does the electrical length of separation change but also the phase of the induced currents. An antenna has therefore an effective bandwidth over which it can be used.

4.8 Worked examples

1. An open-ended rectangular waveguide operated in the dominant mode acts as a radiator. To estimate the radiated field the open-end of the waveguide may be considered as an aperture with the field distribution appropriate to the dominant mode.

Show that the electric field on the axis of waveguide a distance z from the end is approximately given by

$$\mathrm{E} = \frac{1}{\pi^2 z}\left(1 + \frac{\omega}{\beta c}\right)(\omega\mu_0\beta abP)^{\frac{1}{2}}$$

in which P is the power carried by the mode and the remaining constants refer to the waveguide.

In practice a reflection will take place from the end of the waveguide, giving a reflection coefficient Γ. How could the estimate be improved using this fact?

Solution. The electric field is to be calculated from (4.49) with $\boldsymbol{n} = \boldsymbol{a}_1 = \boldsymbol{a}_z$ and, since $\boldsymbol{r}_2$ is in the aperture, $\boldsymbol{r}_2 \cdot \boldsymbol{a}_1 = 0$ (or all elements have equal delay to the observation point). By reference to chapter 3 we may write

$$\mathrm{H}_s = \boldsymbol{a}_x \mathrm{H}_0 \sin(\pi\xi/a)$$

$$\mathrm{E}_s = -\boldsymbol{a}_y \frac{\omega\mu_0}{\beta}\mathrm{H}_0$$

Hence, in magnitude and direction,

$$\mathbf{E}_1 = \frac{k}{4\pi r_1}\boldsymbol{a}_z \times \int_0^a \int_0^b \sqrt{\left(\frac{\mu_0}{\varepsilon_0}\right)}\,[\boldsymbol{a}_z \times (\boldsymbol{a}_z \times \mathrm{H}_s\,\boldsymbol{a}_x) - \boldsymbol{a}_z \times \mathrm{E}_s \boldsymbol{a}_y]\,\mathrm{d}\xi\,\mathrm{d}\eta$$

$$= -\frac{k}{4\pi z}\boldsymbol{a}_y \int_0^a \int_0^b \left[\sqrt{\left(\frac{\mu_0}{\varepsilon_0}\right)}\,\mathrm{H}_0 + \frac{\omega\mu_0}{\beta}\mathrm{H}_0\right] \sin\frac{\pi\xi}{a}\,\mathrm{d}\xi\,\mathrm{d}\eta$$

$$= -\boldsymbol{a}_y \frac{1}{4\pi z}\frac{\omega}{c}\left[\sqrt{\left(\frac{\mu_0}{\varepsilon_0}\right)} + \frac{\omega\mu_0}{\beta}\right]\mathrm{H}_0 \frac{2ab}{\pi}$$

$$= -\boldsymbol{a}_y \frac{ab}{2\pi^2 z}\omega\mu_0\left(1 + \frac{\omega}{\beta c}\right)\mathrm{H}_0$$

The power in the mode is given by

$$P = \frac{1}{2}\,\mathrm{Re}\int_0^a\int_0^b \mathrm{H}_0^2 \frac{\omega\mu_0}{\beta}\sin^2\frac{\pi\xi}{a}\,\mathrm{d}\xi\,\mathrm{d}\eta$$

$$= \frac{1}{2}\mathrm{H}_0^2\frac{\omega\mu_0}{\beta}\frac{ab}{2}$$

Hence $$\mathrm{H}_0 = 2\left(\frac{\beta P}{\omega\mu_0 ab}\right)^{\frac{1}{2}}$$

Substituting $$\mathbf{E}_1 = -\boldsymbol{a}_y\frac{1}{\pi^2 z}(\omega\mu_0\beta ab)^{\frac{1}{2}}\left(1+\frac{\omega}{\beta c}\right)P^{\frac{1}{2}}$$

If the reflection coefficient of the dominant mode is known, then a better estimate of the dominant-mode field in the aperture can be made:

$$\mathrm{H}_s = \boldsymbol{a}_x \mathrm{H}_0 \sin\left(\frac{\pi\xi}{a}\right)$$

$$\mathrm{E}_s = -\boldsymbol{a}_y\frac{\omega\mu_0}{\beta}\frac{1+\Gamma}{1-\Gamma}\mathrm{H}_0\sin\frac{\pi\xi}{a}$$

Γ must of course be referred to the electric field and to the aperture plane.

2. The elements of an N-element linear array are excited with currents each of equal magnitude, but of phase changing by an angle $\psi(<\pi)$ between successive elements. The element spacing is $\frac{1}{2}\lambda$. Assuming that the radiators are isotropic, determine the form of the radiation pattern and find the angle of maximum radiation. Indicate how this and the beam width depend on ψ.

What would be the effect of changing the spacing to 2λ?

In a retrodirective array the signal picked up on an element of a regularly spaced linear array is fed to the element situated at the mirror-image position at the opposite end of the array. The delay between all pairs of elements is the same. Ignoring effects arising from mutual coupling and scattering, find the direction in the plane containing the incident wave normal and the array of the maximum of the re-radiated beam.

Discuss in general terms the performance of a planar array consisting of 5×5 small electromagnetic horns, each row being connected as for a retrodirective array independently of the other rows. Consider only the case when the incident-wave normal is parallel to a plane containing a row and the normal to the array.

(Cambridge University)

Solution. The case referred to is analysed in § 4.7.2(iii), where, if $ka = \pi$, it is shown that

$$M = \frac{1}{N}\left|\frac{\sin \frac{1}{2}N(\pi \cos \chi + \psi)}{\sin \frac{1}{2}(\pi \cos \chi + \psi)}\right|$$

The maximum occurs when $\pi \cos \chi + \psi = 0$, and the adjacent nulls in the radiation pattern are given by $\pi \cos \chi + \psi = \pm 2\pi/N$. If we assume $|\psi| < \pi$, and remembering $0 < \chi < \pi$, a maximum occurs for one value of χ only, given by

$$\cos \chi_m = -\frac{\psi}{\pi}$$

If $\psi > 0$ the maximum lies in $\frac{1}{2}\pi < \chi_m < \pi$. If $\psi < 0$, it lies in $\frac{1}{2}\pi > \chi_m > 0$. The nulls occur for $\chi = \chi_m + \delta$ where

$$\pi \cos (\chi_m + \delta) + \psi = \pm 2\pi/N, \qquad \pi \cos \chi_m = -\psi$$

giving
$$-\psi \cos \delta - \pi \sin \chi_m \sin \delta + \psi = \pm 2\pi/N$$

If $2\pi/N$ is small, δ is small and we have

$$\delta \approx \mp \frac{2}{N} \operatorname{cosec} \chi_m$$

We deduce that the beam width is a minimum when $\chi_m = \frac{1}{2}\pi$, and increases as the beam is swung away from this angle.

If a is made equal to 2λ, $ka = 4\pi$. The general effect is to introduce many more null positions.

Taking the configuration of fig. 4.21 and χ as defined in fig. 4.17, a plane wave approaching *from* the direction $\chi = \chi_0$ will induce a signal in the radiator at $y = sa$ proportional to $\exp(+jkas \cos \chi_0)$. Thus the cross-feed will induce in that radiator a current I proportional to $\exp[jka(N+1-s)\cos \chi_0]$. The re-radiated signal will depend on

$$\sum_{s=1}^{N-1} \exp[jka(N+1-s)\cos \chi_0] \cdot \exp[jkas \cos \chi]$$

$$= \exp[jka(N+1)] \sum_{s=1}^{N-1} \exp[jkas(\cos \chi - \cos \chi_0)]$$

$$= \exp[jka(N+1)] \frac{\sin[\frac{1}{2}ka\,N(\cos \chi - \cos \chi_0)]}{\sin[\frac{1}{2}ka(\cos \chi - \cos \chi_0)]}$$

The maximum of the re-radiated signal will occur for $\chi = \chi_0$.

When the elements become directive we have to multiply this last expression by the radiation pattern of the individual antenna, so that only directions within this pattern can be dealt with. This restriction is of course the price paid for higher gain.

4.9 Summary

An elementary source for electromagnetic waves has been defined, the short electric or Hertzian dipole. This has been used to find the field produced by a half-wave dipole. The power radiated and the radiation resistance have been calculated.

The use of the small dipole to calculate the field arising from any pattern of current has been formalised, and illustrated by application to a small current-carrying loop.

A reciprocity theorem has been proved and applied to the problem of relating the properties of an antenna in reception to those of the same antenna in transmission.

The same reciprocity theorem has been used to find a formula whereby the distant field arising from a source distribution may be found in terms of the electric and magnetic fields on a closed surface enclosing the source. This has been applied to antennae which have an aperture over which the fields can be estimated.

The analysis of arrays made up of a number of individually fed antennae has been illustrated by examples. It has been shown that the radiation pattern is the product of an array factor, depending only on the geometry of the array, and the radiation pattern of the individual element of the array.

Formulae

Short oscillating electric dipole (fig. 4.1)
Distant field

$$\mathrm{H}_\phi = \mathrm{I}\,\mathrm{d}l\frac{\mathrm{j}k}{4\pi r}\mathrm{e}^{-\mathrm{j}kr}\sin\theta$$

$$\mathrm{E}_\theta = \sqrt{\left(\frac{\mu_0}{\varepsilon_0}\right)}\,\mathrm{H}_\phi$$

Radiated power

$$P = \frac{1}{2}\sqrt{\left(\frac{\mu_0}{\varepsilon_0}\right)}\frac{2\pi}{3}\left(\frac{\delta l}{\lambda}\right)^2|\mathrm{I}|^2$$

Radiation resistance

$$R_\mathrm{a} = \sqrt{\left(\frac{\mu_0}{\varepsilon_0}\right)}\frac{2\pi}{3}\left(\frac{\delta l}{\lambda}\right)^2 = 190\left(\frac{\delta l}{\lambda}\right)^2$$

Small loop (fig. 4.10)

$$E_\phi = \sqrt{\left(\frac{\mu_0}{\varepsilon_0}\right)}\left(\frac{ka}{2}\right)^2 I \sin\theta \frac{e^{-jkr}}{r}$$

$$H_\theta = -\sqrt{\left(\frac{\varepsilon_0}{\mu_0}\right)} E_\phi$$

$$R_a = \frac{\pi^5}{6}\sqrt{\left(\frac{\mu_0}{\varepsilon_0}\right)}\left(\frac{2a}{\lambda}\right)^4 = 1.92\times 10^4\left(\frac{2a}{\lambda}\right)^4$$

Half-wave dipole (fig. 4.4)

Assumed current $I = I_0 \cos(\pi z/2l)$

Distant field

$$H_\phi = \frac{jI_0}{2\pi r} e^{-jkr} \frac{\cos\left(\frac{1}{2}\pi\cos\theta\right)}{\sin\theta}$$

$$E_\theta = \sqrt{\left(\frac{\mu_0}{\varepsilon_0}\right)} H_\phi$$

$$R_a = 73\ \Omega$$

Arbitrary current distribution (fig. 4.8)

Distant field $$\mathbf{H} = \frac{jk}{4\pi r_1}\exp(-jkr_1)\mathbf{K}\times \boldsymbol{a}_1$$

where $$\mathbf{K} = \int \mathbf{J}(\boldsymbol{r}_2)\exp(jk\boldsymbol{r}_2\cdot\boldsymbol{a}_1)\,dv$$

4.10 Problems

1. A radiator approximates to an electric dipole of length 250 m at a frequency of 60 kHz. Assuming the current is maintained constant over the length, evaluate the radiation resistance. Find the maximum electric field strength at a distance of 100 km when the radiated power is 1 kW.

2. The radiator in question 1 is now assumed to have the current decreasing linearly from the value at the centre to zero at the ends. Find the new estimate of the radiation resistance and the new field strength under the same conditions as question 1.

3. Show that an input power of 1 W to a half-wave dipole gives a maximum electric field strength at a distance of 1 km of about 10 mV m^{-1}.

4. A long straight wire carries a current given by $I(z) = I_0 \exp(-jkz)$, the co-ordinate z being taken as zero at one end of the wire and l at the other.

Obtain an expression for the distant electric field and show graphically how the electric field varies with direction at constant radius.

5. Explain what is meant by the radiation resistance of an antenna.

A small circular loop of radius a carries a current I. If $a/\lambda = 0.01$ calculate the radiation resistance of the antenna. Comment on the result.

Draw an approximate equivalent circuit of the antenna and briefly discuss how the latter might be fed in practice.

(Southampton University)

6. Determine (i) the open-circuit voltage, (ii) the available power from a half-wave dipole 20 km away from a similar, parallel, dipole transmitting 1 kW at 150 MHz. It may be assumed that the line joining the dipoles is perpendicular to the dipoles.

7. A radiator is modelled by a plane current sheet occupying an aperture defined by $-a < x < a$, $-b < y < b$, $z = 0$, in terms of Cartesian co-ordinates. The current everywhere flows in the y-direction and is of intensity $K \cos \omega t$ A m^{-1}. Obtain an expression for the radiation pattern.

Explain how this result is of relevance in predicting the diffraction of a beam of coherent light limited by a rectangular aperture. A beam of light of wavelength 632.8 nm from a laser emerges through a square aperture of 1 mm side. Assuming the aperture is uniformly illuminated, estimate the pattern which would be seen on a screen placed normal to the beam at a distance of 50 m.

[The electric field strength radiated by a Hertzian dipole of strength I δl is

$$\mathrm{E} = \mathrm{j}\frac{\mathrm{I}\,\delta l}{4\pi}\omega\mu_0\frac{\exp(-\mathrm{j}kr)}{r}\sin\theta$$

where $k = \omega\sqrt{(\mu_0\varepsilon_0)}$ and θ is the angle between the directions of the radius vector and the dipole.]

(Bristol University)

8. Show that a dipole situated parallel to and above a perfectly conducting plane is equivalent to the dipole together with a similar dipole at the image point carrying current in the reverse direction.

Two such half-wave dipoles are situated at a distance $\frac{1}{4}\lambda$ above an earth plane, and are separated by a distance of N wavelengths, $N \gg 1$. The line of centres is perpendicular to the dipoles. Show that in the vertical plane containing the centres of the dipoles the field strength is proportional to $\sin(\frac{1}{2}\pi \sin\theta)\cos(\pi N \cos\theta)$, θ being the angle of elevation. Hence show that, near the vertical, nulls occur at an angular separation of approximately $1/N$ radians.

9. Using the principle of multiplication of patterns, sketch the following radiation patterns:

(a) The horizontal pattern of four vertical radiators spaced one-half wavelength apart and fed with equal currents, but with 180-degree phasing between adjacent radiators.

(*b*) The horizontal pattern of four vertical radiators spaced one-quarter wavelength apart and having a progressive phase shift of 90 degrees between radiators.

(*c*) The free-space vertical patterns of each of the arrays of parts (*a*) and (*b*):

(i) in the plane of the array,

(ii) in the plane perpendicular to the plane of the array.

(The verticai radiators may be assumed to be half-wave dipoles in (*c*).)

(Southampton University)

5

Transmission-line theory

In the previous chapters we have considered the types of electromagnetic waves which can exist in association with certain configurations of sources and conductors. In three important instances those waves were essentially one dimensional in so far as the propagation was concerned: plane waves in space, the principal wave on transmission lines, and the propagating waveguide modes. In the latter two cases the variation of the fields transverse to the axis of the transmission line or waveguide was of secondary importance: the important fact was that forward and backward waves could be sustained. The presence of backward as well as forward waves evidently affects such things as transmitted power, distribution of electric field, and losses in a transmission line. In transmission-line theory, techniques are developed for conveniently studying such effects, and the results are applicable to any form of one-dimensional wave motion.

Historically, transmission lines were first analysed by considering them as distributed circuits. This approach is in contrast to the field theory developed in chapter 3 and is in some ways easier to appreciate. The reason for this greater simplicity is that 'external' variables are considered, namely voltage, current and power. The detailed field distribution is subsumed into parameters such as inductance and capacitance. It is however important to realise that field theory is necessary both to justify the results of the circuit approach and to calculate the circuit parameters for specific geometries. The present chapter therefore commences with the alternative approach to the transmission-line problem, and considers the justification of the method. The subsequent analysis is carried out in circuit terms. The application to waveguides, which are not amenable to circuit treatment, is deferred to §7.3.

5.1 Uniformly distributed networks

We start by investigating the appropriate description of a uniform line in circuit terms. Consider a length of uniform two-conductor line connected between a source and a load. Within any short length we can discern

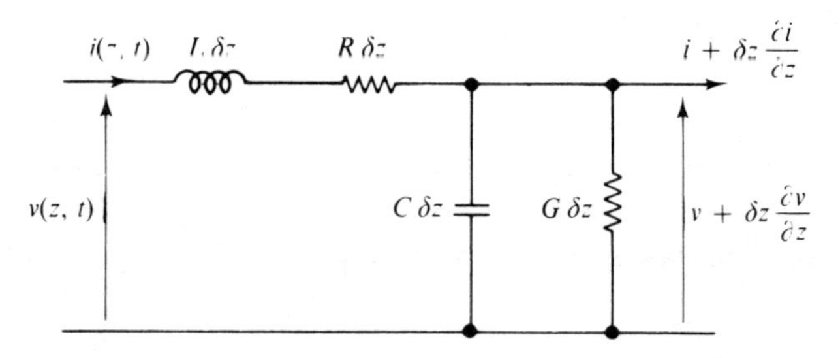

Fig. 5.1. Circuit equivalent to elementary length of transmission line.

both energy storage and energy dissipation: the latter occurs both in the conductors and in the dielectric, and the former as both magnetic and electrostatic energy. A circuit displaying these properties is shown in fig. 5.1. The parameters L, C, R and G will be measured per unit length, and the circuit is thought of as the equivalent to a length δz of the line. The stored energies are $\frac{1}{2}Li^2\,\delta z$, $\frac{1}{2}Cv^2\,\delta z$, and the energy dissipation $(Ri^2+Gv^2)\,\delta z$, which are compatible with ideas of energy distributed along the line. It may be remarked that a length of real line will be symmetrical with respect to the ends, but that the circuit is not. This could be remedied by making either a T or a π circuit with the same elements, but since the limit $\delta z \to 0$ is taken, the symmetry automatically appears. The voltage and current must be described as functions of z and t, so that the terminal conditions are as shown in fig. 5.1. Expressing the change in voltage and current in circuit form we have

$$v-\left(v+\delta z\frac{\partial v}{\partial z}\right)=\left(Ri+L\frac{\partial i}{\partial t}\right)\delta z$$

$$i-\left(i+\delta z\frac{\partial i}{\partial z}\right)=\left(Gv+C\frac{\partial v}{\partial t}\right)\delta z$$

Hence

$$\frac{\partial v}{\partial z}=-\left(Ri+L\frac{\partial i}{\partial t}\right) \tag{5.1}$$

$$\frac{\partial i}{\partial z}=-\left(Gv+C\frac{\partial v}{\partial t}\right) \tag{5.2}$$

Differentiating (5.1) with respect to z and (5.2) with respect to t we have

$$\frac{\partial^2 v}{\partial z^2}=-\left(R\frac{\partial i}{\partial z}+L\frac{\partial^2 i}{\partial z\,\partial t}\right)$$

$$=R\left(Gv+C\frac{\partial v}{\partial t}\right)+L\left(G\frac{\partial v}{\partial t}+C\frac{\partial^2 v}{\partial t^2}\right)$$

Hence v satisfies

$$\frac{\partial^2 v}{\partial z^2} = RGv + (RC + LG)\frac{\partial v}{\partial t} + LC\frac{\partial^2 v}{\partial t^2} \tag{5.3}$$

This is often called the *telegrapher's equation*. Elimination of v by a similar process shows that i satisfies the same equation. We shall consider later general solutions of this equation in the time domain, but we now restrict ourselves to sinusoidal variations in time. All the equations are linear with constant coefficients so that we may use the device of phasors discussed in chapter 1. We make the substitutions

$$v(z, t) \to \mathrm{V}(z)\,\mathrm{e}^{\mathrm{j}\omega t}$$

$$i(z, t) \to \mathrm{I}(z)\,\mathrm{e}^{\mathrm{j}\omega t}$$

Equations (5.1) and (5.2) become

$$\frac{\mathrm{dV}}{\mathrm{d}z} = -(R + \mathrm{j}\omega L)\mathrm{I} \tag{5.4}$$

$$\frac{\mathrm{dI}}{\mathrm{d}z} = -(G + \mathrm{j}\omega C)\mathrm{V} \tag{5.5}$$

Either by substitution in (5.3) or directly from (5.4) and (5.5),

$$\frac{\mathrm{d}^2\mathrm{V}}{\mathrm{d}z^2} = (R + \mathrm{j}\omega L)(G + \mathrm{j}\omega C)\mathrm{V} \tag{5.6}$$

This equation has a solution of the form

$$\mathrm{V}(z) = \mathrm{A}\,\mathrm{e}^{-\gamma z} + \mathrm{B}\,\mathrm{e}^{\gamma z} \tag{5.7}$$

where $$\gamma = [(R + \mathrm{j}\omega L)(G + \mathrm{j}\omega C)]^{\frac{1}{2}}$$

Equation (5.6) is analogous to (1.28) which characterised the propagation of a plane electromagnetic wave in a homogeneous medium. Thus we see that generally $\mathrm{V}(z)$ comprises the sum of two components; $\mathrm{A}\,\mathrm{e}^{-\gamma z}$ representing a voltage waveform travelling in the direction z, and $\mathrm{B}\,\mathrm{e}^{\gamma z}$ a voltage waveform travelling in the opposite direction.

Substitution in (5.4) gives

$$\mathrm{I}(z) = -\frac{1}{R + \mathrm{j}\omega L}\frac{\mathrm{d}}{\mathrm{d}z}[\mathrm{A}\,\mathrm{e}^{-\gamma z} + \mathrm{B}\,\mathrm{e}^{\gamma z}]$$

$$= Z_0^{-1}[\mathrm{A}\,\mathrm{e}^{-\gamma z} - \mathrm{B}\,\mathrm{e}^{\gamma z}] \tag{5.8}$$

where $$Z_0 = [(R + \mathrm{j}\omega L)/(G + \mathrm{j}\omega C)]^{\frac{1}{2}} \tag{5.9}$$

These equations may be compared with equations (1.29) to (1.34) and (1.58). They describe waves travelling in the two directions along the line, with propagation constant $\gamma = \alpha + j\beta$.

The impedance Z_0 is that seen by a progressive wave on a long line. It is called the *characteristic impedance* of the line, and may be compared with the wave impedance ζ of (1.34).

5.1.1 Comparison between circuit formulation and electromagnetic theory

Uniform lossless two-conductor systems have been considered from the point of view of electromagnetic theory in chapter 3. It was shown that in such systems a principal wave exists for which it is possible to define at any given point a voltage between lines and a 'circuit' current along those lines. The wave propagated with the velocity of electromagnetic waves in the unbounded medium, and the ratio of voltage to current for a wave in the forward direction was independent of position along the line. These results are evidently consistent in form with the results obtained in the last section, although numerical comparisons have to be made for given geometries. For example, let us consider a coaxial line of inner conductor radius a and outer conductor of inner radius b, as shown in fig. 3.2. The inductance and capacitance per unit length may be shown to be given by

$$L = (\mu/2\pi)\ln(b/a)\,\mathrm{H\,m^{-1}}$$

$$C = 2\pi\varepsilon/\ln(b/a)\,\mathrm{F\,m^{-1}}$$

Since both conductor and medium are lossless, we evidently have to put $R = 0$, $G = 0$ in the equations of the last section before making a comparison. We then have

$$\gamma = j\omega\sqrt{(LC)} = j\omega\sqrt{(\mu\varepsilon)}$$

$$Z_0 = \sqrt{(L/C)} = \sqrt{(\mu/\varepsilon)}\ln(b/a)/2\pi$$

The former is correct for plane waves in the unbounded medium, and the expression for Z_0 agrees with that found in (3.32). An investigation into the way of calculating L and C from static fields shows that agreement is found in all cases.

In deriving (5.1) and (5.2) the possibility of any mutual inductance coupling points in an axial direction was ignored. This was evidently correct, but it is only in the light of principal wave theory that it can be seen to be so: the principal wave has no axial electric or magnetic field components which would give axial coupling. Logically, an approach via circuit theory, which takes no account of propagation of electric and magnetic fields, would not be expected to produce a wave theory with finite velocity. The distributed circuit approach does in fact give the correct answer for the principal (*TEM*) wave, but does not predict the possibility of waveguide modes.

The case of slight losses in transmission lines was discussed in chapter 3 by the use of perturbation theory. The loss in the conductor was estimated on the assumption that the local field was of the same form as it would have been with lossless conductors. The result was to give a loss proportional to $|H|^2$ or, in terms of current appropriate to the present discussion, $|I|^2$. This is evidently the same as would be given by a series resistance, and is comparable with the parameter R of the present discussion. Losses arising because of the medium between conductors were also seen to be simply accounted for by shunt conductance, as in the present circuit model.

The correct setting for the circuit model can thus be seen. It is more flexible than the approach through electromagnetic theory: for example, it is easy to allow if necessary that R, G, L, C be functions of frequency; the system is not constrained to be perfectly straight; geometries quite intractable to electromagnetic theory can be handled, providing the limitations of the method are clearly understood.

5.1.2 The primary parameters *R, G, L, C*

These parameters have been introduced on heuristic grounds. In practice they must be calculated or measured. In the telegrapher's equation they were implicitly assumed to be constant, but in (5.4) and (5.5) they can be taken to be functions of frequency.

In general R will increase approximately as the square root of the operating frequency due to skin effect, mentioned in chapter 1. If we neglect the field within the conductors, which decreases with increasing frequency, then L is nominally independent of frequency. C depends upon the permittivity of the medium separating the conductors,but can normally be assumed constant. G comprises the d.c. leakage conductance between the conductors plus the equivalent a.c. conductance associated with dielectric losses in the insulating material, which are generally frequency dependent.

The transmission line shown in fig. 5.2(a) is referred to as 'balanced' since the two conductors are identical, and the line would normally be placed symmetrically above an earth plane, or inside a screening tube. Transmission lines which are asymmetrical in cross-section (e.g. coaxial

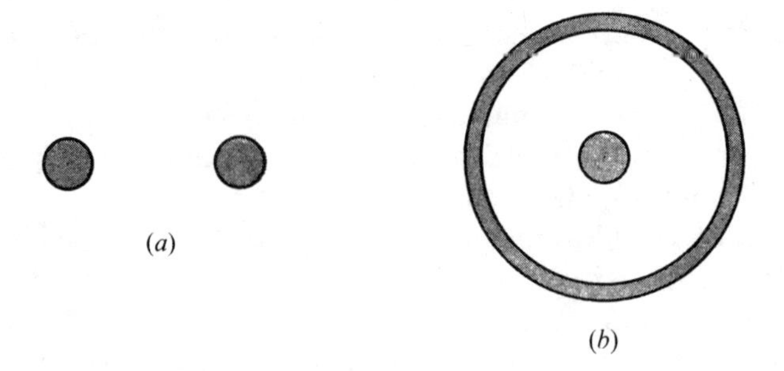

Fig. 5.2. Transmission-line geometry. (*a*) Balanced. (*b*) Unbalanced.

cables; fig. 5.2(*b*)) are referred to as 'unbalanced'. Although the resistance of the two conductors, the inner and the outer, of a length of coaxial cable are generally unequal, we may still use the designation R for the total resistance of a unit length of line.

The junction between balanced and unbalanced lines requires a care not indicated by the circuit model, and illustrates a situation where the field considerations have to be used.

5.2 Waves on long lines

The equations for voltage and current on a transmission line

$$V(z) = A\,e^{-\gamma z} + B\,e^{\gamma z}$$

$$I(z) = Z_0^{-1}[A\,e^{-\gamma z} - B\,e^{\gamma z}]$$

are, as has been pointed out, of the same form as for plane waves or waves on guiding structures. In the present section some of the consequences of the presence of forward and backward waves will be considered. The results are appropriate to waves in space or guided waves, or indeed to any form of one-dimensional wave motion.

A pure forward wave is described by the terms

$$V(z) = A\,e^{-\gamma z} = A\,e^{-\alpha z}e^{-j\beta z}$$

$$I(z) = Z_0^{-1}A\,e^{-\gamma z}$$

It corresponds to a source at one end of an infinite uniform line with no source anywhere else. It is convenient to take the source at $z = 0$ and then these equations are valid for $z > 0$ providing no source exists in

$z > 0$. $V(z)$ is a phasor quantity, so that the voltage which would be measured by a meter is

$$|V(z)| = A\,e^{-\alpha z}$$

in which A has been presumed real.

The magnitude of the voltage decreases exponentially away from the source, with an attenuation constant α nepers per metre or, from (1.60), 8.686α dB m^{-1}. The expression for the instantaneous voltage is

$$\mathrm{Re}\,[V(z)\,e^{j\omega t}] = A\,e^{-\alpha z}\cos(\omega t - \beta z)$$

corresponding to a wavelength $\lambda = 2\pi/\beta$ and a phase velocity ω/β as discussed in chapter 1.

We observe that we have for the ratio of voltage to current

$$V(z)/I(z) = Z_0$$

Z_0 is independent of z, and we may therefore interpret this equation by saying that an impedance Z_0 is presented to the line on the left of z by the part of the line on the right. We may in circuit terms consider the possibility of providing a suitable component with impedance Z_0 and using it to terminate a line. The part of the line between source and this load will then behave as though it were extending to infinity. The significance of the characteristic impedance Z_0 is that it can replace an infinite length of line. A line terminated in the characteristic impedance is said to be *matched.*

Similar remarks can be made about the backward wave described by the term $B\exp(-\gamma z)$. Only the direction of propagation and current flow is reversed. Such a wave, with the co-ordinate origin of the previous paragraph, would be produced by a source at some positive value of z. In the region between the two sources both forward and backward waves would exist. This however is not the usual reason for a backward wave being present: such a wave arises because of a discontinuity in the line giving rise to a reflection of the incident wave. If for example the line is short-circuited at some point, the voltage at that point must be identically zero for all times. This can only be achieved by allowing both forward and backward waves. The backward wave will have a definite phase relationship to the forward wave, and interference between the two will result. We will now examine the pattern of voltage distribution when both waves are present.

5.2.1 Standing waves

The measured voltage V_P at a point P, co-ordinate z, is given by

$$V_P = |V(z)| = |A\,e^{-\gamma z} + B\,e^{\gamma z}|$$

$$= |A\,e^{-\gamma z}||1 + k\,e^{2\gamma z}| \tag{5.10}$$

where $k = B/A$, the reflection coefficient. We can without loss of generality take $|k| \leqslant 1$. It will in general be complex, and we write

$$k = K e^{j\theta} \quad (K \leqslant 1)$$

It is simplest to consider first a lossless line for which $\alpha = 0$. In this case (5.10) becomes

$$\begin{aligned} V_P &= A|1 + k\,e^{2j\beta z}| \\ &= A|1 + K e^{j\theta}\,e^{2j\beta z}| \\ &= A\{[1 + K\cos(2\beta z + \theta)]^2 + K^2 \sin^2(2\beta z + \theta)\}^{\frac{1}{2}} \\ &= A[1 + K^2 + 2K\cos(2\beta z + \theta)]^{\frac{1}{2}} \end{aligned} \tag{5.11}$$

This expression is illustrated graphically in fig. 5.3. As P varies along the line, $\cos(2\beta z + \theta)$ will lie in the range ± 1. When it is $+1$, V_P is at a maximum of $A(1+K)$, and when it is -1, V_P is at a minimum of $A(1-K)$. The variation of V_p along the line is referred to as a *standing-wave pattern*. The pattern is periodic, with a period in z given by $2\beta z = 2\pi$, or $z = \frac{1}{2}\lambda$. The positions of the maxima depend on θ. We see that in principle it is possible to obtain values for K and θ from this type of measurement.

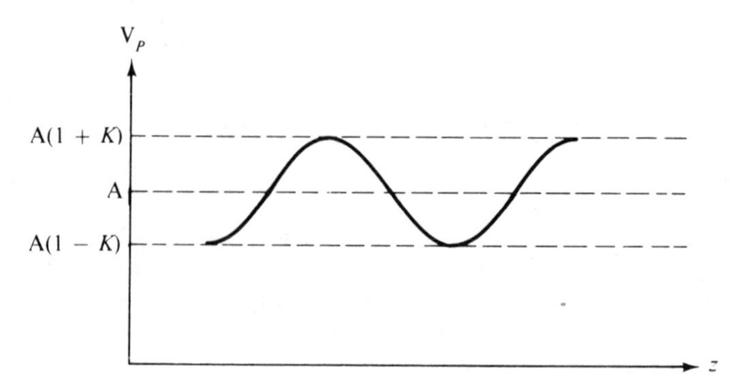

Fig. 5.3. Standing-wave pattern.

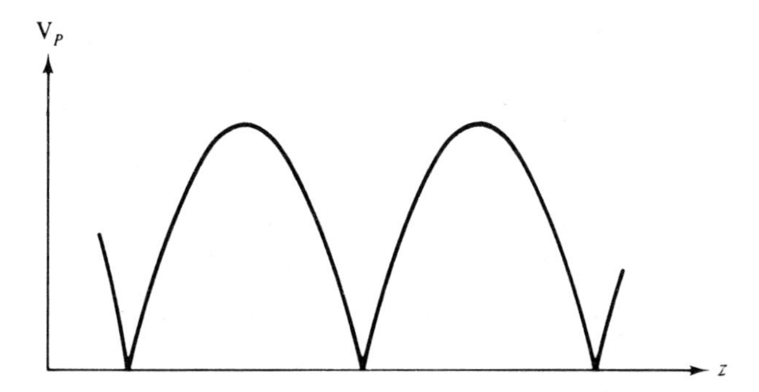

Fig. 5.4. Standing-wave pattern for complete reflection.

The current in the line can be found in a similar way, giving

$$\begin{aligned} I_P &= Z_0^{-1} A|1-k\ e^{2j\beta z}| \\ &= Z_0^{-1} A[1+K^2-2K\cos(2\beta z+\theta)]^{\frac{1}{2}} \end{aligned}$$

The pattern is the same except that current minima correspond to voltage maxima, and vice versa.

If $K = 1$, the voltage falls to zero at the minima, as shown in fig. 5.4. Such a pattern is characteristic of a pure standing wave, whereas the pattern depicted in fig. 5.3 consists of a progressive wave together with a standing-wave component. It is evident that the pattern of fig. 5.4 would be that occurring for a short-circuit placed across the line at a voltage null.

The pattern on a lossy line is somewhat more complicated. We now have

$$\begin{aligned} V_P &= |A\exp(-\gamma z)||1+K\exp[2\alpha z+j(2\beta z+\theta)]| \\ &= A\exp(-\alpha z)\{[1+K\exp(2\alpha z)\cos(2\beta z+\theta)]^2 \\ &\qquad + K^2\exp(4\alpha z)\sin^2(2\beta z+\theta)\}^{\frac{1}{2}} \\ &= A\exp(-\alpha z)[1+K^2\exp(4\alpha z)+2K\exp(2\alpha z)\cos(2\beta z+\theta)]^{\frac{1}{2}} \end{aligned} \tag{5.12}$$

If α is small enough the difference in behaviour over a few wavelengths between (5.11) and (5.12) will be negligible, and we have a similar state of affairs to that shown in fig. 5.3.

To see what happens when α is more significant, take the place from which the reflection originates as $z = 0$ and more backwards to a source at $z = -l$. If αl is large enough, the term in K will have decayed away,

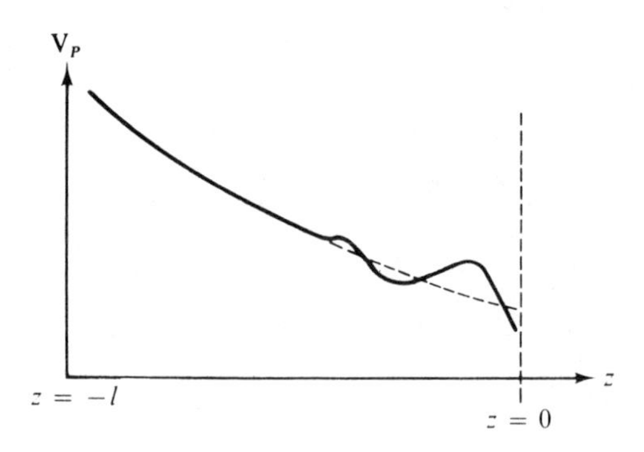

Fig. 5.5. Standing-wave pattern on lossy line.

giving a pattern of the type in fig. 5.5. In this case the reflections have been attenuated to a negligible level by the time the source has been reached. The pattern of fig. 5.5 is found as long as the attenuation in one wavelength is not too great.

5.2.2 Measurement of standing-wave pattern

The concept of measuring voltage across lines is only viable at low frequencies when the wavelength is very long. More often we are dealing with high frequencies corresponding to wavelengths less than a few metres. It will be observed that to find K it is not necessary to measure voltage absolutely, since it is only the ratio of the minimum to the maximum that is required.

A standing-wave detector for coaxial geometry is shown in fig. 5.6. A probe penetrates into the line through a longitudinal slot in the outer. Provided that the penetration is not too great, the field in the line is not

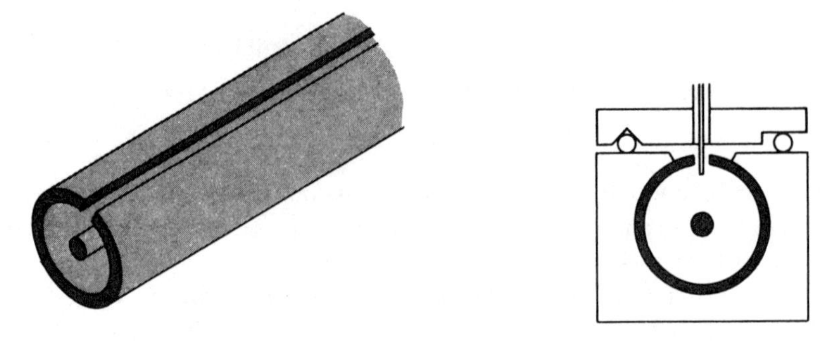

Fig. 5.6. A coaxial standing-wave detector.

greatly perturbed, and a signal proportional to the radial electric field is picked up. This can be rectified and measured, and is of course proportional to the voltage between inner and outer. The chief problem is the mechanical one of keeping the penetration constant as the probe carriage is moved along the line. The theory of such measurements will be dealt with later in this chapter.

Having explored the way in which the voltage and current vary along a line, we will now turn to the problem of determining properties as seen at the terminals.

5.3 The input impedance of a terminated line

We start with the voltage and current given by (5.7) and (5.8).

$$V(z) = A\,e^{-\gamma z} + B\,e^{\gamma z}$$

$$I(z) = Z_0^{-1}[A\,e^{-\gamma z} - B\,e^{\gamma z}]$$

where

$$\gamma = [(R + j\omega L)(G + j\omega C)]^{\frac{1}{2}}$$

$$Z_0 = [(R + j\omega L)/(G + j\omega C)]^{\frac{1}{2}}$$

If the primary parameters and frequency are given, the secondary parameters γ, Z_0 can be calculated. We note

$$R + j\omega L = Z_0\gamma \tag{5.13}$$

$$G + j\omega C = Z_0^{-1}\gamma \tag{5.14}$$

For lowloss lines at high frequency, $\omega L \gg R$ and $\omega C \gg G$. When this is the case, to the first approximation,

$$\gamma = j\beta = j\omega\sqrt{(LC)}$$

$$Z_0 = \sqrt{(L/C)}$$

Thus the line is non-dispersive and its characteristic impedance is purely resistive. For other conditions $\gamma = \alpha + j\beta$ will be complex, with no simple variation with frequency, and Z_0 will also be complex.

In the case of lowloss high-frequency lines it is possible to measure the velocity of propagation ω/β and Z_0 directly. For other cases the primary parameters must be measured. The obvious way of determining, for example, the capacitance per unit length, C, is by taking a length of line open-circuited at both ends, and measuring its capacitance on a bridge. The validity of this approach needs some justification, so we first of all apply our theory to the determination of the input impedance of a length of open-circuited line.

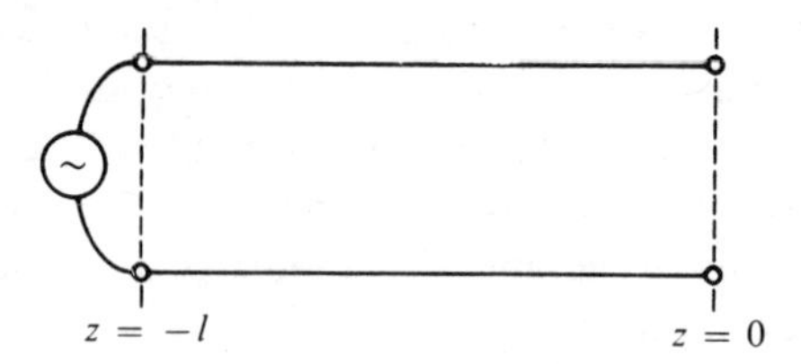

Fig. 5.7. Length of line open-circuited at far end.

5.3.1 Termination in an open-circuit

It is convenient to take the termination at co-ordinate point $z = 0$ and the input to the line at $z = -l$, as shown in fig. 5.7. The open-circuit is then described by the equation

$$\mathrm{I}(0) = Z_0^{-1}[\mathrm{A} - \mathrm{B}] = 0$$

We therefore have

$$\mathrm{A} = \mathrm{B}$$

Hence

$$\mathrm{V}(z) = \mathrm{A}[\mathrm{e}^{-\gamma z} + \mathrm{e}^{\gamma z}]$$

$$\mathrm{I}(z) = Z_0^{-1}\mathrm{A}[\mathrm{e}^{-\gamma z} - \mathrm{e}^{\gamma z}]$$

The impedance seen by the source at $z = -l$ is equal to

$$Z_{1\mathrm{o}} = \frac{\mathrm{V}(-l)}{\mathrm{I}(-l)} = Z_0 \frac{\mathrm{e}^{\gamma l} + \mathrm{e}^{-\gamma l}}{\mathrm{e}^{\gamma l} - \mathrm{e}^{-\gamma l}}$$

This may be put either in the form

$$Z_{1\mathrm{o}} = Z_0 \frac{1 + \mathrm{e}^{-2\gamma l}}{1 - \mathrm{e}^{-2\gamma l}} \tag{5.15}$$

or, using hyperbolic functions,

$$Z_{1\mathrm{o}} = Z_0 \coth \gamma l$$

Let us assume that l is small enough so that $|\gamma l| \ll 1$. We may then expand the exponentials in (5.15) to give

$$Z_{1\mathrm{o}} \approx Z_0[1 + (1 - 2\gamma l)]/[1 - (1 - 2\gamma l)]$$

$$\approx Z_0/\gamma l$$

Using (5.14)

$$Z_{1\mathrm{o}}^{-1} = Z_0^{-1}\gamma l = l(G + \mathrm{j}\omega C)$$

The measurement of admittance of a length l of line will therefore give a result $l(G + \mathrm{j}\omega C)$ provided that $|\gamma l| \ll 1$. Assuming the losses are small, this condition becomes

$$\beta l = 2\pi l/\lambda \ll 1$$

or

$$l \ll \lambda/2\pi$$

For a longer length this result will not hold: if for example we make $\gamma l = \alpha l + \mathrm{j}\beta l$, with $\alpha l \ll 1$ and $\beta l = \frac{1}{2}\pi$

$$\begin{aligned} Z_{1o} &= Z_0[1+\mathrm{e}^{-2\alpha l}\,\mathrm{e}^{\mathrm{j}\pi}]/[1-\mathrm{e}^{-2\alpha l}\,\mathrm{e}^{\mathrm{j}\pi}] \\ &\approx Z_0[1-(1-2\alpha l)]/[1+(1-2\alpha l)] \\ &\approx \alpha l Z_0 \end{aligned}$$

The length of such a line is related to wavelength by

$$\beta l = 2\pi l/\lambda = \tfrac{1}{2}\pi$$

$$l = \tfrac{1}{4}\lambda$$

Thus a length of line $\frac{1}{4}\lambda$ long at a given frequency will not appear as an open-circuit but as a very low resistance.

5.3.2 Termination in a short-circuit

We may similarly treat the case when a short-circuit at $z = 0$ gives

$$\mathrm{V}(0) = \mathrm{A} + \mathrm{B} = 0$$

or

$$\mathrm{B} = -\mathrm{A}$$

At other values of z,

$$\mathrm{V}(z) = \mathrm{A}[\mathrm{e}^{-\gamma z} - \mathrm{e}^{\gamma z}]$$

$$\mathrm{I}(z) = Z_0^{-1}\,\mathrm{A}[\mathrm{e}^{-\gamma z} + \mathrm{e}^{\gamma z}]$$

For the impedance Z_{1s} seen by a source at $z = -l$

$$\begin{aligned} Z_{1s} &= Z_0[\mathrm{e}^{\gamma l} - \mathrm{e}^{-\gamma l}]/[\mathrm{e}^{\gamma l} + \mathrm{e}^{-\gamma l}] \\ &= Z_0[1-\mathrm{e}^{-2\gamma l}]/[1-\mathrm{e}^{-2\gamma l}] \\ &= Z_0 \tanh \gamma l \end{aligned} \tag{5.16}$$

For a short length of line $|\gamma l| \ll 1$, (5.16) gives, approximating as before,

$$Z_{1s} \approx Z_0 \gamma l$$

Equation (5.13) then enables us to write

$$Z_{1s} = l(R + \mathrm{j}\omega L)$$

We again find that by choosing a short length of line we can measure the parameters R, L.

Examples

1. A length of coaxial cable 1.1 m long is measured at 1 kHz. It is found that the open-circuit measurement gives a capacity of 60 pF and the short-circuit an inductance of 0.38 μH.

(*a*) Determine the characteristic impedance and the velocity of propagation.

(*b*) What would be the results of the two measurements at a frequency of 30 MHz?

Solution. (*a*) The line is very short when measured in wavelengths. Hence we may say

$$L = \frac{0.38}{1.1} \times 10^{-6} = 3.45 \times 10^{-7}\ \mathrm{H\,m^{-1}}$$

$$C = \frac{6}{1.1} \times 10^{-11} = 5.45 \times 10^{-11}\ \mathrm{F\,m^{-1}}$$

Hence

$$Z_0 = (L/C)^{\frac{1}{2}} = (0.633 \times 10^4)^{\frac{1}{2}} = 79.6$$

$$v = (LC)^{-\frac{1}{2}} = 10^9(18.8)^{-\frac{1}{2}} = 2.31 \times 10^8\ \mathrm{m\,s^{-1}}$$

(*b*) For any length of line, assumed lossless, (5.15) and (5.16) give

$$Z_{1o} = Z_0[1 + \mathrm{e}^{-2\mathrm{j}\beta l}]/[1 - \mathrm{e}^{-2\mathrm{j}\beta l}]$$

$$= -\mathrm{j}Z_0 \cot \beta l$$

or

$$Y_{1o} = \mathrm{j}Y_0 \tan \beta l$$

and

$$Z_{1s} = \mathrm{j}Z_0 \tan \beta l$$

Now

$$\beta = \omega/v = (2\pi \times 3 \times 10^7)/(2.31 \times 10^8)$$

$$= 0.816$$

$$\beta l = 0.816 \times 1.1 = 0.898\ \mathrm{rad}$$

$$\tan \beta l = 1.254$$

Therefore

$$Y_{1o} = \mathrm{j} \times 1.58 \times 10^{-2}\ \mathrm{S}$$

$$Z_{1s} = \mathrm{j} \times 99.9\ \Omega$$

Measured as capacity and inductance this gives 83.8 pF and 0.53 μH respectively.

2. A telephone line has the following primary constants per loop mile: $R = 20\ \Omega$, $L = 3$ mH, $C = 0.06\ \mu$F and $G = 10\ \mu$S.

Calculate the characteristic impedance and the propagation coefficient of the line at an angular frequency of 5000 rad s^{-1}. How does the magnitude of the characteristic impedance vary between very low and very high frequencies?

Solution. We have $Z_0 = [(R + \mathrm{j}\omega L)/(G + \mathrm{j}\omega C)]^{\frac{1}{2}}$

$$= \left[\frac{20 + \mathrm{j} \times 5 \times 10^3 \times 3 \times 10^{-3}}{10^{-5} + \mathrm{j} \times 5 \times 10^3 \times 0.06 \times 10^{-6}}\right]^{\frac{1}{2}}$$

$$= 288.6\angle -25.6°$$

and

$$\gamma = [(R + \mathrm{j}\omega L)(G + \mathrm{j}\omega C)]^{\frac{1}{2}}$$

$$= 4 \times 10^{-2} + \mathrm{j} \times 7.6 \times 10^{-2}$$

At very low frequencies, $|Z_0| \approx [R/G]^{\frac{1}{2}} = 1414\ \Omega$ and at high frequencies, $|Z_0| \approx [L/C]^{\frac{1}{2}} = 223.6\ \Omega$.

5.3.3 Determination of characteristic impedance and propagation constant from open- and short-circuit impedances

We note from (5.15) and (5.16) that

$$Z_{1\mathrm{s}} Z_{1\mathrm{o}} = Z_0^2$$

$$Z_{1\mathrm{s}}/Z_{1\mathrm{o}} = \tanh^2 \gamma l = [(1 - \mathrm{e}^{-2\gamma l})/(1 + \mathrm{e}^{-2\gamma l})]^2 \qquad (5.17)$$

These results do not depend on the condition $|\gamma l| \ll 1$ and provide a way of determining Z_0 and γ from measurements on any length of line.

In (5.17), γl will be complex unless the line is lossless. The equation may either be manipulated using the relation

$$\tanh(A + \mathrm{j}B) = [\tanh A + \mathrm{j} \tan B]/[1 + \mathrm{j} \tanh A \tan B]$$

or by solving for $\mathrm{e}^{-2\gamma l}$, giving

$$\mathrm{e}^{-2\gamma l} = \frac{1 - (Z_{1\mathrm{s}}/Z_{1\mathrm{o}})^{\frac{1}{2}}}{1 + (Z_{1\mathrm{s}}/Z_{1\mathrm{o}})^{\frac{1}{2}}}$$

If we put the right-hand side into polar form $r\,\mathrm{e}^{\mathrm{j}\theta}$ we find

$$\mathrm{e}^{-2\alpha l} = r$$

$$-2\beta l = \theta \pm 2n\pi \qquad (n = 0, 1, 2, \ldots)$$

Example

Open- and short-circuit termination measurements on a telephone cable 50 km long, at an angular frequency of 5000 rads s^{-1}, gave the following results:

$$Z_{1o} = 328\angle -29.2°\ \Omega$$

$$Z_{1s} = 1548\angle 6.8°\ \Omega$$

Estimate the values of the characteristic impedance, the propagation coefficient and the primary constants of the line at the given frequency. It may be assumed that the line is less than one wavelength long at this frequency.

Solution.
$$Z_0 = [Z_{1o}Z_{1s}]^{\frac{1}{2}} = 713\angle -11.2°\ \Omega$$

$$\tanh \gamma l = [Z_{1s}/Z_{1o}]^{\frac{1}{2}} = 2.17\angle 18° = 2.06 + \mathrm{j}0.67$$

Using the equation for $\exp(-2\gamma l)$ we have

$$\exp(-2\gamma l) = \frac{-1.06 - \mathrm{j}0.67}{3.06 + \mathrm{j}0.67} = \frac{1.25\angle 212.3°}{3.13\angle 12.4°}$$

$$= 0.399\angle 199.9° = 0.399\angle 3.49\ \text{rad}$$

Hence
$$-2\alpha l = \ln(0.399) = -0.919$$

$$\alpha = 0.009\ \text{nepers km}^{-1}$$

$$= 0.08\ \text{dB km}^{-1}$$

We have to find β from

$$-2\beta l = 3.49 \pm 2n\pi$$

Since we are told that the line is less than one wavelength long $0 < \beta l < 2\pi$. Hence

$$2\beta l = 2\pi - 3.49$$

$$\beta = 0.028\ \text{rad km}^{-1}$$

Also
$$\gamma = \alpha + \mathrm{j}\beta = 0.03\angle 72°$$

We can estimate the primary line constants as follows:

$$R + \mathrm{j}\omega L = \gamma Z_0 = 0.029\angle 72° \cdot 713\angle -11.2°$$

which gives $R = 10.1\ \Omega\ \text{km}^{-1}$ and $L = 3.61\ \text{mH km}^{-1}$

Finally,
$$G + \mathrm{j}\omega C = \frac{\gamma}{Z_0} = \frac{0.029\angle 72°}{713\angle -11.2°}$$

which gives $G = 4.8\ \mu\text{S km}^{-1}$ and $C = 0.0081\ \mu\text{F km}^{-1}$.

We note that this method cannot work if the attenuation is too great since both Z_{1o} and Z_{1s} will then be close to Z_0.

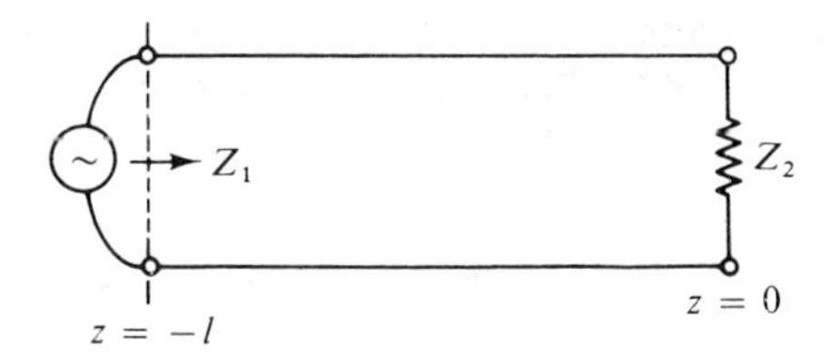

Fig. 5.8. Length of line terminated in load Z_2.

5.3.4 Input impedance with arbitrary termination

We have derived above the expressions for the input impedance of a length of line terminated either in open-circuit or short-circuit. We now consider the case when the load may have any impedance Z_2 as shown in fig. 5.8.

The condition at $z=0$ is now

$$\frac{\mathrm{V}(0)}{\mathrm{I}(0)} = Z_0 \frac{\mathrm{A}+\mathrm{B}}{\mathrm{A}-\mathrm{B}} = Z_2$$

Solving we find

$$\frac{\mathrm{B}}{\mathrm{A}} = \frac{Z_2 - Z_0}{Z_2 + Z_0} = k \tag{5.18}$$

The input impedance Z_1 is given by

$$Z_1 = Z_0 \frac{\mathrm{A}\,\mathrm{e}^{\gamma l} + \mathrm{B}\,\mathrm{e}^{-\gamma l}}{\mathrm{A}\,\mathrm{e}^{\gamma l} - \mathrm{B}\,\mathrm{e}^{\gamma l}}$$

$$= Z_0 \frac{1 + k\,\mathrm{e}^{-2\gamma l}}{1 - k\,\mathrm{e}^{-2\gamma l}} \tag{5.19}$$

Alternatively we may write

$$Z_1 = Z_0 \frac{(Z_2 + Z_0)\,\mathrm{e}^{\gamma l} + (Z_2 - Z_0)\,\mathrm{e}^{-\gamma l}}{(Z_2 + Z_0)\,\mathrm{e}^{\gamma l} - (Z_2 - Z_0)\,\mathrm{e}^{-\gamma l}}$$

$$= Z_0 \frac{Z_2 + Z_0 \tanh \gamma l}{Z_2 \tanh \gamma l + Z_0} \tag{5.20}$$

Although we have taken an impedance Z_2 to terminate the line as though it were prescribed, this is essentially a circuit concept: it is often of more importance to measure the impedance Z_2 of a particular configuration of components in association with a line. It may also be noted that the terms 'open' and 'short' circuit have a rather different significance at high frequencies to that used in lumped-circuit analysis. For instance, an open-wire line supporting a *TEM* wave is theoretically short-circuited by a perfectly conducting sheet of infinite extent placed normally to the line. Similarly, a coaxial cable is short-circuited by connecting a perfectly conducting disc between inner and outer conductors. These conditions can be approximated reasonably satisfactorily in practice but in neither type of line can an 'open-circuit' be obtained by simply leaving the line open. As indicated in chapter 1, the boundary conditions for the electromagnetic field existing in the medium surrounding the conductors must be satisfied in all cases, and this may require rather involved physical arrangements.

5.4 The transmission matrix

When a length of transmission line is used as a connecting link we are often concerned only with terminal behaviour. Since we use the line as a two-port network, any of the appropriate circuit representations can be used. The most convenient is the transmission-matrix form, expressing the input variables in terms of the output variables: that is, in the V, I notation of fig. 5.9, we have to express V_1, I_1 in terms of V_2, I_2.

We have, as usual,

$$V(z) = A\,e^{-\gamma z} + B\,e^{\gamma z}$$

$$I(z) = Z_0^{-1}(A\,e^{-\gamma z} - B\,e^{\gamma z})$$

Our variables may be expressed in the form

$$V_2 = V(0) = A + B$$

$$I_2 = I(0) = Z_0^{-1}(A - B)$$

$$V_1 = V(-l) = A\,e^{\gamma l} + B\,e^{-\gamma l}$$

$$I_1 = I(-l) = Z_0^{-1}(A\,e^{\gamma l} - B\,e^{-\gamma l})$$

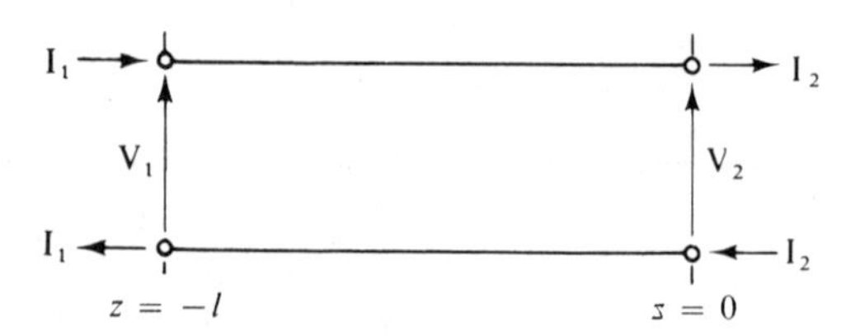

Fig. 5.9. A length of line as a two-port network.

The first two equations can be solved for A and B, giving

$$A = \tfrac{1}{2}(V_2 + Z_0 I_2)$$

$$B = \tfrac{1}{2}(V_2 - Z_0 I_2)$$

Substituting these expressions in the third and fourth equations we have

$$V_1 = V_2 \cosh \gamma l + Z_0 I_2 \sinh \gamma l$$

$$I_1 = Z_0^{-1} V_2 \sinh \gamma l + I_2 \cosh \gamma l$$

In matrix format we may write

$$\begin{pmatrix} V_1 \\ I_1 \end{pmatrix} = \begin{pmatrix} \cosh \gamma l & Z_0 \sinh \gamma l \\ Z_0^{-1} \sinh \gamma l & \cosh \gamma l \end{pmatrix} \begin{pmatrix} V_2 \\ I_2 \end{pmatrix} \tag{5.21}$$

We may derive (5.20) from this result by noting that the termination Z_2 implies that

$$V_2 = Z_2 I_2$$

When this relation is substituted in (5.21), (5.20) results.

When the transmission line is lossless, γ is equal to $j\beta$, and we may express the matrix of (5.21) in terms of trigonometrical functions, with Z_0 real:

$$\begin{pmatrix} \cos \beta l & jZ_0 \sin \beta l \\ jZ_0^{-1} \sin \beta l & \cos \beta l \end{pmatrix} \tag{5.22}$$

5.4.1 Expressions for voltage and current in terms of sending-end conditions

The voltage and current at the sending end are easily available for measurement, and we can find the voltage and current further up the line in terms of these measurements. Since the line termination is now inaccessible, it is more satisfactory to take the origin of z at the source, as in fig. 5.10.

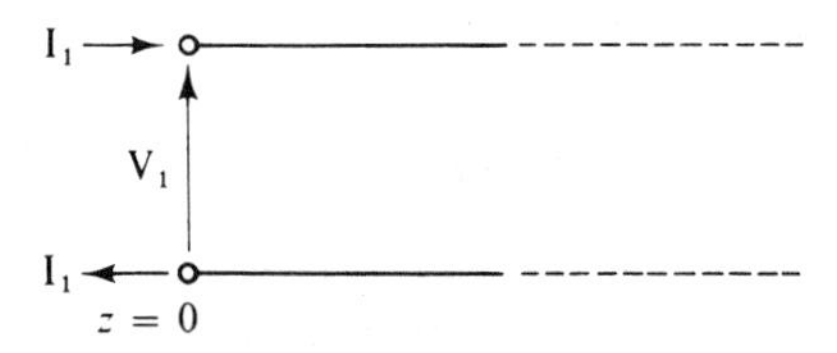

Fig. 5.10. Co-ordinates for measurements at the sending end of a line.

We have

$$V_1 = A + B$$

$$I_1 = Z_0^{-1}(A - B)$$

Solving for A, B,

$$A = \tfrac{1}{2}(V_1 + I_1 Z_0)$$

$$B = \tfrac{1}{2}(V_1 - I_1 Z_0)$$

Hence
$$V(z) = \tfrac{1}{2}(V_1 + I_1 Z_0)\,e^{-\gamma z} + \tfrac{1}{2}(V_1 - I_1 Z_0)\,e^{\gamma z}$$

By collecting similar exponential terms and using hyperbolic functions, we can rewrite this as

$$V(z) = V_1 \cosh \gamma z - I_1 Z_0 \sinh \gamma z$$

The corresponding expression for the current $I(z)$ is

$$I(z) = I_1 \cosh \gamma z - V_1 Z_0^{-1} \sinh \gamma z$$

These expressions are somewhat misleading in that they appear to imply that $V(z)$ and $I(z)$ can grow indefinitely along the line, since the hyperbolic functions increase as $z \to \infty$. This cannot be the case, and in fact for a very long line $V_1 \approx I_1 Z_0$. This can be seen from (5.19):

$$Z_1 = Z_0 \frac{1 + k\,e^{-2\gamma l}}{1 - k\,e^{-2\gamma l}}$$

As l increases, providing $\alpha \neq 0$, $|e^{-2\gamma l}| = e^{-2\alpha l} \to 0$ so that $Z_1 \to Z_0$. Thus for an infinitely long line with any loss, however small, only a forward wave is propagated.

5.5 Lowloss line

The remainder of this chapter will be concerned with lines for which the attenuation over a few wavelengths may be neglected in almost all circumstances. On the occasion when it is necessary to bring in attenuation we can assume it is so small that approximations such as $\exp(-\alpha l) = (1 - \alpha l)$ may be made.

In terms of the primary parameters, α can be found as follows, assuming $R \ll \omega L$, $G \ll \omega C$.

$$\gamma = \alpha + j\beta = (R + j\omega L)^{\frac{1}{2}}(G + j\omega C)^{\frac{1}{2}}$$

$$= j\omega\,\sqrt{(LC)}\left(1 + \frac{R}{j\omega L}\right)^{\frac{1}{2}}\left(1 + \frac{G}{j\omega C}\right)^{\frac{1}{2}}$$

$$\approx j\omega\,\sqrt{(LC)}\left(1 + \frac{R}{2j\omega L} + \frac{G}{2j\omega C}\right)$$

whence $$\alpha = \tfrac{1}{2}[R\sqrt{(C/L)} + G\sqrt{(L/C)}]$$
$$= \tfrac{1}{2}(RZ_0^{-1} + GZ_0)$$
$$\beta = \omega\sqrt{(LC)}$$

We first look at particular cases of the impedance presented by lengths of terminated line.

5.5.1 The quarter-wave transformer

Equation (5.20) becomes, when α is zero,

$$Z_1 = Z_0\left(\frac{Z_2 + jZ_0 \tan \beta l}{jZ_2 \tan \beta l + Z_0}\right) \tag{5.23}$$

Consider a line for which the length at the frequency in question is $\frac{1}{4}\lambda$. We then have $\beta l = 2\pi l/\lambda = \frac{1}{2}\pi$, so that $\tan \beta l$ is very large and (5.21) becomes

$$Z_2 = Z_0^2/Z_1$$

The line thus acts as an impedance inverter. It was mentioned earlier that whereas a good approximation to a short-circuit was possible on a high-frequency line, this was not the case for an open-circuit. We see that we can use a quarter-wave transformer terminated in a short-circuit to provide the open-circuit.

The losses in the line, however small, will prevent the attainment of a perfect open-circuit. We can proceed either from (5.19) or (5.20). In the latter case we use the result,

$$\tanh \gamma l = \tanh(\alpha l + j\beta l)$$
$$= \frac{\tanh \alpha l + j \tan \beta l}{1 + j \tanh \alpha l \tan \beta l}$$

If αl is small, $\tanh \alpha l \to \alpha l$

and $$\tanh \gamma l = \frac{\alpha l + j \tan \beta l}{1 + j\alpha l \tan \beta l}$$

For the line in question $\tan \beta l \to \infty$ so that $\tanh \gamma l \to 1/\alpha l$. We also have $Z_2 = 0$. Hence

$$Z_1 = Z_0 \tanh \gamma l \to Z_0/\alpha l$$

Thus in place of an open-circuit we have a high resistive impedance.

As an alternative to using the hyperbolic function we may use (5.19). When $Z_2=0$, $k=-1$, so that

$$Z_1=Z_0\frac{1-e^{-2\gamma l}}{1+e^{-2\gamma l}}$$

In this case $2\gamma l=2\alpha l+2j\beta l=2\alpha l+j\pi$.

Hence

$$Z_1\approx Z_0\frac{1+e^{-2\alpha l}}{1-e^{-2\alpha l}}$$

$$\approx Z_0\frac{2-\alpha l}{2\alpha l}$$

$$\approx\frac{Z_0}{\alpha l}$$

5.5.2 A line one half-wavelength long

We may treat the more general case when $l=\frac{1}{2}n\lambda$, $\beta l=n\pi$ and $\tan\beta l=0$. From (5.23) we see that $Z_1=Z_2$, so the impedance measured at intervals of a half-wavelength from the termination has the same value as Z_2. It must be emphasised that these simple relationships only obtain on a lossless line and furthermore are only correct at one frequency.

5.5.3 Reactance presented by short- and open-circuited lines

When Z_2 is made a short-circuit or (assuming it is possible) an open-circuit, we have

$$Z_1=jZ_0\tan\beta l=jX_1\quad\text{for } Z_2=0$$

and

$$Z_1=-jZ_0\cot\beta l=jX_1\quad\text{for } Z_2=\infty$$

These functions are plotted in fig. 5.11.

If in fig. 5.11 l is thought of as fixed, the abscissa is in fact a scale of frequency, so that the figure depicts reactance plotted against frequency. For very low frequencies we find the results previously established: that we have either ωL or $1/\omega C$ for the short- or open-circuit termination respectively.

If we consider the short-circuit line, there is, in circuit terms, a resonance when the line is an odd number of quarter-wavelengths long which is similar to that for a parallel L-C circuit. Let us consider the behaviour of Z_1 for frequencies making l nearly $\frac{1}{4}\lambda$.

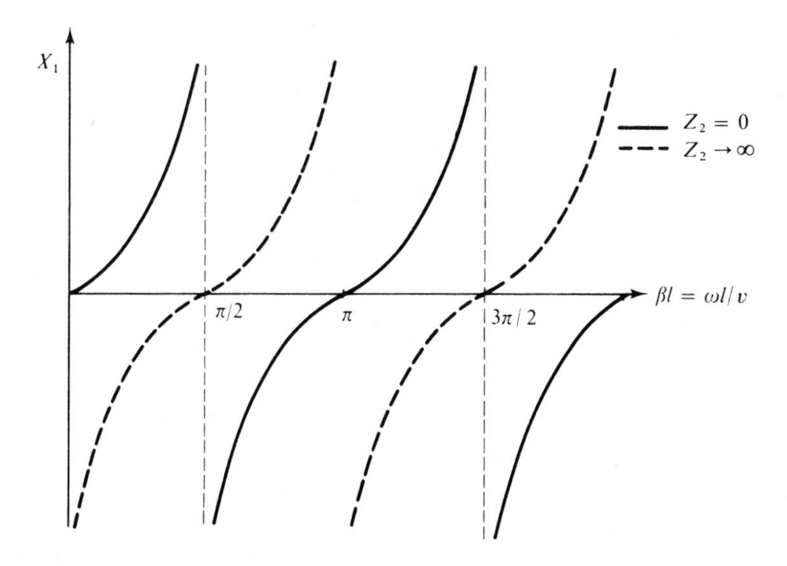

Fig. 5.11. Input reactance for a length of line terminated in an open-circuit or a short-circuit.

If we denote by ω_0 the frequency at which l is equal to one quarter-wavelength we have

$$\beta l = \omega_0 l/v = \tfrac{1}{2}\pi$$

At a neighbouring frequency $\omega_0 + \delta\omega$,

$$\beta l = (\omega_0 + \delta\omega)l/v = (1 + \delta\omega/\omega_0) \times \tfrac{1}{2}\pi$$

We have therefore

$$\begin{aligned}\exp(-2\gamma l) &= \exp(-2\alpha l)\exp(-\mathrm{j}\pi - \mathrm{j}\pi\delta\omega/\omega_0)\\ &\approx -(1-2\alpha l)(1-\mathrm{j}\pi\delta\omega/\omega_0)\\ &\approx -1 + 2\alpha l + \mathrm{j}\pi\delta\omega/\omega_0\end{aligned}$$

Hence the input impedance is given by

$$\begin{aligned}Z_1 &= Z_0(1-\mathrm{e}^{-2\gamma l})/(1+\mathrm{e}^{-2\gamma l})\\ &\approx Z_0(\alpha l + \mathrm{j}\pi\delta\omega/2\omega_0)^{-1} \qquad (5.24)\end{aligned}$$

We see that, when $\delta\omega = 0$, $Z_1 = Z_0/\alpha l$, agreeing with the previous result. We compare this expression with that for a parallel L-C-R circuit, shown in fig. 5.12.

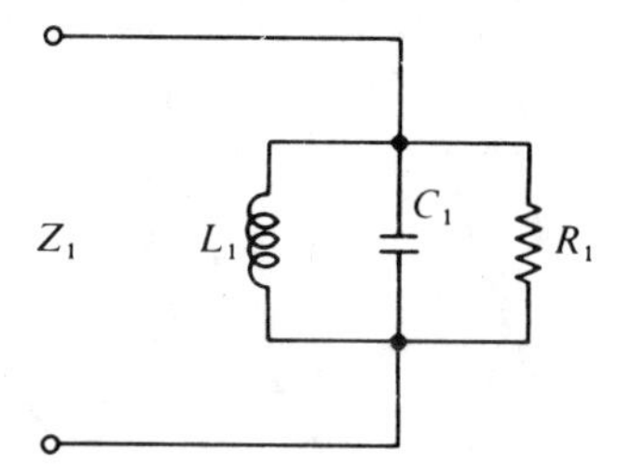

Fig. 5.12. Circuit equivalent to a line $\frac{1}{4}\lambda$ long.

For this circuit

$$\frac{1}{Z_1}=\frac{1}{\mathrm{j}\omega L_1}+\mathrm{j}\omega C_1+\frac{1}{R_1}$$

$$=\mathrm{j}\sqrt{\left(\frac{C_1}{L_1}\right)}\left(\frac{\omega}{\omega_0}-\frac{\omega_0}{\omega}\right)+\frac{1}{R_1}$$

$$\approx \mathrm{j}\sqrt{\left(\frac{C_1}{L_1}\right)}\frac{2\delta\omega}{\omega_0}+\frac{1}{R_1}$$

In this expression $\omega_0=(L_1C_1)^{-\frac{1}{2}}$ is the same frequency as that contained in (5.24). Comparing expressions we must have

$$\frac{1}{Z_1}=\frac{1}{Z_0}\left(\alpha l+\mathrm{j}\frac{\pi\delta\omega}{2\omega_0}\right)\equiv\frac{1}{R_1}+\mathrm{j}\sqrt{\left(\frac{C_1}{L_1}\right)}\frac{2\delta\omega}{\omega_0}$$

Therefore $\quad R_1=\dfrac{Z_0}{\alpha l}, \qquad \sqrt{\left(\dfrac{C_1}{L_1}\right)}=\dfrac{1}{Z_0}\dfrac{\pi}{4}, \qquad \sqrt{(L_1C_1)}=\dfrac{1}{\omega_0}$

giving $\quad C_1=\dfrac{\pi}{4Z_0\omega_0}, \qquad L_1=\dfrac{4Z_0}{\pi\omega_0}$

The Q-factor for the circuit is given by

$$Q=\frac{R_1}{\omega_0 L_1}=\frac{\pi}{4\alpha l}$$

To take a particular example, the attenuation of air-dielectric coaxial line 25 mm in diameter at 300 MHz is about 0.008 dB m^{-1}. The length will be 0.25 m, giving for Q a value of 3400. This is much higher than the Q of any lumped circuit at that frequency. This technique for making lowloss reactive elements is used in the design of filters at high frequencies.

5.5.4 Resonators

A length of line short-circuited at both ends can be regarded as comprising two equal elements and will thus make a resonator at the frequency for which the length is one half-wavelength. Such a resonator can be made of coaxial line without any mechanical problems of mounting the inner conductor, and losses would be solely due to the finite conductivity of the conductors. Suitable methods of feeding are indicated in fig. 5.13. In either case the Q-factor would need modifying to allow for losses introduced by the coupling. Such a resonator would also resonate at higher frequencies for which the length is λ, 2λ, ... The Q would decrease as the effective resistance increases on account of skin effect.

It may be added that a value for Q could also be calculated by the method adopted in chapter 3 for the waveguide resonator, when allowance would also (properly) be made for the finite conductivity of the closing plate at each end.

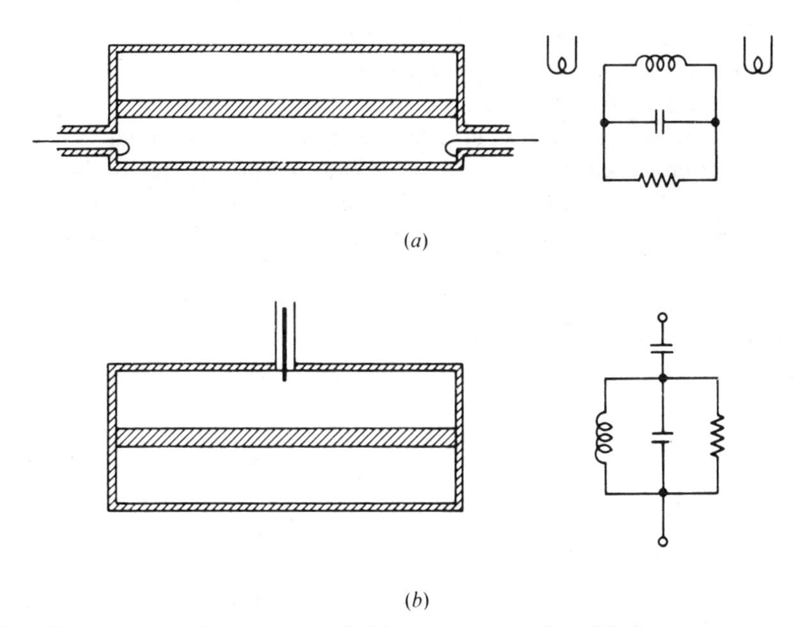

Fig. 5.13. A coaxial resonator. (*a*) Two-port connection. (*b*) One-port connection.

5.6 Surges on lossless lines

We have shown that, in the steady state, reflections occur at mismatched terminations. The same is true of transient disturbances.

On a lossless line we have seen that a forward wave is described by

$$V(z) = A \exp[-j\omega\sqrt{(LC)}]$$

$$I(z) = A\sqrt{(C/L)} \exp[-j\omega\sqrt{(LC)}]$$

The phase velocity ω/β is independent of frequency, so that transients are propagated without change of shape. This can also be seen from (5.3), which reduces to

$$\frac{\partial^2 v}{\partial z^2} = LC\frac{\partial^2 v}{\partial t^2}$$

This is the one-dimensional wave equation, the solutions to which may be written

$$v = f(t - z/c) + g(t + z/c) \tag{5.25}$$

where $c = (LC)^{-\frac{1}{2}}$ and g, f are arbitrary functions. We then find by substitution in either (5.1) or (5.2)

$$i = \sqrt{(C/L)}[f(t - z/c) - g(t + z/c)] \tag{5.26}$$

$f(t - z/c)$ represents a forward wave whose waveform at $z = 0$ is $f(t)$, and $g(t + z/c)$ a backward wave of waveform $g(t)$ at $z = 0$. $Z_0 = \sqrt{(L/C)}$ is sometimes known in this context as the *surge impedance*.

As an example of surges consider the situation shown in fig. 5.14. In this diagram is shown a source of steady e.m.f. E and of internal impedance Z_0 connected through a switch to a length l of open-circuit line. With the line carrying no current or voltage, the switch is closed. It is obvious that eventually there will be a steady e.m.f. E between lines. How is this state arrived at? At the instant at which the switch is closed a forward transient wave is initiated which then moves forward as indicated in fig. 5.15(a), (b) and (c). During the period covered by these

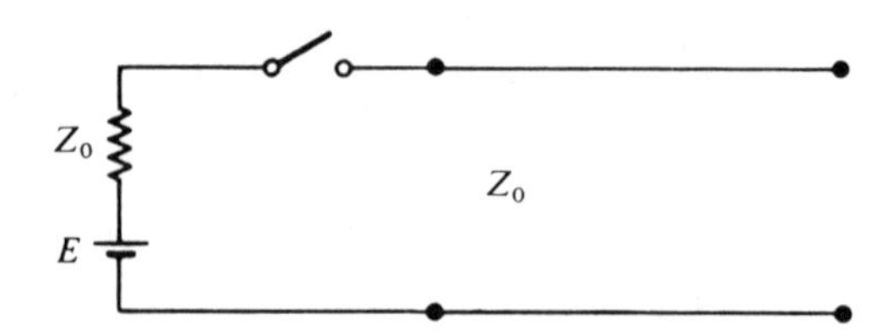

Fig. 5.14. Line with switched input for surge calculation.

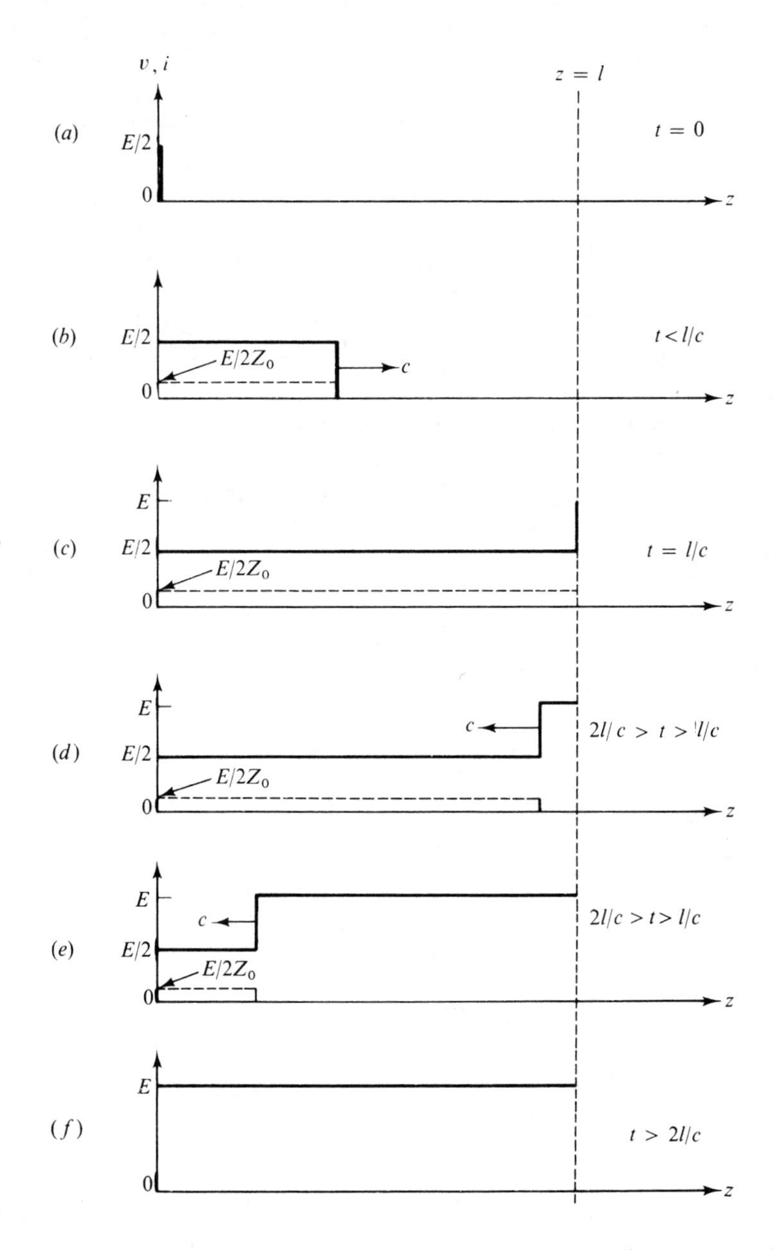

Fig. 5.15. Spatial distribution of voltage and current on a line at successive times. (*a*), (*b*) Before reflection. (*c*) At reflection. (*d*), (*e*) After reflection. (*f*) Final state.

diagrams no backward wave exists on the line, so that from (5.25) and (5.26) $v = iZ_0$. The source therefore sees an impedance Z_0, and the voltage of the wave is $\frac{1}{2}E$. When the wavefront, travelling with velocity c, reaches the open-circuit, a reflection must be produced which ensures that the current at $z = l$ is identically zero. This condition is satisfied by a wave travelling backwards, also of height $\frac{1}{2}E$, for which the current will be $E/2Z_0$ in the direction of the wave, i.e. $-E/2Z_0$. The situation then develops as in figs. 5.15(*d*) and (*e*). The wavefront reaching the source will not give a reflection since the line is there terminated in Z_0. The current then falls to zero and a steady state is reached. If the source had been other than Z_0 in impedance a continually decreasing sequence of reflections would have occurred. We note that at the open-circuit the reflection coefficient (5.18) is unity, corresponding to a reflected wave equal in amplitude and sign to the incident wave. The voltage and current waveforms as functions of time at different points are illustrated in fig. 5.16. This process may be conveniently represented on a lattice diagram, as shown in fig. 5.17(*a*).

The initial surge $\frac{1}{2}E$ is shown in fig. 5.17(*a*) travelling from the input. At the termination this surge is totally reflected and travels back along

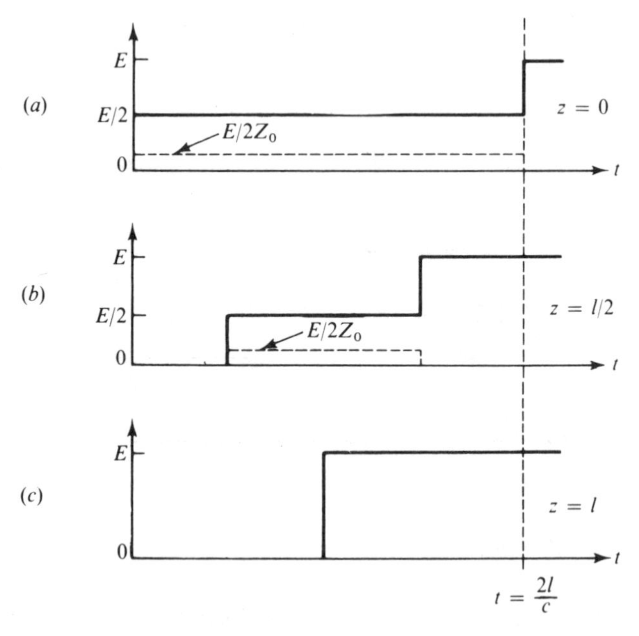

Fig. 5.16. Waveforms in line at different points. (*a*) Beginning of line. (*b*) half-way. (*c*) At load.

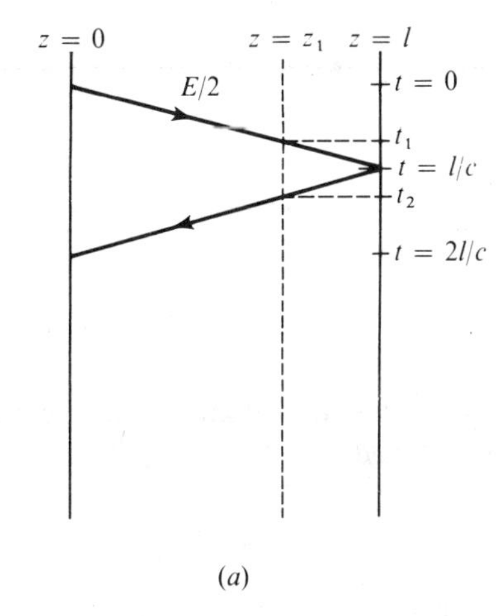

(*a*)

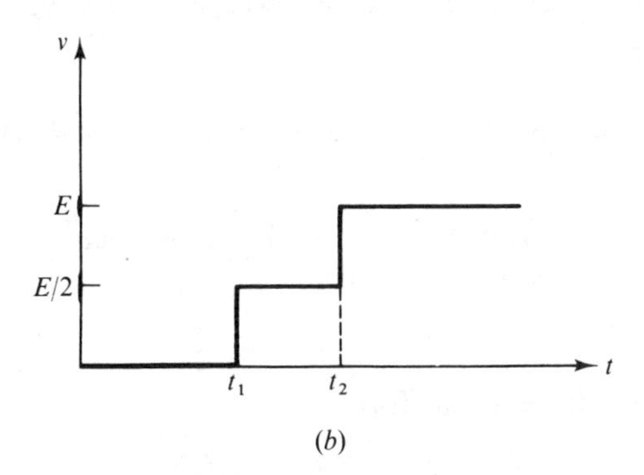

(*b*)

Fig. 5.17. Alternative method for surge calculation. (*a*) Lattice diagram. (*b*) Time waveforms.

the line to the input. Thus at a point z_1 the voltage will vary with time as shown in fig. 5.17(*b*). Since the reflected surge sees a correct termination at the input of the line, there is no further reflection, and the line remains charged to the voltage *E*.

Lattice diagrams can be used for more complicated problems on lossless lines involving multiple reflections.

Example

Sketch the lattice diagram for the circuit shown in fig. 5.18 and plot the time variation of the voltage (*a*) at the input, and (*b*) half-way along the line.

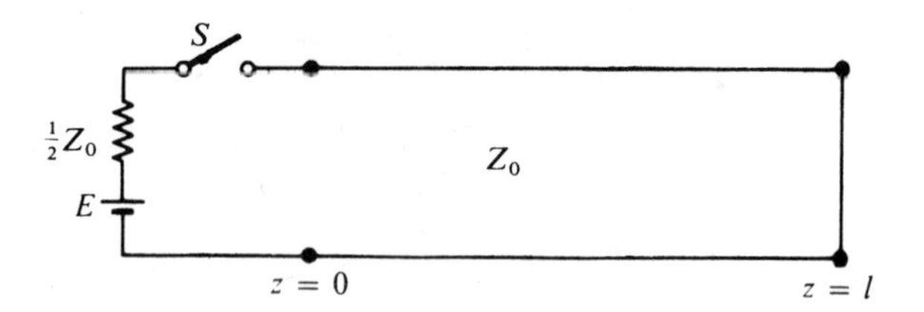

Fig. 5.18. Circuit for surge calculation.

Solution. The voltage reflection coefficient at the short-circuit is -1, and at the input, with S closed, is

$$k_v = \frac{\frac{1}{2}Z_0 - Z_0}{\frac{1}{2}Z_0 + Z_0} = -\frac{1}{3}$$

The initial surge voltage will be $EZ_0/(Z_0 + \frac{1}{2}Z_0) = 2E/3$. We can sketch the lattice diagrams shown in fig. 5.19(a).

The voltages at the input and half-way along the line are shown in Figure 5.19(b).

5.7 Reflection theory for steady-state sine waves on lossless lines

We consider a line terminated at $z = 0$ by an impedance Z_2. The voltage at any point on the line is given by (5.7).

$$\mathrm{V}(z) = \mathrm{A}\,\mathrm{e}^{-\gamma z} + \mathrm{B}\,\mathrm{e}^{\gamma z}$$

This may be written in the form

$$\mathrm{V}(z) = \mathrm{V_i}(z) + \mathrm{V_r}(z)$$

where $\mathrm{V_i}(z)$ is the incident or forward wave at the point z and $\mathrm{V_r}(z)$ is the reflected or backward wave.

We define the voltage reflection coefficient $k_v(z)$ by the equation

$$k_v(z) = \frac{\mathrm{V_r}(z)}{\mathrm{V_i}(z)} = \frac{\mathrm{B}}{\mathrm{A}}\,\mathrm{e}^{2\gamma z}$$

Using (5.18) we may write this in the form

$$k_v(z) = k_v\,\mathrm{e}^{2\gamma z} \tag{5.27}$$

where

$$k_v = k_v(0) = \frac{(Z_2 - Z_0)}{(Z_2 + Z_0)} \tag{5.28}$$

The reflection coefficient $k_v(z)$ will in general be complex, but, provided the termination Z_2 is passive ($\mathrm{Re}(Z_2) > 0$), then $|k_v| \leqslant 1$.

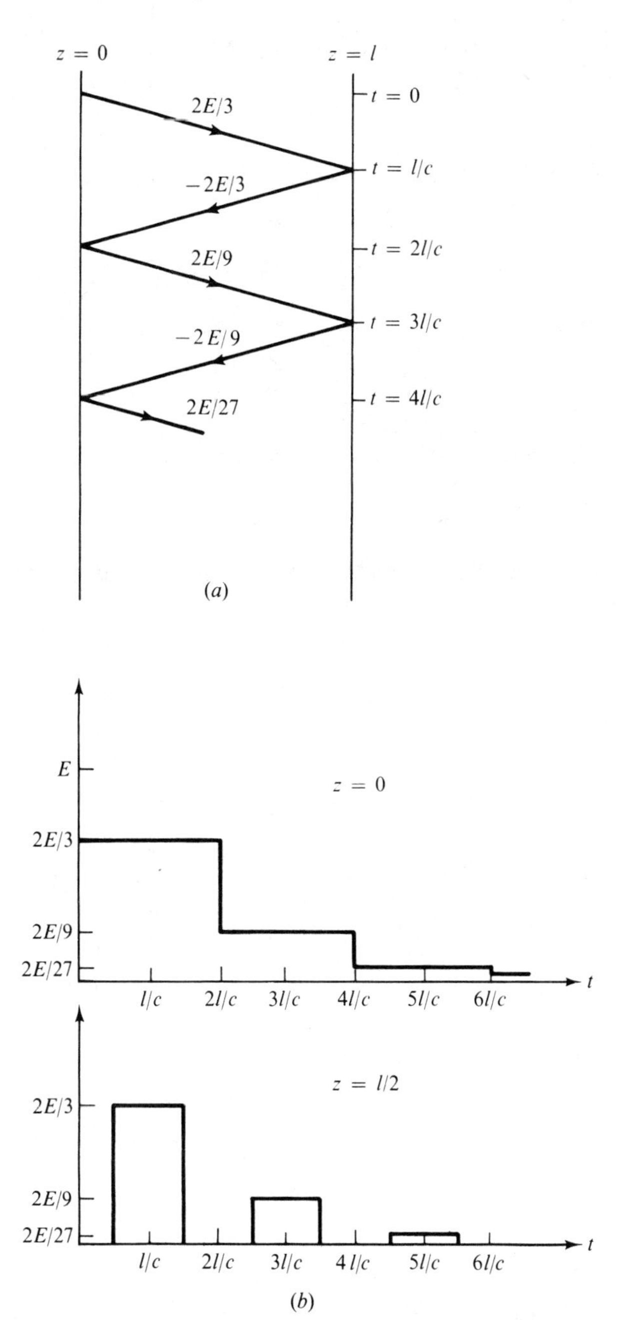

Fig. 5.19. Solutions to problem of fig. 5.18. (*a*) Lattice diagram. (*b*) Time waveforms.

Equation (5.8) for the current enables us also to write

$$\mathrm{I}(z)=\mathrm{I_i}(z)+\mathrm{I_r}(z)$$

and to define a current reflection coefficient

$$k_i(z)=\frac{\mathrm{I_r}(z)}{\mathrm{I_i}(z)}=-\frac{\mathrm{V_r}(z)}{\mathrm{V_i}(z)}=-k_v(z)$$

We note that

$$k_\mathrm{i}=k_i(0)=-\frac{Z_2-Z_0}{Z_2+Z_0}=\frac{Y_2-Y_0}{Y_2+Y_0}$$

For a lossless line we substitute $\gamma=\mathrm{j}\beta$ and then

$$\mathrm{V}(z)=\mathrm{A\,e}^{-\mathrm{j}\beta z}(1+k_v\,\mathrm{e}^{-2\mathrm{j}\beta z})$$

where k_v is given by (5.28). Since k_v is complex we write $k_v=K\,\mathrm{e}^{\mathrm{j}\theta}$. We then have

$$\mathrm{V}(z)=\mathrm{A\,e}^{-\mathrm{j}\beta z}[1+K\,\mathrm{e}^{\mathrm{j}(\theta+2\beta z)}]$$

and, similarly, we can show that

$$\mathrm{I}(z)=Z_0^{-1}\,\mathrm{A\,e}^{-\mathrm{j}\beta z}[1-K\,\mathrm{e}^{\mathrm{j}(\theta+2\beta z)}]$$

Taking the moduli of these expressions, as in (5.11), we find

$$|\mathrm{V}(z)|=\mathrm{A}[1+K^2+2K\cos(2\beta z+\theta)]^{\frac{1}{2}} \tag{5.29}$$

$$|\mathrm{I}(z)|=Z_0^{-1}\,\mathrm{A}[1+K^2-2K\cos(2\beta z+\theta)]^{\frac{1}{2}} \tag{5.30}$$

From (5.29) it can be seen that the magnitude of $\mathrm{V}(z)$ is a maximum when $2\beta z+\theta=0,\ \pm2\pi,\ldots,$ and a minimum when $2\beta z+\theta=\pm\pi,\ \pm3\pi,\ldots$

We now define the *voltage standing-wave ratio* (VSWR) by

$$S=\frac{\mathrm{V_{max}}}{\mathrm{V_{min}}}$$

$$=\frac{1+K}{1-K} \tag{5.31}$$

Inspection of (5.29) and (5.30) shows that at a point for which $|\mathrm{V}(z)|$ is a maximum, $|\mathrm{I}(z)|$ is a minimum and $\mathrm{V}(z)$ and $\mathrm{I}(z)$ are in phase with each other. The impedance at this point, Z_1, defined as the input impedance of the line to the right of the point z and given by $\mathrm{V}(z)/\mathrm{I}(z)$, is hence purely resistive. The value is given by

$$Z_1 = \frac{[V(z)]_{\max}}{[I(z)]_{\min}} = Z_0 \frac{1+K}{1-K}$$
$$= SZ_0 \tag{5.32}$$

Similarly for points at which $|V(z)|$ is a minimum we may show

$$Z_1 = Z_0/S \tag{5.33}$$

Example
A length of line is terminated in an impedance equal to $Z_0(1.5+0.5j)$. Find the reflection coefficient at the termination, the value of the VSWR, the positions of maximum and minimum voltage and the value of the impedance at these points.

Solution. We have $Z_2 = (1.5+0.5j)Z_0$, hence

$$k_v = \frac{Z_2 - Z_0}{Z_2 + Z_0} = \frac{0.5+0.5j}{2.5+0.5j}$$
$$= \frac{0.707\angle 45°}{2.54\angle 11.3°} = 0.28\angle 33.7°$$

Therefore $K = 0.28$, $\theta = 33.7°$ and the VSWR

$$S = \frac{1+K}{1-K} = 1.78$$

Maxima occur when $2\beta z + \theta = 0 \pm 2\pi, \ldots$ Measure l towards the source as shown in fig. 5.20, so that $z = -l$. In this case θ is positive, so

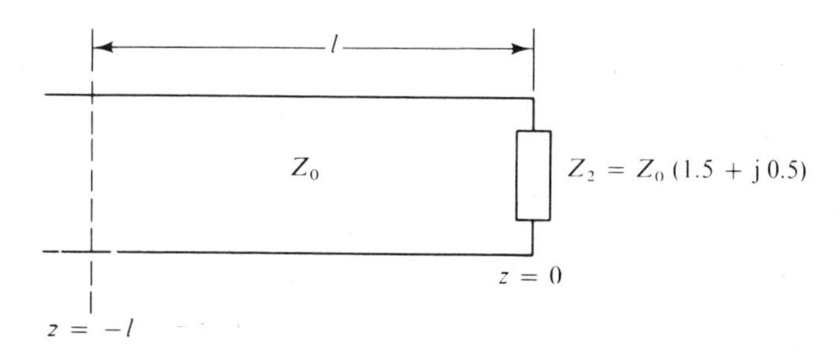

Fig. 5.20. Illustrating calculation on line terminated in $Z_0(1.5+0.5j)$.

that the first position of a maximum will be

$$-2\beta l + \theta = 0$$

$$l = \frac{\theta}{2\beta} = \lambda \frac{\theta}{4\pi} = 0.047\lambda$$

The first minimum will be given by

$$-2\beta l + \theta = -\pi$$

$$l = \frac{\theta}{2\beta} + \frac{\pi}{2\beta} = \lambda\left(\frac{\theta}{4\pi} + \frac{1}{4}\right) = 0.297\lambda$$

The value of Z_1 will be respectively

$$Z_0 S = 1.78 Z_0$$

$$Z_0 / S = 0.56 Z_0$$

The variation of Z_1 with position can be illustrated graphically as shown in fig. 5.21. The axes represent the real and imaginary parts of the normalised impedance Z_1/Z_0. With the angles and distance in the diagram, the points P, Q represent the complex numbers $1 + k_v(z)$, $1 - k_v(z)$ for the point $z = -l$. As l varies, P and Q describe a circle of

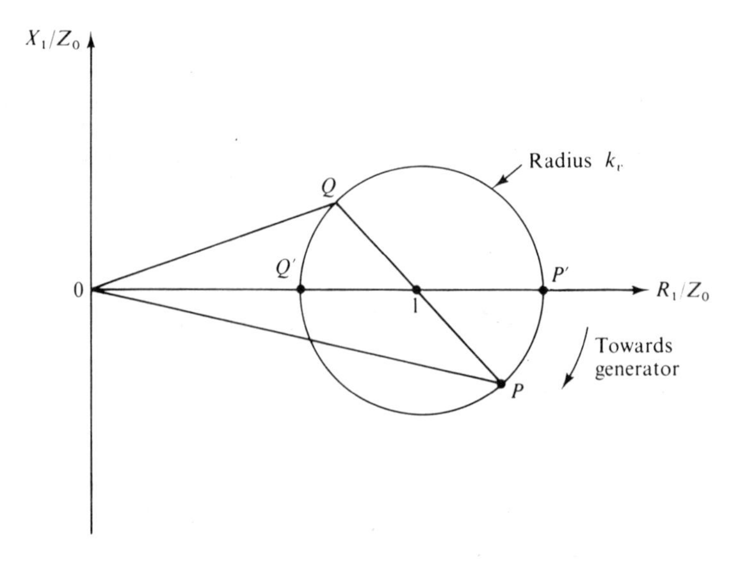

Fig. 5.21. Graphical construction for impedance calculations.

radius $|k_v|$. The diagram shows the previous result, that Z_1/Z_0 is resistive at the points P, Q and of value S or S^{-1}. This form of graphical illustration can be developed, but it is less useful than an alternative polar form in which the complete range of $|Z_1|$ from zero to infinity, $\mathrm{Re}\,(Z_1)>0$, is contained in a limited area.

5.7.1 Derivation of a polar impedance/admittance diagram: the Smith chart

It will be obvious from the earlier parts of this chapter that all expressions contain impedances normalised to the characteristic impedance, for example, (5.20) relates Z_1/Z_0 to Z_2/Z_0, and (5.28) expresses k_v in terms of Z_2/Z_0. We shall presume that this normalisation is carried out, and of course for the lossless line being considered Z_0 is real.

We consider a length of line terminated in a normalised impedance $\mathfrak{z}_2$. The normalised impedance $\mathfrak{z}$ presented by the line at some other point is equal to

$$\mathfrak{z}_2=\frac{\mathrm{V_i}(z)+\mathrm{V_r}(z)}{\mathrm{V_i}(z)-\mathrm{V_r}(z)}=\frac{1+k_v(z)}{1-k_v(z)} \tag{5.34}$$

Equation (5.34) applies also at the point of termination when

$$\mathfrak{z}=\frac{1+k_v}{1-k_v} \tag{5.35}$$

from which is derived

$$k_v=\frac{\mathfrak{z}_2-1}{\mathfrak{z}_2+1}$$

k_v and $k_v(z)$ are related by (5.27) in the form

$$k_v(z)=k_v\,\mathrm{e}^{2\mathrm{j}\beta z} \tag{5.36}$$

We are thus concerned with the relation

$$\mathfrak{z}=\frac{1+k}{1-k} \tag{5.37}$$

where $\mathfrak{z}=r+\mathrm{j}x$, $k=u+\mathrm{j}v$. Equation (5.36) shows that the variations in $k_v(z)$ in which we are interested are simply an increase (or decrease) of argument, easily accomplished if we adopt a polar plot for k, or Cartesian axes in u, v. To connect $\mathfrak{z}$ with k, we find the relation between u and v for which either r or x can have a prescribed value.

From (5.37) we have

$$r+\mathrm{j}x=\frac{1+(u+\mathrm{j}v)}{1-(u+\mathrm{j}v)}$$

$$=\frac{(1+u+\mathrm{j}v)(1-u+\mathrm{j}v)}{(1-u)^2+v^2}$$

$$=\frac{(1+u)(1-u)-v^2+\mathrm{j}2v}{(1-u)^2+v^2} \tag{5.38}$$

Equating real parts gives

$$r=\frac{1-u^2-v^2}{(1-u)^2+v^2}$$

This can be rearranged in the form

$$\left(u-\frac{r}{1+r}\right)^2+v^2=\frac{1}{(1+r)^2}$$

For a given r, this is the equation of a circle in the complex k-plane, centre $(r/(1+r), 0)$ radius $1/(1+r)$. We can draw a family of loci corresponding to different constant normalised resistances, as shown in fig. 5.22(a). When $r=0$, the radius is unity, and the centre is at the origin. As r tends to infinity, the radius decreases to zero and the centre moves towards the point $u=1$, $v=0$. Thus, within a circle of unit radius we can encompass the range $0<r<\infty$. It will be noticed that all the circles pass through the point (1, 0). Similar rearrangement of the imaginary part of (5.38) yields for a given value of x, the normalised reactance, the equation

$$(u-1)^2+(v-x^{-1})^2=x^{-2}$$

in which x is to be regarded as constant. These loci also form a set of circles in the k-plane with centre $(1, x^{-1})$ and radius $|x|^{-1}$. All circles pass through (1, 0). Figure 5.22(a) shows typical r and x loci. The sets of circles are orthogonal, and the whole range of possible values of r and x are contained within the unit circle.

This graphical representation makes possible a rapid transference between $\mathfrak{z}_1$ and $k_v(z)$ in (5.34) or (5.35): given a value say for $\mathfrak{z}_2$, the intersection of the circles corresponding to r_2 and x_2 locates the point representing k_v. If necessary this can be read off either in polar or Cartesian form. More often, it is $k_v(z)$ at some other point which is required. This can be found by modifying the argument of k_v in accordance with (5.36) by moving round a circle of radius $|k_v|$. The value of

(5.34) can then be found by reading off the circles passing through the point. The alteration in argument of $k_v(z)$ derives from a change in position along the transmission line. One complete revolution about the origin corresponds to a change of 2π in $2\beta z$, or a change of $\frac{1}{2}\lambda$ in z. It is therefore convenient, as in fig. 5.22(*a*), to mark the angles in fractions of a wavelength as well as in degrees. This process is best illustrated by examples, but before proceeding to these it must be noted that the representations given in terms of $\mathfrak{z}$ and k_v are equally applicable to an admittance interpretation.

Consider for example (5.34)

$$\mathfrak{z}_1 = \frac{1+k_v(z)}{1-k_v(z)}$$

The normalised admittance seen at the point z is

$$\mathscr{y}_1 = \frac{Y_1}{Y_0} = \frac{1}{\mathfrak{z}_1} = \frac{1-k_v(z)}{1+k_v(z)} = \frac{1+k_i(z)}{1-k_i(z)}$$

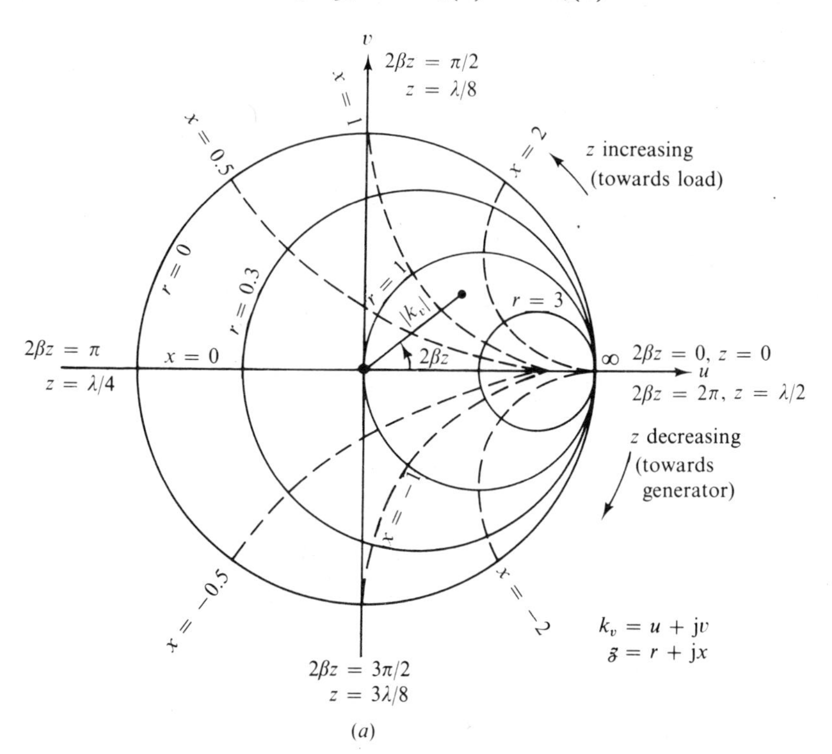

Fig. 5.22. Graphical form for relation between impedance and reflection coefficient. (*a*) Impedance form.

where $k_i(z)$ is the current reflection coefficient. There is thus a complete identity between the pairs of variables ($\mathscr{z}$, k_v) and ($\mathscr{y}$, k_i). We may use the graphical representation derived in the last section for either. Formally the revised labelling will be as shown in fig. 5.22(*b*). The Cartesian axes now represent real and imaginary parts of the current reflection coefficient rather than of the voltage reflection coefficient. It is useful to note that the centre of the chart, where $k = 0$, corresponds to a normalised impedance or admittance of unity. The circle $g = 1$ on the diagram has a special significance in that any point on this circle corresponds to an admittance equal to the characteristic admittance Y_0 in parallel with a susceptance. For use in problem solving a much more complete set of curvęs than that shown in fig, 5.22 is required. A more representative version is shown in fig. 5.23, and the examples in the remainder of this chapter are done using a chart of this variety.

Examples

1. Find the length of short-circuited lossless line which presents a normalised reactance of (a)+2j, (b)−0.5j at the input end.

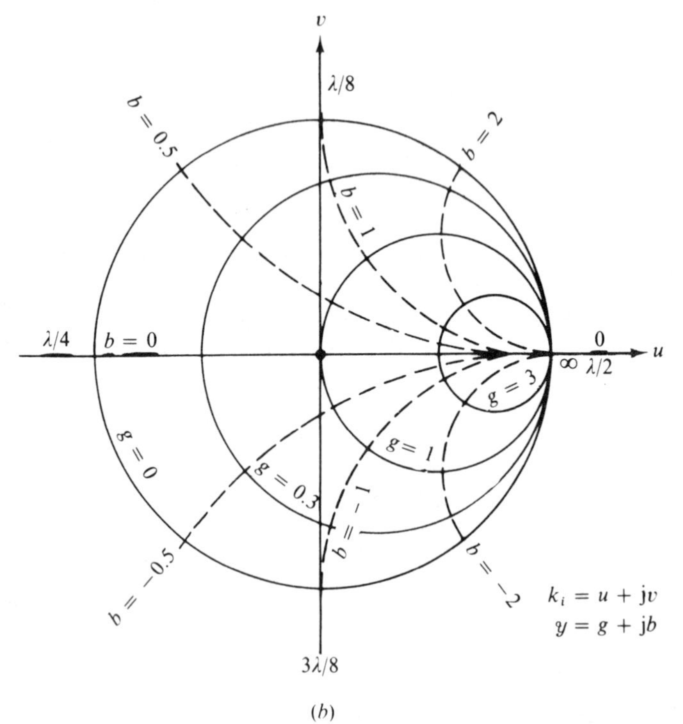

(*b*)

Fig. 5.22 (*contd.*) (*b*) Admittance form.

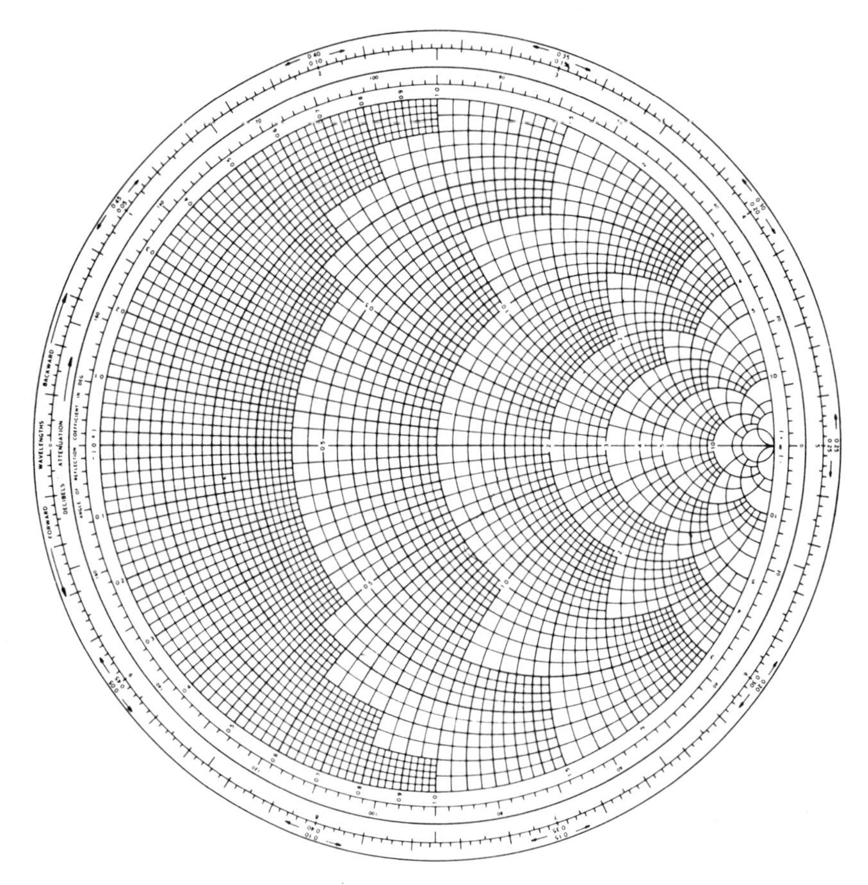

Fig. 5.23. The Smith chart.

Solution. (*a*) Since the line is terminated as shown in fig. 5.24(*a*) in a short-circuit for which $r=0$, $x=0$, we first locate the corresponding point A in fig. 5.24(*b*). The reflection coefficient can be read off as $k_v = 1\angle 180° = -1$. At any other value of z, $k_v(z) = k_v \exp(2j\beta z)$. Since we move back towards the source, z is negative. A distance l towards the source will *decrease* the argument of k_v by $2\beta l$, corresponding to a clockwise rotation about the origin. The point corresponding to a reactance of 2j is the point B, and the angle required is $\angle AOB$. We thus see that the required length is 0.176λ.

(*b*) The reactance -0.5j corresponds to C in fig. 5.24(*b*), and the angle is found to correspond to a length of 0.426λ. It will be noted that $-0.5\text{j} = (2\text{j})^{-1}$ so that the impedances are reciprocal: -0.5j is the impedance that would be presented by a line $\frac{1}{4}\lambda$ long terminated in 2j. Thus 0.426λ should be $(0.176+0.25)\lambda$.

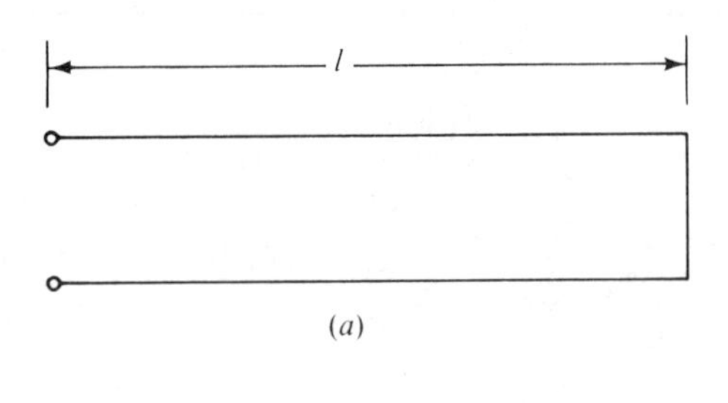

(a)

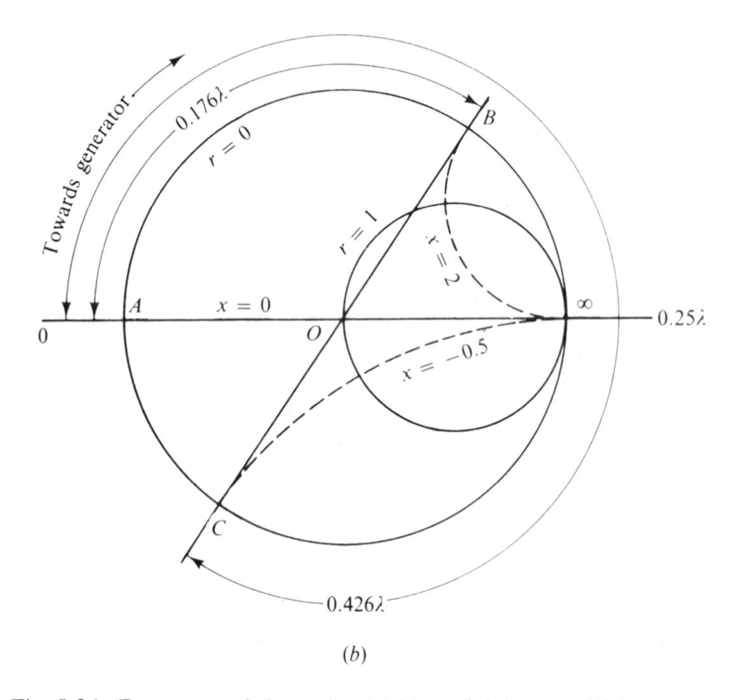

(b)

Fig. 5.24. Reactance of short-circuited line. (a) Circuit. (b) Smith chart.

2. Find the length of short-circuited line that presents a normalised admittance of 2j at the input.

Solution. Referring to fig. 5.25, the short-circuit is $g = \infty$, the point A. As before we turn clockwise until the point B for which $b = 2$ is reached. The length is $(0.25 + 0.176)\lambda$. This agrees with the previous example, since a normalised admittance of 2j is the same as a normalised impedance of $(2\text{j})^{-1}$.

3. An air-spaced coaxial line 2 m long has a characteristic impedance of 50 Ω.

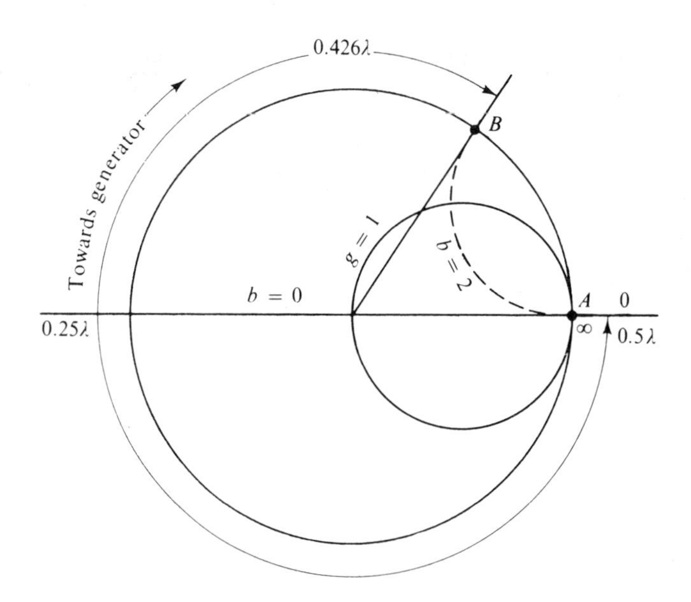

Fig. 5.25. Smith chart for calculation of admittance of short-circuited line.

It is supplied from a source of frequency 100 MHz. The VSWR with a particular termination is found to be 2.5, and a voltage minimum occurs 50 cm before the termination. Determine the impedance of the termination and the input impedance to the line. Express these also as admittances.

Solution (*a*) At 100 MHz in air, $\lambda = 3$ m. The minimum is therefore $\frac{1}{6}\lambda$ away from the termination, as shown in fig. 5.26(*a*). At this minimum the impedance is resistive of normalised value $(2.5)^{-1} = 0.4$. This minimum is thus located at the point A in fig. 5.26(*b*). The load is found by following from A a circular path in an *anticlockwise* direction (towards the termination, or forward) through $\frac{1}{6}\lambda$, to point B. The normalised impedance at B is then read off to be $1.06 - \mathrm{j}1.0$, or $Z_2 = (53 - \mathrm{j}50)\Omega$. Since the input is $\frac{1}{2}\lambda$ from the minimum, the normalised input impedance is also 0.4, or $Z_1 = 20\ \Omega$.

The admittances may be found by inverting the impedances, or by using the fact that inversion on the diagram corresponds to the diametrically opposite point, or by starting from an admittance basis. The latter method is portrayed in fig. 5.26(*c*). The voltage minimum corresponds to a normalised admittance of 2.5, point A' in fig. 5.26(*c*). Moving anticlockwise through $\frac{1}{6}\lambda$ we arrive at B', which is diametrically opposite to B in fig. 5.26(*b*).

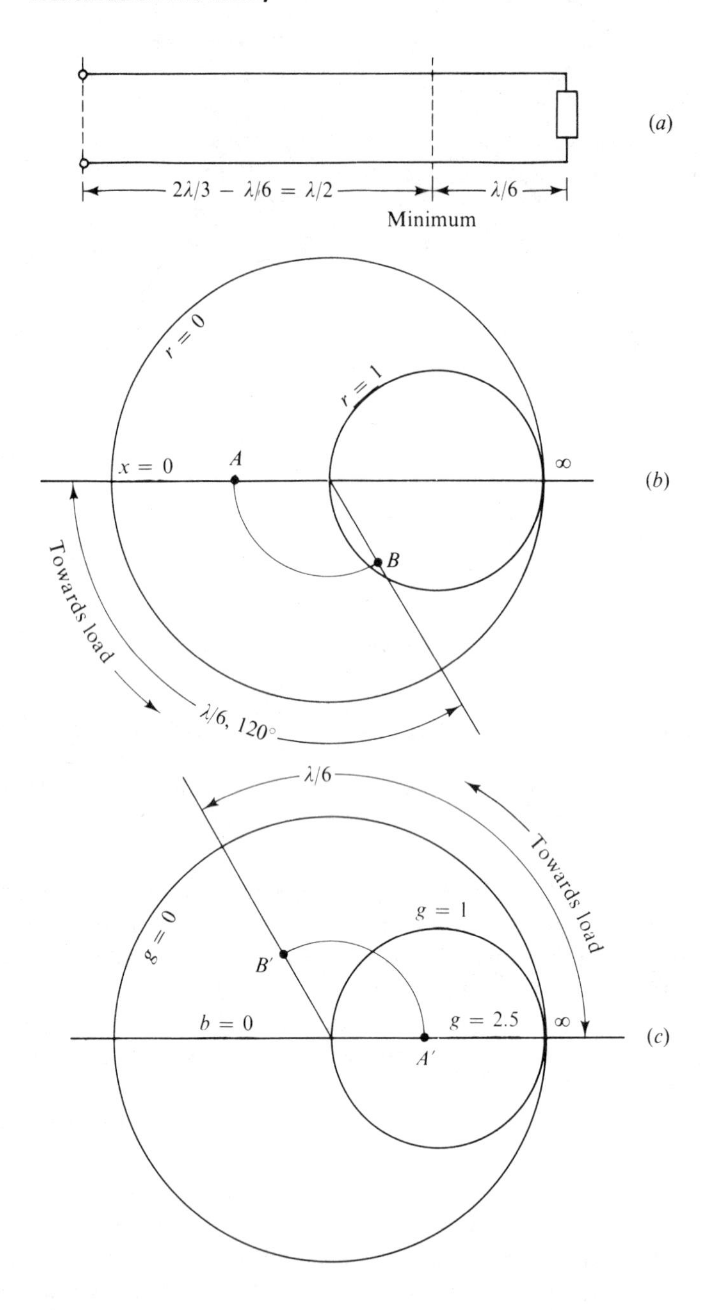

Fig. 5.26. Calculation of impedance from standing-wave data. (*a*) Transmission-line circuit. (*b*) Smith chart for impedance calculation. (*c*) Smith chart for admittance calculation.

5.8 Impedance matching

Impedance matching between a source and a load is normally done to achieve maximum power transfer, and in circuit terms occurs when the load impedance is the complex conjugate of the source impedance. On transmission lines the condition is normally more restrictive, in that the aim is to make both source and load have impedances equal to the characteristic impedance of the line, thus giving a VSWR of unity over most of the line. This has the following advantages.

(*a*) In high-power applications, such as radio transmitters, voltage and current maxima can cause breakdown due to discharge or excessive local heating respectively.

(*b*) Maximum power is transferred since we are assuming lossless conditions on the line.

(*c*) Reflections are minimised, which is particularly important when transmitting digital signals.

The way in which impedance matching is performed is by the insertion of a short length (or lengths) of line in between the given load and the main transmission line. The simplest example is perhaps the quarter-wave transformer, for which it was shown earlier that input and terminating impedances are related by $Z_1Z_2 = Z_0^2$. A line of impedance Z_{01} can be matched to a load of resistive impedance Z_2 by the insertion of a quarter-wave of line of characteristic impedance Z_{02} such that $Z_{01}Z_2 = Z_{02}^2$.

Since in general both resistance and reactance have to be modified, two variables are usually necessary. It has been discussed earlier how reactive elements can be made from lengths of short-circuited line. The shorts can be made variable in position with a sliding plunger, so that variable reactive components are readily obtainable. Such 'stubs' placed across a line form a useful means of providing the variables required.

5.8.1 Single stub, variable position, matching

The configuration is shown in fig. 5.27(*a*), in which both the distance l and the parallel susceptance at P are regarded as variable. If the line to the left of P is to be terminated in Z_0, then, since a variable susceptance is available in parallel at P, the line just to the right of P must look like an admittance of Y_0 in parallel with any susceptance. We thus choose l so that the load presents at P an admittance of this form.

Example

The termination of a line has a normalised admittance of $2+0.5\text{j}$. Show that it is

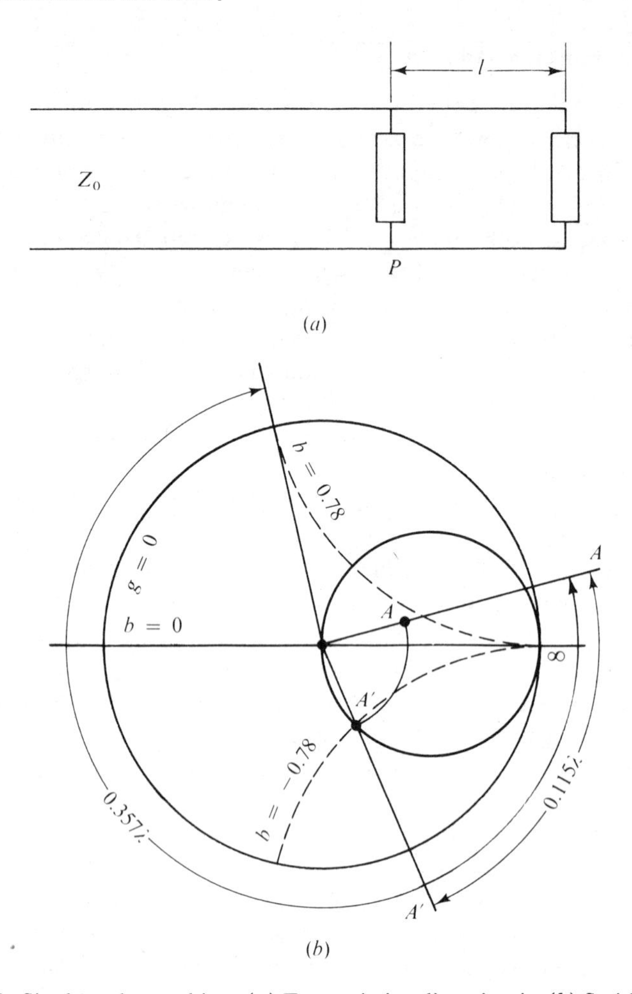

Fig. 5.27. Single stub matching. (*a*) Transmission-line circuit. (*b*) Smith chart.

possible to match this to the line by means of a suitable susceptance placed across the line at a suitable point near the termination.

Solution. The point A in fig. 5.27(*b*) corresponds to the load $2+0.5\text{j}$. By moving clockwise (towards generator) through 0.115λ the normalised admittance becomes $(1-\text{j}0.78)$ at A' so that a susceptance $+0.78$ connected across the line will then present to the left of P a normalised admittance equal to unity. The susceptance can be arranged from a length of short-circuited line of appropriate length. By a subsidiary calculation on the chart this is found to be 0.357λ.

Matching is therefore achieved by placing a short-circuited stub of length 0.357λ at a distance 0.115λ from the load.

5.8.2 Two-stub matching

The method of the last section requires that the position and length of the stubs be variable, depending on load and operating frequency. Changing the position of the stub is particularly inconvenient. Matching with two variable stubs fixed in position is an alternative, although not all loads can be matched. The configuration is shown in fig. 5.28(*a*).

The variable susceptances are at P, Q. The distance between them has to be fixed, and is often taken as $\frac{1}{8}\lambda$. It is convenient to distinguish the locations either side of the variables, shown on the diagram by the addition of + or − as a suffix. At P_- the admittance has to be unity. At P_+ therefore it can be $(1+\mathrm{j}b_1)$, since a parallel susceptance of $-b_1$ can be obtained from the stub. The locus of possible admittance at P_+ is shown by the locus C_1, fig. 5.28(*b*). At Q_- the admittance is that at P_+ transformed forward by, in this case, $\frac{1}{8}\lambda$. Thus the locus C_1 is moved anticlockwise through $\frac{1}{8}\lambda$ to become C_2. Any point on this locus can be matched by variation of the stub at P. Now touching C_2 is a circle of constant conductance, C_3, which for the present numerical distance has the value $g=2$. Any point not inside the circle $g=2$ can by the addition or subtraction of susceptance be brought onto the circle C_2. Hence any load can be matched providing that when transformed to Q through the distance QR it does not come within the circle $g=2$. Alternatively, given QR we may transform the circle $g=2$ towards the load, when the load may not lie within the transformed circle. To be definite let us take QR to be 0.2λ. The circle $g=2$, transformed 0.2λ towards the load (anticlockwise), becomes the circle C_4. Loads of admittance within this circle cannot be matched.

In practice the variation of load will be limited and the distance can be chosen to make sure that matching is possible over the required range.

As an example consider the situation just depicted (a separation of 0.125λ between P and Q, and 0.2λ between Q and R) with a load of admittance $(2-0.3\mathrm{j})$. This is the point A in fig. 5.28(*c*), which when transformed back to Q becomes A' in the figure. This point can be brought onto C_2 by adding either positive or negative susceptance, arriving at B_1 or B_2. Transforming these back to P, they become B'_1, B'_2 on C_1, and by adding suitable susceptance they can both be brought to the centre.

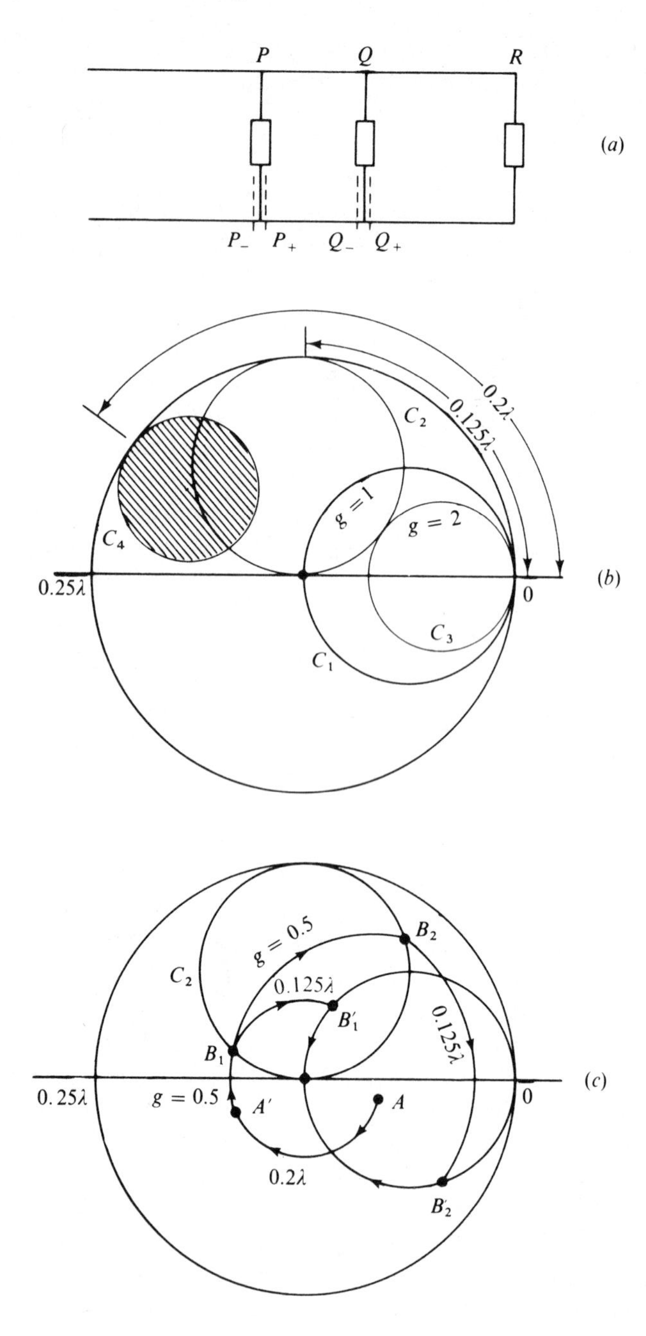

Fig. 5.28. Two stub matching. (*a*) Transmission-line circuit. (*b*) Outline Smith chart. (*c*). Smith chart for problem.

The co-ordinates of the various points are

A	$2-0.3j$
A'	$0.51-0.16j$
B_1	$0.51+0.15j$
B_2	$0.51+1.87j$
B'_1	$1.0+0.7j$
B'_2	$1.0-0.7j$

(*a*) To go from $A' \to B_1$, stub Q must add $0.31j$; to go from $B'_1 \to 0$, stub P must add $-0.7j$.

(*b*) Alternatively, to go from $A' \to B_2$, stub Q must add $2.03j$; to go from $B'_2 \to 0$, stub P must add $+0.7j$.

5.9 VSWR and power

Since any reflection gives rise to a standing wave, we would expect a relationship to exist between $|k_v|$, or S, and the efficiency of transfer of power to the load.

The power delivered to a load is equal to Re (VI*). This is given by the expression

$$\begin{aligned}
&\mathrm{Re}\,[Z_0^{-1}(\mathrm{A\,e}^{-\mathrm{j}\beta z}+\mathrm{B\,e}^{\mathrm{j}\beta z})(\mathrm{A^*\,e}^{\mathrm{j}\beta z}-\mathrm{B^*\,e}^{-\mathrm{j}\beta z})]\\
&\quad=\mathrm{Re}\,[Z_0^{-1}(\mathrm{AA^*}-\mathrm{BB^*}+\mathrm{A^*B\,e}^{2\mathrm{j}\beta z}-\mathrm{AB^*\,e}^{-2\mathrm{j}\beta z})]\\
&\quad=\mathrm{Re}\,\{Z_0^{-1}[|\mathrm{A}|^2-|\mathrm{B}|^2+\mathrm{A^*B\,e}^{2\mathrm{j}\beta z}-(\mathrm{A^*B\,e}^{2\mathrm{j}\beta z})^*]\}\\
&\quad=Z_0^{-1}(|\mathrm{A}|^2-|\mathrm{B}|^2)\\
&\quad=Z_0^{-1}(|\mathrm{V_i}|^2-|\mathrm{V_r}|^2)
\end{aligned}$$

The available power from the source is $Z_0^{-1}|\mathrm{V_i}|^2$. Hence the ratio of power delivered to incident power is given by

$$P_\mathrm{l}/P_\mathrm{i}=1-(|\mathrm{V_r}|^2/|\mathrm{V_i}|^2)=1-|k_v|^2$$

This expression can be related to the voltage standing wave ratio by use of the equation

$$S=(1+|k_v|)/(1-|k_v|)$$

$$|k_v|=(S-1)/(S+1)$$

Finally therefore

$$P_\mathrm{l}/P_\mathrm{i}=4S/(S+1)^2$$

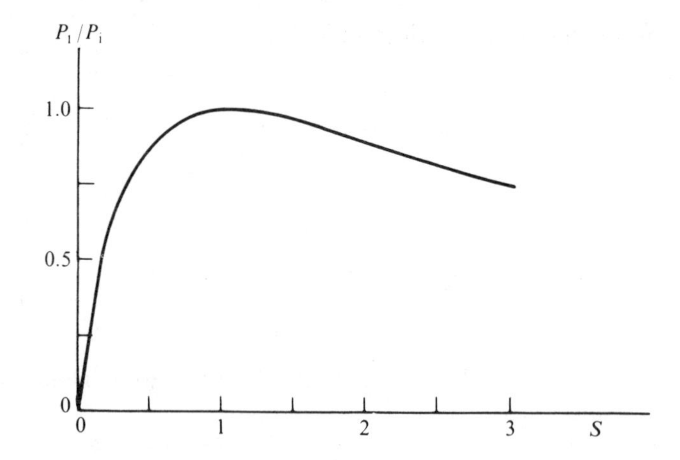

Fig. 5.29. Actual power/available power as a function of standing-wave ratio.

Figure 5.29 shows this relationship graphically. It will be noticed that the power reflected is small for small S. A VSWR of 1.1 for example, gives $P_l/P_i = 0.998$, so that from the point of view of maximum power transfer it is not necessary to strive practically for VSWR very near unity. A VSWR much different from unity indicates that the line will be sensitive to changes in frequency as the electrical length changes. This is a situation to be avoided as far as possible.

5.10 Worked example

A lossless coaxial line of characteristic impedance 100 Ω is connected to a load of admittance $(8-j7)10^{-3}$ S. Find:

(*a*) the VSWR and position of the voltage maximum or minimum nearest to the load;

(*b*) the characteristic impedance and position of a suitable quarter-wave matching transformer;

(*c*) the position and length of a short-circuited matching stub;

(*d*) a suitable two stub matching arrangement.

Solution. (*a*) The normalised admittance y is equal to $0.8-j0.7$. This is marked as the point A in fig. 5.30(*a*). Transforming A backwards (towards the source) by moving on a circle centred on the origin in a clockwise direction, a purely conductive admittance of 0.45 will be reached at A_1 after a distance $(0.5-0.368)\lambda = 0.132\lambda$, or at A_2 by going a distance of $(0.25+0.132)\lambda$, an admittance of 2.2. The VSWR is therefore given by

$$S = 2.2\ (=1/0.45)$$

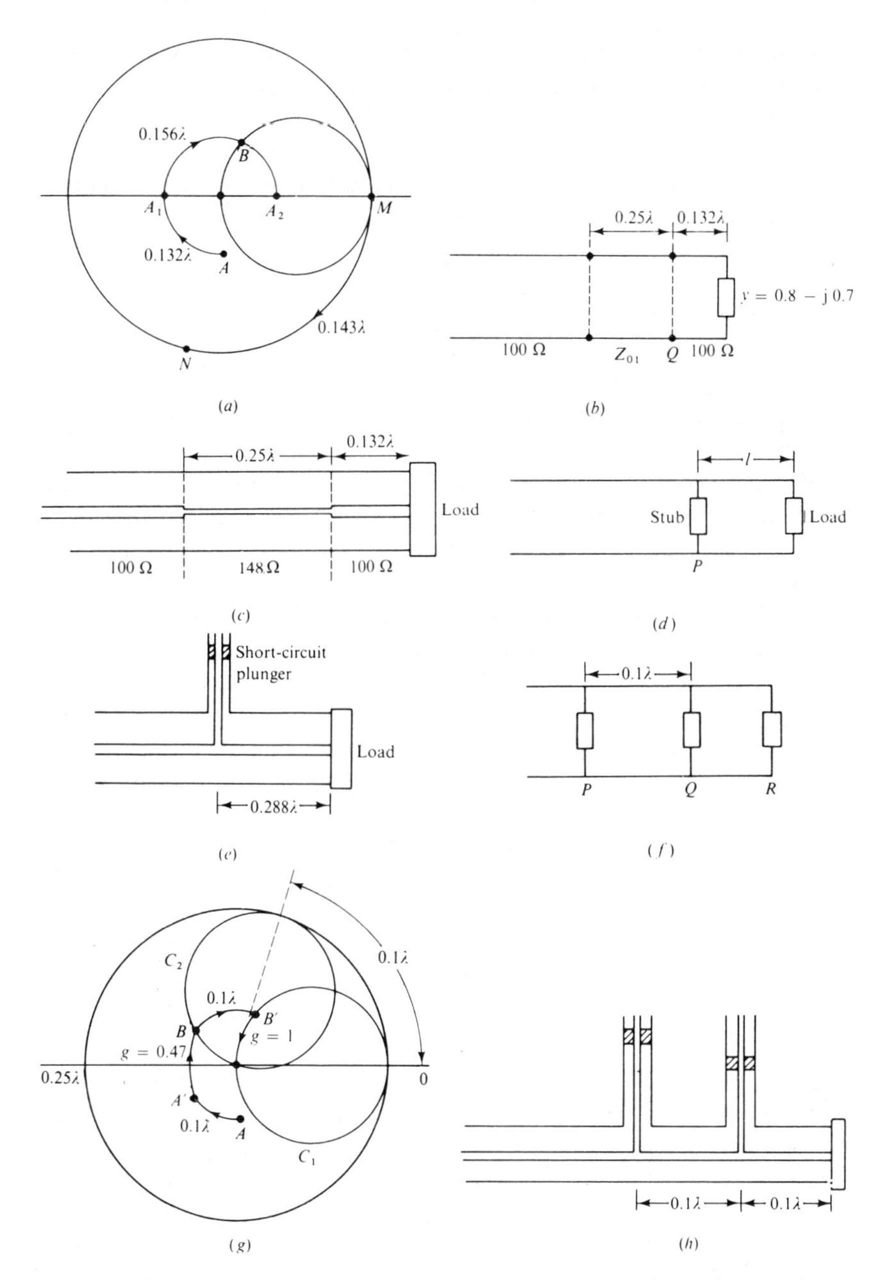

Fig. 5.30. The problem of 5.10. (*a*) Smith chart for VSWR determination. (*b*) Matching by quarter-wave transformer. (*c*) Design of coaxial transformer. (*d*) Single stub matching. (*e*) Design for single stub matching. (*f*) Two stub matching. (*g*) Smith chart for two stub matching. (*h*) Design for two stub matching.

The nearest maximum or minimum is at 0.132λ from the load, when the admittance is 0.45 and the voltage is at a maximum.

(b) Consider the configuration of fig. 5.30(b). At the point Q a purely resistive impedance of $100/0.45 = 220\ \Omega$ is presented by the load and its line. This can be matched to the required $100\ \Omega$ by a $\frac{1}{4}\lambda$ length of different line of characteristic impedance Z_{01} such that

$$100 = Z_{01}^2/220$$

This gives $Z_{01} = 148\ \Omega$.

In coaxial geometry such a line would be constructed by correctly choosing the diameters, giving a structure such as that shown in fig. 5.30(c). Depending on the wavelength, suitable dielectric spacers might be necessary.

Matching could also be achieved by using the longer length on the load, to give an impedance of $45\ \Omega$ (A_2). This would probably be undesirable since the longer length of line would be more frequency-sensitive.

(c) We now consider the configuration of fig. 5.30(d). The length l is to be chosen so that the transformed load admittance at P is of the form $(1 + jb_1)$. From fig. 5.30(a) the shortest distance possible corresponds to the point B, $y = 1.0 + j0.8$, requiring $l = (0.132 + 0.156)\lambda = 0.288\lambda$. The susceptance to be added at P is -0.8. To find the length of short-circuit line needed to give this susceptance we have to transform the point $M(y = \infty)$ to the point N on fig. 5.30(a), giving $(0.393 - 0.25)\lambda = 0.143\lambda$. A cross-check on this figure is that, since an inductive susceptance is required, the length of short-circuit line must be less than one quarter-wavelength.

A possible structure is shown in fig. 5.30(e). The adjustable stub will not be precisely of the length calculated, since the perturbing effects of the junction have to be included. This will usually be equivalent to a small extra parallel capacitance.

(d) We now consider the configuration of fig. 5.30(f). We will arbitrarily select the distance PQ as 0.1λ, and then see if QR can be chosen to allow matching. This is illustrated in fig. 5.30(g). As discussed in an earlier example, the admittances which can be matched at P lie on the circle C_1 which, when transformed forwards (anticlockwise) by 0.1λ to Q, gives the locus C_2. Any admittance with a conductance less than about 3 can be brought onto C_2 by susceptance added at Q. Thus any convenient length may be chosen for QR which does not transform A to within the circle $g = 3$. Assuming 0.1λ is possible, A transforms (backwards, clockwise) to A', $y = 0.47 - j0.16$. An added susceptance of 0.48

at Q will bring A' to B on C_2, where $y = 0.47 + j0.32$. B will then transform to B' on C_1, $y = 1.0 + j0.92$. At P therefore the added susceptance requires to be -0.92. Finally therefore we have the configuration of fig. 5.30(f) with $PQ = QR = 0.1\lambda$, susceptance of stub at Q equal to $+0.48$, susceptance of stub at $P - 0.92$. Lengths for these stubs can be worked out as before. A possible structure is shown in fig. 5.39(h).

5.11 Summary

The consequences of having both forward and backward waves have been investigated. The impedance-transforming properties of a length of line have been determined. It has been shown that resonant systems can be constructed from lengths of transmission line, and can be given equivalent circuits.

The voltage standing-wave ratio (VSWR) has been defined and shown to be related to the relative magnitude of the reflected wave and to the terminating impedance. The use of VSWR to measure impedances has been illustrated. A graphical method for handling transmission-line problems has been derived.

The importance of obtaining the matched condition has been discussed, and various ways of achieving this condition demonstrated.

It has been shown that transients propagate without distortion on a lossless line.

Formulae

Waves on a transmission line

$$V(z) = A\,e^{-\gamma z} + B\,e^{\gamma z}$$

$$I(z) = Z_0^{-1}(A\,e^{-\gamma z} - B\,e^{\gamma z})$$

$$\gamma^2 = (R + j\omega L)(G + j\omega C)$$

$$Z_0^2 = (R + j\omega L)/(G + j\omega C)$$

Input impedance of line terminated in Z_2

$$Z_1 = \frac{Z_2 + Z_0 \tanh \gamma l}{Z_2 \tanh \gamma l + Z_0}$$

$$= Z_0 \frac{1 + k\,e^{-2\gamma l}}{1 - k\,e^{-2\gamma l}}$$

where $$k = \frac{Z_2 - Z_0}{Z_2 + Z_0}$$

Transmission matrix

$$\begin{pmatrix} V_1 \\ I_1 \end{pmatrix} = \begin{pmatrix} \cosh \gamma l & Z_0 \sinh \gamma l \\ Z_0^{-1} \sinh \gamma l & \cosh \gamma l \end{pmatrix} \begin{pmatrix} V_2 \\ I_2 \end{pmatrix}$$

Lowloss line $R \ll \omega L$, $G \ll \omega C$

$$\alpha \approx \frac{1}{2}\left(\frac{R}{Z_0} + GZ_0\right) \qquad \beta \approx \omega \surd(LC)$$

Transient waves on lossless line

$$v(z, t) = f(t - z/c) + g(t + z/c)$$

$$i(z, t) = \surd(C/L)\,[f(t - z/c) - g(t + z/c)]$$

Voltage standing-wave ratio $\quad S = \dfrac{1 + |k|}{1 - |k|}$

5.12 Problems

1. In a certain air-filled coaxial line having $Z_0 = 80\ \Omega$ it is found that $S = 2$ and that two adjacent voltage minima occur 0.15 m and 0.45 m from the termination. Find:
   (i) the frequency of the signal generator;
   (ii) the value of Z_T, the terminating impedance;
   (iii) the fraction of the incident power absorbed by the load.
   What can be inferred about the value of Z_T if, for the same value of S, the minima occur at different distances from the termination? Evaluate Z_T for the particular case of the first minimum occurring 0.3 m from the end.

(London University.)

2. An air-spaced lossless transmission line is terminated in a load impedance that produces a voltage reflection coefficient of $0.5\angle 60°$. The characteristic impedance of the line is $100\ \Omega$. Find the value of the load impedance, the position of the voltage maximum nearest the load, and the voltage standing-wave ratio on the line. Also determine one position and one length of a short-circuited stub, having the same characteristic impedance, which will render a matched condition between this stub and the source, if the frequency of the latter is 100 MHz.

(Bristol University, 1968.)

3. A 30 m length of loss-free air-spaced line has a characteristic impedance of $80\angle 0°\ \Omega$. It is connected at one end to a resistance of $80\ \Omega$. If the other end is connected to a resistance of $160\ \Omega$, determine the magnitude of the impedance which would be measured across the line at its mid-point at a frequency of 12.5 MHz.

4. Show that the voltage (V_x) and the current (I_x) at distance x from the input to a transmission line are given respectively by

$$V_x = A\,e^{-\gamma x} + B\,e^{+\gamma x}$$

and

$$I_x = \frac{A\,e^{-\gamma x}}{Z_0} - \frac{B\,e^{+\gamma x}}{Z_0}$$

where A and B are related to the voltage and the current at the input and Z_0 and $\gamma(=\alpha + j\beta)$ are constants of the line at the operating frequency.

A coaxial transmission line is exactly one half-wavelength long. The input impedance is $(1 + j0)\Omega$ with a short-circuit applied, and $10^4(1 + j0)\ \Omega$ with line of open-circuit. The characteristic impedance is $(100 + j0)\ \Omega$, and the supply frequency is 150 MHz. Determine the attenuation coefficient α.

(Southampton University.)

5. A lossless open-wire feeder connects a transmitter to an antenna. The operating frequency is 90 MHz, the characteristic impedance of the line is 300 Ω and the input impedance of the antenna is resistive and equal to 75 Ω.

Determine the closest point to the load at which a single short-circuited stub may be connected in order to match the line from the source to that point, and find the minimum length of stub required.

If the antenna is replaced by an adjustable resistive load, estimate the maximum and minimum load resistances that can be tolerated if the voltage standing-wave ratio on the main feeder is not to exceed 1.4 : 1, with the stub connected as previously.

(Bristol University, 1975.)

6. Two long transmission lines, each having a characteristic impedance of 400 Ω, are connected by a coaxial cable of 50 Ω characteristic impedance. If a short-duration pulse of magnitude 10 kV travels along the first line towards the junction, determine the magnitude of the first and second pulses entering the second line, stating any assumptions made.

6

Propagation in line systems

Propagation along transmission lines forms a vital part of a communication system. Investigation into current practice reveals different types of transmission lines such as open-wire pairs mounted on telegraph poles, twisted pairs buried in ducts, various forms of coaxial line both with and without frequent repeater amplifiers. On the sea bed at one extreme is the telegraph cable consisting of one conductor with a sea-water return and at the other extreme the highly sophisticated coaxial lines with repeaters designed to withstand the rigorous environment. It is the purpose of this chapter to bring out the relevance of the theory developed earlier to the understanding of the rôles played by these various forms of transmission line.

The theory has concentrated almost entirely on signals described by a simple sinusoidal variation in time. An unchanging signal, be it d.c. or a steady sine wave, does not convey information: information is conveyed by the way in which the signal alters, so that in order to understand the relevance of the characteristics developed for single sine waves it is necessary to consider the characteristics of the actual information-carrying signals which have to be handled.

It is important to realise that the way in which a simple private single-line link might be designed would be quite inappropriate to an addition to an existing communication network. The mere size of a system and the capital locked up in its previous development demand additions to be compatible with the existing system. Such additions are also measured in economic terms: existing equipment has to be used where possible, and the viability of a particular theoretical development may have to await the development of the necessary technology to a point where economic feasibility can be realised. For example, the theory of pulse code modulation was developed as long ago as 1938 but had to await the development of semiconductors before commercial realisation was possible.

6.1 Signals

Various types of signals handled by a modern communications network can be distinguished, such as speech on the telephone network, high-quality speech and music for radio transmission, or television. These are all *analogue* signals, in the sense that the actual waveform of the signal is directly related to the physical sound or brightness from which it originates. To this class we have to add *digital* signals, where the actual signals appearing on a transmission line are pulses of the same shape. The information content is determined by the presence or absence of pulses in the received sequence. The oldest example of such a coded signal is the telegraph signals of dots and dashes, more modern are teletype coding and pulse code modulation (PCM). It is to be noted that the analogue signal is determined completely by the original signal from which it comes, whereas the rate of transmission of digital signals can be as slow or as fast as the transmission system will allow. In fact the early telegraph was used on long distances rather than speech simply because it could be slowed down to a rate that the transmission lines would accommodate.

We shall now briefly consider some characteristics of these signals.

6.1.1 Frequency spectra

It is a fundamental mathematical result that, within wide limits, any function of time can be regarded as being made up of a number of sine waves, each with its appropriate amplitude, frequency and phase. For periodic waveforms this result takes the form of a Fourier series

$$f(t) = \sum_{n=0}^{\infty} (a_n \cos n\omega t + b_n \sin n\omega t)$$

$$= \sum_{-\infty}^{\infty} c_n \exp(\mathrm{j}n\omega t) \tag{6.1}$$

For a non-periodic signal of finite energy the result is given by the Fourier integral theorem, which enables us to write

$$f(t) = \int_{-\infty}^{\infty} F(\omega) \exp[\mathrm{j}\theta(\omega) + \mathrm{j}\omega t] \, \mathrm{d}\omega \tag{6.2}$$

The rules by which the coefficients c_n, or the function $F(\omega)\exp[\mathrm{j}\theta(\omega)]$, are determined from a given time function are not relevant to our considerations in this book: the important thing is that they exist. A further type of function, typical of the electrical signal present in communications systems, is one for which the signal is not of finite energy (as in speech, it may go on for ever) and which is not periodic. In such a case we find that (6.2) can be modified, with $F(\omega)$ being a statistically determined quantity.

Although in some important cases we can calculate $F(\omega)$, it is often necessary to measure it. The way in which this is done is by putting the signal through a filter whose bandwidth is narrow, say Δf, so that only frequencies in the range $f_0 \pm \frac{1}{2}\Delta f$ are transmitted. The output from this filter can be measured, and if the statistical properties of the signal are independent of time, the output averaged over a long time will give a value for $F(f)\,\Delta f$. In this way $F(f)$ may be determined over the range of frequencies. As examples of such experiments some power spectra are shown in fig. 6.1.

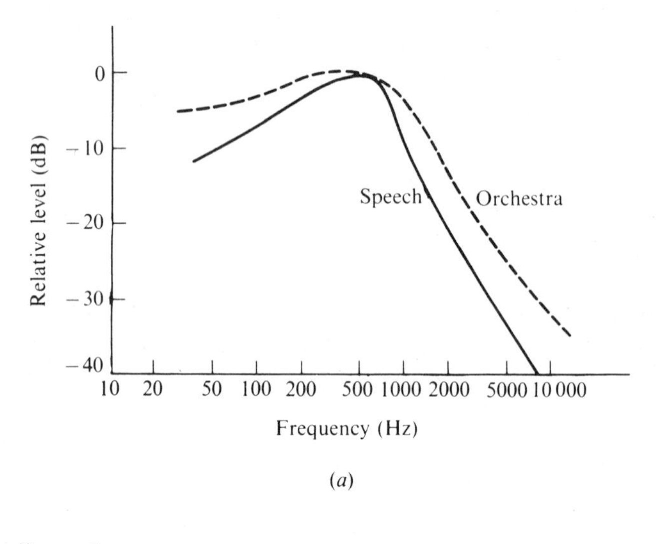

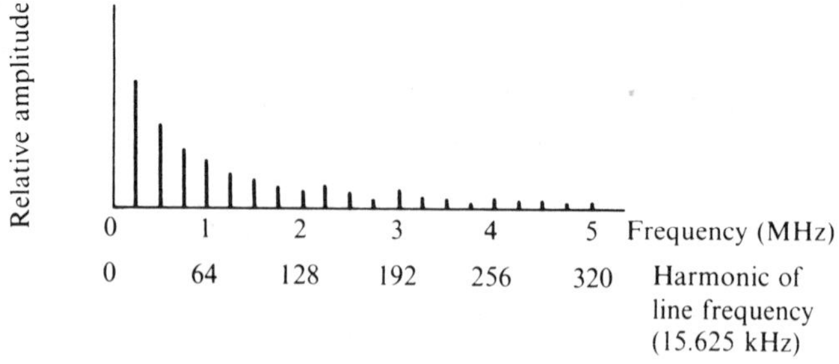

Fig. 6.1. Power spectra. (*a*) Sound. (*b*) Television.

Just as a signal can be expressed in terms of sine waves, so the response of a filter or an amplifier is expressed in terms of its response to a sine wave. Such a two-port will be characterised by its transfer function. Knowledge of the transfer function enables us to say that a sine wave of angular frequency ω will be changed in amplitude by $A(\omega)$ and its phase increased by $\phi(\omega)$. The function $A(\omega)\exp[\mathrm{j}\phi(\omega)]$ is a real function of the variable $\mathrm{j}\omega$.

We can deduce the conditions which are desirable in order that a transmission link shall pass a signal without distortion. A component $F(\omega)\exp[\mathrm{j}\theta(\omega)]\,\mathrm{d}\omega$ of the signal (equation (6.2)) will become after transmission $F(\omega)A(\omega)\exp[\mathrm{j}\theta(\omega)+\mathrm{j}\phi(\omega)]$. Apart from the condition $A(\omega)=1$, the best that can be done is to have $A(\omega)$ constant over the frequency range for which $F(\omega)$ is significant. The phase function $\phi(\omega)$ will correspond to a delay T independent of frequency if ϕ is equal to $-\omega T$. These conditions of constant amplitude and linear phase refer to the modulus and phase of a single function of $\mathrm{j}\omega$, and cannot be independently specified. However, a good approximation may often be made. Auxiliary circuits called *equalisers* are used to compensate for the inherent properties of transmission lines and filters.

6.1.2 Base-band signals

The term *base-band* is applied to the signal originating directly from the physical process via a transducer. For example, the output from a microphone picking up speech is base-band. Whatever processing a signal undergoes in transmission it is necessary to restore it eventually to base-band to feed the final transducer reconverting the electrical signal into its original form. Such a signal will usually have a frequency spectrum extending from some low frequency to a much higher one, and this range must be handled by any transmission system in order to convey the signal without distortion.

It is not necessary that a signal is kept in base-band form in its progress through a transmission system. Apart from anything else, it would not be possible to pass two such signals simultaneously over the same line. One method of transmitting many signals over the same line is by *frequency-division multiplex* (FDM).

6.1.3 Frequency-division multiplexing

It is possible to shift a signal occupying a frequency range $f_1<f<f_2$ to another range $f_1+f_c<f<f_2+f_c$ by the process of modulation. The basic

process by which this is achieved is one of multiplication: circuits can be devised which form the product of signals presented at two inputs. Consider the case when a signal represented by $a \cos \omega_m t$ is multiplied by a carrier $\cos \omega_c t$. The output from the modulator is

$$v(t) = a \cos \omega_m t \cos \omega_c t$$

$$= \tfrac{1}{2}a[\cos(\omega_c + \omega_m)t + \cos(\omega_c - \omega_m)t]$$

The first term represents the original signal with its frequency shifted up to $f_c + f_m$, the second the original signal shifted to $f_c - f_m$. These two terms are called the upper and lower sidebands respectively. A suitable filter will remove the lower sideband and leave us with each component of the original signal shifted up by f_c.

The process of recovering the base-band signal, termed *demodulation* is similarly accomplished by multiplication:

$$\tfrac{1}{2}a \cos[(\omega_c + \omega_m)t] \cos \omega_c t = \tfrac{1}{4}a[\cos \omega_m t + \cos(2\omega_c + \omega_m)t]$$

The second term represents a signal at twice the carrier frequency which is easily filtered out, leaving the original base-band signal.

The most common example of FDM in a communication system is illustrated in fig. 6.2, in which is shown the first stage in assembling telephone signals into a form suitable for junction transmission.

We note that a line carrying a large number of such channels has at any one time a signal not identifiable with any single one of the original signals. The original signals can only be recovered with the aid of suitable filters and demodulators. We note too that N base-band signals each of bandwidth B require a bandwidth NB for transmission by FDM.

6.1.4 Time-division multiplexing

An alternative way in which several signals may be transmitted along one line is by allocating successive time slots to the different signals. For an analogue signal, of necessity in 'real' time, this is only possible if the nature of signal is such that it can be reconstituted from periodic samples: restricting the signal to time slots prohibits us from transmitting the whole signal. For a digital signal it is always possible by a suitable choice of transmission rate.

It can be shown that, provided a signal has a frequency spectrum limited to the range 0 to B, it is only necessary to have $2B$ samples per second in order to be able to reconstruct the original signal. This

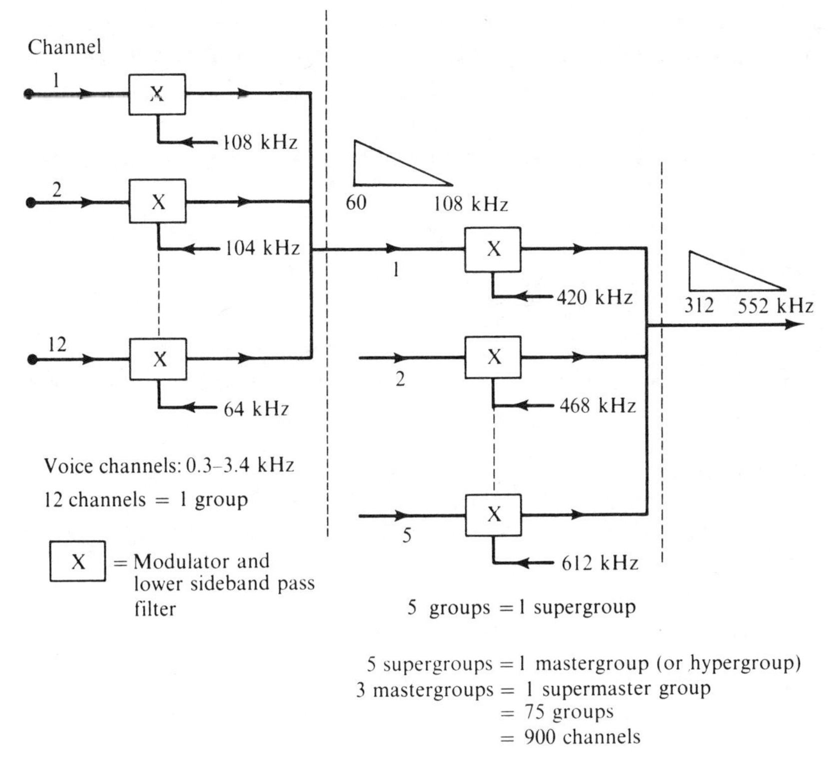

Fig. 6.2. Assembly of individual voice channels in FDM.

sampling process is usually carried out by taking the value of the signal over a very short interval at the appropriate times. Each value is coded as a binary number and transmitted in digital form. We thus have a signal consisting of similar-shaped pulses, as discussed earlier. For an example, telephonic speech is limited to an upper frequency of 3.4 kHz, and is then sampled at 8 kHz. The pulses transmitted along a line are initially 0.375 μs in duration.

6.1.5 Modulated carriers

An FDM signal consists of the sum of many independent signals. With TDM, the signal consists of a train of pulses derived from many different independent signals. It is at times necessary to transmit the multiplexed signal by modulating a carrier with the combined signal, as for example in a microwave transmission link, in which a carrier at 4 GHz may be modulated with a signal comprising may independent telephone channels together with a television link. It is the purpose of this book to discuss primarily the transmission aspects of such links, so the problems imposed on the handling of such composite signals in amplifiers will only be mentioned. One point of interest is the fact that the nature of a composite signal can only be described statistically, and frequently a system is assessed by its response to randomly varying signals.

6.2 Random signals: noise

Unwanted signals arising from many sources are present on all communication links. Known as *noise*, such signals may arise, for example, from:

natural phenomena, such as lightning;
man-made interference, such as ignition surges;
interference between physically adjacent but distinct links (cross-talk);
fluctuations occurring because of the discrete nature of the charge carriers in electric currents (thermal noise and shot noise).

These signals are random in origin, and may be thought of as a random train of individual pulse-like disturbances. Each type of random signal will be characterised by a power spectrum, which will usually have to be determined experimentally. Thermal and shot noise however arise from fundamental physical processes, and are amenable to calculation. Thermal noise is a consequence of the thermal motion of the charge carriers in a conductor when in equilibrium. Shot noise is associated with the fluctuations in a current made up of a stream of charges which, although coming at an average rate which is constant, are originally emitted at random times. Since thermal noise is inevitably present, its magnitude is a matter of importance, not only because of its 'nuisance' value, but as a standard by which to assess other forms of noise.

6.2.1 Thermal noise

A small fluctuating voltage exists across the terminals of a resistor which is in thermal equilibrium with its surroundings. The mean fluctuation is of course zero, since no current flows. The mean-square fluctuation across a resistance R, in the frequency interval between f and $f+\delta f$, is given by the Nyquist formula

$$\delta(\overline{v^2}) = 4kTR\,\delta f \qquad (6.3)$$

In this formula k is Boltzmann's constant, 1.38×10^{-23} J K^{-1}, and T is the absolute temperature. At 290 K, $kT = 4\times10^{-21}$ W Hz^{-1}. Since $\delta(\overline{v^2})$ is proportional to δf and independent of the absolute frequency (at least for radio frequencies), the noise is said to be *white*, by analogy with white light.

The effect of several independent noise sources acting simultaneously is found by addition of the mean-square voltages. (If the sources are not independent this result is not true.) Thus the total noise deriving from a resistance R at the input to an amplifier of voltage gain-frequency response $A(f)$ is given by

$$\overline{v^2} = \int_0^{\infty} [A(f)]^2 4kTR \, \mathrm{d}f \qquad (6.4)$$

Another way of presenting (6.3) is in the form of available power. The noisy resistor has an equivalent circuit of R in series with an e.m.f. equal to $[\delta(\overline{v^2})]^{\frac{1}{2}}$. The maximum power which can be taken from such a source is obtained when a load of resistance R is connected across it. This power is given by

$$R\left\{\frac{[\delta(\overline{v^2})]^{\frac{1}{2}}}{2R}\right\}^2 = kT\,\delta f \qquad (6.5)$$

When both resistors are at the same temperature, this power is compensated for by a flow the opposite way, maintaining equilibrium. From the figures given above, this available power at 290 K is 4×10^{-21} W Hz^{-1}.

The existence of a definite value for one contribution to the total noise signal makes this value a convenient reference point against which to judge other situations. As an example we may consider the noise introduced by a length of transmission line.

6.2.2 Thermal noise on a transmission line

Consider the situation of fig. 6.3. We will calculate the noise power delivered into a matched load from a length of transmission line at temperature T_l connected to a matched source at temperature T_s. Providing the line is lowloss Z_0 will be resistive, and the output of the line will be matched to the load. If the temperatures T_s and T_l are equal, the load sees a resistance Z_0 at the common temperature, and receives power $kT_s\delta f$. This contains one part from the source subsequently attenuated by a factor $\exp(-2\alpha l)$, equal to $kT_s\delta f \exp(-2\alpha l)$. The contribution from the line is therefore $kT_s[1-\exp(-2\alpha l)]\delta f$.

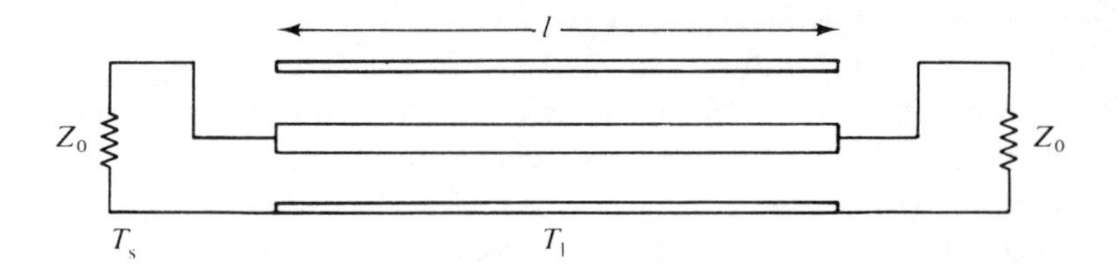

Fig. 6.3. Noise from a source connected by a lossy transmission line.

When the temperatures are unequal we have for the output power

$$\{kT_s \exp(-2\alpha l) + kT_l[1 - \exp(-2\alpha l)]\}\, \delta f \tag{6.6}$$

6.2.3 Degradation of signal by noise

Noise has been referred to as an unwanted signal. In the case of an analogue signal, the wanted signal will eventually be converted to its original form, such as speech or a television picture. The unwanted signal will undergo the same process and in some way impair the final output. On a telephone line it will be heard as a hiss, which can be ignored at low levels but which has a serious effect at high levels. On the television picture it will appear as a variable graininess, or snow. The amount of noise which can be tolerated depends on the final assessment, and is not usually amenable to calculation: the properties of hearing and vision enter into the result. On a digital signal the noise can have the effect of adding a spurious pulse where there should not be one, thus introducing error.

The measurement of noise, and the determination of signal power to give a prescribed signal-to-noise ratio, is thus very necessary. It is however important to remember that the prescribed value necessary to ensure adequate transmission is determined by subjective results, which depend on the use to which the signal is being put. This leads for example to a weighting factor being applied in telephonic systems to allow for the fact that the human ear varies in sensitivity over the range of frequencies covered by speech.

6.3 Propagation characteristics

It is evident that we have to consider propagation of a signal containing more than one frequency along a transmission line, either as a base-band signal, or as a frequency-translated signal.

6.3.1 Phase delay

We have seen that any signal can be regarded as the sum of individual sine waves. An individual sine wave $V_0 \sin \omega t$ will, after transmission along a cable, appear at the other end as

$$V_0 \exp(-\alpha z) \sin(\omega t - \beta z) = V_0 \exp(-\alpha z) \sin[\omega(t - \beta z/\omega)]$$

in which α, β are functions of ω. It thus suffers a change of amplitude determined by the attenuation, and a delay of $\beta z/\omega$. If a sum of sine waves is to be passed without distortion, the attenuation must not vary greatly over the range of frequencies, and the delay for all frequencies should be the same. This is satisfied if the phase velocity $v_p = \omega/\beta$ is independent of frequency, or β is proportional to ω. We thus endeavour to arrange such a characteristic over the required frequency range. For a narrow-band signal at base-band this is not too difficult, but it is more difficult to realise over an extended range. A frequency-shifted signal is restricted to a narrow range of frequencies about a higher frequency, so that we are only concerned with relative delay over the range.

6.3.2 Group delay

Let us consider the transmission of a single channel. We have seen (§6.1.3) that a signal $a \cos \omega_m t$ is translated to become $\frac{1}{2}a \cos(\omega_c + \omega_m)t$. After transmission through a length of transmission line this signal will become

$$\tfrac{1}{2}a \cos[(\omega_c + \omega_m)t - z\beta]$$

in which β is to be evaluated at the angular frequency $(\omega_c + \omega_m)$. Now if $\omega_m \ll \omega_c$ we may express β by means of a Taylor series in the form

$$\beta = \beta_c + \omega_m\left(\frac{d\beta}{d\omega}\right)_c + \frac{\omega_m^2}{2}\left(\frac{d^2\beta}{d\omega^2}\right)_c \cdots$$

in which the suffix c denotes that the function is evaluated for $\omega = \omega_c$. If the range of ω_m is small enough to allow us to neglect the last and higher terms we have to this approximation

$$\tfrac{1}{2}a \cos\left[(\omega_c + \omega_m)t - z\beta_c - \omega_m z\left(\frac{d\beta}{d\omega}\right)_c\right]$$

$$= \tfrac{1}{2}a \cos\left\{\omega_c t - z\beta_c + \omega_m\left[t - z\left(\frac{d\beta}{d\omega}\right)_c\right]\right\}$$

The term $\omega_c t - z\beta_c$ represents a phase-shifted carrier, so that on

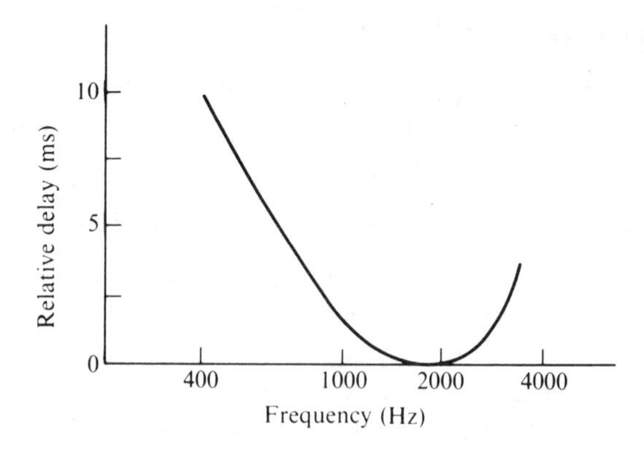

Fig. 6.4. Group-delay characteristic for a 4 kHz band-pass filter.

demodulation by multiplying by $\cos(\omega_c t - z\beta_c)$ and filtering we shall obtain the output

$$\tfrac{1}{4}a \cos\left\{\omega_m\left[t - z\left(\frac{d\beta}{d\omega}\right)_c\right]\right\}$$

Thus, for every ω_m in the narrow band near ω_c, each component of the demodulated signal is delayed by the same amount. This delay is referred to as the *group delay*, and is given by

$$\tau = z\left(\frac{d\beta}{d\omega}\right)_c$$

The total phase response in a transmission network includes major contributions from the filters used to separate channels. The appropriate expression for group delay is then

$$\tau = \left(\frac{d\phi}{d\omega}\right)_c \tag{6.7}$$

In fig. 6.4 is shown a typical group-delay characteristic for a 4 kHz band-pass filter. Compensation for the varying group delay over the pass-band can be effected by two-port networks giving *group-delay equalisation.*

6.3.3 Intercept distortion

It was explicitly stated in the last section that demodulation was accomplished by multiplication by $\cos(\omega_c t - z\beta_c)$. This involves getting the phase of the carrier correct. If it is not correct, so the carrier is say $\cos \omega_c t$, the demodulated output will be

$$\cos[\omega_m(t-\tau)-z\beta_c]$$

Although for one particular value of ω_m the term $z\beta_c$ merely represents a phase shift, when several frequencies are involved (as there will be) it can give rise to distortion. A phase delay ϕ_0 independent of frequency is equivalent to a time delay ϕ_0/ω_m, which varies with frequency. To avoid distortion the phase delay must be made zero, or a multiple of π, or otherwise the method of demodulation suitably chosen. The name *intercept distortion* arises from integration of (6.7) which gives

$$\phi = \omega\tau + \phi_0$$

ϕ_0 is the intercept on the $\omega = 0$ axis.

6.3.4 Group velocity

We have considered in §6.3.2 a single term of the composite signal in a channel. If we allow for two components we have

$$\cos \omega_c t + \tfrac{1}{2}a \cos (\omega_c + \omega_m)t$$

This expression is shown graphically in fig. 6.5. The original signal thus appears as the envelope of a carrier. Since the original signal suffers the group delay, the envelope of the waveform must also suffer the same delay and therefore appears to travel with a velocity

$$v_g = \frac{z}{\tau} = \left[\left(\frac{d\beta}{d\omega}\right)_c\right]^{-1} \tag{6.8}$$

This is known as the *group velocity*.

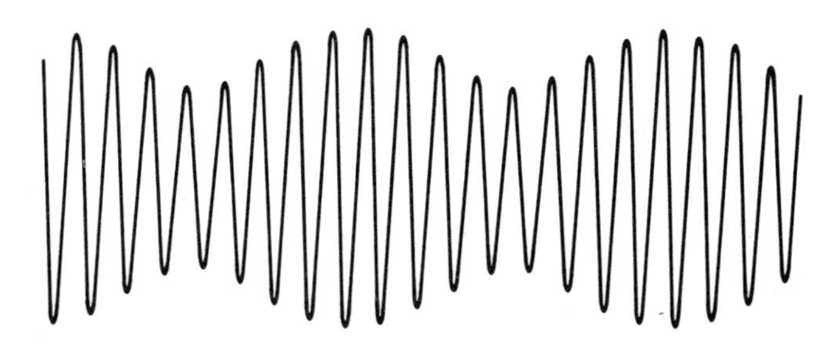

Fig. 6.5. The signal $\cos \omega_c t + \frac{1}{2}a \cos (\omega_c + \omega_m)t$ with $a = 0.5$.

If the phase velocity is truly constant, so that $\omega = \beta v_p$, then the group velocity is also equal to v_p:

$$v_g = \frac{d\omega}{d\beta} = v_p$$

This is the situation for a lossless line, for which $\beta = \omega\sqrt{(LC)}$. A lossless waveguide, however, has an $\omega - \beta$ characteristic which is not linear. It is shown in § 3.4.1 that

$$\beta = \frac{\omega}{c}\left[1 - \left(\frac{\omega_0}{\omega}\right)^2\right]^{\frac{1}{2}}$$

whence

$$v_p = \frac{\omega}{\beta} = c\left[1 - \left(\frac{\omega_0}{\omega}\right)^2\right]^{-\frac{1}{2}}$$

$$v_g = \left(\frac{d\beta}{d\omega}\right)^{-1} = c\left[1 - \left(\frac{\omega_0}{\omega}\right)^2\right]^{\frac{1}{2}}$$

Clearly both v_p and v_g vary with frequency and it is pertinent to ask what range is usable with a given characteristic. This in turn sheds light on the degree of group-delay equalisation which must be achieved. The degree of equalisation which is possible depends on the stability of the system being equalised, and it is this factor which finally limits the frequency range a signal may cover. These considerations are further complicated by the fact that different types of modulation are affected by phase distortion in different ways: for example, amplitude modulation is not affected by intercept distortion.

These considerations are further developed for the case of waveguide propagation in §7.6.

6.4 Transmission-line parameters

It has been shown that the voltage and current on a long line of parameters R, G, L, C are described by

$$V = A\,e^{-\gamma z}$$

$$I = Z_0^{-1} A\,e^{-\gamma z}$$

where

$$\gamma = [(R + j\omega L)(G + j\omega C)]^{\frac{1}{2}}$$
$$Z_0 = [(R + j\omega L)/(G + j\omega C)]^{\frac{1}{2}} \qquad (6.9)$$

The backward wave has been omitted. In these equations the parameters may in fact be functions of frequency: in particular R will increase as $\omega^{\frac{1}{2}}$ because of skin effect; L will also contain a modifying term depending on ω, since at low frequencies flux penetration occurs within the conductors; that part of G due to dielectric loss will increase proportional to ω. Thus in general γ and Z_0 are complicated functions of frequency. Lines are usually designed for low loss, and this assumption enables us to make certain simplifications: if $R \ll \omega L$ and $G \ll \omega C$, we may write

$$Z_0 \approx \sqrt{(L/C)}$$

$$\gamma = \mathrm{j}\omega\left[LC\left(1+\frac{R}{\mathrm{j}\omega L}\right)\left(1+\frac{G}{\mathrm{j}\omega C}\right)\right]^{\frac{1}{2}}$$

$$\approx \mathrm{j}\omega\sqrt{(LC)}\left[1+\frac{1}{2\mathrm{j}\omega}\left(\frac{R}{L}+\frac{G}{C}\right)\right]$$

$$\alpha+\mathrm{j}\beta = \tfrac{1}{2}[R\sqrt{(C/L)}+G\sqrt{(L/C)}]+\mathrm{j}\omega\sqrt{(LC)} \tag{6.10}$$

If we assume $R \propto \sqrt{f}$, $G \propto f$, we have for the attenuation coefficient

$$\alpha = A_1\sqrt{f}+A_2 f \tag{6.11}$$

as exemplified by (3.79) for the coaxial line. The phase coefficient β is that for the lossless line. The shunt conductance G will be of importance at higher frequencies, depending on the nature of the dielectric involved. Coaxial cables use high-quality dielectric, and various ways are devised to reduce the dielectric required to support the wires to a minimum, so that dielectric loss is of little significance. Submarine cables for mechanical reasons use solid dielectric and incur greater loss. On the other hand, twisted pairs use paper dielectric, and losses from this can become appreciable at 100 kHz.

6.4.1 Attenuation on a coaxial line

Equation (3.79) gives for a coaxial line

$$\alpha = \frac{\pi f}{v}\tan\delta + \tfrac{1}{2}R_s\sqrt{\left(\frac{\varepsilon}{\mu}\right)}\left(\frac{1}{a}+\frac{1}{b}\right)[\ln(b/a)]^{-1}$$

The first term arises from dielectric loss, and is independent of the radii a, b. The second term may be written in the form

$$\tfrac{1}{2}R_s\sqrt{\left(\frac{\varepsilon}{\mu}\right)}\frac{1}{b}\left(1+\frac{b}{a}\right)\left[\ln\frac{b}{a}\right]^{-1}$$

Regarded as a function of b/a, with b held constant, this takes the form shown in fig. 6.6. The minimum value occurs for that value of b/a which satisfies the equation

$$\ln(b/a) = 1 + a/b$$

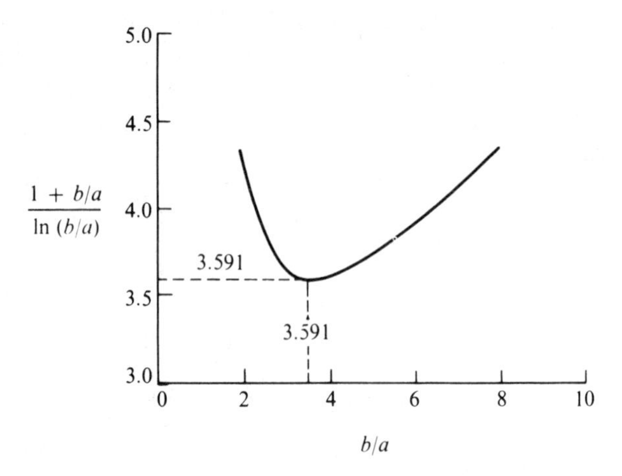

Fig. 6.6. The function $(1+x)/\ln x$, $x=b/a$.

The root is approximately $b/a=3.6$. From (3.32), the characteristic impedance is given by the expression

$$Z_0=\frac{1}{2\pi}\sqrt{\left(\frac{\mu}{\varepsilon}\right)}\ln\frac{b}{a}$$

and for this value of b/a, Z_0 ranges between 51 Ω and 77 Ω, going from solid dielectric ($\varepsilon_r=2.25$) to air. It will be seen that the attenuation coefficient becomes smaller at any frequency if b is increased. Thus diameters of up to 4 cm are used in repeatered submarine cables.

6.4.2 Low-frequency behaviour

At low frequencies G is to all intents and purposes zero, so that $G \ll \omega C$. It is however impossible to satisfy $R \ll \omega L$. Open-wire pairs and twisted pairs carrying base-band speech signals thus have modified types of attenuation and phase coefficient functions. The early submarine telegraph cable consisting of a copper core with thin layers of insulation separating it from a sea-water return has very low inductance. Putting $L=0$, $G=0$ in (6.9) we find

$$\gamma \approx \sqrt{(\mathrm{j}\omega RC)}=\sqrt{(\omega CR)}\frac{1+\mathrm{j}}{\sqrt{2}}$$

$$Z_0 \approx \sqrt{\left(\frac{R}{\omega C}\right)}\frac{1-\mathrm{j}}{\sqrt{2}}$$

Attenuation thus increases with frequency, and it is not possible in practice to transmit speech over such a circuit. Matters would be improved if it were possible to increase the inductance. In the case of open- and twisted-pair lines this was found to be possible by 'loading' the lines with series inductors placed at suitable intervals. If the intervals at which the coils are spaced is a small fraction of the wavelength, the effect may be expected to be as though the extra inductance were uniformly distributed. These loading coils are spaced at about 1850 m intervals in Post Office telephone lines. This is discussed further in § 10.4.

6.4.3 The distortionless line

If it were possible to realise the relation $R/L = G/C$ between the parameters, then we should have

$$\begin{aligned}\gamma &= j\omega\surd(LC)(1+R/j\omega L)^{\frac{1}{2}}(1+G/j\omega C)^{\frac{1}{2}}) \\ &= j\omega\surd(LC)(1+R/j\omega L) \\ &= R\surd(C/L)+j\omega\surd(LC)\end{aligned}$$

For constant values of the parameters, the attenuation and the phase velocity will both be independent of frequency. Signals could thus be propagated without distortion but with attenuation. It is impossible to achieve this relationship in practice, although considerable improvements can be effected. The practice of inserting loading coils is a step towards the distortionless line, although the values used are actually much smaller than would be needed to achieve this condition. Submarine cables have been made containing magnetic cores to give extra uniformly distributed inductance, but are not now used.

6.5 Particular lines

In practice a number of standard forms of transmission line have evolved, and in the following sections a brief description of their properties will be given. We may distinguish various configurations of open-wire lines, coaxial lines and parallel-strip lines. The geometry of these enables electromagnetic calculations to be done when the conductors are perfect. The result of these calculations is to give the characteristic impedance. The velocity of propagation is, as shown in chapter 3, the same as that for plane waves in the unbounded medium. The results of these calculations are summarised in fig. 6.7.

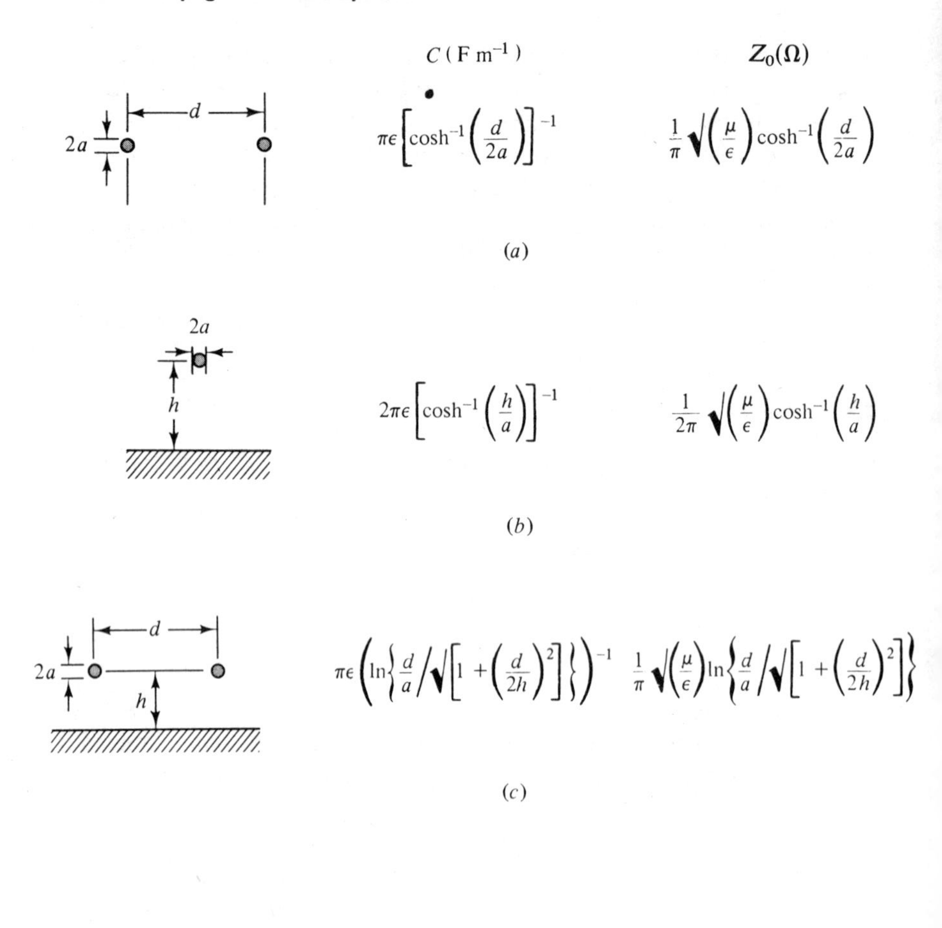

Fig. 6.7. Various transmission lines. (*a*) Open-wire pair. (*b*) Single wire with earth plane. (*c*) Two-wire with earth plane. (*d*) Coaxial.

Losses can be estimated as has been done for the coaxial line, or by using measured values of the resistance. Open-wire lines are normally used at frequencies low enough for resistance measurements to be made. Only for coaxial lines is the practical geometry sufficiently close to the ideal for calculations to be worth refining.

In addition to these ideal geometries, which of course are in practice modified by the need for supports, are the various forms of twisted pairs used in underground ducts in the telephone systems. The characteristics of such lines must be obtained by measurement.

6.5.1 Voice-frequency telephone cables

The type of cable used to connect subscribers to exchanges consists of twisted pairs formed into many-pair cables. Existing cables often use dry paper insulation, but modern designs use plastic. The values given in this section relate to one of this general class of cable, but is not specifically related to a particular manufacture. Although originally designed to operate at voice frequencies, it has subsequently been found possible to operate such lines with digital signals and thus increase their capacity.

The capacitance per unit length over the frequency range is found to be virtually constant at 0.053 μF km^{-1}; the shunt conductance is very small, and contributes less than 1% to the attenuation at the highest frequencies. The variations of inductance and resistance with frequency are dominated by skin effect. At low frequencies, magnetic flux will exist inside the conductors, and thus the inductance will decrease as the frequency increases. When the skin depth is much smaller than the conductor diameter, a new constant level will be reached. This behaviour is illustrated in fig. 6.8. The resistance will increase from its d.c. value once skin effect becomes significant, and thereafter will increase proportional to $\sqrt{f}$. This is illustrated in fig. 6.9.

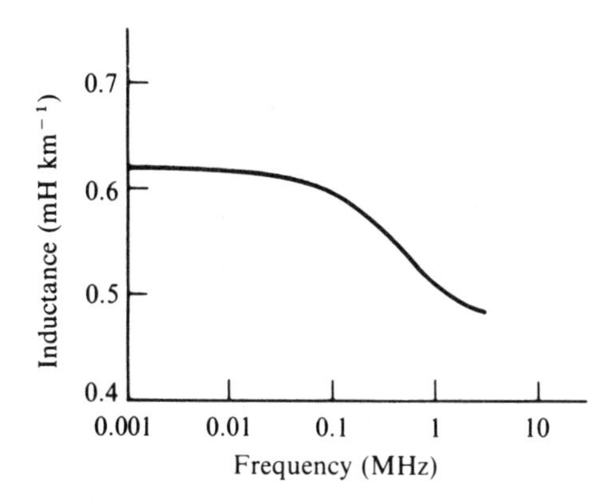

Fig. 6.8. Inductance of twisted-pair cable.

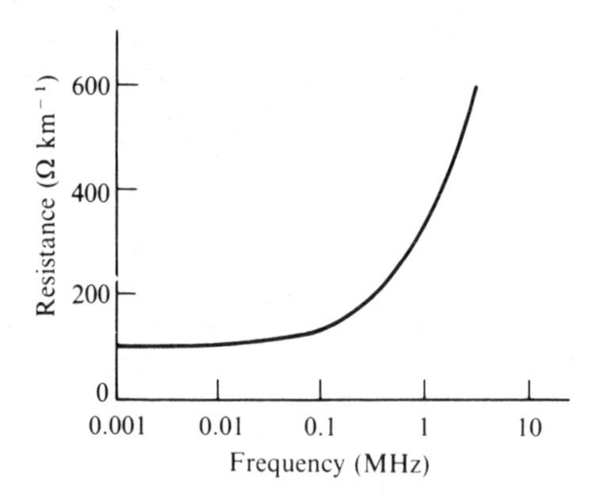

Fig. 6.9. Resistance of twisted-pair cable.

The secondary parameters and attenuation are calculated from the values of R, L and C. At low frequencies when $\omega L \ll R$, Z_0 becomes equal to $\sqrt{(R/\mathrm{j}\omega C)}$. At these frequencies therefore a phase angle of $-\frac{1}{4}\pi$ would be expected, with a large value of $|Z_0|$. At high frequencies, Z_0 will tend to the value $\sqrt{(L/C)}$. This is illustrated in fig. 6.10.

The attenuation characteristic is shown in fig. 6.11. At low frequencies, γ is approximately given by $\sqrt{(\mathrm{j}\omega CR)}$, with R constant. At high frequencies, the attenuation is dominated by skin effect. Therefore at both high and low frequencies, a dependence on $\sqrt{f}$ is to be expected. This is borne out by the graph. The phase velocity is calculated from ω/β and will thus be proportional to $\sqrt{f}$ at low frequencies, levelling off at high frequencies. This is shown in fig. 6.12.

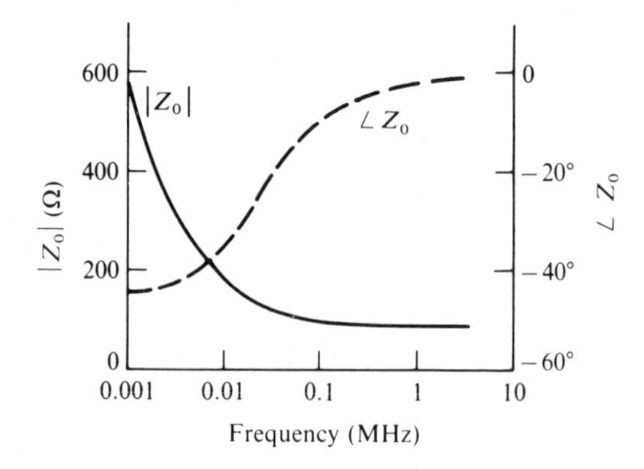

Fig. 6.10. Characteristic impedance of twisted-pair cable.

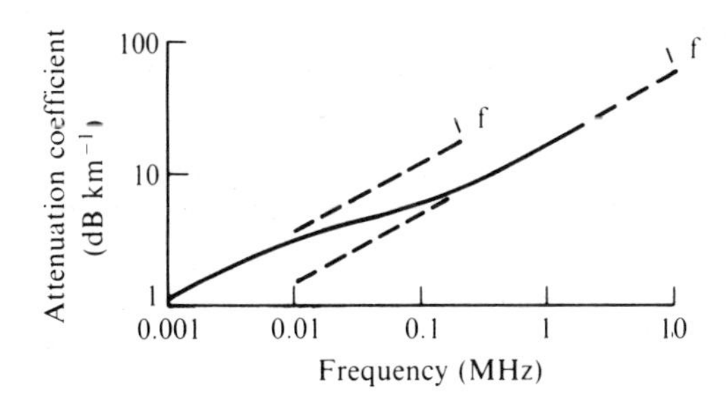

Fig. 6.11. Attenuation characteristic of twisted-pair cable.

It is to be noted that all the parameters change with temperature, because of resistance change and dimensional change. Typically, changes of the order of 0.1% for each degree of temperature are found.

6.5.2 Coaxial land-lines

This section will be restricted to lowloss coaxial cables designed for long runs. There are requirements on the electrical properties, in the form of specifications of characteristic impedance, attenuation and velocity of propagation, which must be met in the manufactured cable. These specifications must be necessarily related to the manufacturing techniques available. A coaxial line is a complicated structure involving an

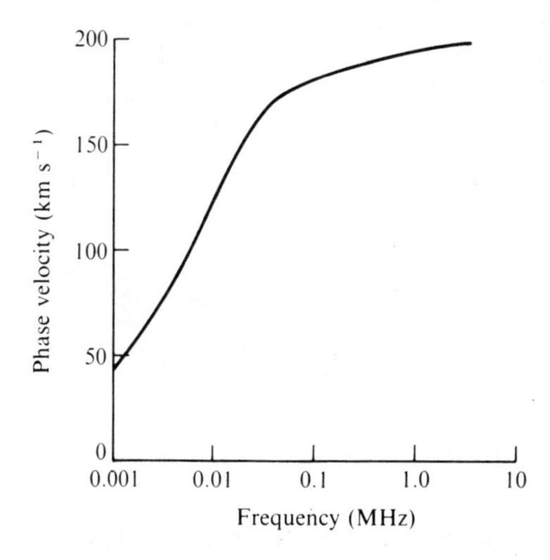

Fig. 6.12. Phase velocity for twisted-pair cable.

inner, an outer, supports and outer cladding, and for communications systems has to be manufactured in long lengths. It is obviously desirable to reduce joints as far as possible, so that the length is determined by problems such as the drawing of the cable into ducts. This typically limits lengths to 500 m. The production must therefore be continuous, and the outer must be a folded structure rather than a seamless tube. The effect of non-uniformity of electrical properties with length will be discussed later. Variations must be kept to a low level and this is done by care in manufacture.

In fig. 6.13(*a*) and (*b*) are shown two methods of making a coaxial cable. These refer specifically to British Post Office designs of inner conductor 2.6 mm, outer 9.5 mm. The inner conductor is drawn to the required size and polyethylene discs sprung onto the inner at 33 mm intervals. The outer is then formed by folding copper tape over the discs.

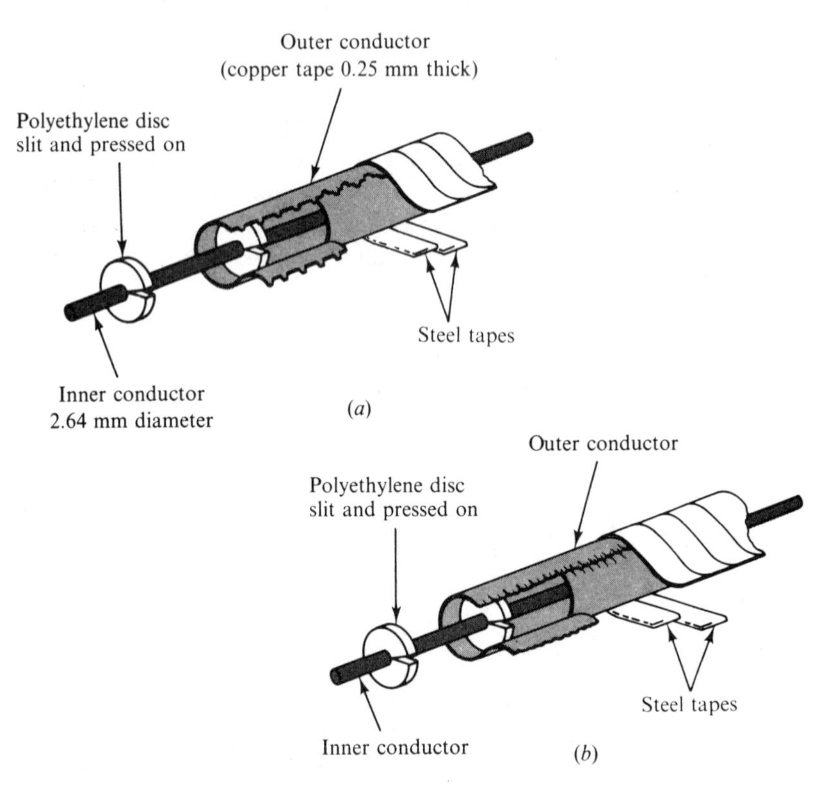

Fig. 6.13. British Post Office 2.6/9.5 mm cable. (*a*) Notched interlacing edges. (*b*) Corrugated edges to avoid overlapping.

The edges forming the longitudinal seam are either corrugated or notched. It is not found necessary to solder the seam. Over the outer is wound a layer of steel tape. The purpose of this tape is to reduce the coupling field penetrating the outer at low frequencies and thus causing cross-talk. It provides shielding at frequencies exceeding 100 kHz.

The characteristic impedance is given by the expression

$$Z_0 = 74.4[1 + 0.0123(1 - j)/\sqrt{f}]$$

in which f is in MHz and Z_0 in ohms. A tolerance of $\pm 0.4\ \Omega$ is allowed.

The attenuation coefficient is given by the expression

$$\alpha = 0.013 + 2.305\sqrt{f} + 0.003f$$

in which α is the attenuation in dB km^{-1} and f is in MHz. This expression is dominated by the second term, arising from skin effect. The allowed variations are 0.5% of the mean attenuation in the range 4–64 MHz, and 1% in the range 64–500 MHz. The temperature coefficient of attenuation is expected to be about 0.002 °C^{-1} at an ambient of 10 °C.

The cable is designed to provide FDM channels up to 60 MHz, and to be capable of digital operation at about 800 Mbit s^{-1}. It is for the latter purpose that the frequency range has to be extended to 500 MHz.

6.5.3 Submarine coaxial cables

The environment of the submarine cable and the necessity of laying it deep in the ocean present special problems, resulting in a design quite different to that for long land-lines. The basis of modern designs (fig. 6.14(*a*)) is a solid-dielectric coaxial line with the strength provided by high-tensile steel strands inside the inner conductor. In manufacture the strength member is first formed. The inner electrical conductor is made by wrapping copper tape (about 0.5 mm in thickness) around the strength member, welding the seam and subsequently swaging the copper into close contact with the steel. The polyethylene dielectric is then extruded onto the conductor and subsequently shaved to size. The outer conductor is formed by wrapping copper tape around the dielectric, with a longitudinal overlap of about 5 mm. Tolerances of these operations are about ± 0.02 mm. A 3 mm sheath of polyethylene completes the light-weight cable for deep-sea laying. The tension during the laying process can rise to 100 kN, and the lay-up of strands in the strength member (some 50 in number) is designed so that application of

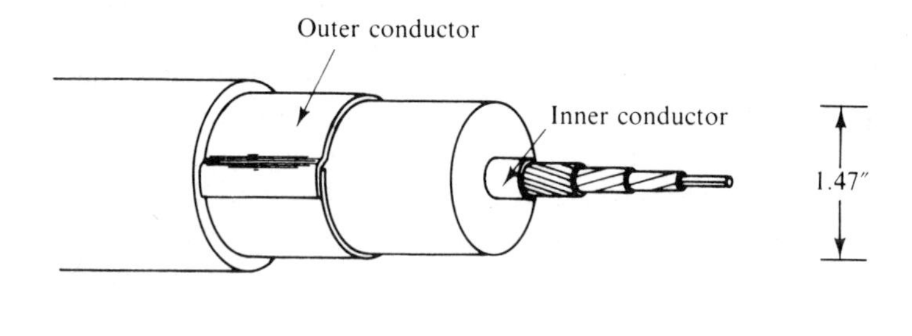

(a)

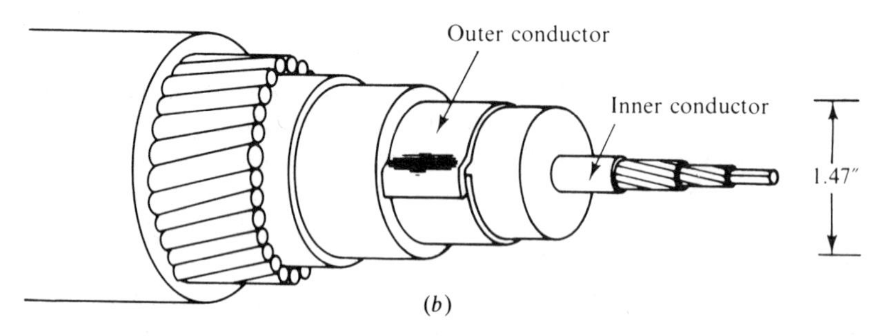

(b)

Fig. 6.14. Submarine coaxial cables. (*a*) Light-weight for deep-sea use. (*b*) Armoured for use on continental shelf.

tension does not cause twist. The pressure of water on the cable may be 90 MN m^{-2}, causing contraction of the circumference by up to 1 mm. Since the total attenuation at the high-frequency end of the band can rise to 16 000 dB, it is essential to be able to predict the cable attenuation to a very high degree of accuracy (0.01%) if repeaters are not to be overloaded.

The electrical characteristics of such cable are very similar to those of other coaxial lines. At the maximum frequency used, the dielectric loss accounts for only a small part of the attenuation, which is predominantly caused by skin effect. For example at 1 MHz, a cable of outer diameter 40 mm will have an attenuation of about 0.5 dB km^{-1}. Such a cable is in practice worked up to 14 MHz, with repeaters at intervals of some 10 km. The cable illustrated in fig. 6.14(*a*) is used in the deep sea where it is unlikely to be disturbed. Nearer the shore, cables are prone to damage from trawlers, and armouring is applied, as shown in fig. 6.14(*b*).

6.5.4 Flexible coaxial cables

The two types of coaxial line discussed earlier have been of special manufacture, designed for a special purpose. The agreement between theoretical performance and actual measurements is very good. It is indeed this agreement which gives confidence that the parameters controlling performance are known, and as a consequence performance in a rigorous environment can be guaranteed.

For many uses such performance is not necessary, and a large variety of flexible coaxial cables is available. It is the purpose of this section to consider typical parameters of such lines.

The use of a solid tube as the outer conductor makes a cable relatively inflexible, so that flexible cables usually have an outer made of copper wires braided into a sleeve, or sometimes a corrugated tube is used. A braided outer has a higher resistance than the equivalent solid conductor and is also less effective as a screen. The inner conductor may be a single copper wire or made of several strands of thinner wire. Strength may be obtained by use of a steel wire copper- or silver-plated to give good conduction. A solid dielectric such as polyethylene extruded onto the inner gives good mechanical properties. Dielectric losses may be minimised by using a cellular plastic or by devising a mounting whereby air spaces are incorporated. For example a polyethylene thread may be loosely wound round the inner, as say a helix of 20 mm pitch. The whole is then inserted into a dielectric tube over which the outer braid is formed. An alternative method in precision cables is to cut a helical slot in a solid dielectric core. Dielectric loss is only of account at higher frequencies: for polyetheylene with $\tan\delta = 10^{-4}$, $\varepsilon_r = 2.25$, the loss at 100 MHz is 1.4×10^{-3} dB m^{-1}, compared with which copper losses are likely to be some fifty times greater. The precise nature of the composite dielectric will however affect the phase velocity.

Nominal attenuation figures published by manufacturers for this type of cable show a loss almost precisely proportional to $\sqrt{f}$ up to 1000 MHz, implying that skin effect is the predominant cause of attenuation. Some idea of the effect on losses of using braided outer or stranded inner can be obtained from figures for different cables. A value for the total series resistance per unit length can be obtained from the attenuation coefficient, using (6.10). The effective braid resistance is found to be several times the resistance of the equivalent solid tube, the ratio rising as diameter increases. It must be remembered that the loss in the outer is smaller than that in the inner.

The screening properties of the outer can be enhanced by the use of double braids, either in contact with each other or separated by a layer of insulation.

6.5.5 Open-wire lines

An open-wire line consists of two parallel conductors supported clear of the ground. In so far as a practical line is actually straight with parallel conductors and the effect of supports is negligible, the characteristics are simply calculable. The conductors are typically of the order 1–2 mm in diameter with spacings of 20–30 cm. The conductors may be copper, but may also be copper alloys to give greater strength. At frequencies for which the skin depth is much less than the radius of the conductor, the inductance and capacity per unit length will have the values calculated for perfect conductors, as shown in fig. 6.7. The resistance will be governed by skin effect. At low frequencies, the inductance will increase slightly because of the uniform distribution of current through the conductors, and for the same reason the resistance will become independent of frequency. Figures 6.15(a) to (d) shows the various parameters for a line with two copper conductors each of diameter 1.5 mm spaced 250 mm apart. The shunt conductance has been taken to be zero, but this will depend greatly on climatic conditions. Attenuation may increase greatly in wet weather. The shape of the curves is very comparable with those presented in §6.5.1.

6.5.6 Microstrip line

The development of solid-state microwave devices together with ease of fabrication of circuits on printed circuit boards has led to very considerable use of transmission lines based on plane parallel conductors. These consist of a substrate which supports thin copper electrodes on either side, with the substrate thickness of the order of 1 mm. Some different forms of construction are shown in fig. 6.16(a) to (d).

The triplate line shown in fig. 6.16(a) is the most difficult to fabricate, since it involves a sandwich of two planar substrates. It is however a true shielded transmission line using the outer planes as the shield and the middle strip as the inner. The interior is wholly filled with dielectric, and therefore the type of analysis performed in chapter 3 is applicable. A *TEM* wave will be propagated with a velocity equal to that of a plane wave in the unbounded medium, and the only parameters to be calculated are characteristic impedance and attenuation. The dielectric has

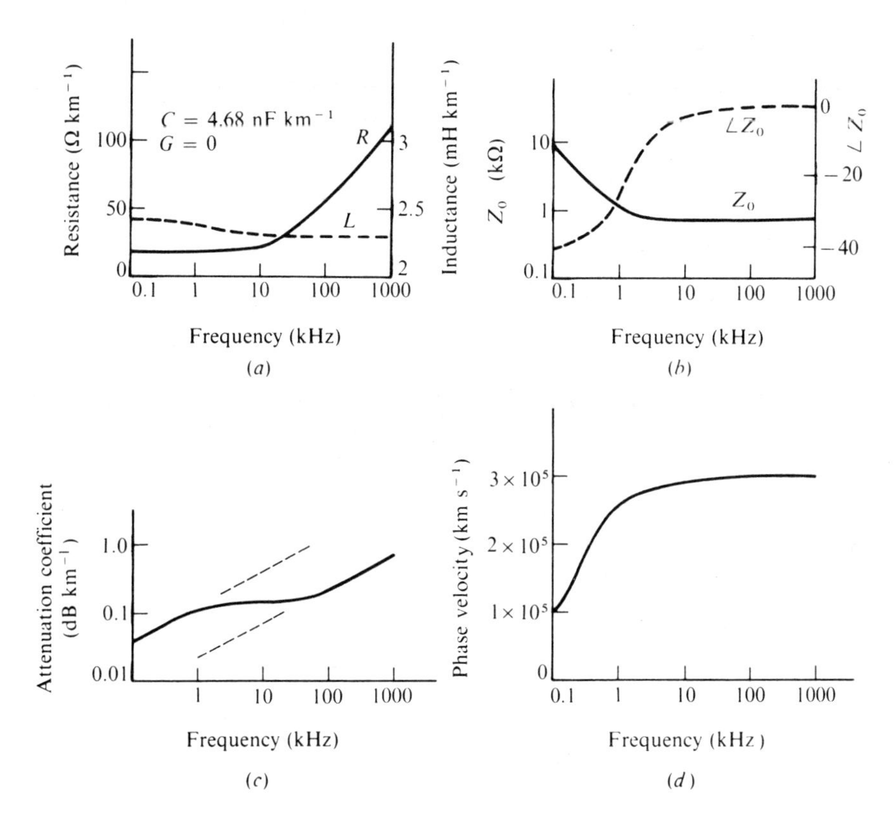

Fig. 6.15. Parameters for open-wire pair. Conductor 1.5 mm diameter, 250 mm spacing. (*a*) Inductance and resistance. (*b*) Characteristic impedance. (*c*) Attenuation coefficient. (*d*) Phase velocity.

the effect of decreasing by $\varepsilon_r^{\frac{1}{2}}$ both the characteristic impedance and the phase velocity as compared with the values for an air-spaced line (where ε_r is the relative permittivity), and it is convenient to show results for an air-spaced line. These are given in fig. 6.17(*b*) for a triplate line dimensioned as in fig. 6.17(*a*). It will be seen that the characteristic impedance is little affected by a reasonable thickness of the centre conductor. For a spacing of 2 mm and thickness 0.1 mm we have $t/b = 0.05$. The attenuation arising from conductor resistance alone is also shown. The attenuation coefficient for a thickness of 2 mm is given by these figures as about 1 dB m^{-1}. To this has to be added dielectric loss, for which the attenuation coefficient is the same as in the case of the coaxial cable, that is, $(c^{-1}\pi f\varepsilon_r^{\frac{1}{2}} \tan\delta)$ neper m^{-1}. At 10 GHz for $\varepsilon_r = 3$ and $\tan\delta = 10^{-3}$, this gives 0.5 dB m^{-1}.

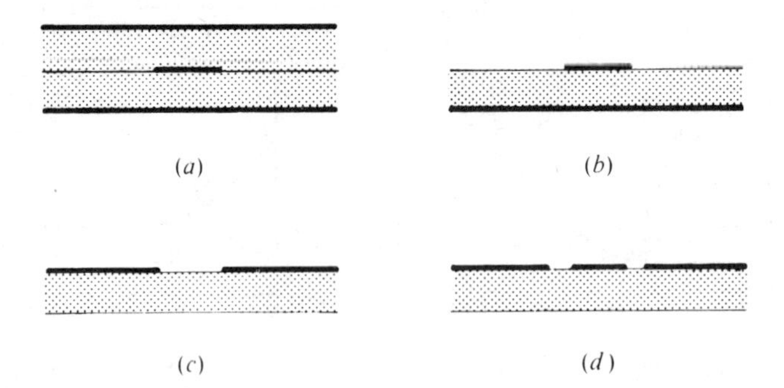

Fig. 6.16. Forms of microstrip line. (*a*) Triplate. (*b*) Microstrip. (*c*) Slot line. (*d*) Coplanar waveguide.

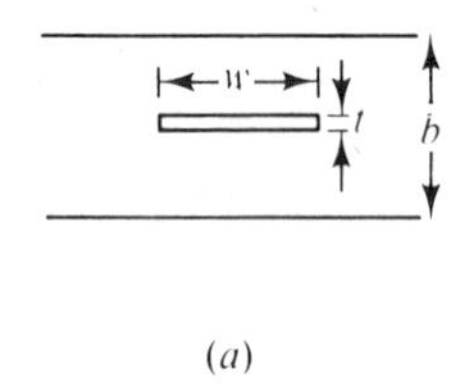

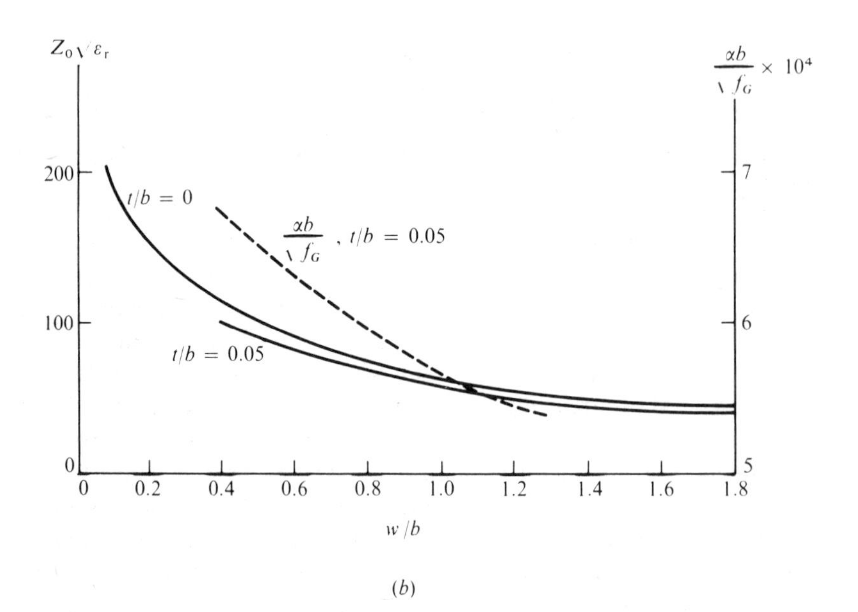

Fig. 6.17. Characteristics of triplate. (*a*) Dimensions. (*b*) Z_0 and α.

As mentioned earlier, triplate line is the most complicated form to fabricate, even though electrically satisfactory. The microstrip line of fig. 6.16(*b*) is much easier to fabricate, although it is more lossy and is also dispersive. It is important to realise that microstrip line does not come in the category of transmission lines discussed in chapter 3: it is not possible to satisfy the boundary conditions at the interface between the substrate top-surface and the surrounding medium (usually air) unless the permittivities are equal, which is not usually the case. This may be seen by considering that *TEM* waves would travel with different velocities in the two dielectrics. However, it is found that a *TEM* wave approximation is reasonably satisfactory, particularly for high-permittivity substrate. This can be understood from the configuration of the lines of electric force as indicated schematically in fig. 6.18(*a*) and (*b*).

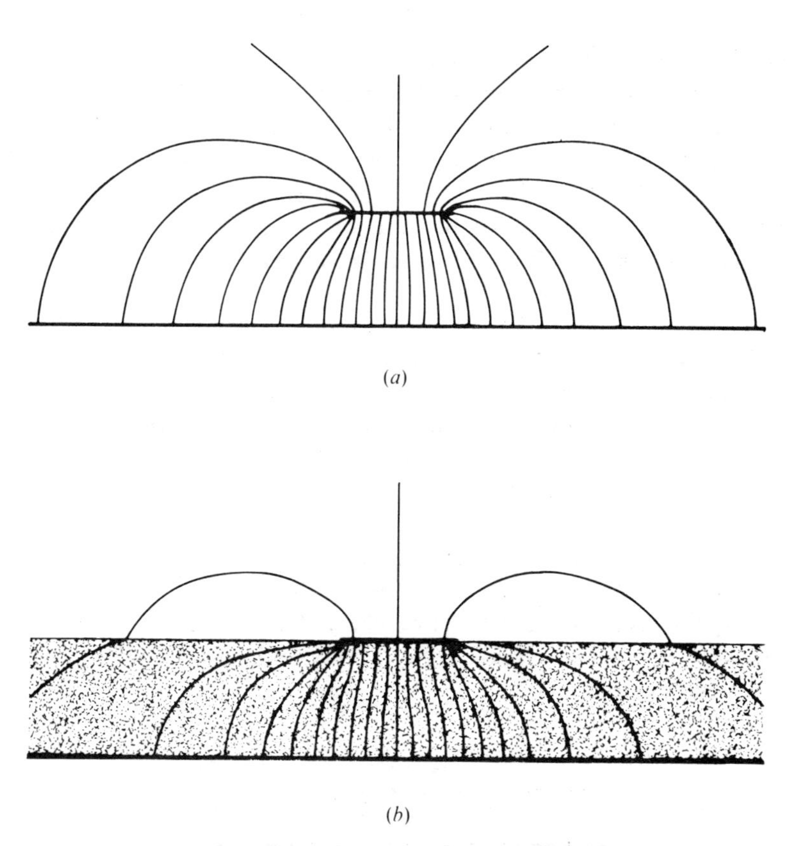

Fig. 6.18. The effect of substrate permittivity. (*a*) Single-dielectric. (*b*) High-permittivity substrate.

The former shows the situation in single-dielectric, the latter the situation with a high-permittivity dielectric substrate.

In calculating the characteristic impedance, it is customary to introduce an effective permittivity, ε_{re}, which will be between those of the substrate and the surrounding air. The characteristic impedance and phase velocity of the actual line have the values which would be found for the conductor geometry immersed in an infinite dielectric of permittivity ε_{re}.

A semi-empirical formula for the characteristic impedance of the microstrip line shown in fig. 6.19 is

$$Z_0 = 60\varepsilon_{re}^{\frac{1}{2}} \ln(4b/d)$$

where

$$\varepsilon_{re} = 0.475\varepsilon_r + 0.67$$

$$d = 0.536w + 0.67t$$

and $w/b < 1.25$, $0.1 < t/w < 0.8$, $2.5 < \varepsilon_r < 6$. This covers a practical range of parameters with about 5% accuracy.

The dispersive nature of propagation along such a line is indicated by an experimentally observed variation of ε_{re} with frequency. Such a result is shown in fig. 6.20. It may be mentioned in passing that triplate line may also be dispersive because of the difficulty of achieving a perfect fit between the two halves of the substrate.

The fact that fields are not confined to the region between the conductors has both useful and detrimental results. Coupling will take place between adjacent circuits, which can be used to advantage in the construction of filters, for example. The fact that conductors are open, and quite likely to be long compared with a wavelength gives rise to losses by radiation as well as through the conductor resistance and dielectric loss. A typical figure for the attenuation coefficient is 0.1 dB per wavelength.

Some common substrates may be mentioned:

Alumina, a ceramic substrate $\varepsilon_r \approx 9.5$ to 10.

Polyguide, copper-clad irradiated plastic (polyolefin) $\varepsilon_r = 2.32$.

Epsilam–10, ceramic-filled Teflon compound, anisotropic, $\varepsilon_r = 10$ or 15 depending on direction.

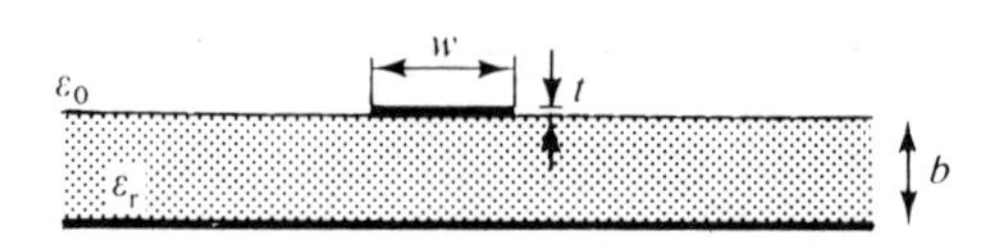

Fig. 6.19. Dimensions of microstrip line.

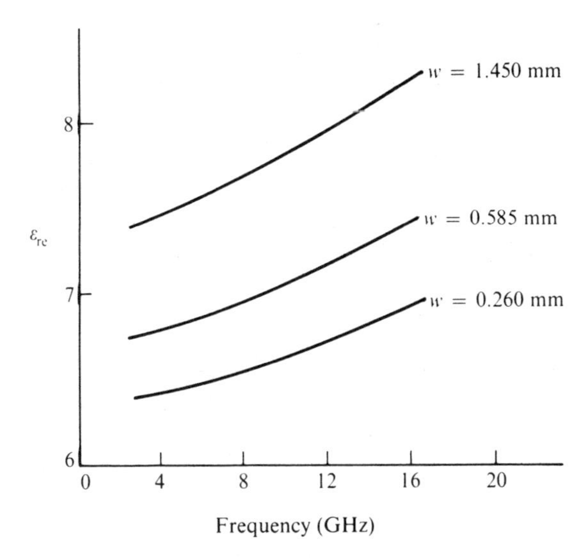

Fig. 6.20. Dispersion in microstrip line.

Sapphire, single crystal $\varepsilon_r = 9.5$ or 11.7 depending on direction.
Quartz, $\varepsilon_r \approx 4.5$ (anisotropic).

The other types of strip line illustrated in fig. 6.16 have properties similar to those discussed for microstrip line, and are covered in the literature.

†6.6 Propagation of pulses on lines

We have been considering the transmission effects from the point of view of signals restricted to narrow bands of frequency. Digital signals consist of a sequence of pulses, so that the propagation of a pulse along a long line is of interest. The results that can be obtained by theory are limited but are of sufficient interest to make presentation worthwhile.

†6.6.1 Mathematical method

We restrict ourselves to the case of a long line terminated in its characteristic impedance, so that the effect of transmission on a sine wave is a multiplication by $\exp[-\gamma(j\omega)z]$. The transfer function for use in the Laplace transform method is therefore $\exp[-\gamma(s)z]$. In order to find the voltage $v_z(t)$ at a point z arising from an input $v_0(t)$ at $z = 0$ we firstly obtain the Laplace transform $\underline{v_0}(s)$ of $v_0(t)$. The Laplace transform of

$v_z(t)$ is then given by

$$\overline{v_z}(s)=\overline{v_0}(s)\exp[-\gamma(s)z]$$

The inverse transform of $\overline{v_z}(s)$ then gives the time function $v_z(t)$. Although this process may be carried through for a given function $v_0(t)$, it is often easier to find first the response to an impulse of unit area, obtaining the impulse response $h(t)$ by

$$h(t)=\mathscr{L}^{-1}\exp[-\gamma(s)z]$$

in which $\mathscr{L}^{-1}$ denotes the inverse transform. The response to $v_0(t)$ can then be obtained by using the convolution integral giving

$$v_z(t)=\int_0^t v_0(u)h(t-u)\,\mathrm{d}u$$

$$=\int_0^t v_0(t-u)h(u)\,\mathrm{d}u$$

†6.6.2 Form for the propagation constant

We have discussed in chapter 5 the propagation of surges in lossless lines. For strictly lossless lines any pulse is propagated without change in shape or amplitude. In the case of the distortionless line ($R/L=G/C$), again a pulse suffers only attenuation in amplitude without change of shape.

The characteristic feature observed when rectangular pulses are propagated along real lines is that the rectangular shape is lost as the sharp edges become rounded. This would not be observed in the distortionless line even though there is attenuation. The reason for the difference of behaviour is that, because of skin effect and dielectric losses, the attenuation constant of a real line increases indefinitely as the frequency increases. To obtain a realistic solution we therefore have to take a model showing this attenuation–frequency characteristic.

A great simplification can be effected by considering pulses which are short. Equation (6.10) gives an expression for the propagation constant which is valid at high enough frequencies, when $\omega L \gg R$, $\omega C \gg G$. If R and G are constant, this expression is of the same form as that for the distortionless line. The difference between the two occurs at low frequencies, for which the expansion is not valid, but if the spectrum of the input pulse is small in content at these low frequencies a reasonably accurate result should be obtained by use of the approximation.

To make the attenuation coefficient an appropriate function of frequency seems easy. We have indeed derived from (6.11), considering skin effect and dielectric losses, the form

$$\alpha = a_1\omega^{\frac{1}{2}} + a_2\omega$$

This does not however give us the corresponding form for the phase coefficient, which must be compatible with the functional form of the attenuation coefficient: $\alpha(\omega)$ and $\beta(\omega)$ are related by being real and imaginary parts of a real function of $j\omega$. We may obtain a suitable form for the skin-effect term by taking

$$\gamma = \alpha + j\beta = 2^{\frac{1}{2}}a_1(j\omega)^{\frac{1}{2}} + j\omega(LC)^{\frac{1}{2}}$$
$$= a_1\omega^{\frac{1}{2}} + j[a_1\omega^{\frac{1}{2}} + \omega(LC)^{\frac{1}{2}}]$$

Although the term $ja_1\omega^{\frac{1}{2}}$ implies variable series inductance proportional to $\omega^{-\frac{1}{2}}$, this need not worry us if the low-frequency response is not of concern. No easy way exists of allowing for dielectric loss. The term proportional to ω arose through taking a loss angle independent of frequency: this is true only from statistical considerations, and no easily tractable circuit model exists. However, the magnitude of dielectric loss is relatively small and we may expect sensible results by taking

$$\gamma = \kappa(j\omega)^{\frac{1}{2}} + j\omega/c \tag{6.12}$$

The constant κ is related to the loss. The constant c is the velocity of propagation, and may be taken to be independent of the frequency.

†6.6.3 The submarine telegraph cable

The results given in this section relate directly to the early submarine telegraph cable, in which a central conductor was separated from sea water by only a thin layer of dielectric, and the sea water used as earth return. The result is of historical interest, as being the earliest result on a distributed system, published by Lord Kelvin in 1855. It also, surprisingly, turns out to be relevant to the modern line with skin-effect losses predominating.

For the cable described, both L and G are very small and we have to a good approximation

$$\gamma(j\omega) = [j\omega RC]^{\frac{1}{2}}$$

giving

$$\gamma(s) = [sRC]^{\frac{1}{2}}$$

The Laplace transform $\bar{h}(s)$ of the impulse response is then given by

$$\bar{h}(s) = \exp[-(s\tau)^{\frac{1}{2}}] \quad \text{where } \tau = z^2RC$$

The inverse transform is given in Abramowitz & Segun, *Handbook of mathematical functions*, section 29.3.82, to yield

$$h(t)=\frac{1}{2\tau\sqrt{\pi}}\left(\frac{\tau}{t}\right)^{\frac{3}{2}}\exp(-\tau/4t)$$

This function is shown in figure 6.21, where it will be observed that the impulse becomes spread out. The extent to which this happens depends on τ and hence z^2.

The response to a step of unit height is given by

$$\int_0^t \frac{1}{2\tau\sqrt{\pi}}\left(\frac{\tau}{u}\right)^{\frac{3}{2}}\exp(-\tau/4u)\,du=\frac{2}{\sqrt{\pi}}\int_{\sqrt{(\tau/4t)}}^{\infty}\exp(-v^2)\,dv$$

This integral can be evaluated in terms of the tabulated *error function*, defined by

$$\text{erf}(x)=\frac{2}{\sqrt{\pi}}\int_0^x \exp(-v^2)\,dv$$

It is convenient to denote the complement of the error function by

$$\text{erfc}(x)=1-\text{erf}(x)$$

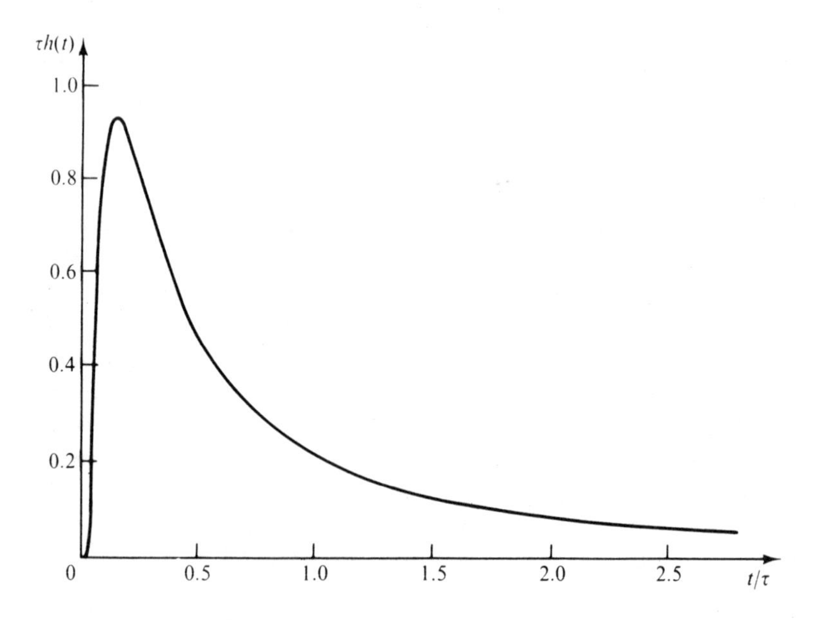

Fig. 6.21. Impulse response of submarine telegraph cable. The function $h(t)=\frac{1}{2}(\tau\sqrt{\pi})^{-1}(\tau/t)^{\frac{3}{2}}\exp(-\tau/4t)$.

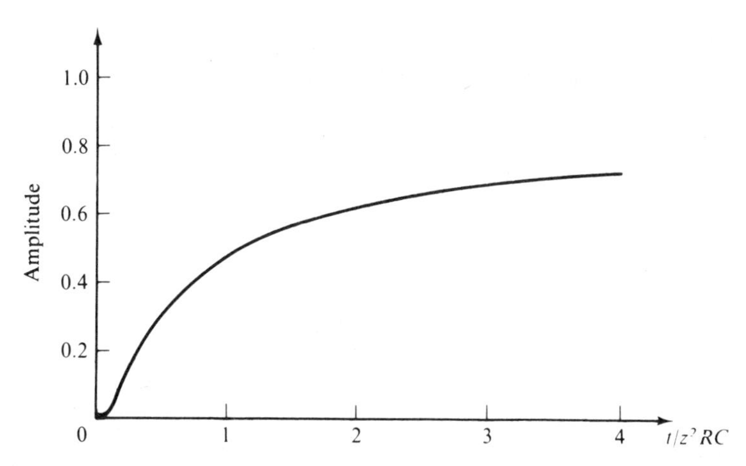

Fig. 6.22. Step response of submarine telegraph cable. The function erfc $[(\tau/4t)^{\frac{1}{2}}]$.

The step response may thus be expressed in the form

$$M(t) = 1 - \text{erf}\,[(\tau/4t)^{\frac{1}{2}}] = \text{erfc}\,[(\tau/4t)^{\frac{1}{2}}]$$

in which $\tau = z^2RC$.

The graphical form of this function is shown in figure 6.22. Certain features may be noted: firstly the delay during which little rise occurs, and secondly the very slow rise towards the final value of unity. The time to rise between 10% and 50% of the final value is found to be $0{\cdot}9z^2RC$. To rise to 70% takes a further $2{\cdot}9z^2RC$. It will be noticed that rise time is proportional to the square of the length of line.

†6.6.4 Lowloss line with skin effect

We will now turn to the case more appropriate to modern lowloss cables. We have to invert, using the expression of (6.12),

$$\bar{h}(s) = \exp\{[-(s/c) - \kappa s^{\frac{1}{2}}]z\}$$
$$= \exp(-sz/c) \times \exp(-\kappa s^{\frac{1}{2}} z)$$

Reference to the theory of the Laplace transform shows that the term $\exp(-zs/c)$ corresponds to a pure delay of z/c. The remaining term is precisely that dealt with in the last section, with $\tau = (\kappa z)^2$. The impulse response is therefore expressed by

$$h(t) = 0 \quad (t < z/c)$$
$$h(t) = (2\tau\sqrt{\pi})^{-1}(\tau/t_1)^{\frac{3}{2}} \exp(-\tau/4t_1)$$

where $t_1 = t - z/c$. The response to a unit step is zero for $t < z/c$ and thereafter is given by the function erfc $(\frac{1}{2}\kappa z t_1^{-\frac{1}{2}})$.

It may be seen from (6.12) that κz is directly related to the attenuation of the length z of the line. If we compare lines by specifying the attenuation w dB at a chosen frequency f_c, then

$$w = 8{\cdot}686 z\alpha(f_c)$$

$$= 8.686 z\kappa(\pi f_c)^{\frac{1}{2}}$$

Hence

$$\tfrac{1}{2}\kappa z t_1^{-\frac{1}{2}} = 0.032 w (t_1 f_c)^{-\frac{1}{2}}$$

The step response thus takes the form

$$M(t) = \text{erfc}\,[0{\cdot}032 w (f_c t_1)^{-\frac{1}{2}}]$$

where w is the loss in decibels at frequency f_c. This response is shown for various values of w in fig. 6.23. The response to a rectangular pulse may be obtained by considering the pulse as a unit step followed later by a unit step in the opposite sense. Results are shown also in fig. 6.23. The pulse width has been chosen to be $1/2f_c$, and the output is given for various values of w, the attenuation at f_c. The distinctive features of the pulse output are (i) the delay in reaching maximum amplitude and (ii) the

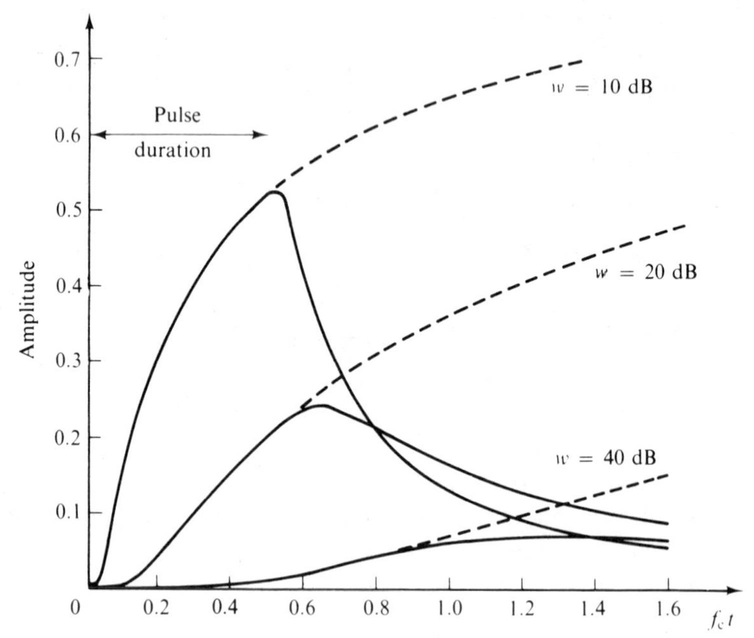

Fig. 6.23. Response of line to a rectangular pulse for different values of attenuation.

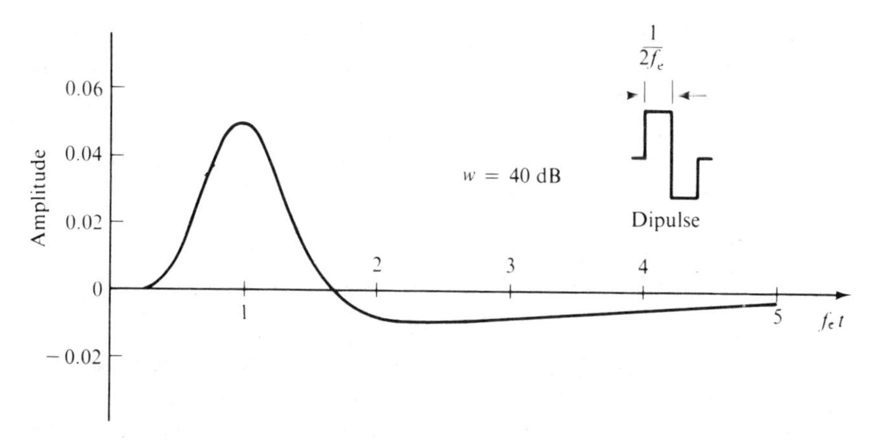

Fig. 6.24. Response of line to a dipulse.

very slowly decaying tail. The latter effect could cause inter-symbol interference in a succession of pulses.

The long tail may be reduced and the pulse confined to a much narrower period by compensating for the high-frequency loss in an equalising circuit. The extent to which this is possible depends on the signal-to-noise ratio, which must be satisfactory at the high-frequency end. A narrower output pulse can also be obtained by using for the input a dipulse, as shown in fig. 6.24 for the case $w = 40$ dB. This has the effect of reducing the low-frequency end of the spectrum. When the high-frequency attenuation is large and the delay longer than the duration of input pulse, the shape of the output pulse is almost independent of the shape of the input pulse. In fig. 6.25 the output for a rectangular pulse of

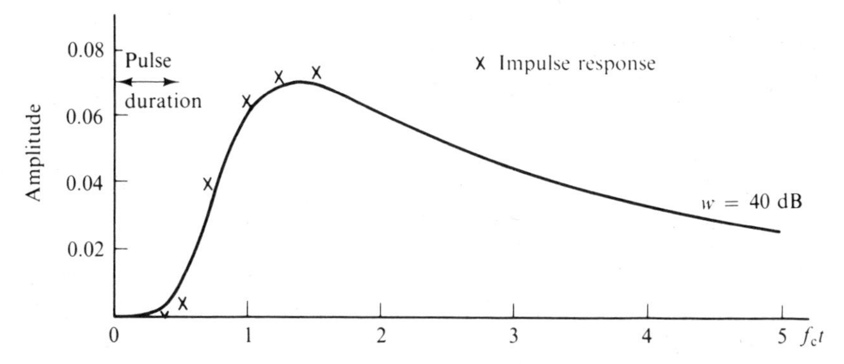

Fig. 6.25. Comparison of response to rectangular pulse with response to impulse.

duration $1/2f_c$ is shown for the case $w = 40$ dB, on a longer time scale than in fig. 6.23. Also shown is the response to an impulse of strength $1/2f_c$ occurring at $f_c t = 0.25$, corresponding to the area and mid-point of the rectangular pulse. It can be seen that these are virtually identical, so that the output shape for any shaped pulse of duration $1/2f_c$ will be very similar.

6.7 Irregularities on lines and testing procedures

It has been mentioned earlier that some degree of variation with length of the electrical characteristics of a transmission line is to be expected. the effect of such variations can be seen by considering irregularities in the value of characteristic impedance. A sudden change in characteristic impedance constitutes a discontinuity in the line. As we have discussed in chapter 5, a discontinuity will cause a reflection to take place. In the first instance a loss of power in the forward wave occurs. Since there will be many irregularities, the reflected wave will itself encounter irregularities and thus produce a forward wave, delayed with respect to the original wave. The total forward wave is thus made up of the wanted signal together with other low-level contributions of various phases. Since the resultant is dictated by the relative phases of the signals, the resultant will change as the frequency is changed. If the irregularities are randomly situated, only a smooth change would be expected. If however irregularities occur in a periodic manner a marked frequency variation would occur. Such periodicities arise through the manufacturing process and by the way in which individual coaxial lines are combined into cables. Joints form a major source of irregularity, but do not occur at precisely the same intervals.

A line can be tested for these defects in various ways. The most obvious is to transmit a single very short pulse along the line and then look for reflected pulses.

6.7.1 Pulse-echo testing

The practical method is to make the line and a simulated line impedance two arms of a bridge. The bridge is then excited with short pulses. The bridge balance prevents most of the input pulse arriving at the bridge detector, which can therefore be of high gain to view subsequent returns. The result of a test carried out on the British Post Office 60 MHz cable is shown in fig. 6.26. The exciting pulse was 10 ns duration.

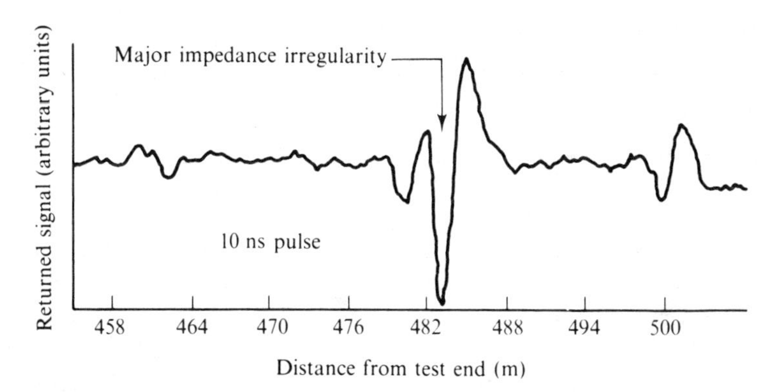

Fig. 6.26. Pulse-echo test on 2.6/9.5 mm coaxial pair (From Still, Stephens & Bundy, *POEEJ* **66**, 174–8).

6.7.2 Time-domain reflectometry

A related test uses, instead of a short pulse, a longer pulse with very sharp rise time. This enables greater discrimination to be made. A rise time of 150 ps occupies a length of 4·5 cm on an air-spaced line, so that resolution, of the order of a few centimetres is possible. In fig. 6.27 is shown a result on the junction between repeater and main transmission cable.

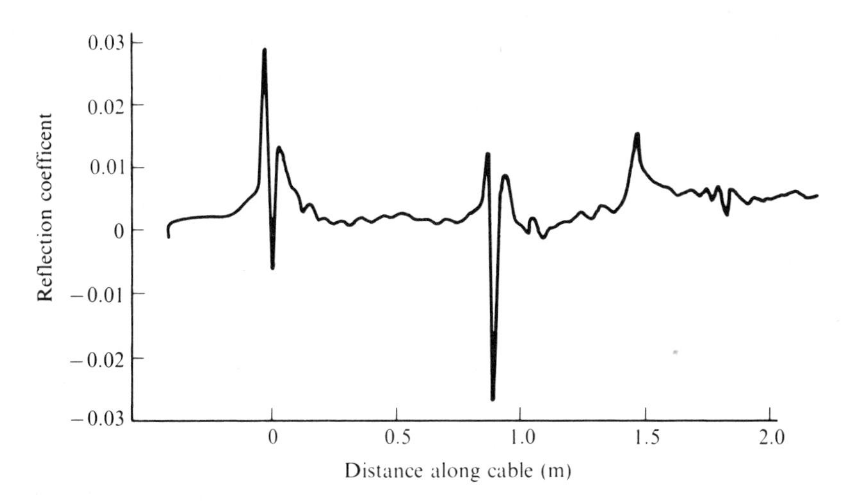

Fig. 6.27. Time domain reflection trace. (From Still, Stephens & Bundy, *POEEJ* **66**, 174–8).

6.7.3 Uniformity of impedance

Periodic irregularities will cause successive reflections. The effect will be most marked when the reflections arrive at the sending ending in phase with each other. Suppose the irregularities occur at the points d, $2d$, $3d, \ldots$ on the line. The delay for an incident wave to arrive at the point nd and for its reflection to return corresponds to a length of $2nd$. The difference in path between two neighbouring reflections is $2d$. If this is a multiple of the wavelength all reflections will add. In terms of frequency, large returns will be expected for frequencies equal to integer multiples of $c/2d$, where c is the phase velocity. This is illustrated by the graph shown in fig. 6.28. The ordinate is the residual gain over a link of 60 MHz cable comprising 36 cable sections and repeaters. The 'rolls' are caused by the junctions between cable and repeater, and have a period about 100 kHz. The phase velocity on the line was known to be $2{\cdot}88 \times 10^8\ \mathrm{ms}^{-1}$, so that the corresponding interval agrees with the repeater spacing of 1400 m.

To test characteristic impedance, a terminated line is matched in a bridge against a standard resistor. The bridge may be excited by various waveforms, and the returned signal will indicate impedance deviations. In figs. 6.29 and 6.30 respectively are shown results for steady excitation of variable frequency and for excitation by a step..

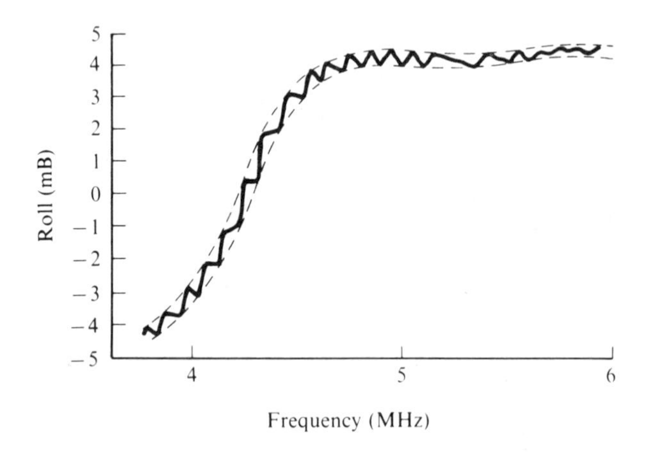

Fig. 6.28. Variation of gain with frequency on a link of 60 MHz cable comprising 36 cable sections and repeaters. (From Still, Stephens & Bundy, *POEEJ* **66**, 174–8).

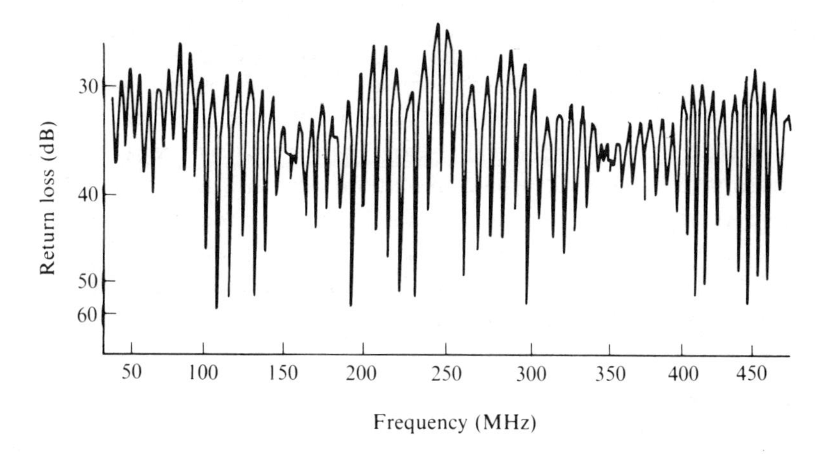

Fig. 6.29. Sweep-frequency return loss. (From Still, Stephens & Bundy, *POEEJ* **66**, 174–8.)

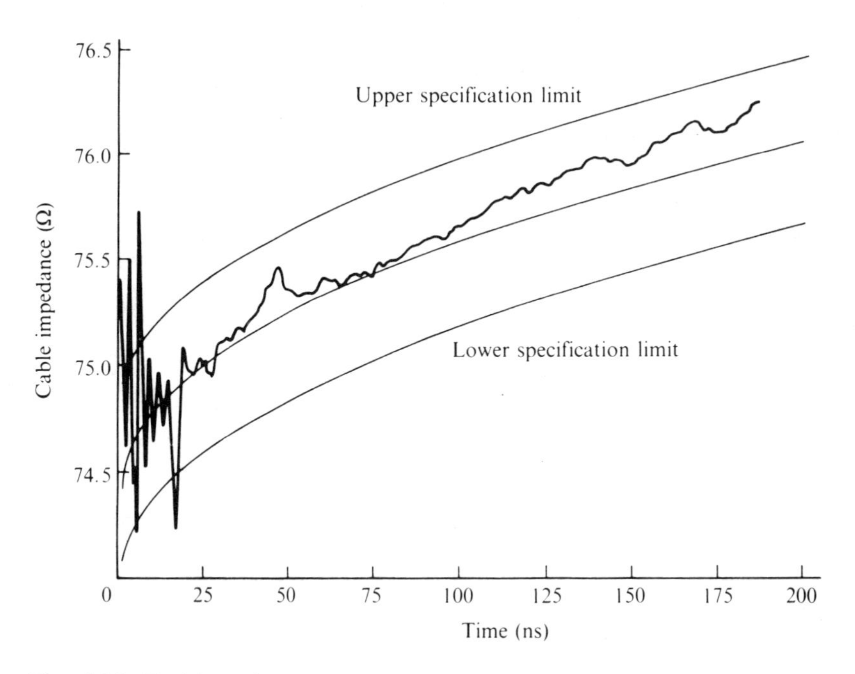

Fig. 6.30. End-impedance trace using step input. (From Still, Stephens & Bundy, *POEEJ* **66**, 174–8.)

6.7.4 Attenuation

Attenuation is measured by comparison of a path containing the line under test with a reference path of variable attenuation. Because of the limitations imposed by the signal-handling capability of repeaters, it is necessary to know and to hold the attenuation of a length of cable to within very precise limits. For the 60 MHz cable the tolerances on a 500 m length of cable vary between $\pm 0{\cdot}07$ dB and $\pm 0{\cdot}39$ dB, at 16 MHz and 484 MHz respectively. Measurement accuracy must be one order of magnitude better than this.

6.7.5 Delay

The delay of a length of line can be measured as follows. Two paths are set up, one containing the line and the other being a reference path of variable attenuation and constant phase delay independent of frequency. The outputs from the two paths are summed, and by adjustment of attenuation and frequency, the frequencies found for which zero output is obtained. Suppose the output from the non-frequency-sensitive path is $\cos(\omega t + \phi_1)$, and the output from the path containing the line is $\cos[\omega(t-\tau)+\phi_2]$. Cancellation will occur whenever ϕ_1 differs from $\phi_2 - \omega\tau$ by an odd multiple of π. If we first find the lowest frequency for which cancellation happens, we shall have

$$\phi_2 - \omega_1\tau = \phi_1 - \pi$$

Since ω increases above ω_1, the next cancellation must be when

$$\phi_2 - \omega_2\tau = \phi_1 - 3\pi$$

In general
$$\phi_2 - \omega_2\tau = \phi_1 - (2n-1)\pi$$

We thus have
$$\omega_n = (2n-1)\pi/\tau + \text{constant}$$

or
$$f_n = (2n-1)/2\tau + \text{constant}$$

Knowledge of n and f_n enables τ to be found. If the length of the line is known then the phase velocity may be determined.

6.8 Long lines with repeaters

The need for incorporating repeater amplifiers into long line links arises because of the attenuation of the line, which, as we have seen, increases predominantly as the square root of the frequency. The loss incurred cannot be made up at the ends because of the presence of noise. We have seen earlier (§ 6.2.2) that a line fed from its characteristic impedance gives over a bandwidth B an available noise output of kTB watts. An input signal of available power P_s will give at the output of the line a signal-to-noise ratio $P_s \exp(-2\alpha l)/kTB$. If this ratio becomes too small, it will be impossible to recover the signal, so that amplification must take place at intervals determined by the attenuation coefficient and the noise level. Various effects follow from a decision to build in repeater amplifiers at fairly frequent intervals, apart from the problems of reliability: we have to concern ourselves with the noise properties of the amplifiers and with equalising at each point the uneven loss characteristic of the cable; the impedance matching of repeater to line at regular intervals affects the transmission curve; the absolute power levels become important, because of the non-linearity inherent in all amplifiers; non-linearity destroys the complete independence of the various channels in an FDM system, so that we have to concern ourselves with intermodulation products and intermodulation noise; it is necessary to feed power to the repeaters along the transmission line, involving filtering. We have in addition a decision to make about the provision of two paths: these can be provided by two separate lines or by the same line with path separation in the amplifiers.

Although most of these problems are matters of circuit design, they do influence the propagation characteristics of a link and we shall look briefly at them.

6.8.1 Noise in amplifiers

In order to specify an amplifier we have to give figures of merit for its various properties. We consider in this section the noise performance. An amplifier produces noise in virtue of its components and the currents flowing in the active devices. It is always necessary to think of an amplifier in conjunction with a particular source impedance, since change of source impedance will change both the signal power entering the amplifier and the noise the amplifier produces. This being the case it is convenient to characterise the amplifier by reference to the source: we imagine the amplifier to be replaced by an ideal one which has the same gain but which is noiseless, and we then ask how much extra noise needs to be delivered by the source to give the same signal-to-noise ratio at the output of the amplifier. (We shall consider only amplifiers for which the gain is sufficiently high to allow us to disregard noise sources after the output.)

Let us consider a source of resistive impedance R_s. We may characterise the source by the maximum power which can be drawn from it, the power delivered into a load of R_s. In terms of open-circuit voltage E_s, this may be expressed as

$$P_s = E_s^2/4R_s$$

If the source is at temperature T_s, the ratio of signal power to noise power across the resistor is equal to P_s/kT_sB, in which B is the bandwidth of a subsequent filter. (It is to be noted that this is true for any load connected across the source,, since the signal and noise are mismatched to the same extent.) The signal-to-noise ratio at the output of the real amplifier connected to this source will be of some value $(S/N)_{out}$. We define the *noise figure n* by the equation

$$n = \frac{P_s/kT_sB}{(S/N)_{out}}$$

We note that the value of n depends on choice of T_s. This is usually taken as 290 K. The same $(S/N)_{out}$ will be obtained if the amplifier is made noiseless and the source considered to have noise power nkT_sB. Since uncorrelated noise powers add, we may regard this as being divided up in the form

$$nkT_sB = kT_sB + (n-1)kT_sB = kT_sB + kT_eB$$

The equivalent temperature T_e is specific to the amplifier (in conjunction with the given source impedance) since if the source were made noiseless the amplifier contribution would remain the same. T_e is the *noise temperature* of the amplifier.

The above definitions are valid for any two-port, although we may then have to take into account noise sources after the two-port. An example is the length of transmission line considered in § 6.2.2. The available noise output per unit bandwidth is given by

$$kT_s\,e^{-2\alpha l} + kT_l(1 - e^{-2\alpha l})$$

In the absence of any source noise ($T_s = 0$), the noise output must be equal to $kT_e\,e^{-2\alpha l}$. Hence

$$T_e = T_l(e^{2\alpha l} - 1)$$

If we consider a situation typical of a length of feeder of attenuation 0·1 dB and $T_l = 290$ K we find $T_e = 6{\cdot}6$ K. Although not of significance in the present context of repeatered lines, this situation is relevant to earth stations for satellite communications when T_s is effectively only a few degrees Kelvin.

Returning now to the problem of a repeater-amplified line system, for normal operation we can assume that each stage of line plus repeater has an overall zero loss, or in other words that the available power gain of the repeater is equal to the attenuation of the preceding section of line. (We shall assume that the various elements of the system are correctly matched, in which case the actual power gain equals the available power gain.) Consider the situation illustrated in fig. 6.31(a). Each amplifier has an available gain g, and a noise figure n. The line has attenuation g and temperature T_0. The noise input p_0 to the first repeater may be assumed to be derived from a matched source also at T_0, and the noise output p, from that repeater will be $gnkT_0B$, in which B is the bandwidth. We can calculate the noise input to the next repeater with the aid of (6.6). In that equation kT_s is now equal to $gnkT_0$ and $\exp(-2\alpha l)$ becomes $1/g$, giving the noise input to the second repeater as

$$[gnkT_0g^{-1}+kT_0(1-g^{-1})]B=[(n+1)kT_0-kT_0g^{-1}]B$$

To find the output at the second repeater we must add to this input

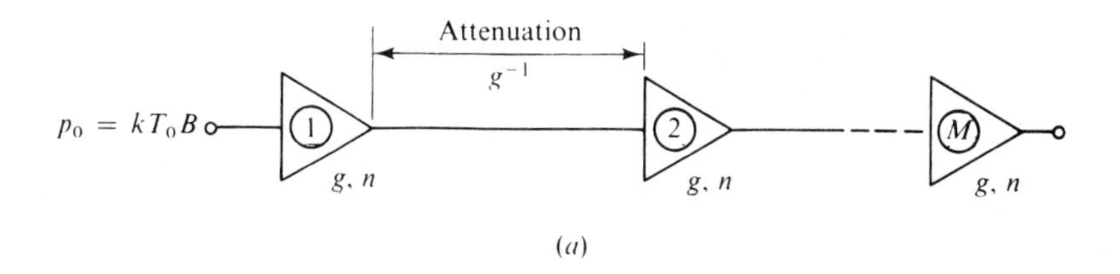

(*a*)

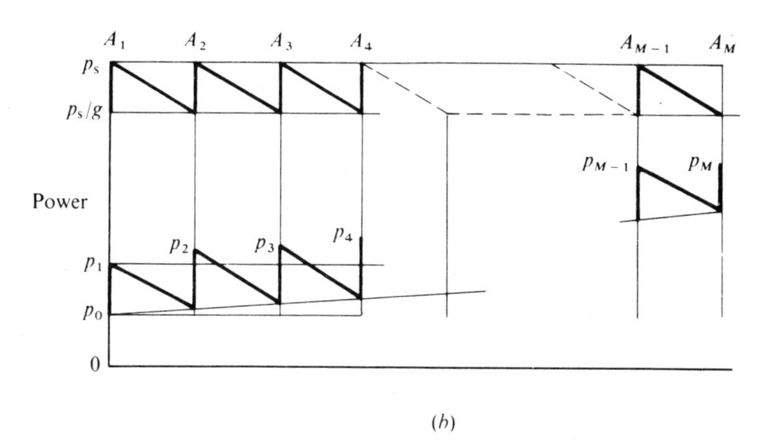

(*b*)

Fig. 6.31. Signal-to-noise ratio on a repeatered line. (*a*) Configuration of line and repeaters. (*b*) Variation of signal-to-noise ratio along the line.

$(n-1)kT_0B$ for the second repeater and multiply by g, giving an output

$$p_2 = 2gnkT_0B - kT_0B$$

Similar calculations show that after M stages the noise output is

$$p_M = MgnkT_0B - (M-1)kT_0B$$

The last term will be negligibly small, so that we may write for the noise output after M stages

$$p_M = nkT_0BgM$$

or, expressed in terms of decibels with respect to 1 W,

$$P_M = N + G + 10\log_{10} M + 10\log_{10} kT_0B$$

where $N = \log_{10} n$ and $G = 10\log_{10} g$.

For a signal output level per channel of p_s, the signal-to-noise power ratio at the output of the first repeater is $p_s/(gnkT_0B)$ whilst at the output of the Mth repeater, it will be $p_s/(gnkT_0BM)$. This is illustrated diagrammatically in fig. 6.31(*b*).

For example, consider a system 2000 km long comprising 100 repeater/line sections. For a channel bandwidth of 4 kHz at 290 K, $kT_0B = 1.6\times10^{-17}$ W. This can be expressed as -168 dB with respect to one watt (dBW) or, more conventionally, with respect to one milliwatt, as -138 dBm.

If each repeater has a gain of 50 dB at the highest frequency and a noise figure of 10 dB, then the total system noise per channel is

$$p_M = MgnkT_0B - kT_0B(M-1)$$
$$= -58 \text{ dBm}$$

If the final signal level is -15 dBm, then $(S/N)_{out}$ is 43 dB. At the output of the first repeater, $(S/N) = -15 - 50 - 10 + 138 = 63$ dB.

Some observations can be made with respect to this example. If we reduce the noise factor of the amplifier, we clearly reduce the output noise. If we double the number of amplifiers, the gain of each one can be reduced to 25 dB (since the line attenuation is proportional to the length) whilst the term $10\log_{10}(2M)$ increases by 3 dB. Thus the overall noise level is reduced by 22 dB. However, we have at least doubled the cost of building and installing the amplifiers, and this economic consideration may outweigh the improved performance.

6.8.2 Intermodulation

A further consequence of using repeater amplifiers in wideband FDM systems is the intermodulation distortion caused by the inevitable non-linearity of the amplifier transfer characteristic. Although negative feedback is used in these amplifiers to improve gain stability and reduce distortion, this can give rise to very rapid deterioration of performance if the amplifier is overloaded. An arbitrary definition of overload is 'that level of absolute output power, in dBm, at which the absolute level of third harmonic distortion increases by 20 dB when the input changes by 1 dB'.

The general non-linear transfer characteristic can be expressed as a simple power series,

$$v_2 = a_1 v_1 + a_2 v_1^2 + a_3 v_1^3 + \cdots$$

where the a_i are constants, v_2 is the output and v_1 the input to the system.

If we let $v_1 = A \cos \omega_1 t + B \cos \omega_2 t$, where A and B are constants, that is v_1 is the sum of two separate frequency signals, then substitution in the expression for v_2 will give rise to harmonic terms and sum and difference terms of the two frequencies ω_1 and ω_2. Thus, considering the first three terms of the general polynomial expression, we have

$$\begin{aligned} v_2 = {} & a_1(A \cos \omega_1 t + B \cos \omega_2 t) \\ & + a_2(A \cos \omega_1 t + B \cos \omega_2 t)^2 + a_3(A \cos \omega_1 t + B \cos \omega_2 t)^3 \end{aligned}$$

Expanding and collecting terms gives the following component magnitudes:

$$\begin{aligned}
\text{d.c.:}\quad & \tfrac{1}{2}a_2(A^2 + B^2) \\
\omega_1:\quad & a_1 A + a_3(\tfrac{3}{4}A^3 + \tfrac{3}{2}AB^2) \\
\omega_2:\quad & a_1 B + a_3(\tfrac{3}{4}B^3 + \tfrac{3}{2}A^2 B) \\
2\omega_1:\quad & \tfrac{1}{2}a_2 A^2 \\
2\omega_2:\quad & \tfrac{1}{2}a_2 B^2 \\
3\omega_1:\quad & \tfrac{1}{4}a_3 A^3 \\
3\omega_2:\quad & \tfrac{1}{4}a_3 B^3 \\
\left.\begin{matrix}\omega_1 + \omega_2 \\ \omega_1 - \omega_2\end{matrix}\right\}:\quad & a_2 AB \\
\left.\begin{matrix}2\omega_1 + \omega_2 \\ 2\omega_1 - \omega_2\end{matrix}\right\}:\quad & 3a_3 A^2 B \\
\left.\begin{matrix}2\omega_2 + \omega_1 \\ 2\omega_2 - \omega_1\end{matrix}\right\}:\quad & 3a_3 B^2 A
\end{aligned}$$

Clearly an increase in level of A and B by 1 dB gives rise to 2 dB increase in each second harmonic level. Similarly each third-order term increases by 3 dB. Thus, the relative magnitude of the harmonics is a function of the level of the fundamental signals. Furthermore, in a cascaded series of similar amplifiers, the intermodulation process becomes very involved, since the distortion products of the first amplifier which fall within the operating bandwidth of the second amplifier will themselves introduce further intermodulation products, and so on throughout the complete link. Further complications arise since in practice the repeater spacing is a random quantity, and this means that the phase of the harmonic components at the output of an amplifier are also randomly varying. In theory, we can distinguish between second and third harmonic components in this respect, and show that (for a simple third-order approximation to the non-linearity of a repeater) the second harmonic distortion accumulates as a summation of mean power, whilst third harmonic distortion accumulates on a voltage basis. In practice, intermodulation measurements are made using either discrete tones or alternatively a white-noise source, which more nearly simulates the condition of a multiplexed voice-channel system.

6.8.3 Operating levels

The average power of a speech waveform is difficult to define, but a useful design specification, based on extensive measurements, is an average power level per channel of -15 dBm. This includes speech and signalling tones and gives 'normal' quality reproduction in a telephone circuit. For an M-channel multiplexed system, therefore, we would expect an average or mean power $P_M = -15 + 10 \log_{10} M$ dBm. We must consider also an additional level representing the probable peak-to-mean power ratio for a given number of channels. This 'multi-channel load factor' must be determined empirically, and fig. 6.32 shows a typical curve of peak-to-mean power ratio against number of channels. Due to the random nature of the signal in a speech channel, the ratio tends to a limiting level of about 11 dB as the number of channels increases. (It is found, in practice, that if the number of channels exceeds about 240, the overall characteristics of the multiplexed signal are very similar to those of normally distributed white noise.)

For example, a system carrying 1000 channels will have a mean power level of $-15 + 10 \log_{10} 1000 = +15$ dB m, and a peak power level of approximately $+26.5$ dB m ($15 + 11.5$ dB from fig. 6.32). It must be

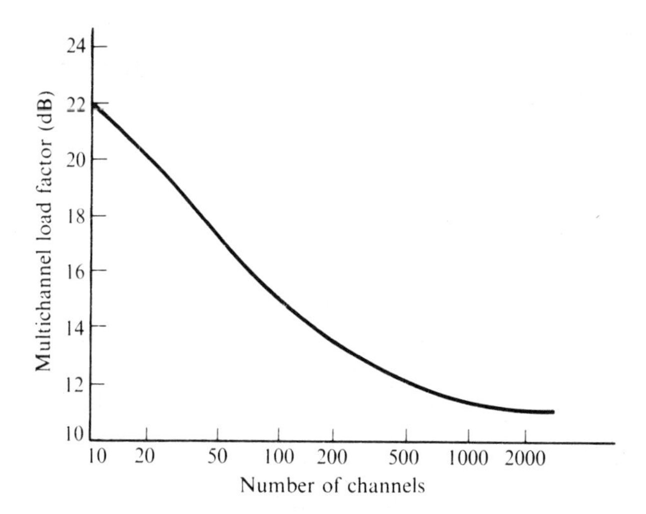

Fig. 6.32. Variation of peak-to-average power ratio with number of speech channels.

emphasised that these figures are subject to statistical variations due to the random nature of the signal. It is not possible to guarantee that a power level is never exceeded, and in practice some form of overload detection and limiting circuit is used to introduce additional attenuation when required. We can represent the various power levels of an operational link on a diagram as shown in fig. 6.33. The overload margin includes an allowance for operating tolerances of amplifiers and equalisers, and for variations of cable attenuation with temperature. The total noise level is the sum of the thermal and intermodulation noise components. The particular division of these noise powers depends

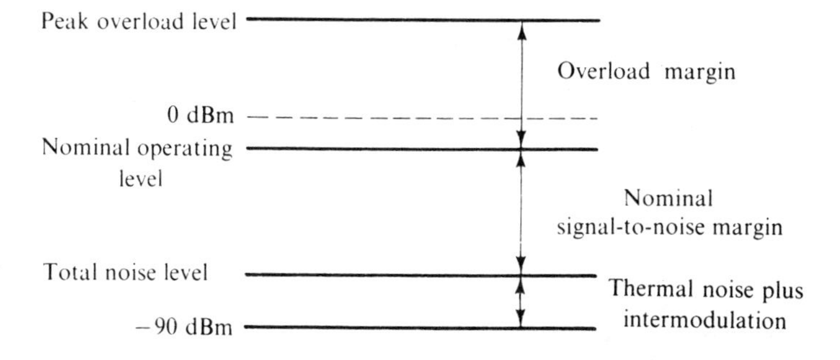

Fig. 6.33. Power levels of different signal components on a repeatered cable.

upon many operational factors, but, basically, the higher the nominal operating level, the greater the contribution of intermodulation distortion to overall noise. Conversely, the higher the operating level, the better the signal-to-thermal-noise power ratio.

6.8.4 Two-wire to four-wire working

The normal telephone transmitter/receiver operates on a single pair of wires. Long-distance links frequently use separate go and return paths since this simplifies the design of the intermediate repeaters, and the transition from two-wire to four-wire working is effected by means of a hybrid transformer, shown in schematic form in fig. 6.34. The balance network, Z_n, is a lumped circuit which ideally has an input impedance equivalent to the impedance of the two-wire line. If the transformers are assumed ideal, the mode of operation is as follows. A signal on the input pair of the four-wire side of the device induces equal signals in the two-wire termination and the balance network, and no signal appears in the output pair of the four-wire circuit. Similarly an input signal from the two-wire system is equally divided between the input and output four-wire circuits, but since the amplifiers in these circuits are non-reciprocal, the signal fed to the return amplifier is simply dissipated. Thus the hybrid transformer permits two-wire to four-wire interconnection at the expense of a 3 dB power loss in each direction. In practice, there is an additional insertion loss of approximately 0.5 dB. The requirement that the balance network impedance exactly equals the line impedance is unrealisable in practice, and this gives rise to two important effects, which we shall discuss briefly.

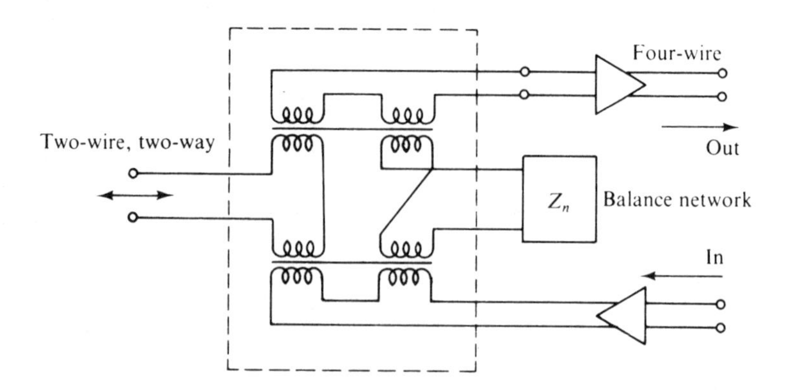

Fig. 6.34. A hybrid transformer.

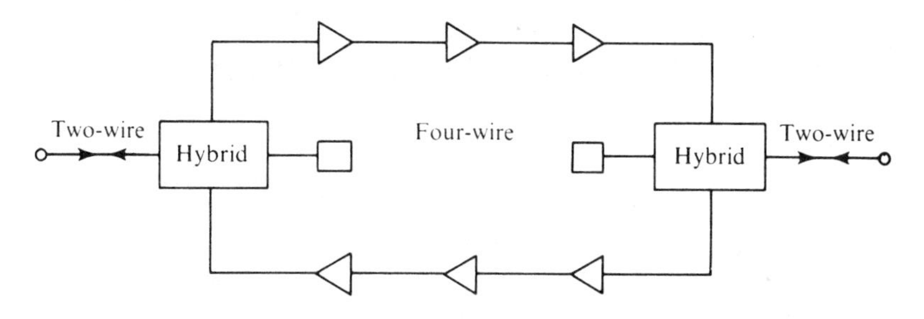

Fig. 6.35. Repeater system incorporating two-wire to four-wire transition.

Figure 6.35 shows the block schematic of a typical repeatered system using a two-wire to four-wire transition. Since the isolation between the go-and-return pairs of the four-wire section is not complete, there is clearly a loop circuit comprising the go-and-return amplifiers in cascade with the two hybrid transformers. As already mentioned, non-ideal terminations give rise to coupling between the go-and-return points of the hybrid transformer, and it can be shown that the effective attenuation of this coupling is expressed as

$$L = 20 \log_{10} |k|^{-1} \text{ dB}$$

where $$k = (Z_n - Z_L)/(Z_n + Z_L)$$

(The expression for k is that used in chapter 5 to describe the reflection coefficient at a mismatched termination.) Clearly, when $Z_n = Z_L$, $k = 0$ and L is infinite. This would allow any degree of amplification in the go and return paths. Generally, the total loop attenuation is given by $2(L+7+A)$ dB where A is the net link attenuation in each direction or the transmission-line loss minus the amplifier gain. The 7 dB represents the total hybrid loss (2×3.5 dB) in each direction. The relative stability of any linear feedback network is a function of the total loop gain, and normally the magnitude of this gain should not exceed unity. If this condition is not met, oscillations at some audible frequency may result, and even if the loop gain is then reduced slightly, unacceptable ringing or slowly decaying responses may result. For satisfactory operation it is normally required that the total loop loss should be greater than 6 dB. Thus, in the worst situation of mismatch between Z_n and Z_L, as $|k| \rightarrow 1$, $L \rightarrow 0$ dB and we have $2(7+A) > 6$ or $A > -4$. This suggests that the maximum net *gain* in either go or return path should not exceed 4 dB.

The second effect of the loop transmission path is the production of unwanted echoes produced by reflections at impedance discontinuities. The subjective effect of echoes is a function of the amplitude and time

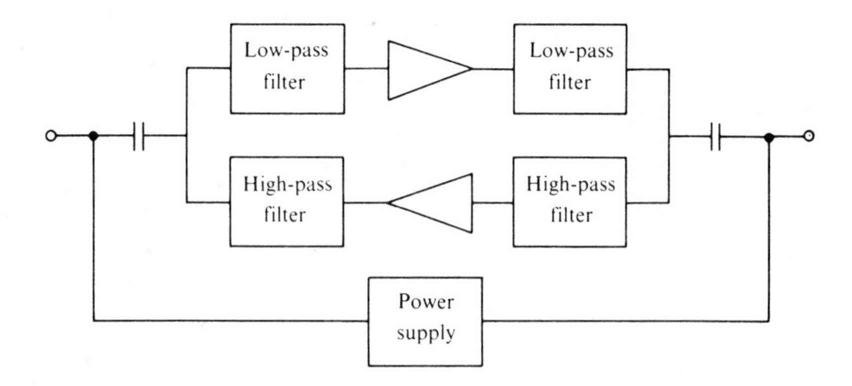

Fig. 6.36. Provision of go-and-return paths in a submarine repeater.

delay of the unwanted signal, and these parameters depend very much on the type of transmission medium used. Clearly, if L and A are increased, the production of echoes is minimised as far as the main loop is concerned. Further complication arises from the 'cross-talk' or coupling occurring between the separate pairs of conductors in a go-and-return path particularly when large differences in signal level occur on adjacent conductors.

In the case of submarine cables, it is uneconomic to provide a four-wire system, and one current technique employed (on the trans-Atlantic CANTAT II cable) divides the available bandwidth of the link, approximately 14 MHz, into two distinct bands, using frequencies up to 6.5 MHz in one direction and frequencies from 6.5 to 14 MHz in the return direction. This permits repeater configurations as shown in fig. 6.36. Submarine-cable systems normally operate with a restricted voice-channel bandwidth of 3 kHz to provide increased overall capacity.

6.8.5 Equalisation

The overall performance of a repeater amplifier system has to be maintained within very close limits in order to satisfy the stringent operating requirements laid down by the International Telecommunications Union. For example, overall gain variation of less than ± 1 dB is expected in a long-haul system having a total link attenuation in excess of 10^4 dB. Equalization is the process of correcting irregularities in the parameters of a given link by means of preset and/or adjustable networks. For example, it is not practicable to space repeaters at exactly equal distances. Since each repeater amplifier has a fixed

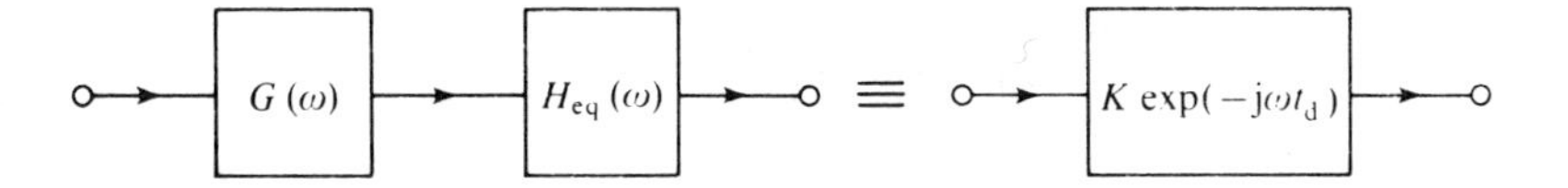

Fig. 6.37. Ideal equalisation.

gain/frequency characteristic designed to compensate for a standard length of cable, additional lumped circuits, called line build-out networks, can be introduced to effectively increase the cable length. The gain characteristic just mentioned is itself equalised with respect to ambient temperature variations by means of a thermistor control circuit within the amplifier. Further circuits afford equalization of the group-delay characteristics of the channel. However, in a long-haul system, the statistical variation of these 'fixed' equalisation networks necessitates the inclusion of automatically adjustable (adaptive) equalisers at discrete intervals. These are controlled by signals derived from continuous measurements made on pilot tones transmitted over the link.

In the ideal situation, an equaliser can be represented by a two-port network with a particular transfer function, such that when cascaded with the transmission system, the overall response is distortionless. Figure 6.37 illustrates this simple idea.

We noted earlier than an ideal communication link has an overall transfer function $G(\omega) = K \exp(-j\omega t_d)$, i.e. has a constant gain K, and uniform time delay t_d. As shown in fig. 6.37, the cascaded system and equaliser should approximate to this ideal. Thus

$$G(\omega) \cdot H_{eq}(\omega) = K \exp(-j\omega t_d)$$

and

$$H_{eq}(\omega) = K \exp(-j\omega t_d)/G(\omega)$$

Whilst appearing to be straightforward, the practical realisation of the necessary network is often very difficult indeed, and normally some compromise solution is reached, such that the overall characteristic is operationally acceptable, but not ideal.

One versatile form of adjustable equaliser is the tapped delay line or transversal filter. The operation of this filter can be explained by reference to fig. 6.38, which shows a simple three-stage device.

The output $y(t)$ can be expressed as a function of the input $x(t)$ and delayed versions of the input.

Thus $y(t) = C_0 x(t) + C_1 x(t-\tau) + C_2 x(t-2\tau)$, where the C_i are weighting coefficients.

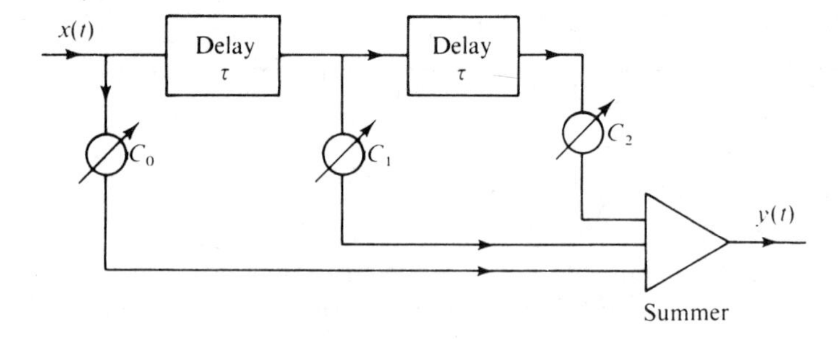

Fig. 6.38. Transversal filter for equalisation.

In terms of frequency spectra this equation becomes

$$Y(\omega) = C_0X(\omega) + C_1X(\omega)\,e^{-j\omega\tau} + C_2X(\omega)\,e^{-2j\omega\tau}$$

$$= X(\omega)\,e^{-j\omega\tau}[C_0\,e^{j\omega\tau} + C_1 + C_2\,e^{-j\omega\tau}]$$

By suitable choice of the weighting coefficients we can produce an approximation to the required transfer function. The greater the number of taps, the more flexible the overall equaliser becomes. One obvious advantage of this type of network is the relative ease with which it could be incorporated in an automatic or adaptive equaliser system.

6.9 A telephone FDM system

In previous sections we have outlined some of the problems associated with establishing a repeatered communication link. As an illustration of a practical system, we shall consider, very briefly, the development of the 60 MHz FDM coaxial cable system introduced by the British Post Office. This represents the culmination of many years of development work and has its roots in the gradual enhancement of the existing systems as new devices and techniques have been developed.

6.9.1 System specification

The system is based on a 2.6/9.5 mm (inner-conductor outer-diameter/outer-conductor inner-diameter) coaxial cable which is capable of operating satisfactorily at frequencies up to 500 MHz, although the present requirement is for a 60 MHz bandwidth. The large available bandwidth presumes future development of a high-capacity digital transmission system. Eighteen of these coaxial cables (plus twelve 0.63 mm diameter copper quads) are contained in a single sheath. Each cable is able to support twelve hypergroups (or $12 \times 900 = 10\,800$ 4 kHz channels). Thus the combined capacity of the composite cable using a pair of coaxial lines for each 10 800 channels is $9 \times 10\,800 = 97\,200$ channels. In fact one pair of cables is reserved as a standby, and so 86 400 circuits in each direction are actually available.

A brief specification is:

maximum length between terminals	278 km
power-feeding stations spacing	66 km
dependent repeater spacing	1.5 km
overall gain/frequency response spread	<0.5 dB

overall insertion loss change <0.5 dB for $0-20$°C ambient.

The overall noise per telephone channel is required to be less than the equivalent of 1.5 pW km^{-1}.

6.9.2 Repeater amplifiers

These are unidirectional transistorized four-stage amplifiers with overall negative feedback designed to give a nominal gain of 27.75 dB at 61.16 MHz. The gain/frequency characteristic is shaped to compensate for the cable attenuation. There are two types of repeater – non-regulated and regulated. The former have preset adjustable attenuations of -1 to $+2$ dB from the nominal, in 0.5 dB steps. Approximately one in five repeaters are regulated in that the gain is automatically adjusted to compensate for parameter changes, as follows. A pilot signal at 61.16 MHz is transmitted over the system. If the level of this signal at a terminal or main station varies by more than ± 0.2 dB it is corrected there by an automatic-gain-controlled pilot amplifier. At the same time, command signals at either 14 kHz or 17 kHz are sent back along the section to the previous main station. These signals can traverse the complete section without amplification via the power-feeding circuits. At the main station, the 14 or 17 kHz tones control a variable-frequency oscillator which in turn modulates a 2.9 MHz regulating pilot tone. This latter is transmitted over the section and causes the gain of each regulated repeater to be changed by the same amount. The 2.9 MHz tone is suppressed at each main station so that each section is independently regulated.

In addition, there are line build-out networks of 2.5 and 5 dB attenuation to allow for non-uniform spacing of the repeaters. Thus a spacing of 1100–1600 m is possible, although in practice the spacing is 1500 m ± 55 m.

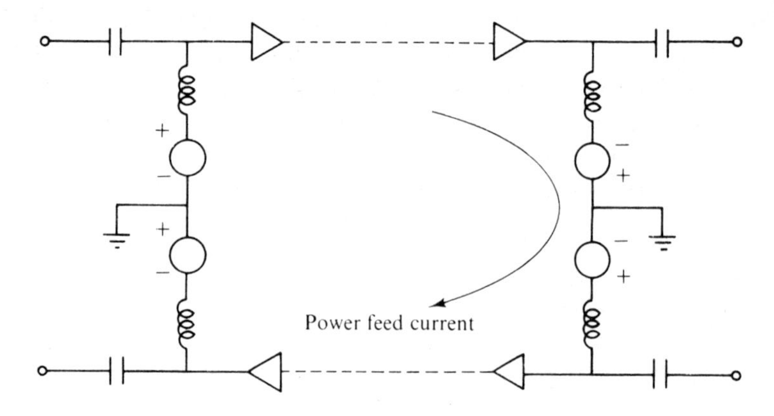

Fig. 6.39. Power supply for repeater amplifiers.

The compensation strategy used is a combination of pre- and post-regulation. This is designed to provide ±1.4 dB change for ±10 °C variation about a mean temperature of 10 °C. This technique ensures that an amplifier mid-way between regulating stages operates at a constant level, and the overall change between summer and winter is minimised.

6.9.3 Power supply

Each repeater requires approximately 21 V d.c. for normal operation, which means that a 66 km section with 1.5 km spacing between repeaters must be supplied with roughly 1000 V. In practice, the section is supplied by two 500 V constant-current sources of opposite polarity, as shown in fig. 6.39. The generators are designed to give a constant current of 110 mA and, under these conditions, an accidental connection between inner and outer conductor, by an operator for example, results in a redistribution of voltage around the circuit, and, ideally, a condition of zero volts at the fault.

6.10 Pulse code modulation on audio cables

Time-division multiplexing of up to thirty separate telephone channels on a single audio pair is now possible using the technique of pulse code modulation (PCM). The principles of operation can be explained briefly with reference to fig. 6.40. A band-limited analogue signal (fig. 6.40(a)) can be represented by a sequence of equally spaced samples of the continuous waveform if these samples are taken at intervals less than or equal to one half of the period of the highest frequency components of the signal. Thus a voicc channel, band limited to 4 kHz, can be accurately sampled at a frequency of 8 kHz, or sampling intervals of 125 μs.

The sampled values are quantised, that is they are rounded to the nearest whole value in a given scale (fig. 6.40(b)) and translated into a suitable code for transmission over the line. In practice, a non-linear

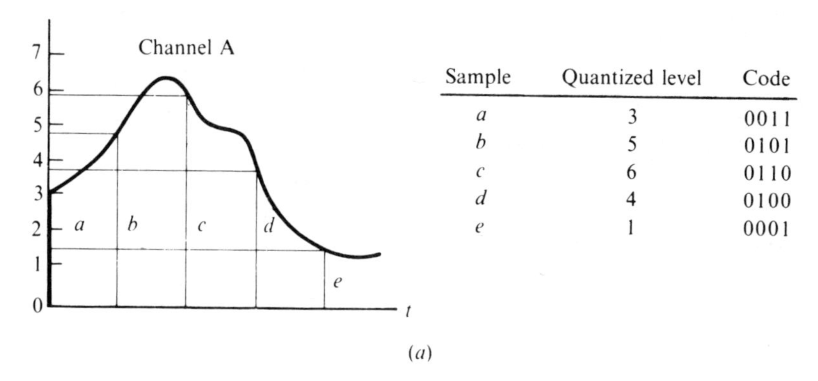

Sample	Quantized level	Code
a	3	0011
b	5	0101
c	6	0110
d	4	0100
e	1	0001

(a)

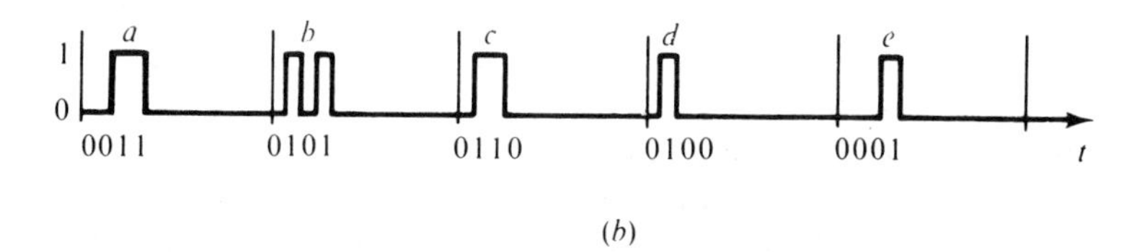

(b)

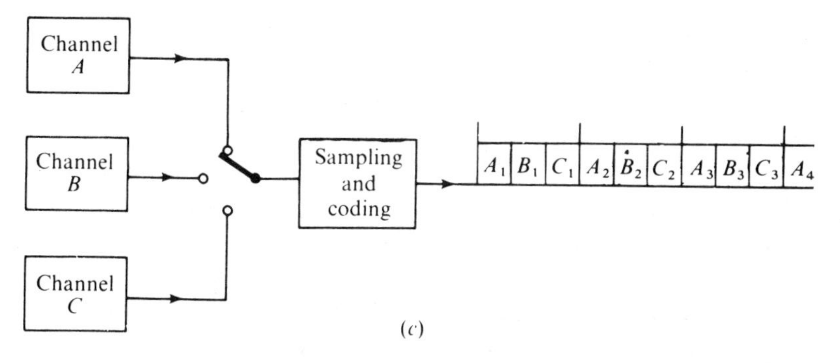

(c)

Fig. 6.40. Principles of pulse code modulation (PCM). (a) Sampling. (b) Quantised signal. (c) Multiplexing.

quantising characteristic is used to provide smaller increments in level for small-amplitude signals, thus improving the signal-to-quantising-error ratio for low levels. By interleaving samples from other audio channels, a composite coded pulse stream is transmitted over the audio line (fig. 6.40(*c*)). Each channel, using 8-bit binary coding (giving $2^8 = 256$ levels) sampled at 8 kHz, produces pulses at a rate of 64 kbit s^{-1}, and so a 30-channel system has an overall bit rate of 2.048 Mbit s^{-1}. Thus each pulse can be 488 ns in duration, but to allow for adequate separation in time, a 50% duty cycle is used, and so each transmitted pulse is 244 ns long. We saw in § 6.6 the problems of transmitting short-duration pulses over transmission lines, and in order to maintain adequate fidelity over a long link, it is necessary to regenerate the transmitted signal at frequent intervals – nominally every 2000 yards – along the line. However, the significant process in this procedure is that of regeneration (as opposed to amplification in analogue cable systems). As long as the pulse train can be unambiguously detected at the repeater by means of a suitable threshold circuit, this is the only information required to re-fabricate the original signal completely. Thus, the regeneration process, in principle, removes the effects of any noise and distortion associated with the transmission. In practice, there will be some error introduced, but in terms of peak signal to r.m.s. noise power ratio a binary PCM system only requires about 20 dB for acceptable distortion levels, whilst an analogue FDM system requires something of the order of 60 to 70 dB.

At the receiving end of the link, the incoming bit stream is separated sequentially into the individual channels, which are then decoded and, using a non-linear characteristic complementary to that at the input, converted into pulse samples of varying amplitude. Finally, these samples are passed through a low-pass filter to produce an analogue output.

Clearly, the operation of this system depends crucially on accurate timing. The clock frequency at the regenerators and receiving terminals is derived from the transmitted bit stream, using a high-Q tuned circuit, but timing jitter can result from the randomly varying phase characteristic of the fundamental component of the pulse train. The deterioration of the pulse shape can give rise to inter-symbol interference (analogous to adjacent channel interference in a FDM system) and this is reduced by equalisation applied prior to the threshold detector in the regenerator (and receiver). The presence of noise inevitably causes errors in the threshold circuit, and fig. 6.41 shows the relationship between error rate and signal-to-noise ratio for a typical threshold

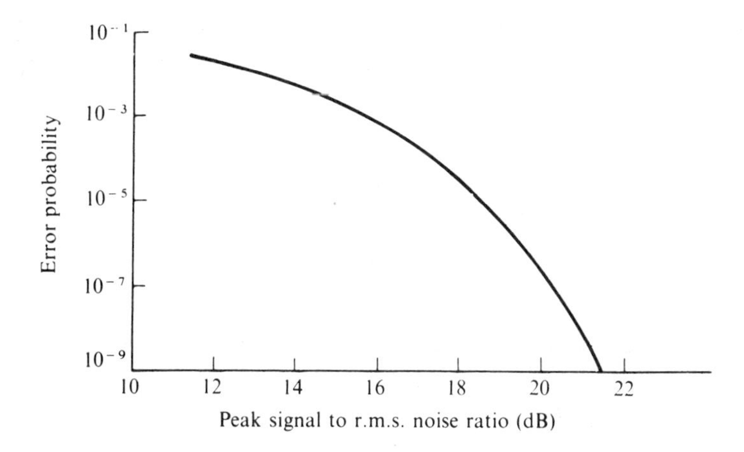

Fig. 6.41. Dependence of error on signal-to-noise ratio for PCM system.

detector. The actual threshold level is derived from the mean pulse amplitude received from the line, thus providing some overall attenuation equalisation.

6.10.1 Advantages of PCM

The overall impact of PCM systems has already been very significant, and extensive developments are currently planned. The following are some advantages afforded by PCM over analogue systems.

(*a*) The transmission of the coded signal is virtually distortionless and independent of the length of the link.

(*b*) Higher levels of interference can be tolerated from noise, cross-talk, etc.

(*c*) Although regenerative repeaters must be closely spaced on audio lines, they are relatively cheap and robust.

(*d*) Since the signals are transmitted in digital form, it is possible to envisage exchanges operating directly with this and other digital data, and large integrated computer-controlled switching systems become possible.

(*e*) If we choose to use a coaxial cable or waveguide system as the transmission link, it is possible to develop hierarchical TDM systems that will operate at bit rates of several hundred megabits per second.

6.11 Worked example

A coaxial line has a copper inner of diameter 3 mm and an aluminium outer of inside diameter 10 mm. It is filled with dielectric of relative permittivity 2.5 and negligible loss. Obtain expressions for the characteristic impedance and propagation constant valid at frequencies for which the skin depth is much less than the radius of the inner.

Solution. If skin effect is ignored, the problem has been solved earlier. It was shown in the problem of § 2.2.5 that a wire of radius a has an impedance because of skin effect given by

$$\frac{1}{2\pi a}\sqrt{\left(\frac{\omega\mu}{2\sigma}\right)}(1+\mathrm{j})$$

To find the total series impedance per unit length we have to allow for both conductors and the inductance per unit length. The latter for a coaxial line is $(\mu/2\pi)\ln(b/a)$. Hence the series impedance per unit length is equal to

$$Z_1=\mathrm{j}\omega\frac{\mu_0}{2\pi}\ln(b/a)+\frac{1}{2\pi}(1+\mathrm{j})\sqrt{(\tfrac{1}{2}\omega\mu_0)}\left(\frac{1}{a\sqrt{\sigma_\mathrm{i}}}+\frac{1}{b\sqrt{\sigma_0}}\right)$$

$$=\mathrm{j}\omega\frac{\mu_0}{2\pi}\ln(b/a)\left[1+\frac{(1-\mathrm{j})}{\ln(b/a)}(2\omega\mu_0)^{-\frac{1}{2}}\left(\frac{1}{a\sqrt{\sigma_\mathrm{i}}}+\frac{1}{b\sqrt{\sigma_0}}\right)\right]$$

The shunt admittance per unit length is $\mathrm{j}\omega C$, or

$$Y_2=\mathrm{j}\omega\frac{2\pi\varepsilon}{\ln(b/a)}$$

Hence

$$Z_0=\left(\frac{Z_1}{Y_2}\right)^{\frac{1}{2}}$$

$$\approx\frac{1}{2\pi}\sqrt{\left(\frac{\mu_0}{\varepsilon}\right)}\ln(b/a)\left[1+\frac{1}{2}\frac{(1-\mathrm{j})}{\ln(b/a)}(2\omega\mu_0)^{-\frac{1}{2}}\left(\frac{1}{a\sqrt{\sigma_\mathrm{i}}}+\frac{1}{b\sqrt{\sigma_0}}\right)\right]$$

in which use has been made of the binomial theorem since the extra term is small. Putting in the given figures we find

$$Z_0=45.7\left[1+(1+\mathrm{j})\frac{12.5}{\sqrt{f}}\right]$$

The frequency at which the skin depth is one-third of the inner radius may be found to be 1.7×10^4 Hz, so that the approximation is certainly good above 0.5 MHz, which is a likely requirement. To find the attenuation, we use $\gamma=(Z_1Y_2)^{\frac{1}{2}}$. Making use of the above calculation,

$$\gamma=\alpha+\mathrm{j}\beta=\mathrm{j}\omega\sqrt{(\mu_0\varepsilon)}\left[1+(1-\mathrm{j})\frac{12.5}{\sqrt{f}}\right]$$

Hence $$\alpha = 2\pi\sqrt{(\mu_0\varepsilon)} \times 12.5\sqrt{f}$$
$$= 41.3 \times 10^{-8}\sqrt{f} \text{ nepers m}^{-1}$$
$$= 3.59 \times 10^{-6}\sqrt{f} \text{ dB m}^{-1}.$$

It may be checked that the circuit approach gives the same expression for attenuation coefficient as was given in § 6.4.1, and also that the design data given in § 6.5.2 agree with theory.

6.12 Summary

This chapter has applied the results of previous chapters to two-conductor transmission lines used as communications links.

The nature of signals and noise has been considered with a view to specifying appropriate parameters such as phase and group velocities, group delay and attenuation.

Values of the circuit parameters for various types of transmission line were given. Results were obtained for the propagation of pulses along a line whose attenuation is proportional to $\sqrt{f}$.

The particular problems associated with long lines containing repeaters were considered. Finally practical systems used for frequency-division multiplex and pulse code modulation were briefly described.

Formulae

Thermal noise from a resistor in bandwidth B

$$p_n = kTB$$

$$\overline{v_n^2} = 4kTRB$$

Noise factor $$n = \frac{(\text{Signal power/noise power})_{\text{in}}}{(\text{Signal power/noise power})_{\text{out}}}$$

Phase velocity $$v_p = \omega/\beta$$

Group velocity $$v_g = \frac{d\omega}{d\beta}$$

Group delay $$\tau = z\frac{d\beta}{d\omega} = \frac{d\phi}{d\omega}$$

6.13 Problems

1. Explain how thermal noise sets a limit to the sensitivity of a communication system and discuss two sources of noise which may occur in an amplifier.

A microphone has a resistive internal impedance of 50 Ω and is matched by means of a transformer to the resistive input impedance of a high-gain amplifier which has a bandwidth of 20 kHz. If the input transformer can be assumed to be ideal, calculate the signal-to-noise ratio which might be expected at the output when the microphone generates an e.m.f. of 10 μV (r.m.s.). The temperature may be taken as 17 °C and Boltzmann's constant is 1.38×10^{-23} J K^{-1}.

(London University, 1974.)

2. 'Minimum attenuation and distortion occur on a uniform transmission line when the primary constants satisfy the relationship $L/R = C/G$.' Discuss this statement with reference to a telephone cable with and without lumped loading.

A line has the following primary constants per loop mile: $R = 40$ Ω, $L =$ 1 mH, $C = 0.05$ μF, G is negligible. Loading coils with inductance L_d of 100 mH and negligible resistance are added at intervals l miles apart, giving rise to an attenuation coefficient which is nominally constant up to a frequency $f_c = [\pi^2(L_d + lL)lC]^{-\frac{1}{2}}$. Determine the minimum value of l if $f_c = 4$ kHz and estimate the attenuation coefficient at 5 krad s^{-1} before and after loading, assuming the additional inductance to be uniformly distributed.

(Bristol University, 1972.)

3. State the principles involved in (*a*) frequency-division and (*b*) time-division multiplex systems of transmission. Explain the causes of inter-channel interference in each system.

A number of speech signals, each limited to a bandwidth of 300–3400 Hz, are to be transmitted by a time-division, pulse-position modulation system in which the pulse width is 1.0 μs. The pulse shift is ± 1.3 μs and the minimum adjacent channel pulse separation is 1.4 μs. Estimate, choosing a suitable sampling frequency, the maximum number of channels which may be transmitted.

(London University, 1973.)

4. Describe the different types of transmission line used in communication systems and state the advantages and disadvantages of each.

The primary coefficients of a coaxial cable per loop kilometre, measured at 0.5 MHz, are: $R = 32$ Ω, $L = 250$ μH, $C = 0.050$ μF and G is negligibly small. The cable is used with intermediate repeaters in a wideband telephone system covering the frequency range 0.5–12.5 MHz. If the cable resistance is assumed to vary as the square root of the operating frequency, and the attenuation in any channel is not to exceed 54 dB between repeaters, calculate the approximate maximum allowable spacing for the repeaters. Justify the approximation.

(London University, 1973.)

5. Outline the design problems associated with a long-haul coaxial cable telecommunication system with particular reference to the requirements for repeaters and equalisers.

Show that the noise power per channel, P_R, of a system comprising N identical repeater sections is given by

$$P_R = N_F + G - 138 + 10 \log_{10} N \text{ dB m},$$

where N_F and G are the noise factor and available gain of a repeater in dB.

A certain system designed to carry 600 channels, has 1000 dB attenuation at the maximum operating frequency and is 1000 km long. The repeater amplifiers selected have a 5 dB noise figure, and an overload level of 22 dB m at their output. If the equivalent peak power of a 600 channel system is taken to be 25 dB m, and the maximum allowable output noise must be more than 55 dB down on the normal operating level, find the minimum number of repeaters required. (Bristol University, 1976.)

6. Service on a long open-wire pair telephone line is interrupted due to an unknown fault. Impedance measurements made at one end of the line yielded the following results:

Frequency (kHz)	2.5	5.0	7.5	8.5	10	12.5
|Input impedance| (Ω)	260	600	1700	2000	1680	860
kHz	15	20	22.5	25	27.5	30
Ω	450	680	1350	1800	1400	830
kHz	32.5	35	37.5	40	42.5	45.0
Ω	600	740	1150	1520	1500	1000
kHz	47.5	50	52.5	55	57.5	60
Ω	780	680	800	1150	1500	1300

Estimate the approximate position of the fault and state whether it is due to an open-circuit or a short-circuit. (Southampton University, part question.)

7. A coaxial cable of characteristic impedance 50 Ω, has a solid copper inner conductor of diameter 0.9 mm and a braided copper outer conductor of inside diameter 2.95 mm. Measurements at 100 MHz give an attenuation coefficient of 0.13 dB m^{-1}. If the skin resistance of copper is taken to be $8.1\sqrt{f_G}$ mΩ, where f_G is the frequency in GHz, estimate the increase in the outer conductor resistance attributable to the use of braided rather than solid copper.

7

Waveguide systems

We shall consider in this chapter some aspects of the use of waveguides, as opposed to transmission lines. As with transmission lines, we have two broad areas to consider: the short link when the losses in the guide although important are not critical, and the long link for communication purposes which is only viable if the attenuation is adequately small. For the short link we may well be interested in power-handling capacity, for example in feeds to radar antennae. The reason for consideration of waveguides for communication use will be seen to be the promise of very wide bandwidths for multichannel operation on trunk routes.

The choice between coaxial line and waveguide in particular circumstances is dictated by several considerations, and we shall commence by comparing some properties.

7.1 Comparison of coaxial line and waveguide

Among the more important topics we may list
mode of propagation
attenuation
frequency range
power-handling capacity
construction

7.1.1 Mode of propagation

On a coaxial line it is possible to propagate a principal or *TEM* wave at any frequency, and this mode of propagation is used almost exclusively. However, *TE* or *TM* modes can also be supported, but, as with all waveguide modes, these exhibit a cut-off frequency below which transmission is not possible. Normally, the operational frequency of coaxial lines is considerably less than the lowest cut-off frequency, and any *TE* or *TM* modes can only be evanescent.

In waveguides, single-mode propagation is normally preferred, and this occurs in the frequency band between the lowest cut-off frequency (that of the dominant mode) and the cut-off frequency of the next highest mode. Multimode propagation is also possible, since, as pointed out in chapter 3, the modes are nominally independent. The signals in the different modes, although at a common frequency, can thus each be used as separate carriers, provided that separation can be achieved. This must be by use of the polarisation properties of the different modes. Although the modes are nominally independent, there is in practice a degree of mode coupling brought about by imperfections of the waveguide and the presence of bends and terminations. This will cause cross-talk between several channels and, as will be seen later, distortion on a single channel.

7.1.2 Attenuation

In comparing waveguide and coaxial line from the point of view of attenuation, it is necessary to be careful about the basis of comparison. It has been seen previously (§ 6.4.1) that, for a coaxial, losses arising from the resistance of conductors can always be reduced by increasing the size. This is also true for waveguide, as shown for example by (3.80) and (3.81). If a waveguide is required to operate in the dominant mode, its size is limited by the frequency to be handled. Likewise, the size of a coaxial line is limited if at a given frequency only the *TEM* wave is to be propagated. We therefore have to pay attention to the system requirements when making the comparison. This detail analysis is pursued in § 7.2.3.

In general it may be said that coaxial lines exhibit higher attenuation than waveguides where it is physically possible to use either: at low frequencies waveguides become too large, and at high frequencies the coaxial line is impossible to make in the small size necessary if the *TEM* mode alone is to propagate.

Coaxial line must incorporate support for the inner. Over a narrow frequency range these supports can take the form of metal stubs one quarter-wavelength long, as shown in fig. 7.1, but over a wide frequency range dielectric is necessary. This dielectric inevitably involves an extra loss as the frequency increases.

Whereas the attenuation coefficient for a coaxial line increases steadily with frequency (fig. 3.16), that for a waveguide normally exhibits a broad minimum (fig. 3.17). At the lower end this is because cut-off is approached, and at the higher end skin effect predominates. The only known exception to this rule is the TE_{0n} modes in circular waveguide, for which attenuation decreases indefinitely as the frequency

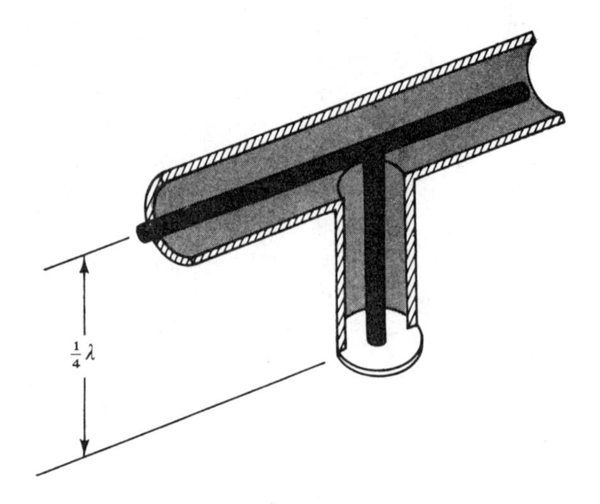

Fig. 7.1. Quarter-wavelength support stub for coaxial line.

increases. The TE_{01} mode forms the basis of the 'long-haul' waveguide for communications, and is discussed in detail later in this chapter.

7.1.3 Frequency range

The range of frequencies which can be used depends both on the way in which attenuation varies with frequency and on the dispersion, which is determined by the variation of the phase constant with frequency.

The phase velocity on lowloss coaxial lines is virtually independent of frequency over the possible operating range. Waveguide on the other hand has a phase constant which is frequency-dependent, since β approaches zero as the frequency approaches cut-off (see, for example, fig. 3.3).

The sidebands of a modulated signal occupy a range of frequencies. If too wide a band is covered, the sidebands will be differentially delayed and severe distortion will result. However, in general the higher the frequency the wider will be the usable band, so that waveguides at millimetric frequencies may be expected to offer bandwidths much in excess of those available from coaxial lines.

In fig. 7.2 is shown for a waveguide mode the variation of group delay as a function of frequency.

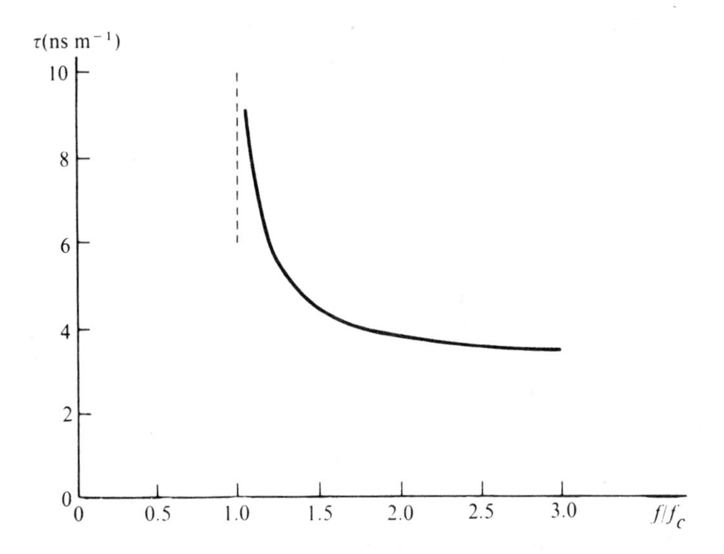

Fig. 7.2. Variation of group delay with frequency for waveguide mode.

7.1.4 Power-handling capacity

The power which can be handled by either coaxial line or waveguide is limited by the allowable magnitude of electric field: beyond a certain value breakdown of the dielectric will occur. The dielectric involved is usually air or other gas. Although the electric field can in theory be reduced by use of solid dielectric, in practice little advantage is gained because of the difficulty of avoiding voids at interfaces. Both coaxial lines and waveguides operated at high power may be pressurised with dry nitrogen (at a pressure of about 3 atmospheres) to increase the power rating. Power densities of the order of $400\,\mathrm{kW\,cm^{-2}}$ are theoretically possible, although these cannot be achieved in practice.

As with attenuation, care must be taken in comparing the power-handling capacity of coaxial line and waveguide because of the need to make a choice of dimensions. This analysis is performed in § 7.2.4.

7.1.5 Construction

Some types of coaxial line have been considered in chapter 6, showing ways devised for supporting the inner to meet various requirements. A method avoiding dielectric has been shown in fig. 7.1. By contrast, waveguide is mechanically self-supporting. It is however necessary to pay attention to the finer details of fabrication to ensure, for example, uniformity and good internal-surface finish. A suitable range of components such as bends has to be devised and made, as well as a range of couplers to join one guide to another with the necessary precision.

7.2 Short-run waveguide systems

In this section we shall consider the use of rectangular waveguide in the dominant mode. This application is that usually encountered in connecting receivers and transmitters to antennae, and what might be called general systems. The use of waveguide as a medium for long-distance communication is considered in a later section.

7.2.1 Choice of waveguide dimensions

It was shown in chapter 3 that a whole range of cut-off frequencies exists for propagation in a rectangular tube, with one mode, TE_{10}, dominant. These cut-off frequencies are given by (3.54), which for an air-filled guide becomes

$$f_{mn} = \tfrac{1}{2}c[(m/a)^2 + (n/b)^2]^{\frac{1}{2}}$$

in which $c = 3 \times 10^8\ \mathrm{ms}^{-1}$ and a is assumed to be greater than b. The four lowest cut-off frequencies are therefore

$$f_{10} = c/2a$$

$$f_{20} = c/a = 2f_{10}$$

$$f_{01} = c/2b = f_{10}(a/b)$$

$$f_{11} = \frac{1}{2}c\left[\frac{1}{a^2} + \frac{1}{b^2}\right]^{\frac{1}{2}} = f_{10}\left(1 + \frac{a^2}{b^2}\right)^{\frac{1}{2}}$$

Of these f_{11} is the greatest. The relative magnitudes of f_{20} and f_{01} depend on the ratio a/b, and they will be equal if $a = 2b$. The range of frequencies over which the dominant mode alone will propagate will therefore have its greatest value if $a/b > 2$.

Other considerations will show that it is undesirable that b should be made less than $\frac{1}{2}a$, and practical waveguide ranges adhere to the ratio 2:1 for $a:b$.

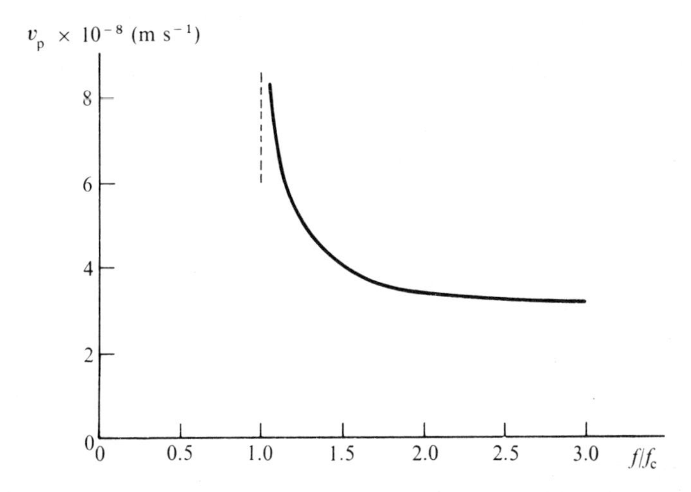

Fig. 7.3. Variation of phase velocity with frequency for waveguide mode.

7.2.2 Frequency coverage

With the choice of 2:1 for the ratio $a:b$, a range of $f_{10}<f<2f_{10}$ is available for single-mode propagation. However, the propagation constant changes very rapidly near cut-off, as shown by (3.74) in the form

$$\frac{1}{v_p}=\frac{\beta}{\omega}=\frac{1}{c}\left(1-\frac{f_{10}^2}{f^2}\right)^{\frac{1}{2}}$$

The phase velocity is shown in fig. 7.3 as a function of f/f_{10}. The rate of change is least at the higher frequencies, and a nominal centre frequency of $1.5f_{10}$ is chosen. This allows a useful range of approximately ±20% in frequency.

†7.2.3 Attenuation

The attenuation coefficient for dominant-mode propagation in air-filled rectangular waveguide was derived in (3.80)

$$\alpha=\frac{R_s}{b}\sqrt{\left(\frac{\varepsilon_0}{\mu_0}\right)}\left[1+\frac{2b}{a}\left(\frac{f_{10}}{f}\right)^2\right]\left[1-\left(\frac{f_{10}}{f}\right)^2\right]^{-\frac{1}{2}} \tag{7.1}$$

in which R_s is the skin resistance. (It may be noted that for copper R_s is about 16 mΩ at a frequency of 4 GHz.) From this expression, as mentioned earlier, it can be seen that b should be made as large as possible in order to decrease the attenuation at any given frequency. The form of the expression was shown graphically in fig. 3.17.

It may be seen that at a given operating frequency the attenuation is determined once it has been decided to use a nominal frequency of $1.5f_{10}$, as discussed in the last section. If the operating frequency is denoted by f, we have

$$f_{10} = c/2a = f/1.5, \qquad b = \tfrac{1}{2}a$$

Hence in terms of the wavelength $\lambda(=c/f)$, $b = 3\lambda/8$. Substituting this expression together with $f/f_{10} = 1.5$ in (7.1), we find

$$\alpha = 5.2\frac{R_s}{\lambda}\sqrt{\left(\frac{\varepsilon_0}{\mu_0}\right)} \text{ nepers m}^{-1} \tag{7.2}$$

We note that R_s is proportional to $f^{\frac{1}{2}}$, so that for a waveguide chosen for the operating frequency f in the way described that attenuation increases as $f^{\frac{3}{2}}$. To fix orders of magnitude we evaluate this expression at 4 GHz in a copper waveguide, to find $\alpha = 0.026$ dB m^{-1}.

The attenuation coefficient for an air-spaced coaxial line given in (3.79) may be expressed in the form

$$\alpha = \frac{1}{2}\frac{R_s}{\lambda}\sqrt{\left(\frac{\varepsilon_0}{\mu_0}\right)}\left(\frac{\lambda}{a}+\frac{\lambda}{b}\right)\Big/\ln(b/a)$$

in which a, b are the radii of inner and outer conductors respectively and λ is the wavelength at the operating frequency. The factor $(R_s/\lambda)\surd(\varepsilon_0/\mu_0)$ is the same as appeared in (7.2). In fig. 7.4 is shown the dependence of the attenuation coefficient on a/λ, b/λ, in the form of contours of constant attenuation coefficient evaluated at 4 GHz. In this figure are also marked lines corresponding to various values of characteristic impedance, determined by the ratio b/a. The attenuation coefficient can evidently be made as small as desired by increasing a and b. This process is however limited by the fact that a waveguide mode commences to propagate at a free-space wavelength λ_c determined by

$$\lambda_c = 2.95(a+b) \tag{7.3}$$

For the purposes of comparison we choose the frequency of this mode to be the same as that for the waveguide, $2f_{10}$. Hence in terms of an operating frequency equal to $1.5f_{10}$

$$\lambda_c = c/2f_{10} = 0.75\,c/f = 0.75\lambda$$

whence (7.3) becomes

$$(a+b)/\lambda \approx 0.25$$

This line is shown on fig. 7.4, and it will be seen that optimum choice of

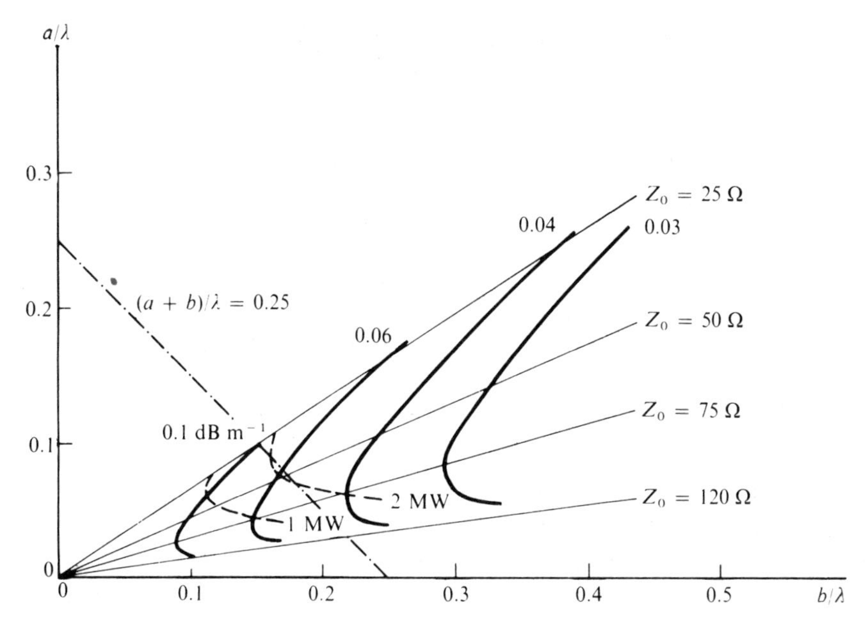

Fig. 7.4. Attenuation and maximum power in copper coaxial line at 4 GHz, air dielectric: λ = operating wavelength, b = outer diameter, a = inner diameter Z_0 = characteristic impedance. Full line: curves of constant attenuation. Pecked line: curves of constant maximum power.

Z_0 gives a value of attenuation coefficient of about twice the value of 0.026 dB m^{-1} determined for the waveguide.

We thus deduce that with the design criteria used the waveguide is superior to the coaxial line from the point of view of attenuation. In practice the problem will be slightly different, in that choice will be made from a fixed range of dimensions. The conclusion is however not significantly altered.

†7.2.4 Power-handling capacity

We consider the dominant TE_{10} mode in a rectangular waveguide, for which the relevant field quantities are given in (3.74)

$$|E_y| = |A|\frac{\omega\mu_0 a}{\pi}\sin\frac{\pi x}{a} \tag{7.4}$$

$$P = \frac{1}{4\pi^2}|A|^2\omega\mu_0\beta a^3 b \tag{7.5}$$

$$\beta = \frac{2\pi f}{c}\left[1-\left(\frac{f_{10}}{f}\right)^2\right]^{\frac{1}{2}}$$

If we denote by E_m the maximum permitted electric field, (7.4) gives

$$|A| = \frac{\pi}{\omega\mu_0 a} E_m$$

Substitution in (7.5) then yields

$$P = \frac{\beta ab}{4\omega\mu_0} E_m^2$$

$$= \frac{1}{4}\sqrt{\left(\frac{\varepsilon_0}{\mu_0}\right)}\left[1 - \left(\frac{f_{10}}{f}\right)^2\right]^{\frac{1}{2}} E_m^2 ab$$

Assuming a nominal frequency of $1.5f_{10}$ and the value $3\,\text{MV m}^{-1}$ appropriate to air we find

$$P/ab = 4.45 \times 10^9\ \text{W m}^{-2}$$

In more suitable units this figure is $445\ \text{kW cm}^{-2}$. It is thus best to keep the area ab as large as possible, which again indicates that b should have the value $\frac{1}{2}a$. Since dimensions are inversely proportional to operating frequency, the maximum power which can be carried is inversely proportional to the square of the frequency. If we use the expressions of the last section for the dimensions a, b of the waveguide in terms of operating wavelength ($a = 0.75\lambda$, $b = 0.375\lambda$) we find

$$P = 1.25 \times 10^9 \lambda^2\ \text{W} \tag{7.6}$$

Thus WG 16 would on this basis handle 1.13 MW. In practice an allowance must be made to give a safe operating margin.

To compare this figure with one for coaxial line, the relevant formulae are contained in chapter 3

$$E_m = |V|/[a \ln(b/a)]$$

$$P = |V|^2/2Z_0$$

$$Z_0 = \frac{1}{2\pi}\sqrt{\left(\frac{\mu_0}{\varepsilon_0}\right)} \ln(b/a)$$

Hence

$$P = \pi\sqrt{(\varepsilon_0/\mu_0)}\, a^2 E_m^2 \ln(b/a)$$

If we consider an air-spaced line at wavelength λ, we have

$$P = \lambda^2 \times 7.5 \times 10^{10} \times (a/\lambda)^2 \ln(b/a)$$

and once again the best choice of a, b has to be considered. In fig. 7.4

are shown contours of constant P at a frequency of 4 GHz. It will be seen that, with the dimensional constraint imposed by the requirement that only a *TEM* mode propagates, P has a maximum of about 2 MW. On the other hand (7.6) for waveguide at 4 GHz gives 7 MW. This comparison is made on the incorrect assumption that the coaxial line has no region of enhanced field strength arising from supports. In practice a factor of perhaps $\frac{1}{4}$ has to be allowed to give safe working.

We thus conclude that waveguide has the better capacity for power handling.

7.3 Waveguide circuits

7.3.1 Equivalent transmission-line terminology

It is possible to draw a very close parallel between source-to-load connections through waveguide and the same connections using transmission line. The analogy is so close that in fact we use the same ideas and terminology. If we put to one side the consideration of the way in which power may be fed into or extracted from a waveguide (which we also did for transmission lines at very high frequencies), the waveguide itself when excited in the dominant mode can only propagate the one mode, with forward and backward waves. As in a transmission line, the backward wave is excited by reflection from discontinuities in the uniform waveguide, either by a load at the end or by obstacles. We have seen in chapter 3 that power flow is calculated entirely on the basis of the transverse electric and magnetic fields, which are fixed, as far as their pattern across the guide is concerned, by the particular cross-section being used. For example, rearrangement of (3.74) enables us to write for the forward wave in rectangular guide

$$E_y = E_i \sin \frac{\pi x}{a} \exp(-j\beta z)$$

$$H_x = -E_i \frac{\beta}{\omega\mu} \sin \frac{\pi x}{a} \exp(-j\beta z)$$

To include a backward wave we have only to include additional terms in $\exp(j\beta z)$, allowing for the fact that since power flow is reversed the magnetic field will be reversed. We may write

$$E_y = \sin \frac{\pi x}{a} [E_i \exp(-j\beta z) + E_r \exp(j\beta z)]$$

$$H_y = -\frac{\beta}{\omega\mu} \sin \frac{\pi x}{a} [E_i \exp(-j\beta z) - E_r \exp(j\beta z)]$$

These formulae are to be compared with those for transmission lines

$$V(z) = A \exp(-j\beta z) + B \exp(j\beta z)$$

$$I(z) = Z_0^{-1}[A \exp(-j\beta z) - B \exp(j\beta z)]$$

The only terminal property that we can measure in a waveguide is the power delivered into a load. The transverse field pattern is not open to observation, and indeed is fixed by the nature of the waveguide. The only important thing about the transverse field is the magnitude, from the axial variation of which we can find out if there is a reflected wave. Observation of the variation of magnitude with axial position can be made in the same way as for a coaxial line: if an axial position can be found in the waveguide wall where a slot can be cut without interfering with wall currents, then a probe can be inserted and used as in the transmission-line standing-wave detector. We can then measure standing-wave ratios and positions of maxima and minima and this information can be used to find reflection coefficients.

The transmission-line theory of chapter 5 has shown that a one-to-one relation exists between reflection coefficients and terminating impedance. Knowledge of the magnitude and phase of the reflection coefficient could be used to derive the value of the ratio of terminating impedance to characteristic impedance. The reflection coefficient in a waveguide can be used in the same way to attribute a value of normalised impedance to whatever physical termination or non-uniformity is giving rise to the reflection. The fact that it is normalised is of no importance, since actual impedance is only of significance when it is possible to connect a 'circuit' impedance across the transmission line.

Consider an obstacle in a waveguide, such as the post illustrated in fig. 7.5(*a*). An incident wave will excite currents in the post, which will in turn generate waves of all modes travelling away from the post, in both directions. Since the guide is operating in the dominant mode, all other modes are attenuated away from the post and there will be left a single backward wave of given amplitude and phase. We may attribute to the post a normalised impedance which would give rise to this wave. There will also be a wave transmitted past the post, again with a given amplitude and phase. Provided the post is thin compared to a guide wavelength it is likely that the reflected and transmitted waves can be interpreted in terms of a normalised admittance in shunt on a transmission line. The post, of given geometry, can be said to be an admittance of this value.

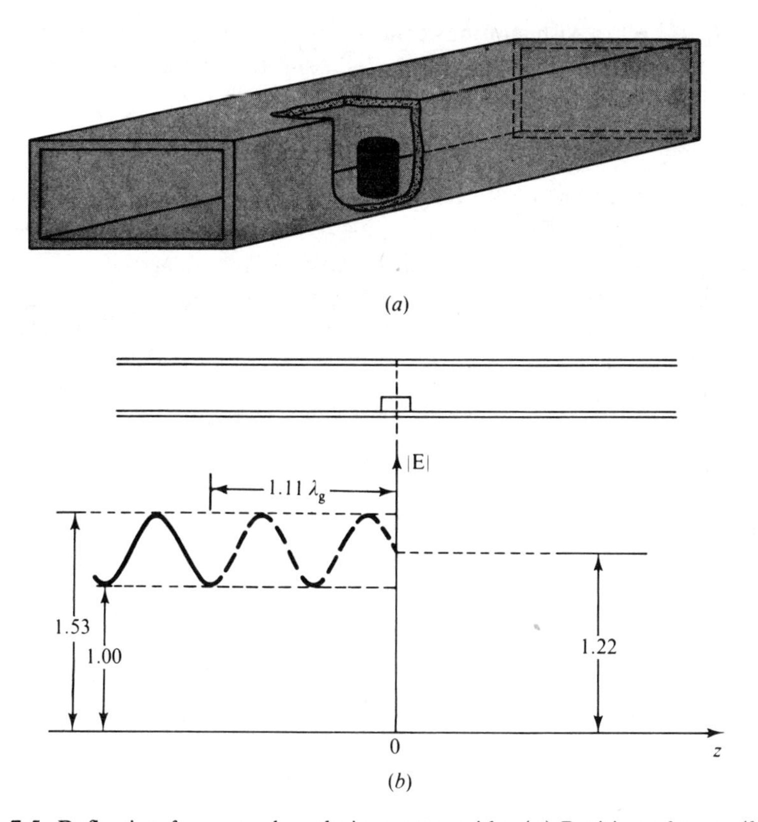

Fig. 7.5. Reflection from an obstacle in a waveguide. (*a*) Position of post. (*b*) Standing-wave pattern.

Example

In fig. 7.5(*b*) is shown the observed standing-wave pattern for electric field strength in the waveguide containing the post of fig. 7.5(*a*). Show that the results are consistent with the post being regarded as an admittance and determine the value of this admittance.

Solution. The measured amplitude of electric field will be proportional to

$$|E_i \exp(-j\beta z) + E_r \exp(j\beta z)| = |E_i \exp(-j\beta z)[1 - k \exp(2j\beta z)]|$$
$$= |E_i|[1 + K^2 - 2K \cos(2\beta z + \theta)]^{\frac{1}{2}}$$

in which $k = K\,e^{j\theta}$ is the current reflection coefficient at $z = 0$. See section 5.7 for transmission lines. A standing-wave ratio S will occur, given by

$(1+K)/(1-K)$, with minima occurring at values of z for which $2\beta z+\theta$ is an even multiple of π. The normalised admittance y associated with the reflection coefficient is given by

$$y=(1+k)/(1-k)$$

In the case of fig. 7.5(b) we take $z=0$ at the centre of the post: even though measurements cannot be made in the immediate neighbourhood we can derive the dominant-mode component by extrapolating from the observed pattern, as shown by the dotted lines. Nearest the post on the generator side is a minimum at $z=-1.11\lambda_g$, or extrapolating nearer the post, at $z=-0.11\lambda_g$. Thus

$$\theta=-4\pi z/\lambda_g=4\pi\times 0.11=1.382$$

Also $$K=(S-1)/(S+1)=0.209$$

Hence $$\begin{aligned} y &= \frac{1+k}{1-k} \\ &= \frac{(1+k)(1-k^*)}{(1-k)(1-k^*)} \\ &= \frac{1-K^2+2K\mathrm{j}\sin\theta}{1+K^2-2K\cos\theta} \\ &= \frac{0.956+\mathrm{j}0.412}{0.965} \\ &= 0.99+\mathrm{j}0.43 \end{aligned}$$

To the accuracy with which such measurements can be made, y is equivalent to a reflectionless resistive load in parallel with a capacitive susceptance of 0.43. The reflectionless load is provided by the continuing waveguide to the right of the post. By interpreting the electric field as ‘voltage’ we expect the amplitude to be continuous at the admittance, which it is. We therefore conclude that the effect of the post is calculable as that of a shunt admittance.

7.3.2 Complex wave amplitudes

When we use transmission-line concepts we invite some association with ‘voltage’ and ‘current’. These cannot be given absolute meaning, as discussed in §3.9, but in any event they are useful only in calculating power. In dealing with waveguides we replace voltage with a wave amplitude whose modulus is related to the power carried by the wave. Consider a forward wave carrying power P_i. We define a wave amplitude, a, such that

$$P_i = \tfrac{1}{2}a \cdot a^*$$

and, as a function of z,

$$a = a_0 \exp(-j\beta z)$$

To a backward wave is attributed an amplitude, b, such that $P_r = \tfrac{1}{2}b \cdot b^*$. The wave amplitudes may be used to define voltage and current variables by the equations

$$\begin{aligned} V &= Z_0^{\frac{1}{2}}(a+b) \\ I &= Z_0^{-\frac{1}{2}}(a-b) \end{aligned} \qquad (7.7)$$

where Z_0 is real and equivalent to the characteristic impedance of a transmission line. Since neither V nor I have a unique physical significance, Z_0 is not uniquely definable and these equations are written in the form

$$v = a + b$$

$$i = a - b$$

The 'normalised variables' v, i have dimensions of $(\text{power})^{\frac{1}{2}}$. We can check that these definitions are consistent: the power flow in the forward direction should be Re $(\tfrac{1}{2}vi^*)$. Substituting

$$\begin{aligned} P &= \tfrac{1}{2}\,\mathrm{Re}\,[(a+b)(a^*-b^*)] \\ &= \tfrac{1}{2}\,\mathrm{Re}\,(|a|^2 - |b|^2 + a^*b - ab^*) \\ &= \tfrac{1}{2}(|a|^2 - |b|^2) \\ &= P_i - P_r \end{aligned}$$

In a situation in which we are given voltage and current in association with a transmission line of given Z_0 we may find the wave amplitudes by inversion of (7.7), giving

$$\begin{aligned} a &= \tfrac{1}{2}(VZ_0^{-\frac{1}{2}} + IZ_0^{\frac{1}{2}}) \\ b &= \tfrac{1}{2}(VZ_0^{-\frac{1}{2}} - IZ_0^{\frac{1}{2}}) \end{aligned} \qquad (7.8)$$

7.3.3 Scattering parameters

The reflection and transmission properties of an obstacle such as the post discussed above are described completely by the complex wave

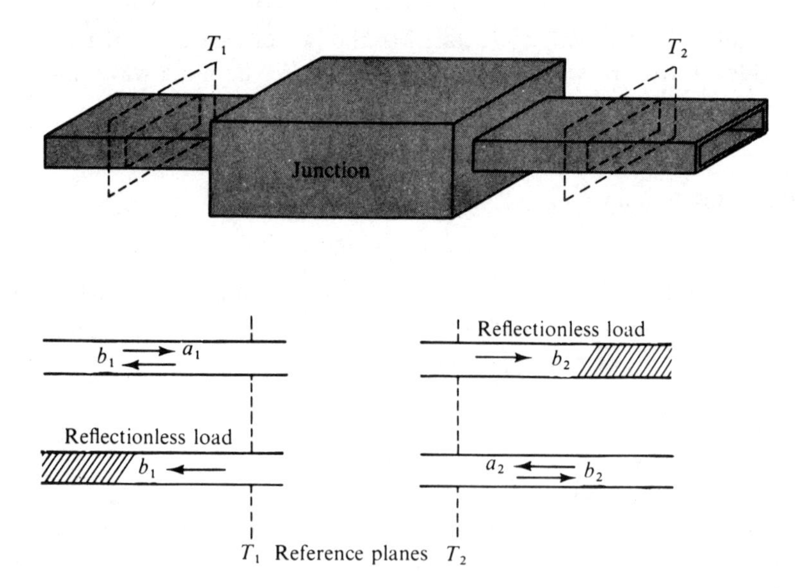

Fig. 7.6. Situation defining scattering parameters of a waveguide junction.

amplitudes of the incident, reflected and transmitted waves. Since the wave amplitude changes with position, unique description is only possible if two reference planes, as in fig. 7.6, are defined. It is then possible to state the ratios of both reflected and transmitted waves to the incident wave. If we denote by a_1, b_1 the incident and reflected wave amplitudes at plane T_1 and b_2 the transmitted wave at T_2, we can measure for a given obstacle the ratios b_1/a_1, b_2/a_1. Further, we can allow a wave to fall on the obstacle from the opposite side. This will give rise to a reflected wave and a transmitted wave. (It must be noted that the direction of the 'incident' wave is always *towards* the obstacle, and therefore in opposite directions on the two sides.) Since the system is linear we can relate the variables shown in fig. 7.7 by two equations

$$\begin{aligned} b_1 &= S_{11}a_1 + S_{12}a_2 \\ b_2 &= S_{21}a_1 + S_{22}a_2 \end{aligned} \qquad (7.9)$$

The parameters S_{ij} are the *scattering parameters* of the two-port formed by the obstacle between the two reference planes. The concept is applicable to any guide carrying waves. We will work out a simple example in transmission lines.

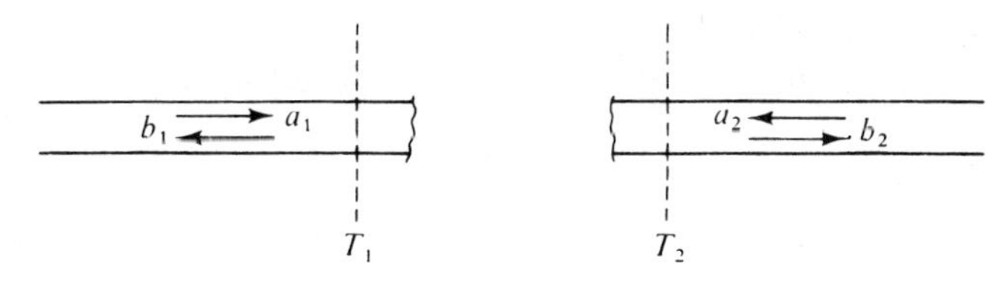

Fig. 7.7. Waveguide junction with waves incident from both sides.

Example

A component of normalised admittance y is placed across a transmission line. Obtain expressions for the scattering parameters referred to the plane through the shunt admittance.

Solution. In this case the reference planes are coincident at the plane of the admittance. The situation is shown in fig. 7.8.

We have in circuit terms

$$\left.\begin{aligned} \mathrm{V} &= \mathrm{A}\,\mathrm{e}^{-\mathrm{j}\beta z} + \mathrm{B}\,\mathrm{e}^{\mathrm{j}\beta z} \\ \mathrm{I} &= Y_0(\mathrm{A}\,\mathrm{e}^{-\mathrm{j}\beta z} - \mathrm{B}\,\mathrm{e}^{\mathrm{j}\beta z}) \end{aligned}\right\} z < 0$$

$$\left.\begin{aligned} \mathrm{V} &= \mathrm{C}\,\mathrm{e}^{-\mathrm{j}\beta z} \\ \mathrm{I} &= Y_0\mathrm{C}\,\mathrm{e}^{-\mathrm{j}\beta z} \end{aligned}\right\} z > 0$$

At $z = 0$ there must be continuity of voltage, so that we have

$$\mathrm{A} + \mathrm{B} = \mathrm{C}$$

There must also be a current balance at $z = 0$, giving

$$Y_0(\mathrm{A} - \mathrm{B}) - Y_0\mathrm{C} = yY_0\mathrm{V}(0) = yY_0\mathrm{C}$$

or

$$\mathrm{A} - \mathrm{B} = (1 + y)\mathrm{C}$$

Hence we find

$$\mathrm{C} = 2\mathrm{A}/(2 + y)$$

$$\mathrm{B} = -y\mathrm{A}/(2 + y)$$

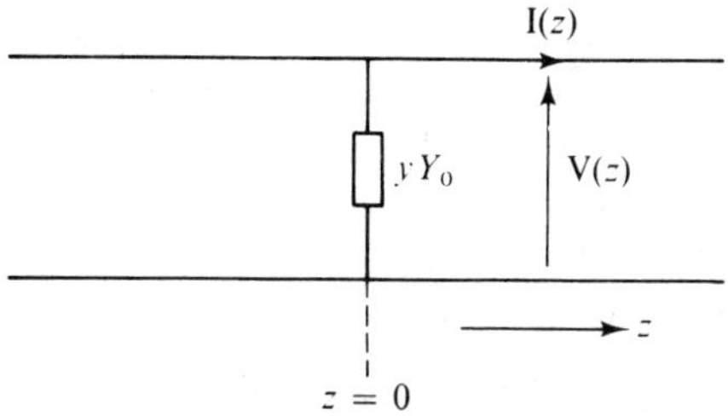

Fig. 7.8. Transmission line with admittance in shunt.

Now the wave amplitudes can be determined using (7.8). Putting $z=0$, so that we are finding amplitudes at the reference plane, we find

$$a_1 = Y_0^{\frac{1}{2}} A$$

$$b_1 = Y_0^{\frac{1}{2}} B$$

$$b_2 = Y_0^{\frac{1}{2}} C$$

The amplitude a_2 is zero, since there is no wave incident from $z>0$. The scattering parameters S_{ij} may from (7.9) be derived as

$$\left.\begin{aligned} S_{11} &= b_1/a_1, \quad a_2 = 0 \\ S_{21} &= b_2/a_1, \quad a_2 = 0 \\ S_{12} &= b_1/a_2, \quad a_1 = 0 \\ S_{22} &= b_2/a_2, \quad a_1 = 0 \end{aligned}\right\} \tag{7.10}$$

Hence for the present case we have

$$S_{11} = B/A = -y/(2+y)$$

$$S_{21} = C/A = 2/(2+y)$$

Since the junction is symmetrical we have also $S_{22}=S_{11}$, $S_{12}=S_{21}$.

†7.3.4 Some properties of the scattering parameters

We may observe that the condition $a_2=0$ used in (7.10) means that side 2 is connected to a passive reflectionless load. The parameter S_{11} may be seen to be the reflection coefficient for a wave incident on side 1 when side 2 is terminated in a reflectionless load. S_{21} is the transmission coefficient under the same condition. S_{22} and S_{12} may be similarly interpreted in terms of a wave incident on side 2 with side 1 correctly terminated.

Unless non-reciprocal media (such as ferrites) are present in the waveguide, the two-port device will be reciprocal. Interpreting this property in the form that a measurement relating a variable at one side to one on the other side is not changed by reversing the device, we must have b_2/a_1 in the first case equal to b_1/a_2 in the second. Thus $S_{21}=S_{12}$.

Many of the two-ports in which we are interested are lossless. In this case the total power dissipated in the junction must be zero. Hence we have

$$\tfrac{1}{2}(|a_1|^2 - |b_1|^2 + |a_2|^2 - |b_2|^2) = 0$$

Direct substitution for b_1 and b_2 give

$$|a_1|^2(1-|S_{11}|^2-|S_{21}|^2)+|a_2|^2(1-|S_{22}|^2-|S_{12}|^2)$$
$$+2\text{Re}\,(S_{11}S_{12}^* a_1 a_2^* + S_{21}S_{22}^* a_1 a_2^*)=0$$

This equality holds for all values (complex) of a_1, a_2, so that the following identities must be satisfied:

$$|S_{11}|^2+|S_{21}|^2=1 \tag{7.11}$$

$$|S_{22}|^2+|S_{12}|^2=1 \tag{7.12}$$

$$S_{11}S_{12}^*+S_{21}S_{22}^*=0 \tag{7.13}$$

From the last of these equations we have

$$S_{11}/S_{21}=-(S_{22}/S_{12})^*$$

and hence

$$|S_{11}/S_{21}|=|S_{22}/S_{12}|$$

From the first and second we have

$$|S_{11}/S_{21}|^2=|S_{21}|^{-2}-1=|S_{22}/S_{12}|^2=|S_{12}|^{-2}-1$$

Therefore

$$|S_{12}|=|S_{21}| \tag{7.14}$$

and

$$|S_{11}|=|S_{22}| \tag{7.15}$$

Since S_{11}, S_{22} are reflection coefficients we know that by suitably choosing the reference planes both quantities may be made real. (This, in transmission-line terms, is placing reference planes where the impedance is resistive.) If we assume this is done (7.15) becomes

$$S_{11}=S_{22}$$

Equation (7.13) will then become

$$S_{11}(S_{12}^*+S_{21})=0$$

or

$$S_{21}=-S_{12}^*$$

Taken together with (7.14) we must have

$$S_{12}=S_{21}=\text{j}K$$

Then (7.11) gives

$$S_{11}=(1-K^2)^{\frac{1}{2}}$$

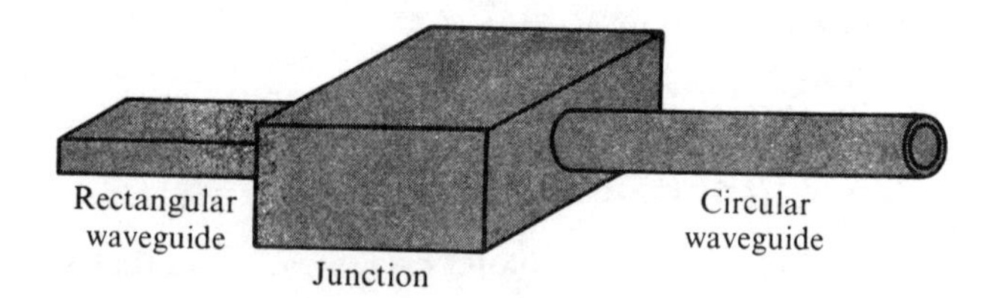

Fig. 7.9. Junction between two different waveguides.

Referred to these particular reference planes a lossless two-port has therefore a scattering matrix

$$S=\begin{pmatrix}(1-K^2)^{\frac{1}{2}} & \mathrm{j}K \\ \mathrm{j}K & (1-K^2)^{\frac{1}{2}}\end{pmatrix} \tag{7.16}$$

We can extend the concept of scattering parameters. Consider for example the system of fig. 7.9, in which a two-port device is shown between two different guides. The waves in each of the guides can be expressed in terms of the incident and reflected waves proper to the particular guide, and therefore all the variables in (7.9) can be defined. Hence the scattering parameters can be found. In the following example are found the scattering parameters of the junction between two transmission lines of different characteristic impedances.

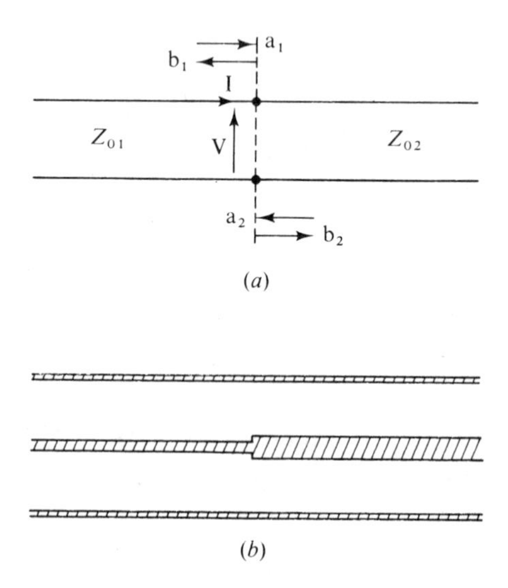

Fig. 7.10. Junction between two transmission lines of differing characteristic impedance. (*a*) Circuit. (*b*) Realisation in coaxial line.

Example

A junction is made between transmission lines of characteristic impedances Z_{01}, Z_{02}, as shown in fig. 7.10(a). Find the scattering matrix with both reference planes in the plane of the junction.

Solution. Using (7.8) we have for the wave incident from $z<0$

$$V = Z_{01}^{\frac{1}{2}}(a_1+b_1)$$

$$I = Z_{01}^{-\frac{1}{2}}(a_1-b_1)$$

For the wave incident from $z>0$ we have

$$V = Z_{02}^{\frac{1}{2}}(a_2+b_2)$$

$$I = -Z_{02}^{-\frac{1}{2}}(a_2-b_2)$$

since the conventional current direction is opposite to the incident wave.

Both voltage and current must be continuous at $z=0$, so that

$$Z_{01}^{\frac{1}{2}}(a_1+b_1) = Z_{02}^{\frac{1}{2}}(a_2+b_2)$$

$$Z_{01}^{-\frac{1}{2}}(a_1-b_1) = -Z_{02}^{-\frac{1}{2}}(a_2-b_2)$$

Eliminating b_2 we find

$$b_1 = (Z_{02}+Z_{01})^{-1}[(Z_{02}-Z_{01})a_1 + 2(Z_{02}Z_{01})^{\frac{1}{2}}a_2]$$

Similarly

$$b_2 = (Z_{02}+Z_{01})^{-1}[2(Z_{02}Z_{01})^{\frac{1}{2}}a_1 - (Z_{02}-Z_{01})a_2]$$

Hence for the scattering matrix we have

$$S = (Z_{02}+Z_{01})^{-1}\begin{pmatrix} Z_{02}-Z_{01} & 2(Z_{02}Z_{01})^{\frac{1}{2}} \\ 2(Z_{02}Z_{01})^{\frac{1}{2}} & Z_{01}-Z_{02} \end{pmatrix} \tag{7.17}$$

We observe that (7.14) and (7.15) are satisfied, but, since the reference planes have not been chosen as required, the matrix is not in the form of (7.16). We notice that knowledge only of the ratio of the characteristic impedances is required.

A practical configuration of such a junction might take the form shown in fig. 7.10(b). A full solution of the field problem would give a scattering matrix differing from that in (7.17) because of the extra modes required in the junction region. At long wavelengths the modification would consist of an additional capacitative susceptance at the junction. At shorter wavelengths a two-port representation would be needed.

It will be noticed that the two examples of this section have referred to transmission lines and impedance, whereas the virtue of scattering parameters is that a knowledge of characteristic impedance is not required: this is solely due to the practical difficulty of performing field calculations for obstacles in waveguides whereas the circuit calculations in lines are very straightforward.

The other extension of scattering-parameter concepts is to multiport devices, which are of common occurrence in microwave systems. In this case we have n ports, each with incident and reflected waves a_i, b_i, $i = 1, 2, \ldots, n$. The equations take the form

$$b_1 = S_{11}a_1 + S_{12}a_2 + \ldots + S_{1n}a_n$$

$$b_2 = S_{21}a_1 + S_{22}a_2 + \ldots + S_{2n}a_n$$

$$-------------------$$

$$b_n = S_{n1}a_1 + S_{n2}a_2 + \ldots + S_{nn}a_n$$

In matrix notation

$$b = Sa$$

The same technique as before can be used to show that where reciprocity applies $S_{ij} = S_{ji}$.

The properties of a lossless multiport junction can also be investigated by equating the total input power to zero:

$$\sum_{i=1}^{n} (a_i a_i^* - b_i b_i^*) = 0$$

By matching all ports except port 1, which ensures $a_2, a_3, \ldots, a_n = 0$ so that $b_i = S_{i1}a_1$, we see that

$$|a_1|^2 - \sum_{i=1}^{n} |S_{i1}|^2 |a_1|^2 = 0$$

or

$$\sum_{i=1}^{n} |S_{i1}|^2 = 1$$

Similarly,

$$\sum_{i=1}^{n} |S_{ij}|^2 = 1 \qquad (j = 1, 2, \ldots, n)$$

If we then consider the more general situation, we have

$$\sum_{i=1}^{n} \left[a_i a_i^* - \sum_{p=1}^{n} S_{ip} a_p \sum_{q=1}^{n} S_{iq}^* a_q^* \right] = 0$$

The coefficient of $a_p a_q^*$ must also vanish if the expression is to be zero for all values of the a_i, so that

$$\sum_{i=1}^{n} S_{ip} S_{iq}^* = 0$$

These relations may be succinctly expressed by the matrix result

$$SS^* = 1$$

or

$$S^{-1} = S^*$$

The scattering-matrix formulation is useful in specifying the properties of multiport junctions, which may be frequently considered as lossless. It can also be used to specify circuit devices such as microwave transistors. To relate the scattering parameters to conventional admittance or hybrid parameters we have to define a characteristic impedance to be associated with each port. Wave amplitudes and actual voltage and current can then be related through the equations

$$V_i = Z_{0i}^{\frac{1}{2}}(a_i + b_i)$$
$$I_i = Z_{0i}^{-\frac{1}{2}}(a_i - b_i)$$

7.3.5 Waveguide impedance elements

In the previous section the effect of a post in a waveguide was considered as an example, and it was pointed out that a normalised admittance could be attributed to such an obstacle. It is desirable to match waveguides in the same way as it is desirable to match transmission lines; for example, at high power the VSWR should be near unity to avoid excessive local stresses, or a receiving aerial should be matched to the waveguide feed to obtain maximum received signal. As with transmission line, we need to introduce admittances in shunt at points along the waveguide. Such admittances can conveniently be produced by posts and by thin diaphragms. To work out the admittance of any particular diaphragm it would be necessary to solve the electromagnetic problem of finding reflection and transmission coefficients for an incident wave in the dominant mode. Such a calculation is usually very difficult, although analytic answers can be obtained under restricted conditions. Data on the admittance of obstacles is therefore frequently experimental, obtained by calculation from measurements of standing waves. In this section we shall give a few details of commonly used admittance

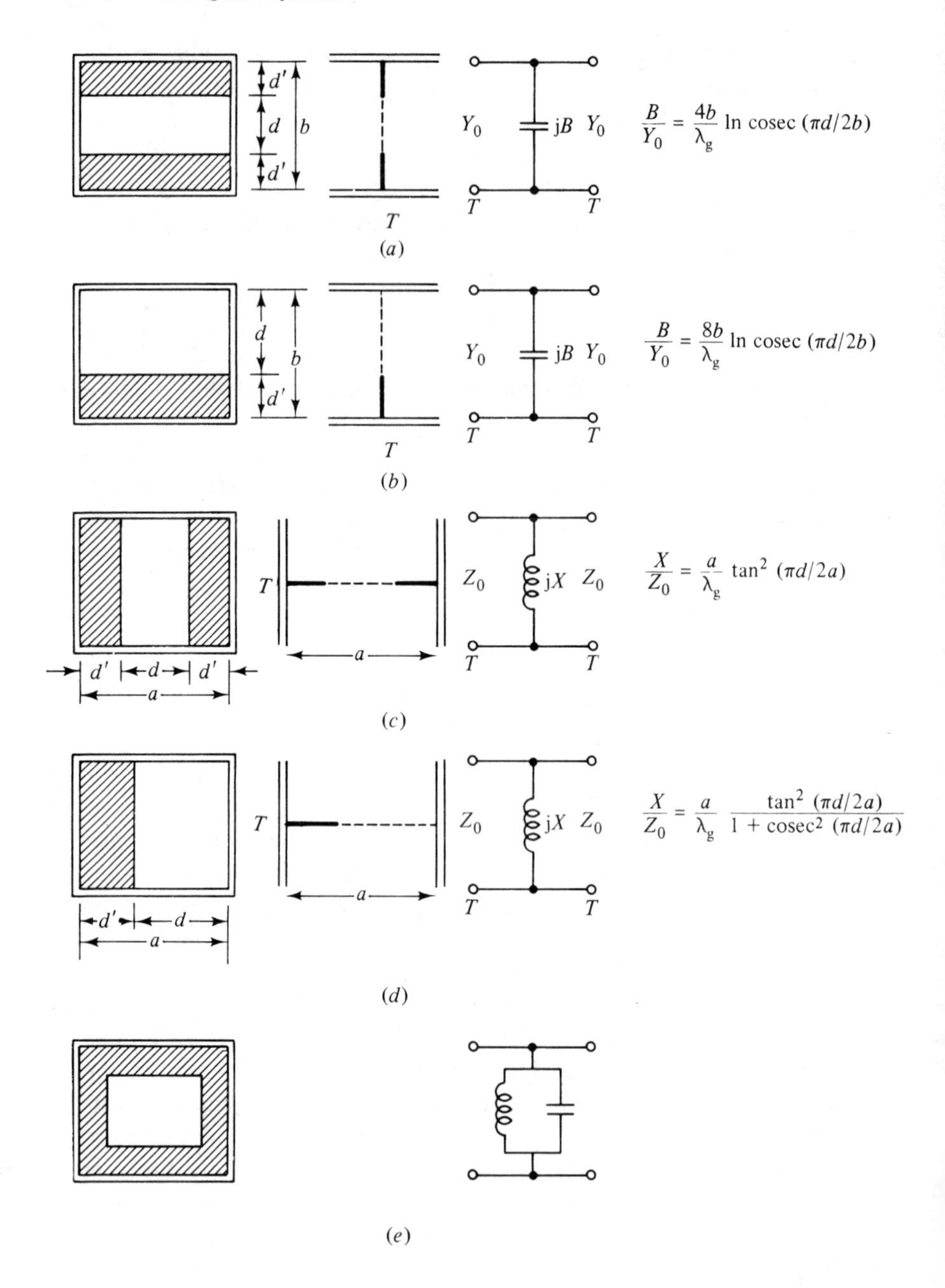

Fig. 7.11. Irises in waveguide as admittances. (*a*) Symmetrical capacitative. (*b*) Asymmetrical capacitative. (*c*) Symmetrical inductive. (*d*) Asymmetrical inductive. (*e*) Resonant.

components, although it must be realised that many others are possible and are used in the appropriate circumstances. The results are all for dominant-mode propagation.

Irises

Thin irises partially blocking the waveguide are frequently used for matching purposes. A selection is shown in fig. 7.11. Those shown in fig. 7.11(a) and (b), in which the aperture extends to the narrow walls of the guide, have a normalised susceptance which is found to be capacitative, and are therefore referred to as capacitative irises. The equations shown in the figure represent the actual susceptance to better than 5% in the normal frequency range. The variation of susceptance with frequency for some particular cases is shown in fig. 7.12.

In fig. 7.11(c) and (d) are shown inductive irises, the aperture extending to the broad faces. As in the capacitive case, the equations are only

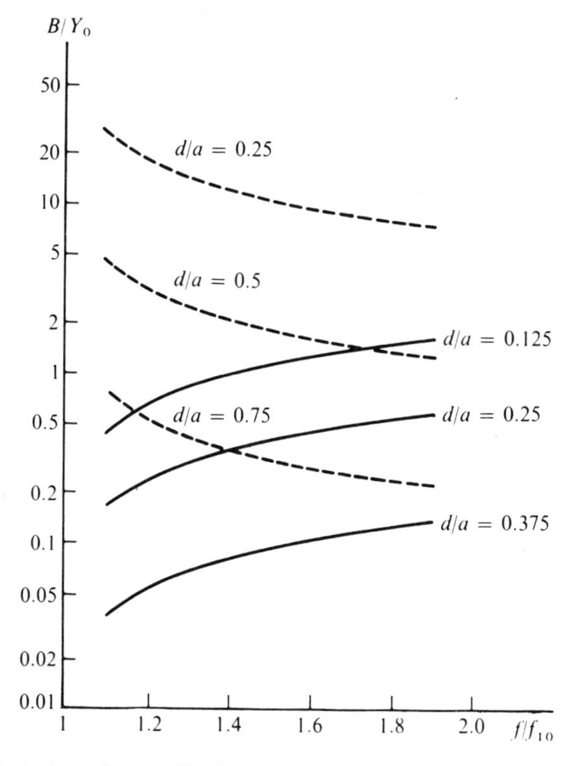

Fig. 7.12. Variation of normalised susceptance of irises with frequency. Full line: symmetrical capacitative ($b = \frac{1}{2}a$). See fig. 7.11(a). Pecked line: symmetrical inductive ($b = \frac{1}{2}a$). See fig. 7.11(c).

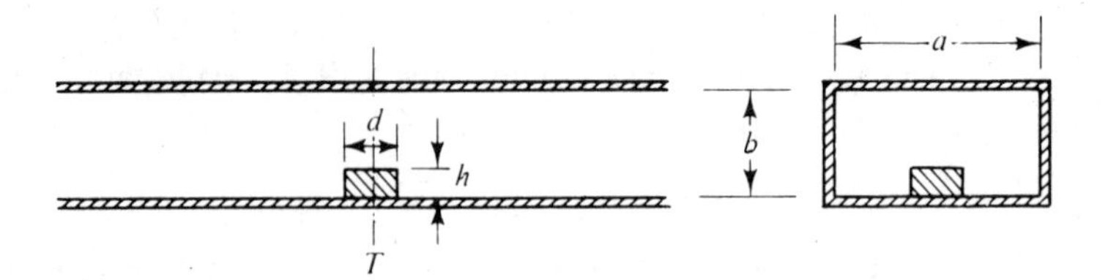

Fig. 7.13. Post in waveguide.

approximate. The fact that one geometry gives capacitative susceptance and the other inductive can be understood by reference to the field pattern of the dominant mode. The electric field is across the narrow dimension, so that narrowing this dimension might be expected to have the effect of increasing this field locally and thus storing more electrostatic energy. There will of course be higher, evanescent modes generated. Since they are evanescent they only contribute to stored energy and hence reactance. By contrast the inductive iris will modify primarily the magnetic field.

The aperture of fig. 7.11(*e*) gives a resonant iris, which might be expected as it combines inductive and capacitative irises.

The use of irises of these types in a waveguide is not suitable at high power on account of the high fields produced in the vicinity of the sharp edges. Alternative devices have to be used for this purpose.

Posts

Another type of obstacle often used is a post on a broad face of the guide, as shown in fig. 7.13. Such an obstacle would be expected, for small penetrations, to be capacitative. It is found that as the height is increased a resonance occurs, giving the effect of a series L–C circuit in shunt across the waveguide. The diameter of a post is likely, for mechanical reasons, to be an appreciable fraction of a wavelength, so that the circuit interpretation has a T-equivalent, as shown in fig. 7.14,

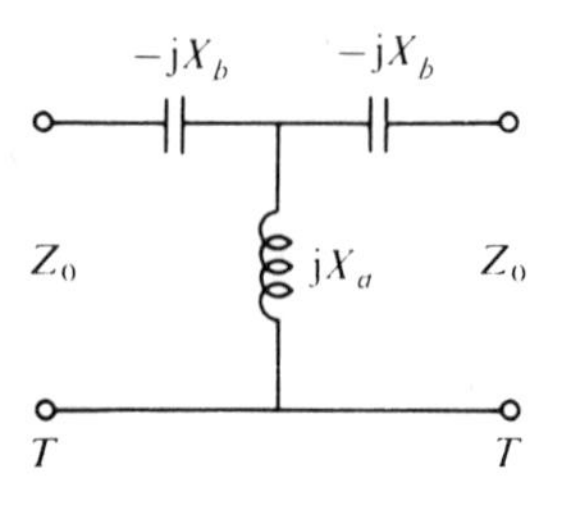

Fig. 7.14. Equivalent circuit for post of fig. 7.13.

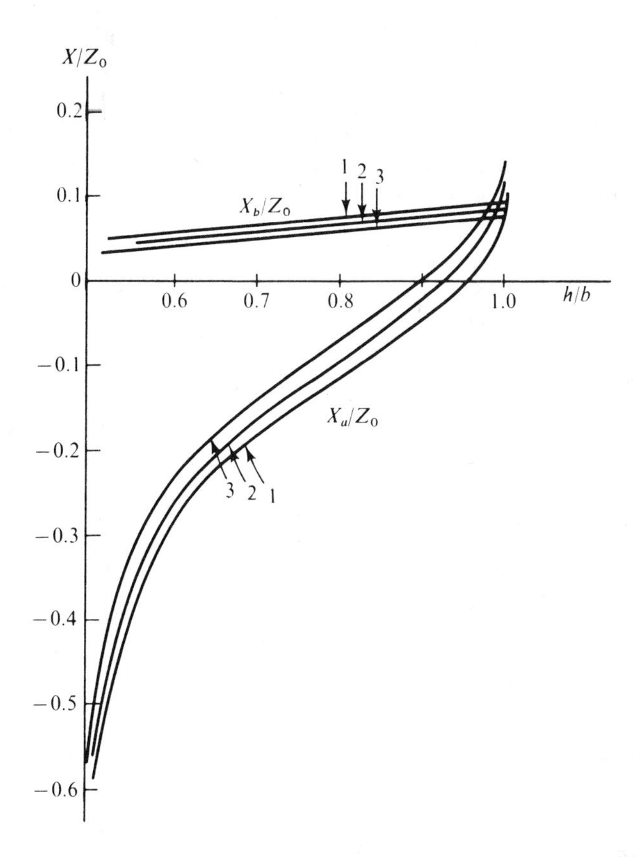

Fig. 7.15. Parameters in equivalent circuit of fig. 7.14 for the post of fig. 7.13 for WG 16. Curve 1 corresponds to $\lambda = 3.4$ cm, $f/f_{10} = 1.34$. Curve 2 corresponds to $\lambda = 3.2$ cm, $f/f_{10} = 1.43$. Curve 3 corresponds to $\lambda = 3.0$ cm, $f/f_{10} = 1.52$.

rather than a simple shunt reactance. Some experimental results for a circular post are given in fig. 7.15.

An adjustable post can be conveniently realised by a screw tapped into the waveguide wall although, since currents will flow between the walls and the post, care has to be taken with the contact. This remark applies also to irises and any insert through a wall, since if currents cross a joint care has to be taken in making that joint.

Example

Measurement of the standing waves in the WG 16 feed to an antenna shows that at a frequency of 9.29 GHz the SWR is 1.7, with a maximum of the electric field 6.2 cm from the commencement of a corner in the waveguide. Determine the size and location of an inductive iris which will bring the SWR to unity.

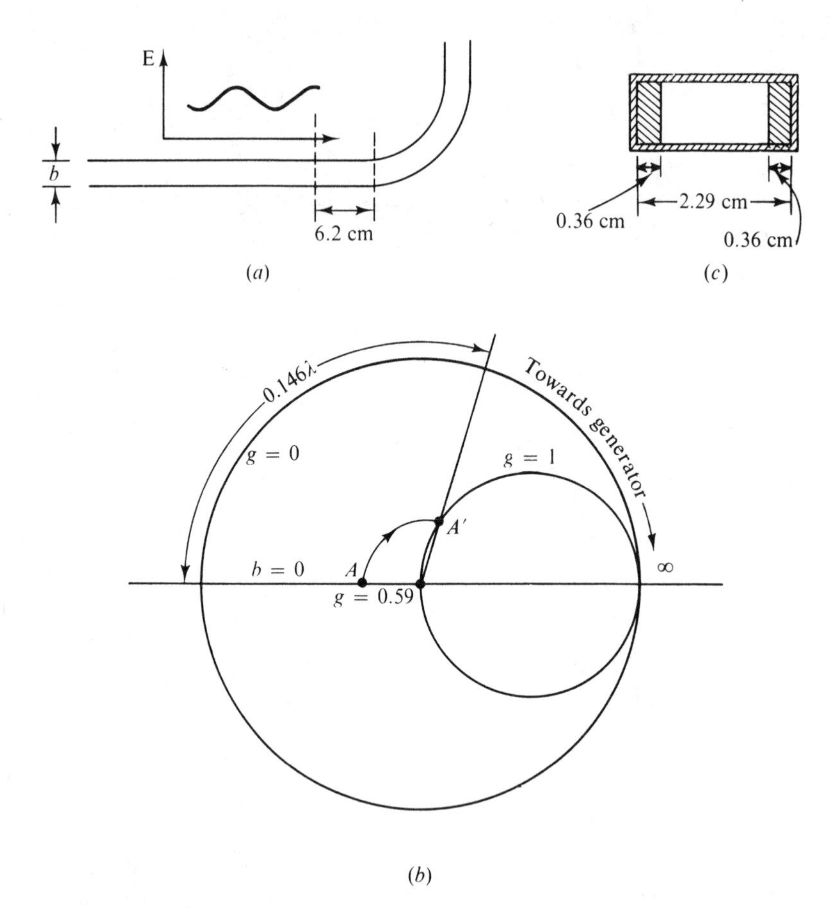

Fig. 7.16. Matching procedure for a waveguide. (*a*) Standing-wave pattern. (*b*) Smith chart. (*c*) Realisation of inductive iris.

Solution. The situation is shown in fig. 7.16(*a*). For WG 16, $a = 2.29$ cm, $b = 1.02$ cm, so that $f_{10} = c/2a = 6.55$ GHz. The guide wavelength λ_g is given by

$$\lambda_g = 2\pi/\beta = c(f^2 - f_{10}^2)^{-\frac{1}{2}} = 4.56 \text{ cm}$$

The maximum therefore occurs at $6.2/4.56 = 1.35\ \lambda_g$ before the reference. In terms of admittance, at this point $Y/Y_0 = 1/1.7 = 0.59$, which is point A on the Smith chart of fig. 7.16(*b*). By transforming A through a distance of $0.146\,\lambda_g$ towards the generator a point A' is reached for which the admittance is $1+j0.55$. This can be reduced to unity by the addition of an inductive susceptance equal to -0.55. The same point A' can also be reached by going towards the antenna a distance of $(1-0.146)\lambda_g = 0.854\,\lambda_g = 3.9$ cm. This point would be feasible since it is far enough away from the bend for the field to be virtually pure dominant mode. If we assume that the susceptance of a symmetrical iris is as in fig. 7.11

$$\frac{B}{Y_0} = -\frac{\lambda_g}{a}\cot^2\frac{\pi d}{2a}$$

we can determine d/a from

$$0.55 = \frac{4.56}{2.29}\cot^2\frac{\pi d}{2a}$$

This gives

$$\frac{\pi d}{2a} = 1.09$$

and hence

$$d = 1.58\text{ cm}$$

This is indicated in fig. 7.16(*c*).

7.4 Waveguide components

The last section has given examples of ways in which impedances could be realised in waveguide circuits. A considerable range exists of other components which are necessary for the working of a waveguide system. We shall consider some of the more important ones.

7.4.1 Variable short-circuits

In transmission-line systems we saw that lengths of line with a short-circuit at one end played a very useful rôle. An adjustable shorting plunger can be reasonably satisfactorily made in circular geometry by the use of spring fingers to make the sliding contact. With waveguide of rectangular section it is not so easy, and less so in any case as the wavelength decreases. The most satisfactory method of making a short is by the use of a non-contact plunger, an example of which is shown in fig. 7.17(*a*). The faces on either side of the gap at the side are regarded as the conductors of a transmission line whose characteristic impedance will depend on the size of the gap. The circuit equivalent of the gap in the plunger of fig. 7.17(*a*) is therefore as shown in fig. 7.17(*b*): the gap around the face of the plunger has an impedance Z_A derived from three quarter-wave sections in sequence, two of characteristic impedance Z_{01} and the other Z_{02}, terminated in a high impedance Z_B. Since a quarter-wave transformer inverts, the impedance Z_A is given by

$$Z_A = \frac{Z_{01}^2}{Z_B}\left(\frac{Z_{01}}{Z_{02}}\right)^2$$

For the parallel-plate transmission line, the characteristic impedance is proportional to the gap, so that with the geometry of fig. 7.17(*a*) $(Z_{01}/Z_{02})^2$ can be made as little as 0.01. Z_B will be large so that Z_A can be very small.

An alternative form of plunger is shown in fig. 7.17(*c*). In this case the transmission line is folded, and the contact between plunger and wall occurs at a high-impedance point with little current flowing.

This technique is used in other cases where mechanical joints form part of the waveguide, as for example in coupling one guide to another. 'Choke' couplers of the type shown in fig. 7.18 avoid the passage of current through the contact. Line *A* gives a high impedance over the

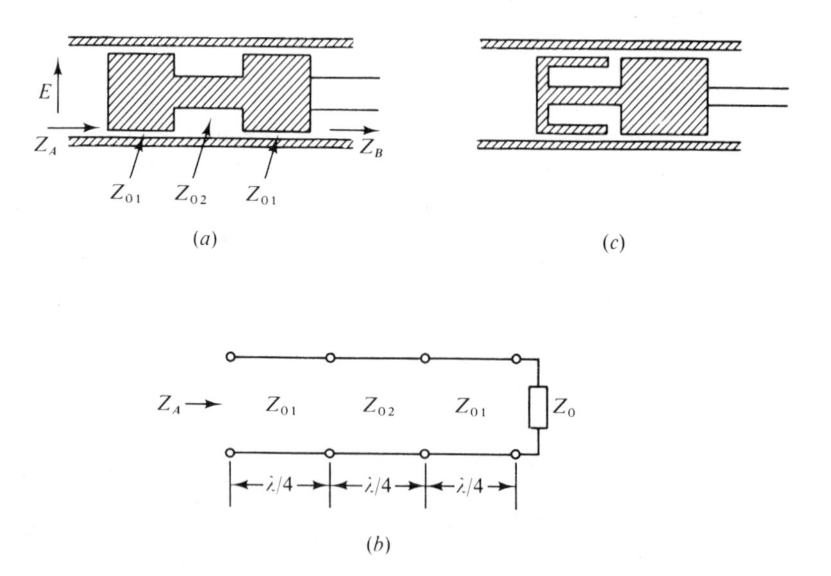

Fig. 7.17. Variable short-circuits. (*a*) Non-contact plunger. (*b*) Equivalent circuit. (*c*) Re-entrant form.

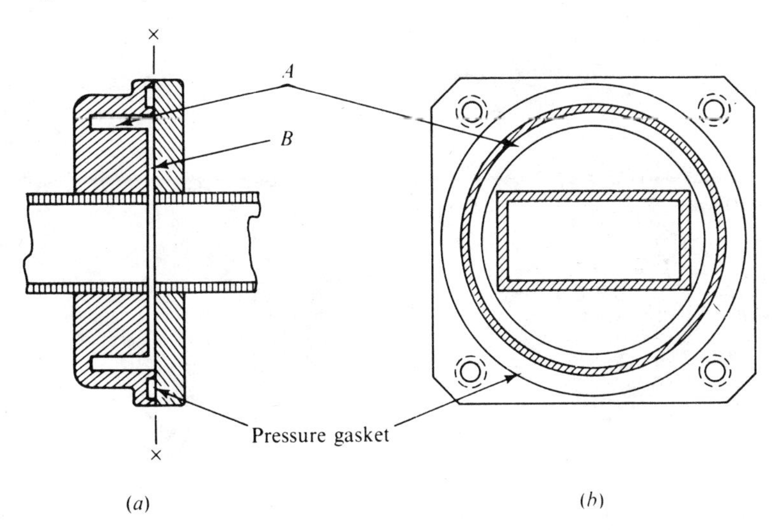

Fig. 7.18. Choke coupler for rectangular waveguide. (*a*) Longitudinal section. (*b*) Transverse section ×–×.

outside of the radial transmission line *B*, producing low impedance across the gap in the waveguide.

7.4.2 Non-reflecting termination

A non-reflecting termination is needed for many measurements. It is theoretically possible to produce such a termination in the form of a thin-film resistive sheet of the correct resistivity across the waveguide followed by a short-circuit one quarter-wavelength away. This arrangement suffers from the disadvantage of being frequency-sensitive both in the requisite resistivity and in the length of the quarter-wave section. A reflectionless termination is better made by arranging for the gradual absorption of the incident wave. A resistive sheet forms a gradual taper in a plane containing the electric field. Providing the taper is made several wavelengths long the reflection is very small. Such a device is shown in fig. 7.19.

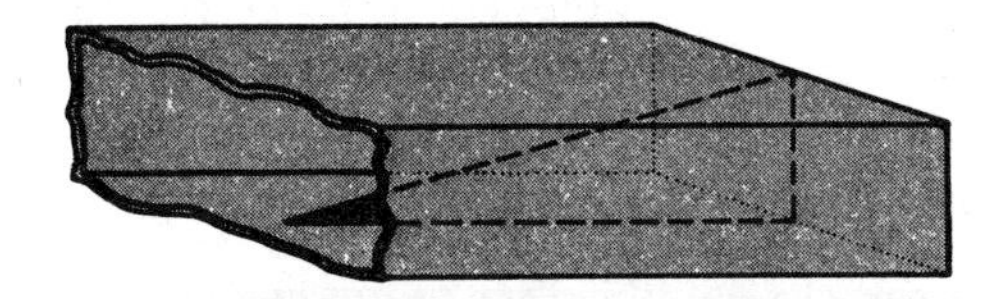

Fig. 7.19. Reflectionless load for rectangular waveguide.

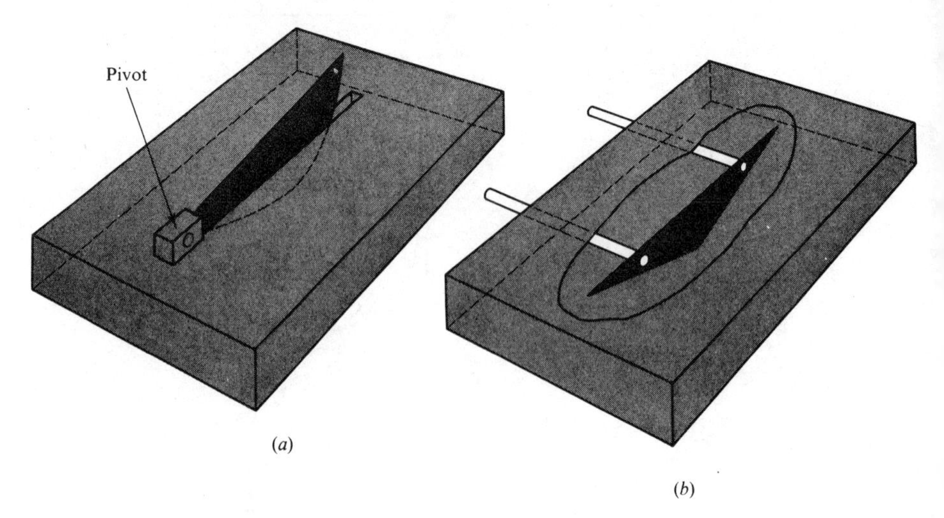

Fig. 7.20. Variable attenuators. (*a*) Penetration into slot. (*b*) Movement across the waveguide.

7.4.3 Attenuators

Fixed attenuators can be made in a way similar to that used for non-reflecting terminations: the difference is in the length and depth of resistive sheet. Variable attenuators can be made using the same principle: a resistive sheet is introduced through a non-radiating slot by an amount which can be controlled. An example is shown in fig. 7.20(*a*). Another method is to mount a sheet of resistive material, as shown in fig. 7.20(*b*) parallel to the guide across the narrow dimensions, and adjust its penetration: at the wall the electric field is zero, so no loss is incurred. At the centre the electric field is a maximum and attenuation will be greater.

Such attenuators need calibrating. An absolute change of attenuation can be obtained by the use of a piston attenuator, as described in §3.8.6. The attenuation constant for the evanescent mode can be precisely calculated, although the insertion loss of the complete attenuator is not known. Such attenuators are usually used on the output of signal generators. At the minimum attenuation the power output is measured directly, and lower levels can then be obtained.

7.4.4 Apertures

It is often necessary to couple small amounts of energy between two waveguides or parts of the same guide. Sometimes the coupling may be into free space. Such coupling may be done by small holes or slots in correctly chosen parts of the waveguide wall, or by placing an iris with a small window across the guide.

The type of coupling obtained by slots or holes in the waveguide wall can be ascertained from the pattern of current flow in the walls. This is shown for the dominant mode in rectangular guide in fig. 7.21(*a*). The introduction of thin slots in the waveguide walls may in some cases cause virtually no modification of the current path (*A* and *B* of fig. 7.21(*b*)), whereas in other positions (*C*, *D* and *E*) severe distortion of the current paths will be caused, and the slots will radiate energy. The non-radiating slot *A* down the centre of the broad face of the waveguide is particularly useful for monitoring the distribution of electric field in the waveguide, since the dominant TE_{10} mode has the maximum electric field E_x in this plane. Central holes in the broad face can be used to couple into another guide through penetration of the electric field. An edge slot such as *B*

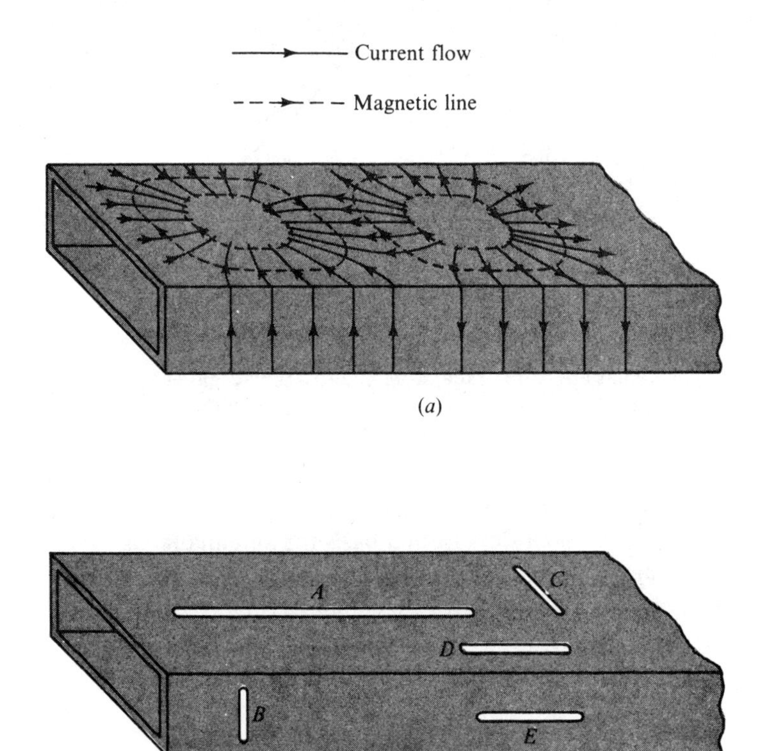

Fig. 7.21. Slots in waveguide walls. (*a*) Current flow. (*b*) Disposition of slots.

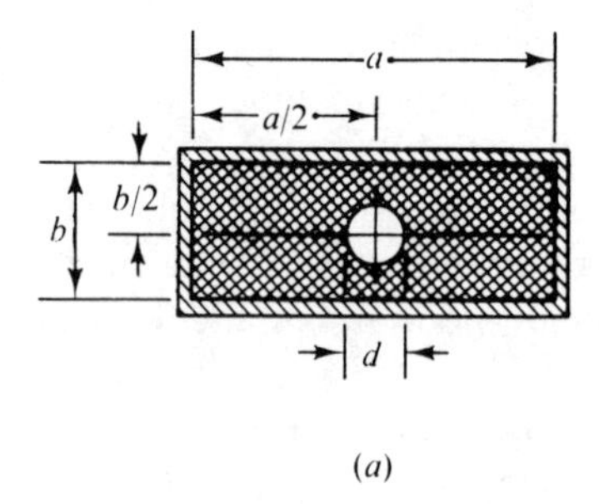

(a)

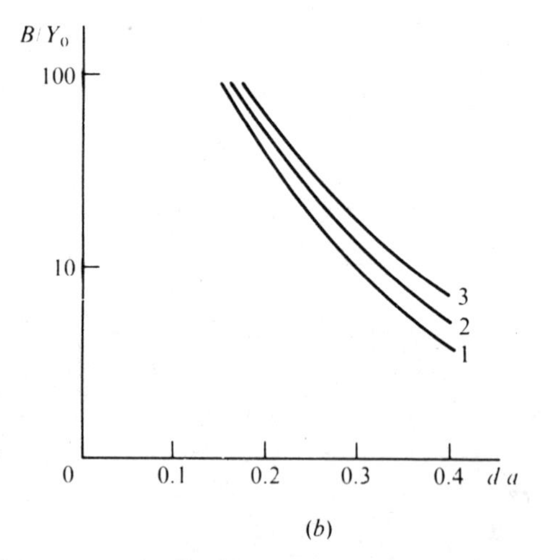

(b)

Fig. 7.22. Circular aperture in diaphragm across a waveguide (*a*) Geometry. (*b*) Shunt susceptance of a zero thickness aperture in WG 16. Curve 1: $f/f_{10}=1.67$. Curve 2: $f/f_{10}=1.45$. Curve 3: $f/f_{10}=1.25$.

causes no radiation, but as such a slot is inclined towards C a radiation field is produced. Such slots form a basis for antenna feeds.

As an example of the small window across a guide, fig. 7.22(*b*) shows values of shunt susceptance for the small circular window shown in fig. 7.22(*a*). The values quoted are calculated and have an accuracy of better than 10%.

7.4.5 Waveguide junctions

The formation of connections between transmission lines is simple to imagine, since each has two wires, and series or parallel connections may be used. Such connections are useful in many ways: as a means of adding components for matching purposes or for power sharing between several loads, for example. Waveguide junctions are also possible. Figure 7.23(a) and (b) shows the E-plane and H-plane T-junctions in rectangular waveguide. An indication of the field pattern in the neighbourhood of the junction is shown in the diagram. Since we have regarded the electric field as equivalent to voltage on a transmission line, the E-plane junction is often referred to as a series junction. In H-plane junctions the disposition of the electric vector indicates that it is a shunt junction. These terms of course are used in a general sense since the large dimensions of the junctions mean that complicated equivalent circuits will be required to represent them.

Various interesting general properties of three-port lossless junctions can be deduced: (i) that a variable short in one arm may always be so positioned that the other two ports are isolated from each other, (ii) that if there is symmetry between two of the ports, a variable short in the third can be so positioned that reflectionless transmission takes place

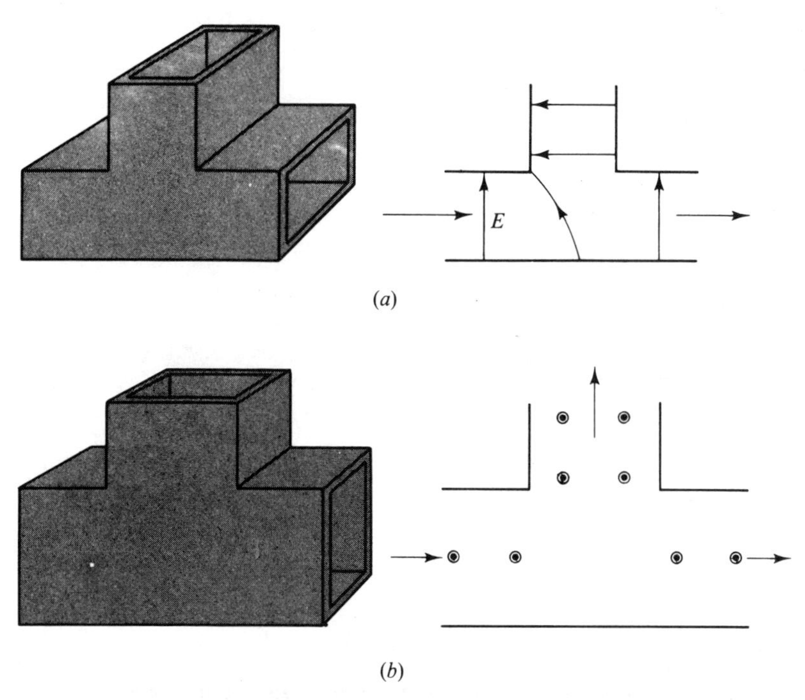

Fig. 7.23. Junctions in rectangular waveguide. (a) E-plane. (b) H-plane.

between those two. A third property concerns 'complete matching': a junction is said to be completely matched if any one port presents a reflectionless impedance to its feed when all the others are terminated in reflectionless loads. It can be shown that for a three-port this state is unrealisable, and only one port at a time can be matched.

Hybrid junctions

A hybrid junction is a four-port device which ideally has the property that, with each port matched, power fed into one port is equally divided between two of the remaining three ports, with no coupling to the fourth port.

One form is the magic-T shown in fig. 7.24(*a*). This may be regarded as composed of an E-plane and an H-plane junction. Power flowing into port 1 will divide equally into guide 2 and 3, but will not couple into port 4 because the electric vector does not contain a component in the appropriate direction. Likewise power flowing into port 4 will divide between guides 2 and 3 and is not coupled to port 1. In the first case the electric fields in guide 2 and 3 are parallel, whereas in the latter case the electric fields are anti-parallel. The junction may be used as a balanced mixer, with signal in to port 1 and local oscillator in port 4. The

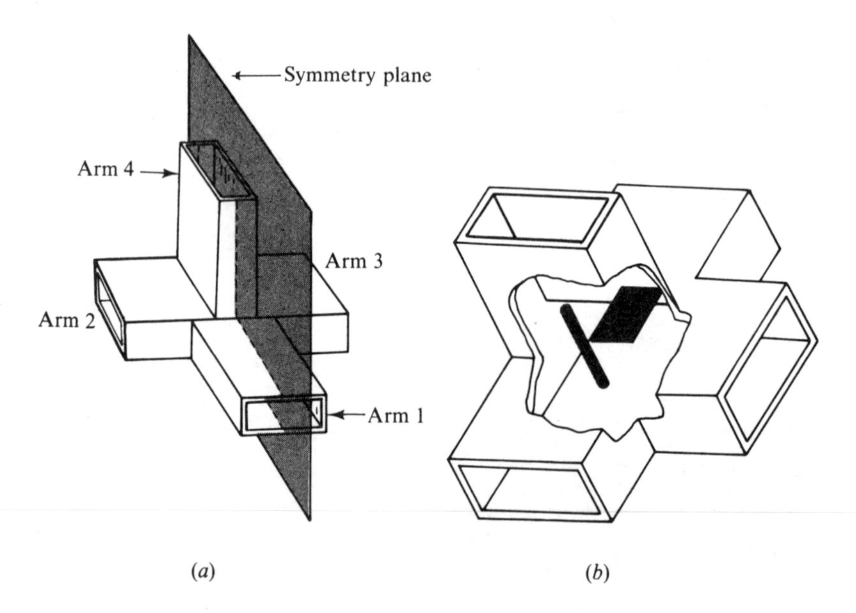

Fig. 7.24. The magic-T. (*a*) Configuration. (*b*) Positions of post and iris. (From Dicke, Purcell & Montgomery, *Principles of microwave circuits*.)

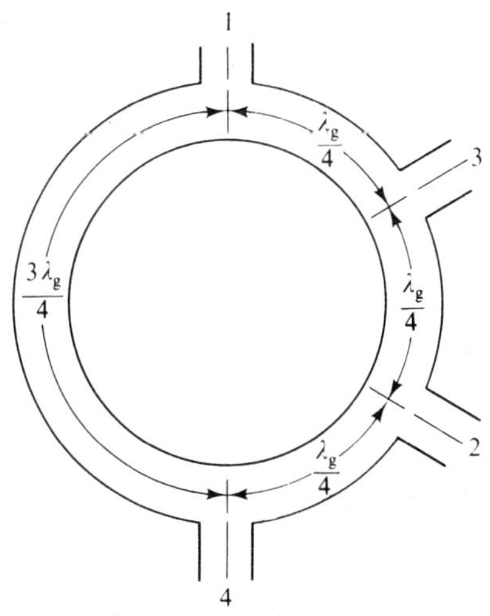

Fig. 7.25. The hybrid-ring.

intermediate-frequency signals appearing at mixers attached to ports 2 and 3 will then be in anti-phase, making possible a degree of cancellation of local oscillator noise. A magic-T constructed as in fig. 7.24(*a*) will require additional matching elements to be incorporated, as shown in fig. 7.24(*b*).

Another junction is the hybrid-ring, shown in fig. 7.25. If it is remembered that one half-wavelength of line has the effect of changing the sign of voltage and current between input and output, it will be seen that the four ports are symmetrical save for a phase inversion between 1 and 4. Feeding from port 1 will therefore cause in a load on port 3 currents through the two paths which are equal and opposite, so that no coupling takes place to port 3. The voltage across it is also zero and the port may be short-circuited. When this is done, using the impedance-inverting property of a quarter-wavelength line, the impedance seen at port 1 is $\frac{1}{2}(Z_0'^2/Z_0)$, where Z_0 is the characteristic impedance of line at the ports and Z_0' the characteristic impedance of the ring. Hence for port 1 to be matched to Z_0, Z_0' must equal $\sqrt{2}\,Z_0$. Similar investigation for the other ports will show that opposite ports are not coupled.

Directional couplers

A four-port network with great application is the directional coupler. Ideally such a coupler may be conceived as having input port 1, output port 2 with auxiliary ports 3 and 4. When the ports are terminated in reflectionless loads, there is no coupling between 1 and 4 or 2 and 3. An input P_1 into port 1 produces an output P_2 in port 2 and an output P_3 in

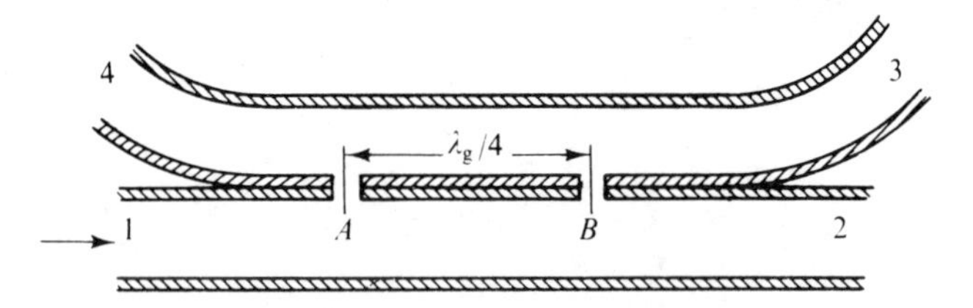

Fig. 7.26. A directional coupler.

port 3. The coupling factor is defined as P_3/P_1. A measure of departure from the ideal is given by the directivity P_3/P_4.

One form for such a device is shown in fig. 7.26. Ports 1 and 2 are the ends of a section of a waveguide feeder. Some of the energy flowing from 1 to 2 will be coupled via the slots A and B into the adjacent waveguide 3 to 4. The coupling at A establishes dominant-mode propagation in the second guide, with power flow in both directions towards 3 and 4. Similarly, energy coupled via B will establish propagation in both directions. Thus energy flowing towards port 3 will be the sum of the two induced waveforms. Because of the extra path length of $2\times\frac{1}{4}\lambda_g$, energy flowing towards port 4 via B will be out of phase with energy flowing in this direction via A. Ideally, the energy coupled to port 4 will be zero for the incident primary flow from 1 to 2. If there is a reflected wave from 2 to 1 we can show, by the same argument, that this will be coupled only to port 4. Thus the coupler separates the incident and reflected power in the waveguide feed.

This type of construction is one of many: it is only necessary to devise coupling which will produce two waves with the required phase difference in the auxiliary guide. Even a single hole can be used, relying on the fact that coupling occurs through both electric and magnetic fields. An array of holes can be used to give broad-band couplers usable over a range of frequencies.

7.4.6 Resonant cavities

It has been mentioned in chapters 3 and 5 that lengths of transmission line or waveguide can be used as resonant cavities of high Q. Such cavities are very convenient in waveguide, since they take the form of a closed section of guide with a small coupling aperture. In fig. 7.27(a) is shown an example of a length of rectangular guide, defined by a short circuit at one end and a small coupling aperture at the other. Such a cavity will resonate at a frequency for which the length is a multiple of

$\frac{1}{2}\lambda_g$. In the neighbourhood of such a frequency, the equivalent circuit is as shown in fig. 7.27(*b*) and (*c*). The coupling aperture behaves as a shunt inductive susceptance of high value, and it is of interest to note that it is possible to choose its value so that the feeder guide sees a matched load at the resonant frequency. This can be seen from the Smith chart in fig. 7.27(*d*). The length of lossy guide can be represented by lossless guide with a resistor of high conductance at the far end, characterised by point *F*. Moving backwards, a distance $\frac{1}{2}\lambda_g$ brings us back to *F*, but slightly before $\frac{1}{2}\lambda_g$ has been traversed, at point *H* the normalised admittance is unity conductance in parallel with a positive susceptance. This susceptance can be cancelled by a shunt inductive

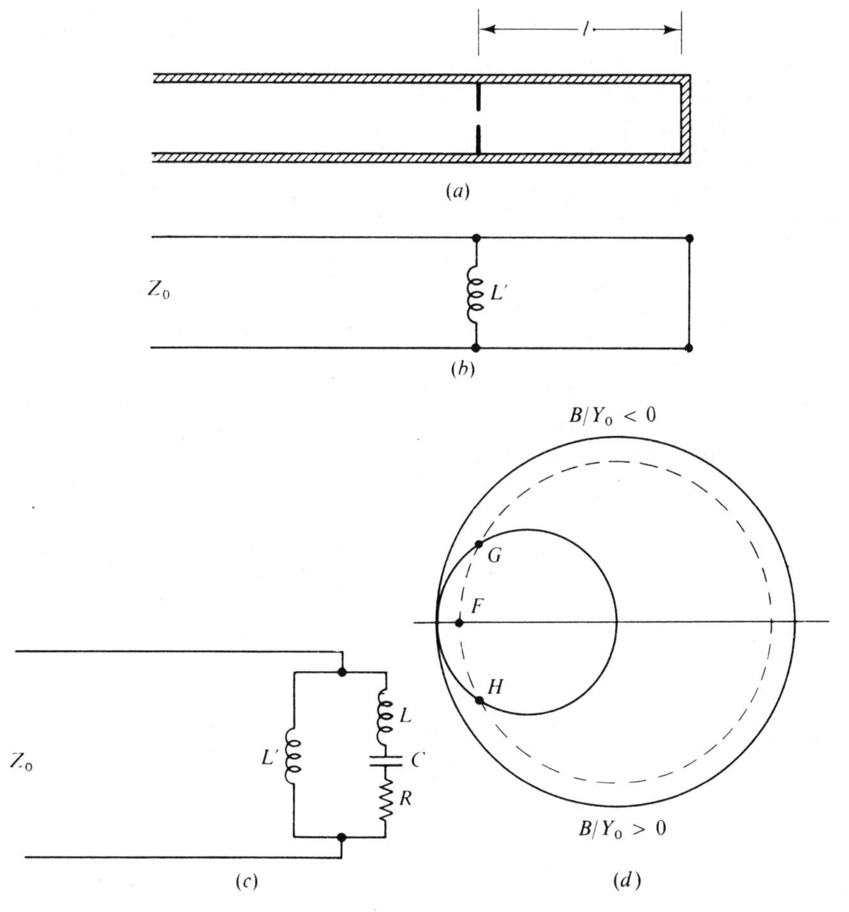

Fig. 7.27. Waveguide resonant cavity. (*a*) Configuration. (*b*), (*c*) Development of equivalent circuit. (*d*) Smith chart.

admittance. Evidently the smaller the resistance, the higher is the Q-factor, and the larger is the required shunt susceptance. Referring back to §7.4.4, it will be realised that the larger the susceptance, the smaller the hole and the less coupling between cavity and feed.

When the coupling iris is replaced by a plate, the cavity is isolated, and the Q-factor is said to be the unloaded Q of the cavity, Q_u. When coupling is allowed, the Q-factor is then the loaded Q, Q_l, which is always smaller than the unloaded Q. The external Q, Q_e is defined by

$$\frac{1}{Q_l}=\frac{1}{Q_u}+\frac{1}{Q_e}$$

It is often required to obtain cavities of high Q, and it becomes important to consider the effects of fabrication. The effects of joints can be seen by reference to the current flow in the end plates of short-circuited lengths of guide. At a short-circuiting plate, the lines of magnetic force will be of the same form as in the propagating mode, which are shown in fig. 7.28(a), (b) and (c) for the TE_{10} rectangular and TE_{11}, TE_{01} circular. The first two have non-zero tangential magnetic fields at the edge of the section, implying that wall currents exist in the plate and cross the joint into the walls. The TE_{01} circular has a magnetic field which is zero at the edge of the section and hence no currents cross the joint. Since good joints are difficult to make, a resonator using the TE_{01} circular mode might be expected to give less trouble. It does also have lower losses, practical difficulties apart. As orders of magnitude, resonators at 10 GHz using the modes listed above might have Qs of 8000, 12 000, 25 000 respectively.

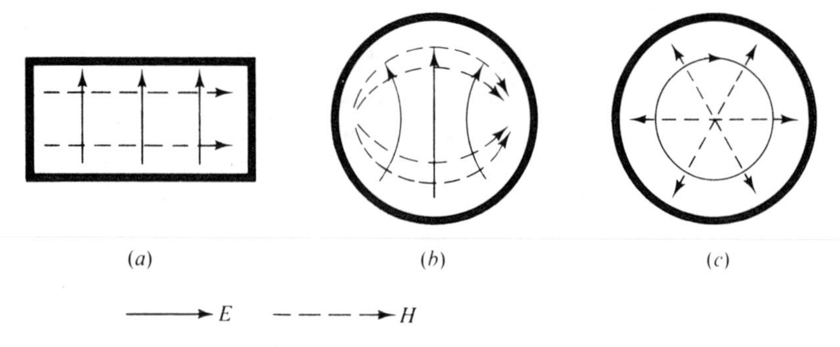

Fig. 7.28. Magnetic fields in propagating modes. (a) TE_{10} rectangular. (b) TE_{11} circular. (c) TE_{01} circular.

7.4.7 Components using ferrites

A range of important components make use of the properties of magnetically saturated ferrites at microwave frequencies. The elementary theory of the radio-frequency behaviour of ferrites is given in § 10.5. The mode of operation of practical devices can be understood in outline if we appreciate that a ferrite will support *TEM* waves that are circularly polarised, and that the permeability is different for the two senses of rotation of the magnetic vector. The simplest device to consider is the Faraday 'plate' which can be used to rotate the plane of polarisation of a wave in a circular guide.

The Faraday 'plate'

Consider the configuration shown in fig. 7.29(a). The dominant mode in the circular guide is TE_{11}, for which the pattern is shown in fig. 7.29(b). Assume that the ferrite rod can be regarded as only perturbing the field. Section 10.5 discusses how in an unbounded ferrite medium the plane of polarisation is rotated due to the different velocities of right- and left-handed circular polarisations. A similar effect will exist in the wave-guide: TE_{11} has a definite orientation of the electric field at the centre, in the region of the rod rather like a *TEM* wave. After propagating down the section of guide containing the rod, a similar pattern will exist but with its plane of polarisation rotated, as shown in fig. 7.29(c). Rectangular-to-circular sections at either end act as filters accepting only one polarisation. The angle θ between these two sections is arranged to be the same angle as that by which the plane of polarisation is rotated. this being so, transmission from port 1 to port 2 will, apart from losses, be complete. Consider now propagation in the reverse direction from port 2 to port 1. Section 10.5 shows that the rotation of the plane of polarisation is in the same direction in space, being fixed not by the direction of travel of the wave but by the direction of magnetisation of the ferrite. The wave coming up to port 1 will not have the appropriate polarisation to pass out, but will be at an angle 2θ to the preferred direction. If, for example, θ is made to be $\frac{1}{4}\pi$, then no wave will emerge at port 1. This device would then be an *isolator*, allowing transmission from port 1 to port 2, but none from port 2 to port 1. Such a device is of great use when interposed between an oscillator and a system: it prevents the system reacting back on the oscillator. As described, the wave entering port 2 would be reflected: it is usually desirable to arrange resistive material near each port to absorb waves whose polarisation is perpendicular to the preferred direction.

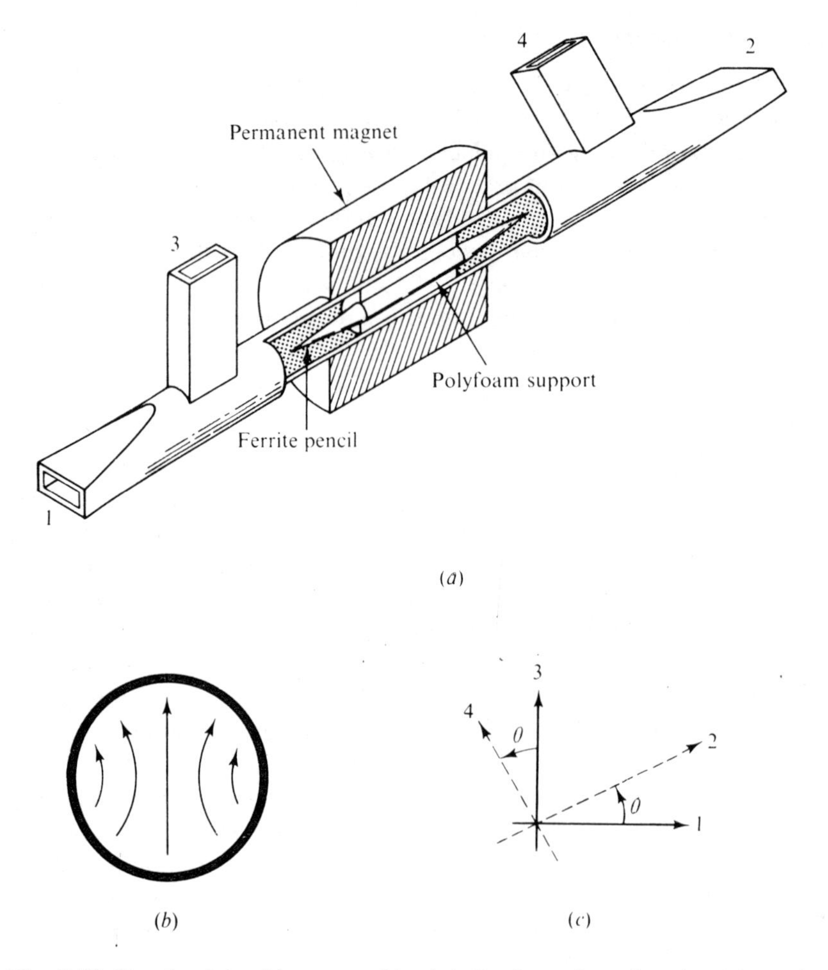

Fig. 7.29. Faraday 'plate' in waveguide. (*a*) Configuration, direction of steady magnetisation parallel to axis of waveguide. (*b*) Electric field of TE_{11} circular mode (*c*) Rotation of plane of polarisation.

The combination of rotation and filters for selecting particular directions of polarisation enable other useful devices to be made. One of these is the *circulator*, using the two further ports shown in fig. 7.29(*a*) enabling the waves polarised perpendicular to the directions preferred by ports 1 and 2 to be extracted. As previously discussed a wave entering port 1 emerges at port 2, and a wave entering at port 2 will be of the wrong polarisation to be accepted by port 1. It will however emerge at port 3. A wave entering port 3 will emerge at port 4 (in the

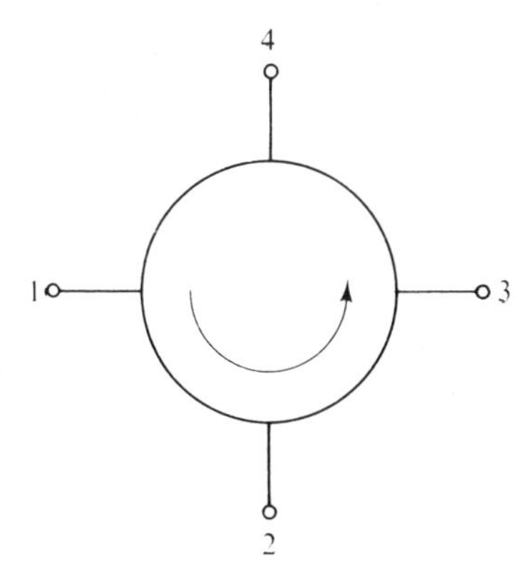

Fig. 7.30. Symbol for four-port circulator.

same way as between ports 1 and 2), being rejected by port 2. Finally a wave entering port 4 will be accepted by port 1. The symbol for such a four-port circulator is shown in fig. 7.30. An isolator results if ports 3 and 4 are terminated in matched loads.

It must be emphasised that the above discussion leaves out of account a number of effects, such as the effect of the ferrite on waveguide modes. Losses also have to be investigated before a useful device can be made.

Ferrites in rectangular guide

It is impossible to discuss exhaustively the many types of isolator and circulator which can be made. However the principles involved when ferrites are used in rectangular guide are somewhat different to those discussed above in which one mode can have alternative polarisations in the same waveguide.

Consider a rectangular waveguide carrying the dominant TE_{10} mode. The magnetic field configuration was given in fig. 3.4, and is repeated in fig. 7.31(*a*). The configuration is for a wave travelling left to right. Along a line AA' the magnetic field direction for the points 1, 2, 3, 4 may be seen to be that shown in fig. 7.31(*b*). For an observer at a fixed point on the guide, the points arrive in succession 1, 2, 3, 4 so that the direction of the field is observed to rotate in the direction shown. On a line BB' the other side of the centre line, the same considerations give the results shown, with the direction of rotation in the opposite sense. A wave travelling from right to left will interchange AA' with BB'.

Consider now such a guide with a ferrite sheet placed and magnetised as shown in fig. 7.32(*a*). The magnetic vector at the sheet is rotating

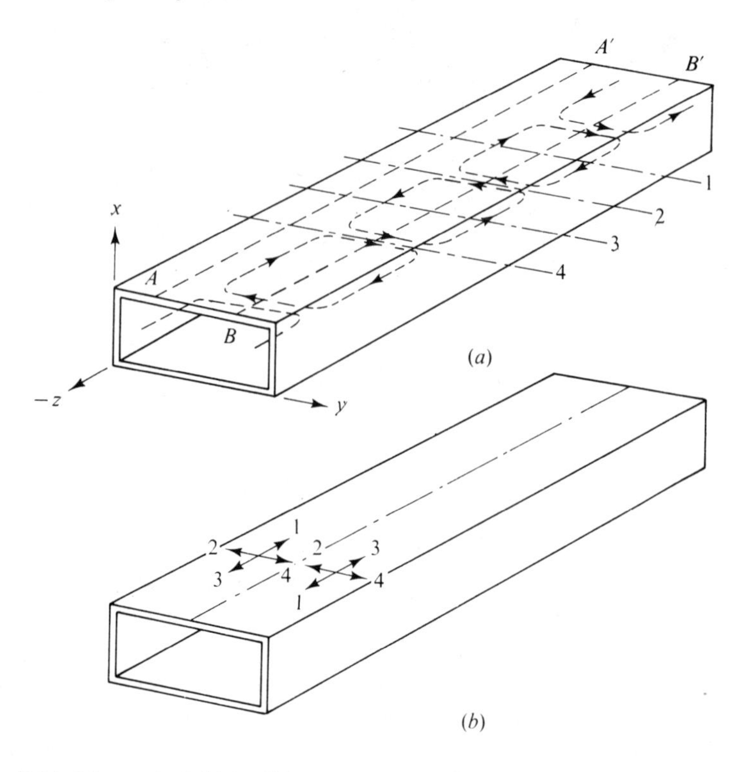

Fig. 7.31. Magnetic field in TE_{10} rectangular. (*a*) Field lines. (*b*) Direction of rotation.

anticlockwise, looking *into* the magnetisation vector, and by the results in § 10.5 the effective permeability is μ_+. If the wave is reversed in direction the ferrite will now experience the opposite sense of rotation of $\boldsymbol{H}$, giving permeability μ_-. Under normal conditions we shall have $\mu_- > \mu_0 > \mu_+$. The ferrite will cause the phase constant β to be smaller for the forward wave and larger for the reverse wave. The phase shift incurred on propagation over a given length will be different for the two directions, and a directional phase-shifter results. The effect may be increased by using symmetrically placed sheets asymmetrically magnetised as shown in fig. 7.32(*b*). If in this case the sheets are magnetised in the same direction, no directionality in phase constant results, but another effect results. Since $\mu_+ < \mu_-$, for the forward wave a concentration of energy towards the right-hand side of the guide (B) will occur. For the reverse wave, the concentration will occur to the left (A). Thus the actual fields are displaced on reversing the direction of propagation. This displacement can for example be used to provide

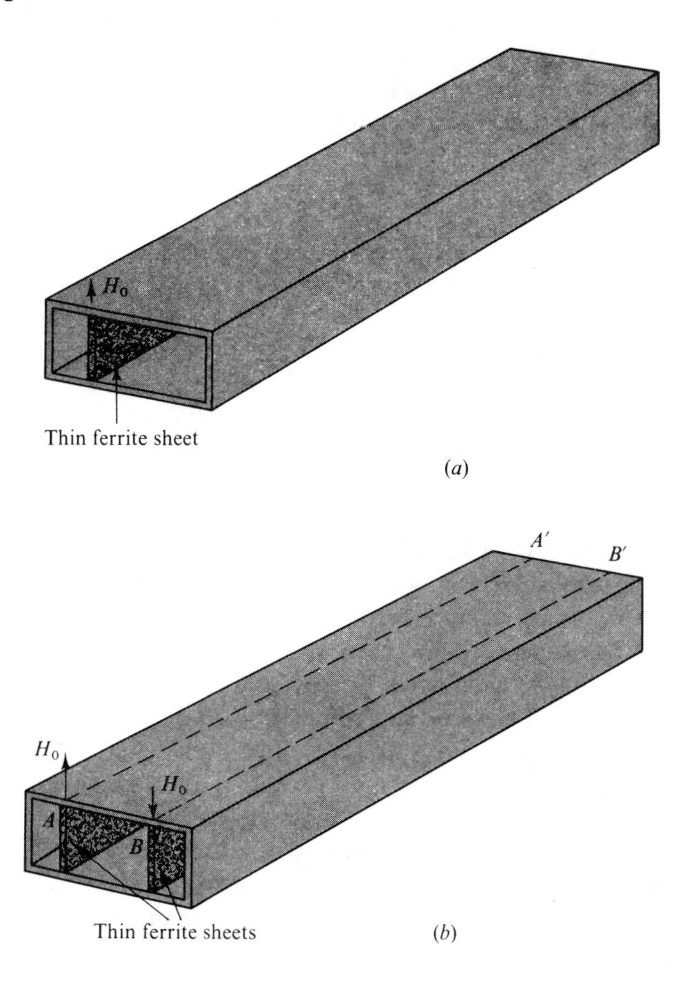

Fig. 7.32. A ferrite sheet in a rectangular waveguide. (*a*) Single sheet. (*b*) Two sheets, polarised in opposite directions.

differential attenuation, or differential coupling into another guide. The effects described in this section form the basis of the non-reciprocal microwave devices now available.

7.5 Waveguide measurements

The performance of a waveguide transmission system can be characterised by measurements of reflection on the feeder (indicated by the presence of standing waves), by monitoring the frequency spectrum of the operational channel, and by signal-to-noise performance tests on the overall system. In this section, we shall look at some of the basic measurement techniques used in waveguide systems.

7.5.1 Standing waves

We saw in chapter 5 the usefulness of standing-wave measurements on transmission systems, and we can use very similar techniques with waveguide systems.

A simple way of indicating the presence of reflections is to use a directional coupler to monitor the incident and reflected power on the feeder. If further information is required, such as the effective terminating impedance, then the field inside the waveguide feed must be monitored to indicate the magnitudes and positions of maxima and minima of standing waves.

It is important to note that the possibility of sampling the field distribution within the waveguide depends critically on the mode (or modes) of propagation existing within the guide. If we are dealing with multimodal propagation then, even in the ideal case when no mode coupling exists, the combined electric and magnetic fields of the separate modes will give rise to a complex structure. Furthermore, the access slots in the waveguide walls will normally not be suitable for more than one mode, and will probably constitute a radiating slot for a different mode. Fortunately, as discussed earlier, the normal or dominant mode in a rectangular waveguide is frequently the only propagating mode, and as long as the slotted section is not in close proximity to a discontinuity, the measurements made on that section will give reliable data concerning the conditions in the waveguide.

We have seen that, for a rectangular waveguide supporting the dominant TE_{10} mode, the current paths in the waveguide walls permit the insertion of a narrow slot along the centre of the broad face without introducing significant distortion of these current paths. Consequently, the internal field structure is not modified and a short probe can be inserted through the slot to sample the electric field, thus giving an indication of the longitudinal variations. Figure 7.33 shows a sketch of a standing-wave detector comprising a coaxial probe and matching section coupled to a semiconductor detector. At microwave frequencies, these slotted sections and detectors have to be very accurately manufactured to ensure minimum interference with the propagation in the waveguide and maximum consistency of measurement over several waveguide wavelengths. In a rectangular waveguide, for example, the

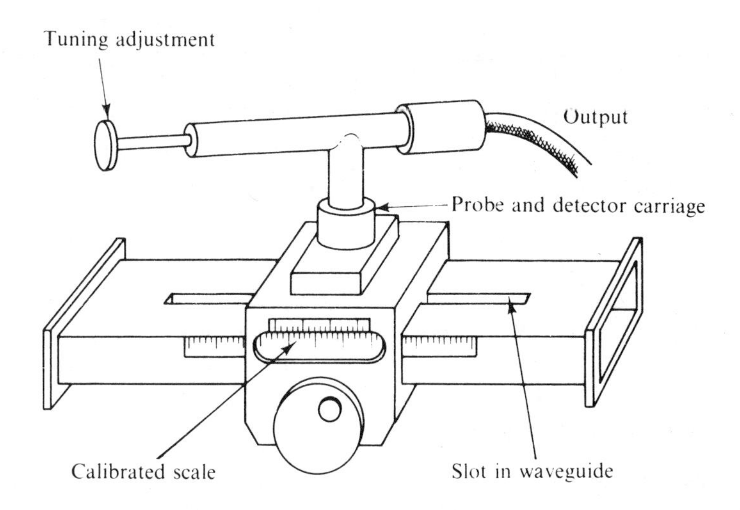

Fig. 7.33. Standing-wave detector in rectangular waveguide.

probe penetration has to be constant to about 0.1% of the narrow dimension.

7.5.2 Wavemeters

Although frequency determination is in principle made by time measurement, it is convenient to have simpler means of making measurements even though they may be less accurate.

We can estimate the frequency of a signal propagating in a waveguide by using the slotted-line device of the previous section. If a standing wave exists, then we can estimate the guide wavelength λ_g and thus from the knowledge of the waveguide the equivalent free-space wavelength and frequency can be determined. A more accurate measure of frequency can be obtained using a tunable cavity wavemeter coupled to the waveguide. Resonant cavities have been briefly discussed in §7.4.6. They can have a very high Q value which corresponds to very good frequency selectivity, and if we arrange for the physical dimensions of the cavity to be adjustable, we can produce a very accurate frequency meter. It has proved possible to obtain Q values of 40 000 at 35 GHz. The TE_{01} mode in circular guide is frequently used. The circular geometry has the advantage that non-contact plungers can be moved directly with a micrometer. Circular plungers have also been used in rectangular guide for this reason. Cavities can be used either as an absorption or as a transmission wavemeter. These arrangements are shown diagrammatically in fig. 7.34.

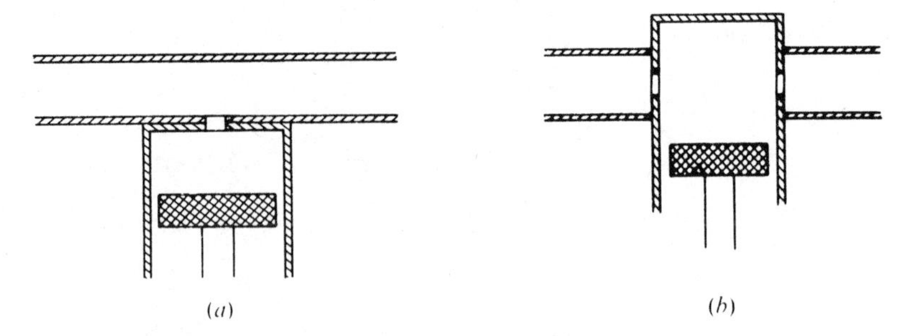

Fig. 7.34. Cavity wavemeters. (*a*) Absoprtion. (*b*) Transmission.

7.5.3 Measurement of power

The absolute determination of power flow is a fundamental measurement. The most usual way of measurement is to arrange for the conversion of the power to heat in a resistive load. The heat produced can be measured in a variety of ways, depending on the power level. At high power a flow calorimeter approach is adopted: for example, a load may take the form of a water-cooled wedge, when power may be found from knowledge of water flow and temperature rise. At lower powers a static calorimeter is more satisfactory, or the load may take the form of a temperature-sensitive resistor, such as a bolometer, or a thermocouple. Such devices enable the temperature rise to be measured directly, or if low-frequency power can be applied to the device simultaneously with the radio-frequency power, operation at constant temperature is possible. In all these methods the main problem is to match the load to the source, so that all power is absorbed. Powers as low as a few milliwatts can be measured to about 0.1%. Powers much lower than this level are measured by comparison against a known source in a radio receiver. The source must be fitted with an adjustable attenuator, and at the highest available level the power must be measurable by absolute means. Lower levels can then be obtained by attenuation, provided the calibration is absolute. A piston attenuator is frequently used. Instruments usually work on these principles, although many other effects have been investigated.

7.5.4 Measurement of scattering parameters

As we mentioned in §7.3.3, scattering parameters are not only a useful and apposite way of characterising microwave circuit elements, but also they can be measured relatively simply. We have seen that the reflection coefficient at a given port is one of the scattering parameters, and we can determine this, in magnitude and phase, using the slotted-waveguide system outlined in a previous section. In order to measure the transmission coefficients S_{ij}, $i \neq j$, we must employ some comparative measure of input and output from the network under specified termination conditions. Complex ratio measurement at microwave frequencies would be extremely difficult, and measuring sets normally use a frequency changer to enable all gain and phase measurements to be made at an intermediate frequency. Figure 7.35 shows a block diagram of a typical system.

The scattering parameters for the two-port are defined by the equations

$$b_1 = S_{11}a_1 + S_{12}a_2$$

$$b_2 = S_{21}a_1 + S_{22}a_2$$

By terminating port 2 in a reflectionless load, a_2 is made zero. S_{11} is then obtained as a reflection coefficient on the generator side and S_{21} by comparing output at port 2 with the incident wave amplitude. By reversing the junction S_{22} and S_{12} can be measured.

The information can be displayed in analogue or digital form and in some more elaborate analysers can be presented in polar form on a Smith chart display on a cathode ray tube. Swept-frequency test sets permit rapid evaluation of system performance over a wide frequency range.

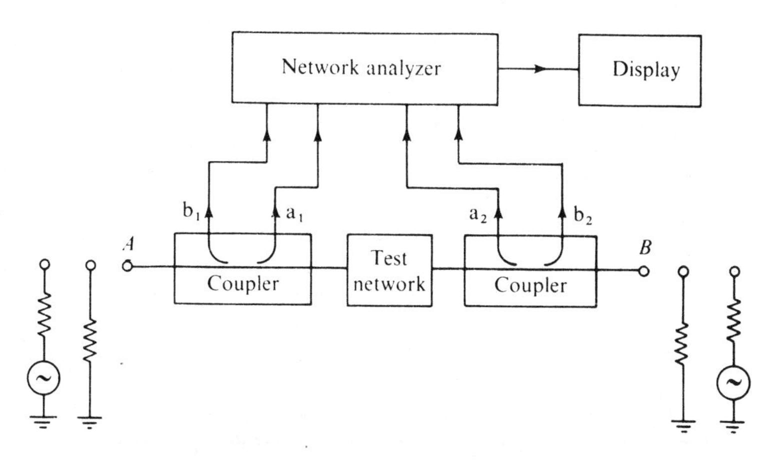

Fig. 7.35. Block diagram of microwave network analyser.

7.6 The long-haul waveguide

The use of waveguide as a communications link was suggested by the low loss of the TE_{01} propagation in circular waveguide. The attenuation of this mode in a circular tube decreases indefinitely as the frequency increases, as has been seen in §3.10.5. An attenuation of the order of 2 dB km^{-1} would make possible runs of the order of 20 km without intermediate amplification. The use of waveguide in this fashion gives rise to problems not encountered in the short-run systems in which dominant-mode propagation is used. The expression for attenuation can be derived from (3.81), giving

$$\alpha = \frac{R_s}{a}\sqrt{\left(\frac{\varepsilon}{\mu}\right)}\left(\frac{f_{01}}{f}\right)^2\left[1-\left(\frac{f_{01}}{f}\right)^2\right]^{\frac{1}{2}}$$

To obtain low attenuations it is necessary that the operating frequency be much greater than the cut-off frequency. This is shown in fig. 3.18 for a 50 mm diameter tube for which the cut-off frequency is 7.32 GHz. To obtain attenuations of the order of 2 dB km^{-1} it is necessary to work at about 50 GHz, giving a ratio of 7 : 1 for f/f_{01}. At such a

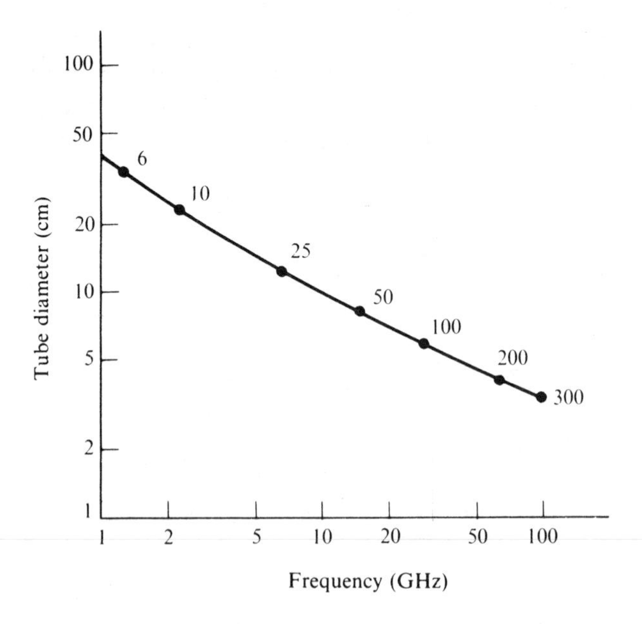

Fig. 7.36. Relation between diameter and frequency for constant attenuation of 1.2 dB km^{-1} in TE_{01} circular mode. The number of propagating modes is also shown.

frequency many other modes will propagate in addition to the preferred mode. In fig. 7.36 is shown the way in which tube diameter and frequency are related in order to keep a fixed attenuation of 1.2 dB km^{-1}. Also shown in the figure are the number of propagating modes. A further factor influencing the choice of tube diameter and frequency is the available bandwidth, which has to be large if such a communication channel is to stand a chance of being economically viable.

7.6.1 Available bandwidth

The matter of propagation of signals along dispersive lines was considered in §6.3.4, where it was shown that signals confined to a narrow range of the frequency spectrum were propagated with the group velocity $(\mathrm{d}\beta/\mathrm{d}\omega)^{-1}$. The group velocity will be a continuous function of frequency, so that limits must be placed on the frequency range which can be used. Some variation can be permitted to be subsequently removed by equalising circuits. An estimate of the bandwidth possible can be obtained as follows.

In the neighbourhood of a frequency ω_0 we may write

$$\beta(\omega) \approx \beta_0 + (\omega - \omega_0)\left(\frac{\mathrm{d}\beta}{\mathrm{d}\omega}\right)_0 + \tfrac{1}{2}(\omega - \omega_0)^2\left(\frac{\mathrm{d}^2\beta}{\mathrm{d}\omega^2}\right)_0$$

The effect of the first and second terms on propagation has been discussed in connection with the phase and group velocities (§6.3). The third term represents a distortion, usually known as delay distortion. The total phase delay from this term over a length of path l will be

$$\delta\phi = \tfrac{1}{2}l(\omega - \omega_0)^2\left(\frac{\mathrm{d}^2\beta}{\mathrm{d}\omega^2}\right)_0$$

The magnitude permissible will depend on the type of modulation. In terms of $\delta\phi$ the frequency range is given by

$$\omega - \omega_0 = \left(\frac{2\delta\phi}{l}\right)^{\frac{1}{2}}\left(\frac{\mathrm{d}^2\beta}{\mathrm{d}\omega^2}\right)_0^{-\frac{1}{2}}$$

If we assume symmetrical sidebands we finally have

$$\Delta\omega < 2\left(\frac{2\delta\phi}{l}\right)^{\frac{1}{2}}\left(\frac{\mathrm{d}^2\beta}{\mathrm{d}\omega^2}\right)^{-\frac{1}{2}}$$

For the case under consideration $\beta = (\omega^2 - \omega_{01}^2)^{\frac{1}{2}}/c$, so that differentiating we find

$$\frac{d^2\beta}{d\omega^2} = -\frac{1}{c}\frac{\omega_{01}^2}{(\omega^2 - \omega_{01}^2)^{\frac{3}{2}}}$$

In order to keep attenuation low we have seen that ω/ω_{01} must be substantially greater than unity, so that we may write

$$\frac{d^2\beta}{d\omega^2} \approx -\frac{\omega_{01}^2}{c\omega^3}$$

Hence
$$\Delta\omega < 2\left(\frac{2\delta\phi}{l}\right)^{\frac{1}{2}}\left(\frac{c\omega^3}{\omega_{01}^2}\right)^{\frac{1}{2}}$$

or
$$\Delta f < 2\left(\frac{c\delta\phi}{\pi l}\right)^{\frac{1}{2}}\frac{f^{\frac{3}{2}}}{f_{01}}$$

$$= 1.95 \times 10^4 \left(\frac{f}{l}\right)^{\frac{1}{2}}\frac{f}{f_{01}}(\delta\phi)^{\frac{1}{2}}$$

To obtain approximate values we will take $\delta\phi$ to be of the order of 1 radian, although the permissible value will depend on the modulation employed, and may well be greater than this. Applying figures appropriate to the 50 mm waveguide with $f_{01} = 7.32$ GHz, $f = 30$ GHz, $l = 25$ km we find Δf is about 90 MHz. This is the value which would be appropriate in the absence of equalisation. It is found that practical systems are sufficiently stable to allow satisfactory equalisation over much larger bandwidths, and current practice uses channels 560 MHz wide. The total usable frequency range is limited at the lower end by increasing difficulty of equalisation and at the higher end by available generators, but the range between 30 GHz and 110 GHz is technically feasible. Such a range must be used to support discrete channels each of bandwidth determined by the criteria discussed earlier in this section.

7.6.2 Propagation in multimode waveguide

The discussion in the previous section has shown that in order to use waveguide as a communication medium it is necessary to work at a frequency where many modes will propagate. In an ideal situation this would not matter since all modes would be independent: in a practical guide however irregularities have the effect of generating many modes. This process of mode conversion has two main effects. Firstly power is removed from the wanted mode into unwanted ones thus increasing the attenuation. Further, unwanted modes can be converted back into the wanted one. Since the different modes have different velocities, it is apparent that this reconversion will give rise to some distortion. Although irregularities in the sense of imperfections have been mentioned as the case of mode conversion, departures from the straightness at bends will cause the same effect, so that care in manufacture will never eliminate the problem. It is therefore necessary to arrange that as little generation as possible of unwanted modes takes place, by proper design and manufacture, and those that are produced are attenuated.

The mechanism giving rise to the low attenuation of TE_{0n} modes was seen in chapter 3 to be that the wall currents were circumferential, being associated with the axial magnetic field only, and that the magnitude of this axial field for a given power flow decreased as the frequency increased. One way of achieving the desired objective is therefore to make the guide wall so that axial currents are suppressed in preference to the circumferential. This should attenuate modes which need axial currents to support them, although it would not discriminate between TE_{0n} modes. Several methods for making such a waveguide wall have been investigated. One method found to be successful is to form the wall as a helix of fine copper wire supported by a material behaving as a lossy dielectric, as shown in fig. 7.37. The attenuation of the TE_{01} modes is slightly worse than the ideal copper tube, but 2 dB km^{-1} is attainable. The lossy dielectric attenuates modes with axial currents.

Another method of discriminating between modes is the dielectric-lined waveguide. Analysis of waveguide consisting of a circular tube

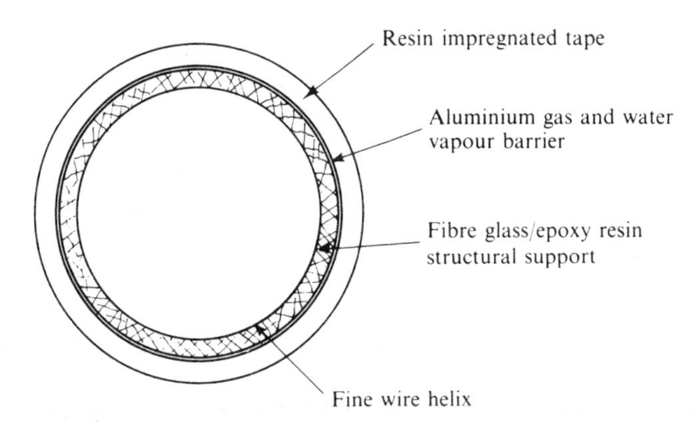

Fig. 7.37. Long-haul waveguide of helix construction.

with a thin layer of dielectric on the wall shows that this has the effect of enhancing the difference of velocities of unwanted and wanted modes.

To illustrate the use of the long-haul waveguide a brief account will be given of the trials conducted by the British Post Office.

7.6.3 A practical system

The British Post Office has undertaken a trial experiment using waveguide communication over a distance of 14 km. The total experiment included design of the waveguide and method of laying, the development of suitable terminal equipment, and the development of suitable oscillators.

Waveguide trials

The waveguide is the 50 mm diameter helix type discussed in the last section. The manufacturing process consists of winding the helix of enamelled copper wire onto a precision mandrel, embedding it in a glass-fibre epoxy-resin jacket to give the required electrical and mechanical characteristics, and completing with an aluminium screen and protective plastic coat. The guide was manufactured in 3 m lengths, each of which was individually checked for straightness. The minimum radius of curvature accepted on each 3 m length was 1 km, and the average was about 2 km.

The route chosen limited necessary bends to a radius greater than 300 m, although some sharper ones were introduced for experimental checks. Similarly, bends suitable for use in a manhole were developed and incorporated. The waveguide was laid inside a carefully aligned duct.

Measurements on the overall attenuation gave figures of between approximately 2 and 3 dB km^{-1} for the range 30 to 110 GHz.

Channel structure

For operation as a communications system the available bandwidth is divided into two bands, 30–70 GHz and 70–110 GHz, to provide paths in opposite directions. Each of these bands is sub-divided into four blocks each about 10 GHz wide into which carriers are channelled. The channel bandwidth depends on various factors. The fastest practicable modulation rate was regarded as 300 Mbit s^{-1} requiring a bandwidth of 560 MHz. Consequently each 10 GHz block is divided into sixteen channels 560 MHz wide, with 1 GHz separating blocks.

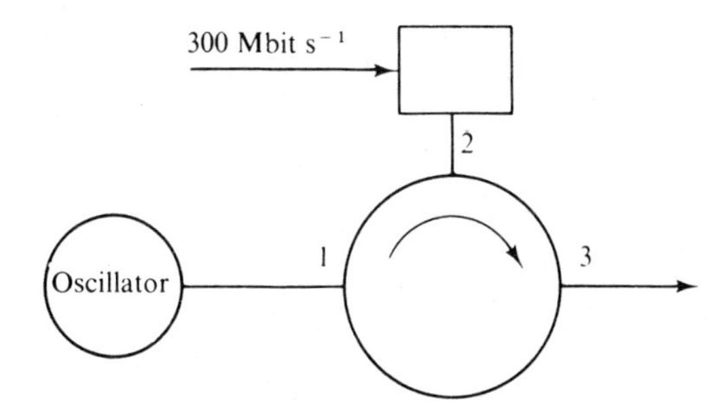

Fig. 7.38. Modulation of carrier with phase-shift-keying.

Modulation
It is necessary to modulate a carrier at about 300 Mbit s^{-1}. The original signal in a digital system consists of a sequence of levels: in a binary system, 0 or 1. The method of modulation is phase-shift-keying (PSK) in which a phase reversal of the carrier takes place when a level changes from 0 to 1. This can be achieved with a circulator and a switch, as indicated in fig. 7.38. The carrier entering at port 1 will appear out of port 2. The switch presents either an open- or a short-circuit depending on the applied level, causing a total reflection of the carrier. The phase of the output will be either 0° or 180° compared with the input. The reflected wave is the modulated carrier required, and issues from port 3.

Oscillator
The solid-state device used to produce the carriers required is the IMPATT diode. The structure is shown in schematic form in fig. 7.39,

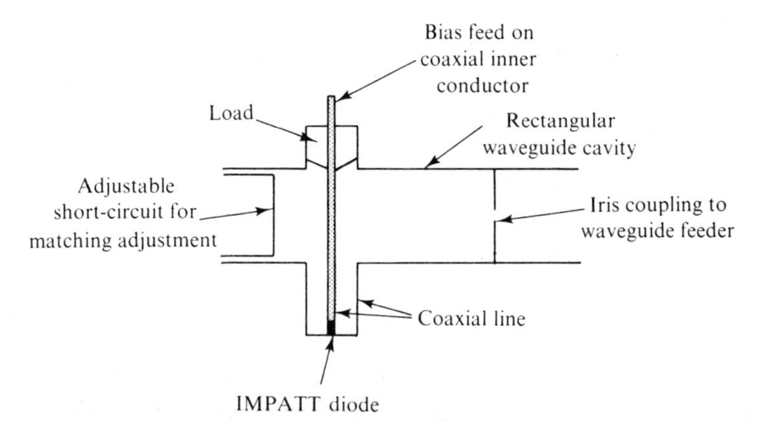

Fig. 7.39. Schematic diagram of oscillators using IMPATT diode.

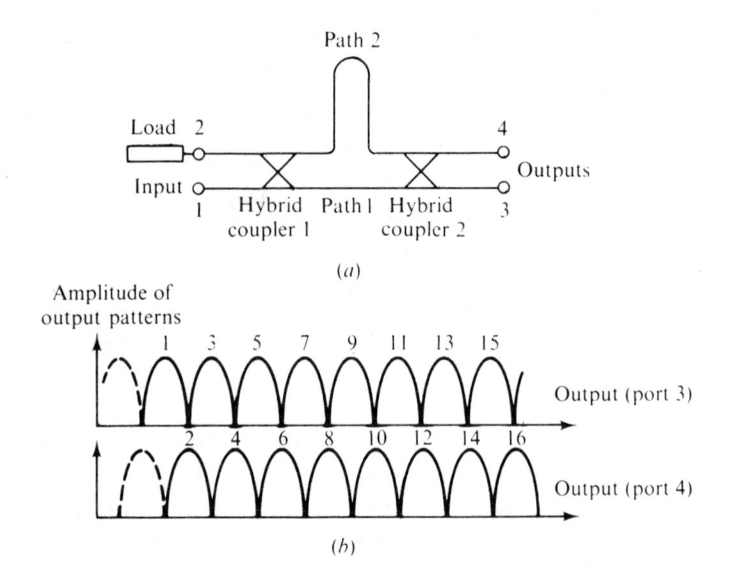

Fig. 7.40. Channel filters. (*a*) Configuration. (*b*) Output spectrum.

and consists fundamentally of a coaxial to rectangular waveguide junction. The inner conductor of the coaxial section acts as the radiator in the rectangular waveguide. Matching to the waveguide is through an iris, the size of which determines the power transferred to the feeder. Conversely, the size of the iris determines the loading placed by the matched feeder onto the oscillator. The variable short-circuit provides a reactance variation.

Filters

It is necessary to separate the various frequencies from one another. A type of structure used is shown in fig. 7.40. To appreciate the action we first recall the properties of a 3 dB directional coupler, such as the magic-T or hybrid-ring of §7.4.5. An input to port 4 splits equally to give equal in-phase outputs at ports 1 and 2; an input to port 3 splits to give equal but anti-phase outputs at 1 and 2; an input to port 1 gives equal and anti-phase outputs to ports 3 and 4; an input to port 2 gives equal and in-phase outputs to ports 3 and 4.

Consider now the configuration of fig. 7.40(*a*). An input to port 1 gives equal in-phase outputs in paths 1 and 2. These outputs recombine in the second hybrid coupler, but the port at which the output appears will depend on the relative phases. If the lengths of paths 1 and 2 differ by an odd multiple of $\frac{1}{2}\lambda_g$ the output will appear at port 4, with cancellation at port 3. If the path lengths differ by an even multiple of $\frac{1}{2}\lambda_g$ an output will appear at port 3, with cancellation at port 4. In this way neighbouring bands of frequencies will be separated, as shown in fig. 7.40(*b*). Repetition of the process by adding further filters in each channel will eventually sub-divided the original spectrum into channels.

7.7 Worked example

A matched load in a waveguide may be made by placing a thin resistive sheet perpendicular to the axis of the guide at a suitable distance in front of a short-circuit. What should be the resistance of the sheet in ohms/square, and at what distance from the short-circuit should the sheet be placed? Calculate these values for the case of a waveguide of width 22.9 mm operated at a frequency of 10 GHz.

Obtain an approximate expression for the reflection coefficient of the system in terms of a small deviation from the frequency at which the match is perfect.

Solution. At any point in a plane transverse to the axis of a waveguide the ratio of the electric to magnetic fields in the plane is a function of the mode and the frequency. Further the fields are mutually perpendicular. If a resistive sheet is placed in the plane, a current sheet will flow and produce a discontinuity in the magnetic fields on either side of the sheet. If the sheet is made of resistance per unit square equal to the ratio E/H for the waveguide, it is only necessary to maintain H zero on the far side of the sheet to obtain a perfect match. This can be done by use of a short-circuited length of guide of length $\frac{1}{4}\lambda_g$. We note that continuity of H_z is automatically ensured by continuity of transverse E.

In the present case of the TE_{10} mode in rectangular guide, the required resistance per square is $\omega\mu_0/\beta$, where $\beta=\surd(\omega^2/c^2-\pi^2/a^2)$. When $a=22.9$ mm and $\omega=2\pi\times10^{10}$, $\beta=158.3$ and $\lambda_g=39.7$ mm. Hence $\omega\mu_0/\beta=499\ \Omega$ and $\frac{1}{4}\lambda_g=10$ mm.

We observe the precise analogy with a system consisting of a transmission line of $Z_0=\omega\mu_0/\beta$ and phase coefficient β terminated in a resistance equal to Z_0 followed by a quarter-wave short-circuit. We may most simply find out what happens when the frequency changes by use of this analogy. It has been shown that the admittance of a length l of short-circuited line is $\mathrm{j}Y_0 \cot \beta l$. The terminating admittance of the line is therefore $G+\mathrm{j}Y_0 \cot \beta l$. The current reflection coefficient is then given by

$$k=\frac{Y_2-Y_0}{Y_2+Y_0}=\frac{G+\mathrm{j}Y_0 \cot \beta l-Y_0}{G+\mathrm{j}Y_0 \cot \beta l+Y_0}$$

Now at frequency ω_0, β takes the value β_0, so that $G=\beta_0/(\omega_0\mu_0)$. At the frequency ω we therefore have

$$k=\left[\frac{\beta_0}{\omega_0\mu_0}+\frac{\beta}{\omega\mu_0}(\mathrm{j}\cot\beta l-1)\right]\Big/\left[\frac{\beta_0}{\omega_0\mu_0}+\frac{\beta}{\omega\mu_0}(\mathrm{j}\cot\beta l+1)\right]$$

We note that when $\omega=\omega_0$, since $\beta_0 l=\frac{1}{2}\pi$, k is correctly zero. At a frequency only slightly different from ω_0, the numerator is small and to a first order we may put $\omega=\omega_0$ in the denominator, giving

$$k\approx\frac{1}{2}\left[1+\frac{\beta\omega_0}{\beta_0\omega}(\mathrm{j}\cot\beta l-1)\right]$$

Since cot βl will be small we have further, to a first order,

$$k\approx\frac{1}{2}\left[1-\frac{\beta\omega_0}{\beta_0\omega}+\mathrm{j}\cot\beta l\right]$$

Let us write $\beta-\beta_0=\delta\beta$, $\omega-\omega_0=\delta\omega$, where, from the definition of β, $\beta_0\,\delta\beta=\omega_0\,\delta\omega/c^2$. Then

$$\begin{aligned}\frac{\beta\omega_0}{\beta_0\omega}&=\frac{\beta_0+\delta\beta}{\beta_0}\left(\frac{\omega_0+\delta\omega}{\omega_0}\right)^{-1}\\&\approx\left(1+\frac{\delta\beta}{\beta_0}\right)\left(1-\frac{\delta\omega}{\omega_0}\right)\\&\approx\left(1+\frac{\omega_0}{\beta_0^2c^2}\,\delta\omega-\frac{\delta\omega}{\omega_0}\right)\end{aligned}$$

Hence

$$1-\frac{\beta\omega_0}{\beta_0\omega}=-\frac{\delta\omega}{\omega_0}\left(\frac{\omega_0^2}{\beta_0^2c^2}-1\right)$$

Further

$$\begin{aligned}\cot\beta l&=\cot\left(l\,\delta\beta+\tfrac{1}{2}\pi\right)\\&=-\tan l\,\delta\beta\\&\approx-l\,\delta\beta\\&=-\frac{\pi}{2}\frac{\delta\beta}{\beta_0}\\&=-\frac{\pi}{2}\frac{\omega_0^2}{\beta_0^2c^2}\frac{\delta\omega}{\omega_0}\end{aligned}$$

Finally therefore

$$k\approx-\frac{1}{2}\frac{\delta\omega}{\omega_0}\left[\left(\frac{\omega_0}{\beta_0c}\right)^2-1+\mathrm{j}\frac{\pi}{2}\left(\frac{\omega_0}{\beta_0c}\right)^2\right]$$

Inserting values we find

$$k = -\frac{1}{2}\frac{\delta f}{f_0}[0.75 + j2.74]$$

7.8 Summary

This chapter has considered the properties and use of waveguide in general systems and as a communication medium. Comparison has been made between waveguide and coaxial line from the points of view of attenuation and power-handling capacity.

The use of transmission-line techniques in waveguide circuits has been developed. Scattering parameters have been defined as an appropriate method of describing waveguide junctions. The properties of thin irises as impedance elements have been given. Brief consideration has been given to methods of measurement.

The advantages and problems of propagation in a multimode circular guide have been considered, and the British Post Office 'long-haul waveguide' communications link described.

Formulae

Complex wave amplitudes

Incident wave a

Reflected wave b

Normalised voltage and current

$$\mathrm{v} = \mathrm{a} + \mathrm{b} = \mathrm{V}Z_0^{-\frac{1}{2}}$$

$$\mathrm{i} = \mathrm{a} - \mathrm{b} = \mathrm{I}Z_0^{\frac{1}{2}}$$

Power

$$P_\mathrm{i} = \tfrac{1}{2}\mathrm{aa}^*$$

$$P_\mathrm{r} = \tfrac{1}{2}\mathrm{bb}^*$$

Scattering parameters for a junction of n ports

$$\mathrm{b}_m = \sum_{p=1}^{n} S_{mp}\mathrm{a}_p \qquad (\text{for } m = 1, \ldots, n)$$

7.9 Problems

1. The TE_{10} mode is used in an air-filled waveguide of rectangular cross-section. The internal dimensions are 23 mm × 10 mm.

If the peak value of electric field which can be used is 2 kV mm^{-1}, find the maximum power which can be delivered to a matched load by the waveguide when the frequency used is 10 GHz.

(London University.)

2. A dominant-mode wave of frequency 10 GHz, propagating in a rectangular guide with broad dimension $a = 25$ mm encounters a sudden transition to a long length of guide for which $a = 12$ mm. What is the form of the dependence of the fields on z in the narrower guide? Calculate the component of the time-averaged Poynting vector in the direction of propagation in both waveguides, and compare the results. What is the magnitude of the reflection coefficient at the transition?

If the narrower guide is of length 20 mm, and is then followed by a transition to the broader guide, describe briefly the effect on the fields.

(Southampton University.)

3. What is Poynting's vector?

The diameter of the inner conductor of a coaxial line is 0.4 cm, the inner diameter of the outer conductor is 1.0 cm, and the characteristic impedance is 55 Ω. The line is used to supply a load with pulses of radio-frequency power of 10^{-6} s duration at the rate of 400 s^{-1}. Calculate the maximum mean power that can be transmitted when the voltage standing-wave ratio is 2 : 1 and the line insulation breaks down if the electric stress at the surface of the inner conductor exceeds 25 kV cm^{-1}.

4. In §3.9.1 it was shown that dominant-mode transmission in a rectangular waveguide could be regarded as in an equivalent transmission line for which $Z_0 = 2\omega\mu a/\beta b$, $v_p = \omega/\beta$, $\beta = [\mu\varepsilon(\omega^2 - \omega_{10}^2)]^{\frac{1}{2}}$.

Show that in circuit terms this transmission line has a distributed series inductance of $2a\mu/b$ per unit length and a distributed shunt admittance consisting of inductance $2a^3\mu/b\pi^2$ in parallel with capacitance $b\varepsilon/2a$, both per unit length. It may be assumed that for a line with distributed series reactance X and shunt susceptance B per unit length, $Z_0 = \sqrt{(X/B)}$, $v_p = \omega/\sqrt{(XB)}$.

5. A two-port network can be characterised by a set of admittance or y-parameters where in matrix form $[I] = [Y][V]$. Show that the y-parameters can be expressed in terms of the normalised scattering parameters $[S]$ by the equation $[Y] = \{[1] - [S]\}\{[1] + [S]\}^{-1}$.

Measurements on a microwave transistor gave the following scattering parameters, referred to 50 Ω characteristic impedance.

$$S_{11} = 0.76\angle 145°,\ S_{12} = 0.013\angle 55°$$

$$S_{21} = 1.12\angle 25°,\ S_{22} = 0.26\angle -164°$$

Estimate the equivalent admittance parameters.

6. The scattering parameters of a two-port waveguide junction referred to planes T_1, T_2 are S_{11}, S_{12}, S_{21}, S_{22}. Show that if the reference plane of port 1 is moved distance l away from the junction the new scattering matrix is

$$\begin{bmatrix} S_{11}\exp(-2j\beta l) & S_{12}\exp(-j\beta l) \\ S_{21}\exp(-j\beta l) & S_{22} \end{bmatrix}$$

8

Microwave radio systems

A radio system can be conveniently sub-divided into three sections: an antenna, associated with a transmitter, which radiates power; an antenna placed in the field of the transmitting antenna which receives some of the radiated power and delivers it to a load; and finally the medium in between the two antennae through which propagation takes place. Some of the electromagnetic problems associated with antennae have been considered in chapter 4, and propagation has been considered in chapters 1 and 2. The purpose of the present chapter is to develop the parameters required to assess, in engineering terms, the performance of a link.

We shall look at some of the characteristics of radio systems, with particular reference to wideband microwave links, since these represent an alternative to the multichannel systems discussed in the previous two chapters. We shall begin by considering the simplest situation, propagation between points in free space. Modern satellite communication links approximately fulfill this condition because of the distances involved and because the major part of the propagation path exists outside the earth's atmosphere.

We shall then discuss the characteristics of electromagnetic propagation in the region of the earth's lower atmosphere, or troposphere including the effect of the earth's surface. It is within this region that the point-to-point relay systems operate. Finally, we shall look briefly at some operational microwave systems.

8.1 Antennae in transmission

8.1.1 Radiation pattern of a transmitting antennae

We have seen in chapters 2 and 4 that at a distant point the field associated with a radiating source is inversely proportional to distance and is a function of direction from the source. For example, both the short dipole and the half-wave dipole discussed in §§4.1 and 4.2 concentrate radiation in the equatorial plane, with zero radiation along the axis of the dipole.

For the more general case it was shown in chapter 4 (4.42) that at a distant point the electric field radiated by an antenna could be expressed in the form

$$\mathbf{E}_a = j\zeta I_0 \mathbf{e}_a e^{-jkr}/r \tag{8.1}$$

in which $\zeta = \sqrt{(\mu_0/\varepsilon_0)}$, $k = \omega/c = 2\pi/\lambda$, I_0 is the current and $\mathbf{e}_a$ is a dimensionless function of direction depending on the way the current is distributed in the antenna. Since the distant radiated wave is quasi-plane, the Poynting vector $\mathbf{E}\times\mathbf{H}^*$ is radial in direction with magnitude $\zeta^{-1}\mathbf{E}\cdot\mathbf{E}^*$, so that the power density is given by

$$P_a = \tfrac{1}{2}\zeta^{-1}\mathbf{E}_a \cdot \mathbf{E}_a^* \tag{8.2}$$

Substituting from (8.1)

$$P_a = \tfrac{1}{2}\zeta I_0^2 \mathbf{e}_a \cdot \mathbf{e}_a^*/r^2 \tag{8.3}$$

The variation with direction is contained within the function $\mathbf{e}_a \cdot \mathbf{e}_a^*$. This variation can be shown in the form of a polar plot in which the radial vector has the magnitude of P_a for a fixed distance from the source and the direction is that of the point at which P_a is measured. To avoid specifying any particular distance it is usual to display P_a/P_m, P_m being the maximum radiated power density in any direction. Such a polar plot is referred to as the *radiation pattern* of the antenna. The locus of P_a/P_m describes a surface, an example of which is shown in fig. 8.1(a) for the short electric dipole. Several variants are useful. It is obviously not straightforward to display a three-dimensional plot, so it is customary to

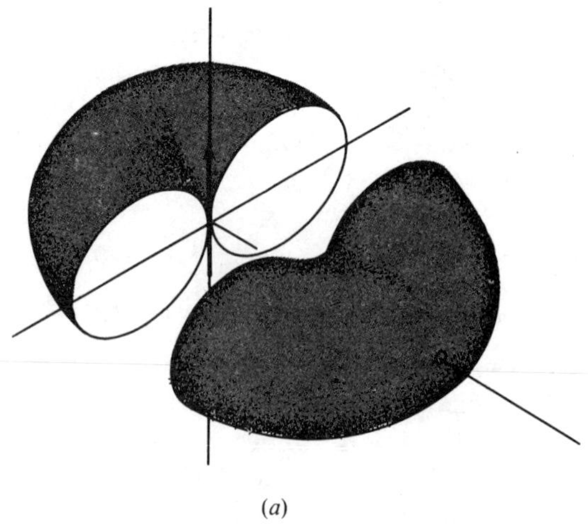

(a)

Fig. 8.1. Radiation pattern of a short dipole. (a) Three-dimensional plot.

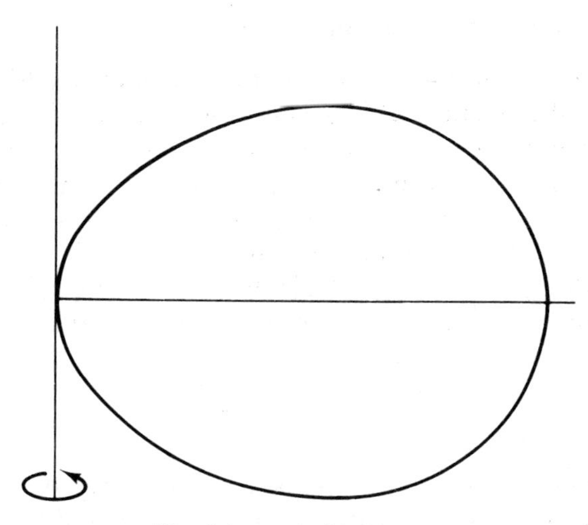

Fig. 8.1 contd. (*b*) Section.

show cross-sections in suitable planes. For example, the radiation pattern of the short dipole is symmetrical about the axis, so it is only necessary to show the single plot of fig. 8.1(*b*). Then again the ratio P_a/P_m from which the radial length is derived may be expressed either linearly or in decibels. It may not be necessary to show P_a/P_m over all directions, in the case for example of a high-gain antenna of the type discussed in §4.6.1. The radiation pattern of such an antenna might be displayed in Cartesian form as in fig. 8.2.

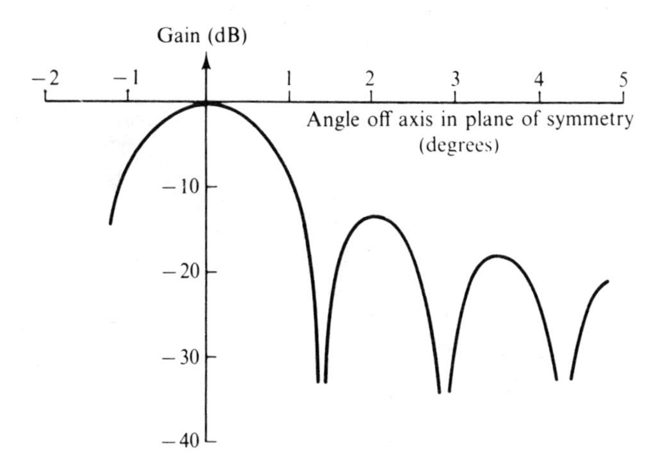

Fig. 8.2. Radiation pattern for a high-gain antenna.

8.1.2 Power gain of a transmitting antenna

The radiation pattern defined in the last section describes the directional properties of the radiated field. The absolute magnitude is described by comparing the actual field with that which would be produced at a similar point from a standard reference antenna transmitting the same total power. The reference antenna is usually taken as the fictional *isotropic radiator*, for which the input power is radiated uniformly in all directions. If the total power radiated by the antenna under test is P_t, and the power density is P_a (given by (8.3)), the *power gain G* is defined by

$$G = \frac{P_a}{(P_t/4\pi r^2)} = \frac{4\pi r^2 P_a}{P_t} \tag{8.4}$$

In this expression P_t must of course be equal to the result of integrating P_a over a sphere of radius r. We note that $r^2 P_a$ does not depend on r, and may in fact be given its own physical interpretation: $P_a\,dS$ is the power crossing an area dS at distance r, dS being normal to $\mathbf{r}$, as shown in fig. 8.3. We may write

$$P_a\,dS = r^2 P_a \frac{dS}{r^2}$$

dS/r^2 is the measure of the solid angle $d\Omega$ subtended by dS at the origin, so that

$$P_a\,dS = r^2 P_a\,d\Omega$$

We may therefore interpret $r^2 P_a$ as the power per unit solid angle (steradian) in the direction of dS, i.e. the power which would be radiated

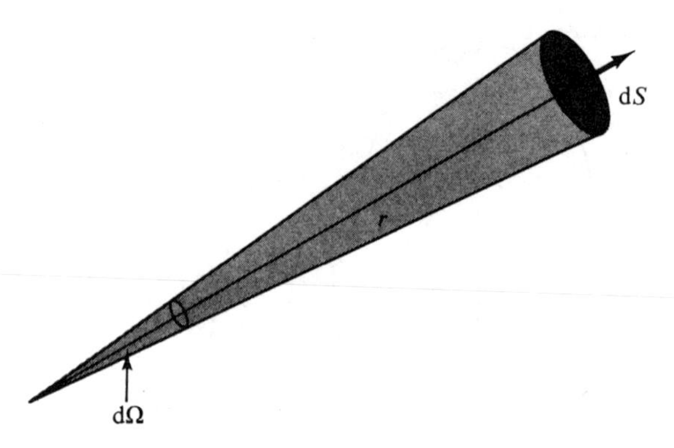

Fig. 8.3. Definition of solid angle.

if the density were constant into a cone subtended by an area of r^2 at distance r. Thus we may write

$$G = 4\pi \frac{\text{power/steradian}}{\text{total power}}$$

Equation (8.4) may be applied to an antenna under experimental investigation. For example the total power may be defined in terms of current and measured radiation resistance R_a, in which case

$$P_t = \tfrac{1}{2} I_0^2 R_a$$

The power density P_a might be measured directly as power P_r received into an area A, so that

$$P_a = P_r / A$$

It might also be convenient to measure the electric field rather than power density, in which case P_a is given by (8.2). Thus a suitable form for G might be

$$G = \frac{4\pi r^2}{\zeta R_a} (\mathbf{E}_a \cdot \mathbf{E}_a^* / I_0^2) \tag{8.5}$$

We note that substitution from (8.1) gives

$$G = \frac{4\pi\zeta}{R_a} \mathbf{e}_a \cdot \mathbf{e}_a^* \tag{8.6}$$

As an example we may consider the short electric dipole for which it was shown in §4.1 that

$$\mathrm{E}_a = \mathrm{j}\zeta I_0 \,\delta l\, k \sin\theta / 4\pi r$$

$$R_a = \zeta \frac{2\pi}{3} \left(\frac{\delta l}{\lambda}\right)^2$$

Applying (8.5), we find

$$G = \tfrac{3}{2} \sin\theta$$

As defined by (8.4), G is a function of direction, and it will be seen that the ratio of G to its maximum G_m is in fact the same as P_a/P_m. The directional properties of G are therefore described by the radiation pattern, and the additional information provided by G is the value of G_m. In practice when a figure for the power gain of an antenna is given, it is usually that for the maximum.

As another example we have that, for a half-wave dipole, (4.22) and (4.24)),

$$|E_a| = |\zeta H_\phi| = \zeta I_0/2\pi r$$

$$R_a = 73\ \Omega$$

Hence $$G_m = 1.64, \quad \text{or } 2.1\text{ dB}$$

8.1.3 Polarisation

In (8.1), $\mathbf{e}_a$ was described as a dimensionless function of direction. Its vector direction determines the direction of the electric field vector at a distant point, and thus the polarisation of the radiated wave. Where this direction is constant the expression $[\mathbf{e}_a \cdot \mathbf{e}_a^*]^{\frac{1}{2}}$ is merely the magnitude of the vector $\mathbf{e}_a$. This is the case for example with the short electric dipole. It is necessary to point out however that $\mathbf{e}_a$ might be of the form

$$\mathbf{e}_a = e_a(\boldsymbol{a}_x + j\boldsymbol{a}_y) \tag{8.7}$$

This expression can be compared with (1.48) describing a circularly polarised plane wave; an antenna for which (8.7) is true would be radiating a circularly polarised wave. In a general case the nature of the polarisation will vary with direction.

Circular polarisation is used for transmitting to and from a satellite, to avoid the variations that would be imposed on a constant direction of polarisation by the changing aspect of the satellite.

8.2 Antennae in reception

8.2.1 Receiving area of an antenna

When an antenna is used in reception the electromagnetic problem is quite different to that of transmission. This has been discussed in §4.5.2 and it was shown that by application of the reciprocity theorem a relation could be established between properties in reception and transmission. We consider a receiving antenna to be in the distant field of the transmitter, so that (4.45) applies. This equation states that the voltage V_0 developed across the antenna terminals on open-circuit is related to the electric field $\mathbf{E}_0$ of the incident wave and the quantity $\mathbf{e}_a$ of (8.1) in the form

$$V_0 = 2\lambda\, \mathbf{E}_0 \cdot \mathbf{e}_a \tag{8.8}$$

In this expression $\mathbf{e}_a$ is evaluated for the direction looking into the oncoming wave. This equation is only valid when the medium of propagation is itself reciprocal, and in the context of the propagation of radio waves this excludes low-frequency ionospheric propagation. The motion of charged particles in the earth's magnetic field corresponds to the ferrites discussed in §7.4.7.

The internal resistance of the source representing the antenna terminals is R_a, so that the power available (into a matched load) will be

$$P_r = \frac{1}{8R_a} |V_0|^2 = \frac{\lambda^2}{2R_a} |\mathbf{E}_0 \cdot \mathbf{e}_a|^2$$

Let us assume that the antenna radiates a plane-polarised wave, so that the direction of $\mathbf{E}_0$ will be normal to the radius vector corresponding to the incoming wave. By definition $\mathbf{e}_a$ is normal to the radius vector along which it is calculated. This situation is shown in fig. 8.4. We see that by rotation of the antenna about the axis representing the incoming wave direction the angle γ may be made zero. Assuming this has been done we then have

$$\mathbf{E}_0 \cdot \mathbf{e}_a = \mathrm{E}_0 \mathrm{e}_a$$

We can express E_0 in terms of the power flow in the incident wave. If this wave has a power density P_w, we have

$$P_w = \tfrac{1}{2}\zeta^{-1}\mathrm{E}_0^2$$

We may then write from (8.8)

$$P_r = \frac{\lambda^2}{R_a} \zeta |\mathrm{e}_a|^2 P_w$$

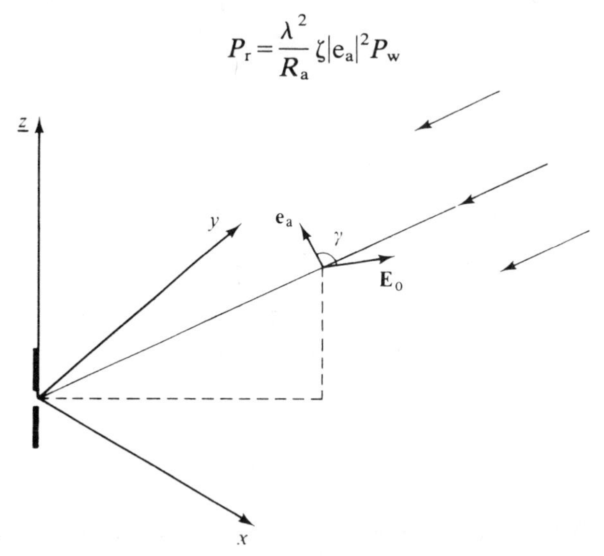

Fig. 8.4. An antenna in an incident plane wave.

or, using (8.6),

$$P_r = \frac{\lambda^2}{4\pi} G P_w = A P_w \tag{8.9}$$

The quantity A has dimension of area, and is termed the *effective receiving area* of the antenna. The expression derived gives the maximum value, corresponding to the load being matched and the antenna oriented correctly to the polarisation of the incoming wave. With this proviso we have the reciprocal relations

$$\begin{aligned} A &= \frac{\lambda^2}{4\pi} G \\ G &= \frac{4\pi}{\lambda^2} A \end{aligned} \tag{8.10}$$

8.2.2 Effective receiving area of an aperture

It might be expected that the effective receiving area is related to the physical size of the antenna: in the case of a parabolic mirror for example an aperture is well defined. Consider the uniformly illuminated aperture of §4.6.3. This problem postulates a rectangular aperture of dimensions $2a \times 2b$, across which is maintained an electric field E_0 in the x-direction, constant in magnitude and phase. The distant electric field was calculated in (4.53). The maximum of the radiation pattern occurs on the line normal to the aperture, where

$$\mathrm{E}_x = E_0 \frac{\mathrm{j}k}{4\pi} 8ab \frac{\mathrm{e}^{-\mathrm{j}kr}}{r}$$

The power density is therefore

$$P_a = \tfrac{1}{2}\zeta^{-1}|\mathrm{E}_x|^2 = \tfrac{1}{2}\zeta^{-1} E_0^2(2abk/\pi r)^2$$

The power crossing the aperture corresponds to that carried by a plane wave and is given by

$$P_t = 4ab \cdot \tfrac{1}{2}\zeta^{-1} E_0^2$$

Applying (8.4) we find

$$G = 4\pi \frac{r^2 P_a}{P_t} = \frac{4\pi}{\lambda^2} 4ab$$

From (8.9) the effective area is seen to be equal to $4ab$, the physical area of the aperture. This equality is dependent on obtaining 'uniform illumination', which is not only difficult to achieve but is undesirable in terms of side lobes. In practice an effective area of some 60% of the physical aperture is more likely.

A parabolic reflector of diameter D has an aperture equal to $\frac{1}{4}\pi D^2$. The maximum conceivable gain is therefore

$$G = \frac{4\pi}{\lambda^2}\frac{\pi D^2}{4} = \left(\frac{\pi D}{\lambda}\right)^2$$

A more practical approximation would be

$$G = 6(D/\lambda)^2 \tag{8.11}$$

The relationship between the gain of an antenna and the width of the main radiation lobe will now be considered.

8.2.3 Relation between beamwidth and gain for a highly directional antenna

If the gain of an antenna is known as a function of direction then the beamwidth is automatically known. The purpose of this section is to point out that if an antenna is designed to have a very high gain in one direction then its beamwidth must be small. This relationship follows from the definition of antenna gain as given by (8.4). Let us suppose that the power density is constant within a cone of semi-angle α about the preferred direction, and zero outside, as shown in fig. 8.5. Then the power density within the cone is given by

$$P_a = \frac{G}{4\pi r^2}P_t \tag{8.12}$$

The area of the cap of the cone shown in fig. 8.5 is exactly equal to $2\pi r^2(1-\cos\alpha)$. When α is much less than unity this becomes $\pi(\alpha r)^2$.

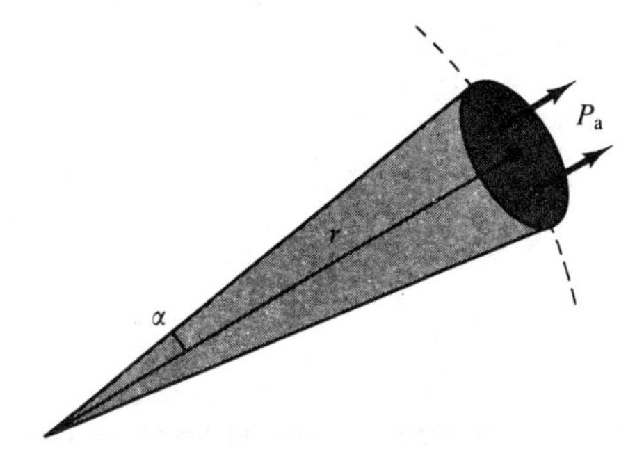

Fig. 8.5. Radiation confined within a cone.

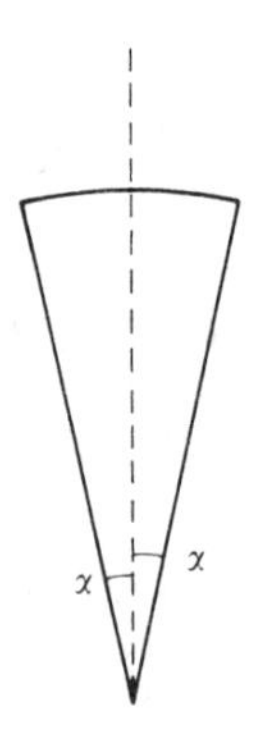

Fig. 8.6. Radiation pattern for power-gain constant in cone of semi-angle α.

The total power crossing the cap is therefore equal to $P_a\pi(\alpha r)^2$, and if outside the cone the power is zero we must have

$$P_t = P_a\pi(\alpha r)^2$$

Using (8.12) we find

$$\alpha^2 = 4G^{-1}$$

or

$$\alpha = 2G^{-\frac{1}{2}} \tag{8.13}$$

In terms of receiving area, using (8.10),

$$\alpha = \lambda(\pi A)^{-\frac{1}{2}} \tag{8.14}$$

This is true only for the axially symmetrical radiation pattern shown in fig. 8.6, and values will depend on the precise shape assumed for the beam, but as an order of magnitude calculation (8.13) and (8.14) are instructive.

If we apply (8.13) to the parabolic mirror whose gain is given by (8.11) we find $2\alpha = 1.6\lambda/D = 94\lambda/D$ degrees. A more practical figure for the angle between directions of half-power is given by

$$2\alpha = 70\lambda/D \text{ degrees}$$

Thus a parabolic mirror 1 m in diameter for $\lambda = 3$ cm gives a half-power beamwidth of about 2°.

It is evident that a tidy relationship can only be established for highly symmetrical beams, but nevertheless the general form of relation is clear: high gain implies a narrow beam, although it may not be equally narrow in different sections.

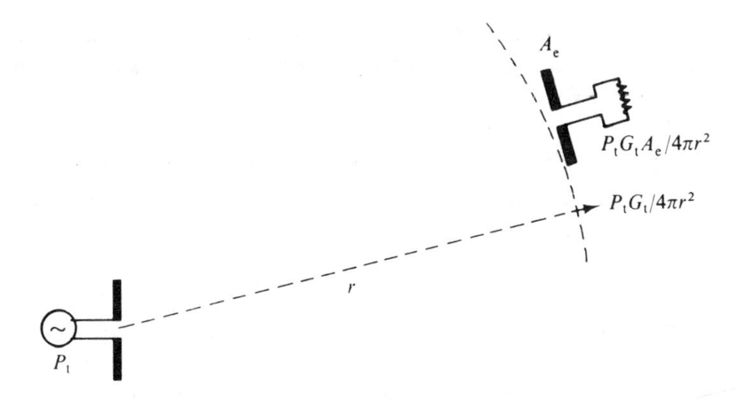

Fig. 8.7. A transmitter–receiver system.

8.3 Transmitter–receiver systems

Consider a transmitter antenna with gain G_t, transmitting a total power P_t watts, as shown in fig. 8.7. Then the power density, distance r from the antenna, is

$$P_a = G_tP_t/4\pi r^2 \text{ Wm}^{-2}$$

A receiving antenna of effective area A_e and power gain G_e at this point, directed towards the transmitter, produces an available power P_aA_e given by

$$P_r = G_tP_tA_e/4\pi r^2$$

Using (8.10) this may be written in terms of the gain of the receiving antenna in the forms

$$P_r = P_tG_tG_e(\lambda/4\pi r)^2 = P_tA_tA_e/\lambda^2 r^2 \tag{8.15}$$

We can interpret P_t and P_r as transmitter power and received power only if no power is dissipated in the antennae themselves. In practice there will inevitably be some losses since the antenna elements are dissipative and carry the excitation currents associated with the radiation. At low frequencies, where it is physically impractical to construct half-wave or resonant antennae, the effective radiation resistance can be very small indeed, and losses in the antenna structure can be very significant. For example, the radiation resistance of a short dipole of length $2l$ with a sinusoidally distributed current was shown in (4.25) to be given by

$$R_a = \frac{2}{3\pi}\sqrt{\left(\frac{\mu}{\varepsilon}\right)}\left(\frac{2l}{\lambda}\right)^2$$

For $2l = \lambda/10$, this gives $R_a = 0.8\Omega$. An antenna may well have an effective loss resistance greater than the radiation resistance, resulting in a low overall efficiency.

G_t and G_e are power ratios and may be expressed in decibels by the formula $10 \log_{10} G$. If they are both expressed in this way, and P_t and P_r are expressed in decibels above a suitable reference, (8.15) may be written in the form

$$P_e = P_t + G_t + G_e - 10 \log_{10}(4\pi r/\lambda)^2 \qquad (8.16)$$

The last term in this equation represents the free-space propagation loss.

We can take account of the losses in the transmitter and receiver antennae by introducing a term $-L$ dB in (8.13). Thus

$$P_e = P_t + G_t + G_e - 10 \log_{10}(4\pi r/\lambda)^2 - L \text{ dB} \qquad (8.17)$$

Example

A microwave radio link, operating at a carrier frequency of 3 GHz uses identical parabolic antennae of circular aperture, radius 1 m, for transmission and reception. The path length is 50 km, and the loss in each terminal is 2 dB. If the input power at the transmitter is 10 W, estimate the overall transmission loss, and the available power at the receiver. Assume ideal free-space propagation and correct alignment of the antennae.

Solution. In this problem $\lambda = 0.1$ m. From (8.11) the gain of each antenna is equal to $6 \times 10^2 = 600$, or 27.8 dB. The propagation loss is equal to

$$20 \log_{10}(4\pi \times 5 \times 10^4/10^{-1}) = 136.0 \text{ dB}$$

The transmitted power is 10^4 mW, or 40 dBm. Hence applying (8.17)

$$P_e = 40 + 2 \times 27.8 - 135 - 2 \times 2 \text{ dBm}$$

$$= -44.4 \text{ dBm}$$

Converting this to an absolute level, $P_e = 3.63 \times 10^{-5}$ mW.

8.4 Antenna measurements

There are several important parameters associated with an antenna array that must be determined experimentally. These include the impedance of the array, the frequency bandwidth, the three-dimensional polar diagram, the gain, and the effective noise temperature, among others. The frequency bandwidth is clearly important in the case of wideband multiplexed systems. Simple resonant antennae such as the half-wave dipole discussed earlicr have relatively narrow bandwidths, depending on the diameter of the conductors (see fig. 4.7). At microwave frequencies, the bandwidth of a large aperture antenna is primarily limited by the characteristics of the feed system illuminating the aperture. The effective noise temperature of an ideal antenna is dependent on the direction in which it is pointing, and its bandwidth, as discussed later in this section.

8.4.1 Impedance measurements

Impedance measurements are conveniently obtained using the techniques discussed in chapter 5. A slotted line or waveguide is connected in series with the antenna feed, allowing measurements of VSWR and the positions of minima on the line to be made. The impedance or admittance of the antenna can then be determined with reference to a specific plane. A suitable reference plane can be established by terminating the feeder in a short circuit, which gives rise to clearly defined nulls at half-wavelength intervals along the feeder. The bandwidth of the system can readily be measured using this technique. (Note that the reference-plane position is a function of the operating frequency.) Clearly, the antenna array must be operated in its 'normal' environment. Any local reflections or interference will modify the near-field distribution of the antenna and this may result in significant changes in the measured impedance.

8.4.2 Polar diagram measurement

Radiation or polar diagrams are measured in a particular plane of interest, or at most in two orthogonal planes, since this provides adequate information in most practical cases. The basic procedure involves two antennae spaced sufficiently far apart to exclude the near-field effects. A minimum distance of $2D^2/\lambda$ is generally required, where D is the equivalent diameter of the antenna aperture. A fixed transmitter antenna is placed in the same horizontal plane as the antenna under test. Both antennae must be carefully sited to minimise possible reflections from local objects and surfaces. The antenna under test is connected to a suitable receiver and indicator. It is then rotated on its axis, and an indication of the polar diagram in the given plane is obtained. The plane of polarisation of the transmitting antenna should be adjustable, since the side lobes of an array may have different polarisation to the main lobe. The polar diagram will also be a function of the operating frequency of the system, and several tests would normally be made at frequency intervals over the operational bandwidth.

This technique is only applicable to relatively small antenna arrays. When investigating large arrays, it may be necessary to arrange that the transmitting antenna moves around the system under test at a constant radius. Or, alternatively, near-field measurements can be made, from which the far-field characteristics can be determined. Another possibility is the construction of a scale model of the array, and the system is tested at a correspondingly higher frequency. Thus, an antenna designed to operate at 100 MHz can be scaled down by a factor of 10, and its performance measured at 1000 MHz.

8.4.3 Gain measurement

In principle, we can determine the gain of an antenna in any direction by measuring the power density in the far field in that direction for a given total transmitted power. Then (8.4) gives the gain directly. The power density can be determined using a standard receiving antenna with known gain, such as the simple half-wave dipole.

An alternative approach is to use two identical antennae, and apply (8.15) to the system. In this equation G_t and G_e are now equal, so that

$$P_r/P_t = (\lambda G/4\pi r)^2$$

Hence

$$G_e = (P_r/P_t)^{\frac{1}{2}}\, 4\pi r/\lambda$$

8.5 Antenna noise

We have seen in chapter 6 that any resistive source produces noise. The available noise in a bandwidth B from such a source is equal to kTB watts. The output terminals from an antenna behave as a source whose resistance is equal to the radiation resistance. In general this source will appear noisy, in part because the antenna itself is constructed of imperfect conductors at ambient temperature and in part because of radiation received by the antenna. The magnitude of the noise can be expressed by means of a noise temperature T_e, which the radiation resistance would need to have in order to produce the observed noise.

Noise radiation falling on the antenna may come from many sources such as man-made interference, thermal radiation from hot bodies, or radiation generated in some physical process in stars or interstellar space.

8.5.1 Thermal radiation

From the point of view of the systems engineer the only quantity necessary for assessment is the noise temperature, which will depend on many variables, such as the direction in which the antenna is pointing. It is perhaps useful however to point out that radiation from a body in virtue of its temperature is of the same nature as the fluctuations expressed by the formula $p_n = kTB$. Provided that the radiating surface is such that it absorbs all radiation falling on it (a black body), it can be shown that the radiation emitted from the surface over all directions in bandwidth δf at frequency f is given by Planck's formula

$$\delta E = 2\pi h f^3[\exp(hf/kT) - 1]^{-1}\, \delta f \ \text{Wm}^{-2}$$

in which f, k, T have their usual significance and h is Planck's constant equal to 6.63×10^{-34} Js. At radio frequencies it is usual that $hf \ll kT$, so that the exponential is equal to $1 + hf/kT$. We then have $\delta E \approx 2\pi kTf^2\, \delta f$. This approximation is known as the Rayleigh–Jeans law. It seems plausible that if an antenna is in the interior of a cavity whose walls are at temperature T the noise temperature of the antenna should also be T. This can be proved by appropriate application of the Rayleigh–Jeans law to such a situation. The same result obviously applies if the only directions in which the antenna can receive, as given by its radiation pattern, are covered by such a body. For example, a high-gain microwave antenna pointing at the earth's surface will register an effective noise temperature of about 290 K, whereas pointing into space the temperature will be much lower.

8.5.2 Measurement of noise temperature

The distinction between discrete and distributed sources is blurred by the relatively large beamwidths of even directive antennae. In fig. 8.8 is shown the type of variation with frequency which may be expected to be found with a lossless antenna. At low frequencies, very high effective noise temperatures are found. The radiation from space comes primarily in the plane of our galaxy (the Milky Way), with a maximum towards the centre of the galaxy. At microwave frequencies, temperatures are much less, typically a few degrees Kelvin, except when radiating bodies fall in the antenna beam. Such maxima are fixed in space, and an antenna of

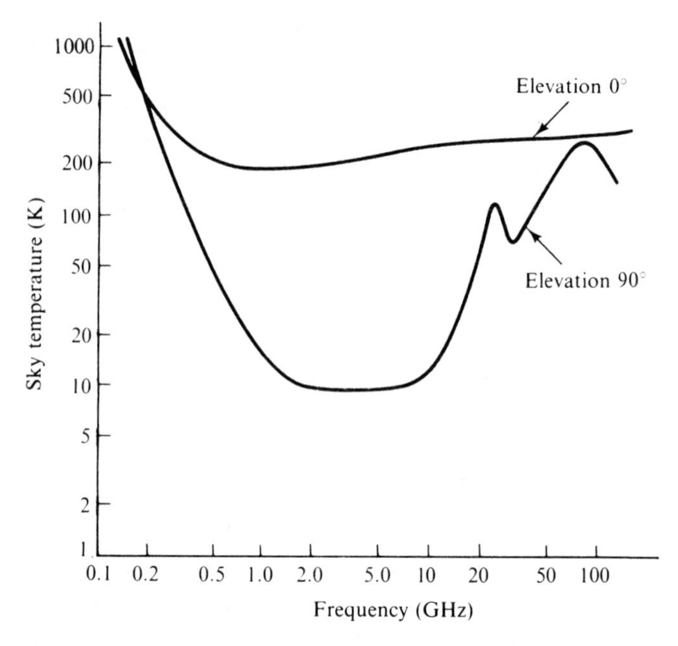

Fig. 8.8. Variation of sky temperature with frequency (typical values).

orientation fixed on earth will encounter temperatures which vary with time. An antenna with its beam at grazing incidence to the earth's surface will pick up thermal radiation from the earth, whereas in the zenith the low average temperature of space will be measured. At frequencies higher than about 10 GHz, appreciable attenuation by the earth's atmosphere exists, and absorbing matter produces in turn thermal radiation. This phenomenon may be compared to the problem discussed in §6.8.1 where a lossy feeder was shown to affect the signal-to-noise ratio. The broad minimum that exists between 1 and 10 GHz makes this range suitable for satellite communication systems.

The effects of the side lobes of the antenna radiation diagram are clearly important in determining the minimum effective noise temperature, and it is sometimes necessary to sacrifice some antenna gain in order to achieve the best possible side-lobe suppression and hence a lower overall effective temperature.

Antenna noise measurements are normally made by comparison with a calibrated noise source. If we are interested in measuring effective temperatures in the range 10–100 K, the local noise source must be cooled with liquid nitrogen to give effective noise temperatures of 80 K or with liquid helium to provide a reference source at a temperature of 4 K.

8.6 Reciprocity of a transmitter–receiver system

The reciprocal relation between an antenna in transmission and the same antenna in reception was derived in chapter 4 and used in the previous sections. It is of interest to note that this result is related to the reciprocity in a circuit sense of the two-port network formed by the terminal pairs of a transmitting antenna and a (different) receiving antenna. Consider the two situations illustrated in fig. 8.9(*a*) and (*b*). The two situations are identical except that in fig. 8.9(*b*) the source position has been changed to the opposite end. Thus if the two-port is reciprocal we must have $V_1'/I_2' = V_2''/I_1''$. We assume that reactances in both circuits have been reduced to zero, so that, in fig. 8.9(*a*), the transmitted power from antenna 1 is

$$P_t' = |V_1'|^2/4R_{a1}$$

The received power into the matched load of antenna 2 will be

$$P_r' = P_t' G_1 A_2 / 4\pi r^2$$

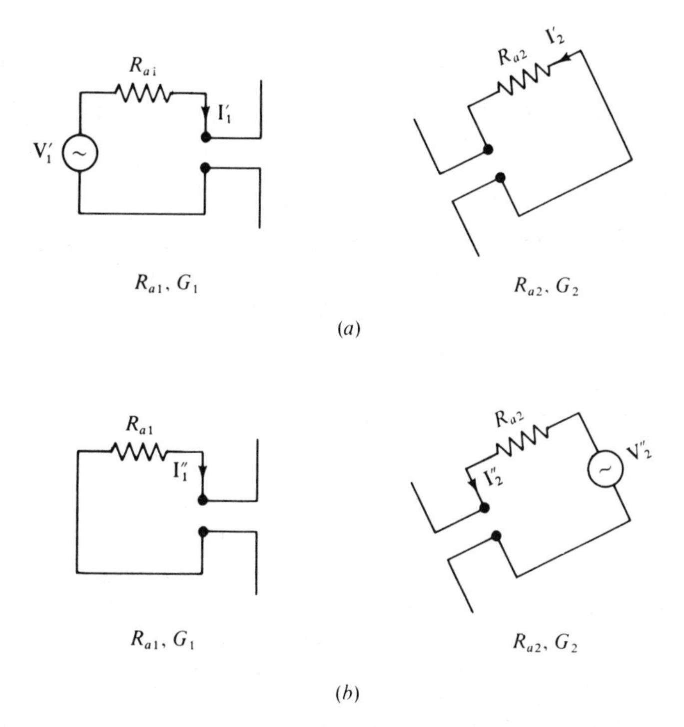

Fig. 8.9 Reciprocity of two-port made by two antennae. (*a*) Antenna 1 transmitting. (*b*) Antenna 2 transmitting.

in which G_1 is the gain of antenna 1 and A_2 the effective area of antenna 2 for the appropriate direction. But P'_r is also equal to $R_{a2}|I'_2|^2$, and hence

$$R_{a2}|I'_2|^2 = \frac{G_1 A_2}{4\pi r^2} \cdot \frac{1}{4R_{a_1}} |V'_1|^2$$

from which $$|I'_2/V'_1|^2 = G_1 A_2/(16\pi r^2 R_{a_1} R_{a_2})$$

Similar analysis for the situation of fig. 8.9(*b*) yields

$$|I''_1/V''_2|^2 = G_2 A_1/(16\pi r^2 R_{a_1} R_{a_2})$$

If we can presume that the two-port is indeed reciprocal we must have

$$G_1 A_2 = G_2 A_1$$

or $$G_1/A_1 = G_2/A_2$$

Since the antennae are quite arbitrary, it follows that the ratio of power gain to receiving area is the same for any antenna. It is necessary to evaluate the ratio for any one antenna, and this we have done in §8.2.1.

8.7 Satellite communication systems

Propagation between the earth and an orbiting satellite takes place in close approximation to propagation between two points in space, as discussed in §8.3. The major part of the path is outside the earth's atmosphere, and highly directional antennae are used. As an example, the International Telecommunications Satellite System (INTELSAT) will be briefly discussed.

The INTELSAT system is based on the use of satellites in a geo-stationary, or geosynchronous, orbit. These orbits are in the earth's equatorial plane, and have a radius such that the period of rotation of the satellite is equal to the period of the earth's rotation. The satellite then appears stationary with respect to the earth's surface. The radius of the orbit may be calculated to be 42 250 km, giving the height of the satellite above the earth's surface as about 36 000 km. In principle three such satellites can provide coverage of the whole surface, as shown in fig. 8.10.

The satellites are active, containing transmitters powered from solar cells. Signals are transmitted from the earth at a frequency near 6 GHz.

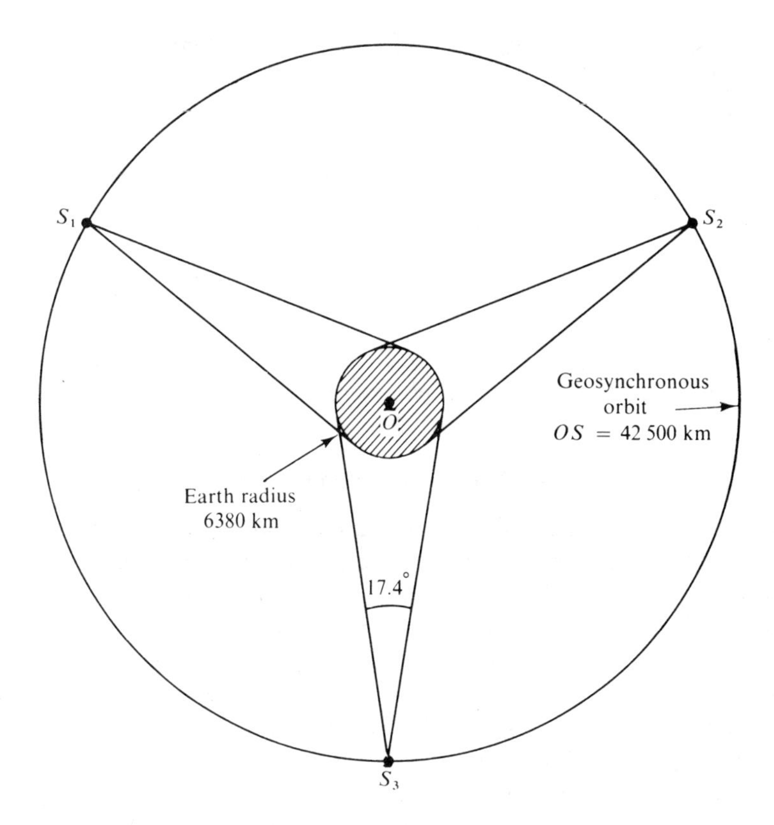

Fig. 8.10. Coverage of the earth's surface by three geosynchronous satellites.

The signals are received by the satellite, amplified and re-transmitted at about 4 GHz to an earth station. The propagation loss over each direction (equation (8.16)) is given by 20 $\log_{10}$ $(4\pi r/\lambda)$ which, for $r = 36\,000$ km and $\lambda = 6$ cm, is equal to about 200 dB. This large loss dictates considerable care in receiver design if a satisfactory signal-to-noise ratio is to be preserved on a round trip. It is also pertinent to point out that the total delay is of the order of 270 ms, which is near the limit for satisfactory two-way telephone communication. A further limitation results from the need to cover the part of the globe visible from the satellite: an arc of some 17° must be covered by the satellite antenna in the equatorial plane. The line-of-sight path is at grazing incidence in latitude 81° so that the polar regions cannot be covered. At latitude 52° the line-of-sight to the satellite is about 30° to the horizon.

Some details of INTELSAT satellites are given in table 8.1

Table 8.1

Satellite	Year of launch	Mass (kg)	Primary power (W)	Effective bandwidth (MHz)	Telephone circuits*	Cost per circuit–year ($)
I	1965	38	40	50	240	32 500
IVA	1975	790	500	800	6 000	1 000
V	1979	967	1200	2300	12 000	800

* A circuit refers to two one-way telephone channels.

8.7.1 Figure of merit for a receiving antenna

If we consider the earth station of a satellite link we can see that there will be two components of noise determining the final signal-to-noise ratio: one component arising from receiver noise (including antenna losses), and one received by the antenna in the form of radiation. This latter will include 'sky' noise, depending on the direction in space towards which the antenna is looking, and also any contribution made by radiation from the earth if the radiation pattern is so placed. This usually happens at low angles to the horizon, or because of side lobes. Consider a satellite at distance r radiating a total power P_s, and received by an earth station with an antenna of gain G_e and noise temperature T_e. The available signal power at the output of the earth-station antenna can be found using (8.15), giving

$$P_r = (\lambda/4\pi r)^2 P_s G_s G_e$$

in which G_s, G_e are the gains of the satellite and the earth-station antennae respectively. The available noise power at the antenna output can be expressed as kT_eB, in which T_e will be taken as 'sky' temperature, excluding for the moment contributions from antenna losses. The best possible signal-to-noise ratio is therefore given by

$$\frac{P_r}{kT_eB} = \frac{P_sG_s}{kB}\left(\frac{\lambda}{4\pi r}\right)^2 \frac{G_e}{T_e} \tag{8.19}$$

In this expression P_s is limited by the on-board power available to power the satellite transmitter and G_s is limited by the need to provide coverage of the visible globe. The wavelength λ is largely fixed by the need for high-gain antennae and a low sky temperature, and r is fixed by the geosynchronous orbit. The only variable available is therefore the earth-station antenna gain G_e, since T_e has a minimum fixed by extra-terrestrial radiation.

The quantity G_e/T_e is a figure of merit for the antenna. It is often quoted in the form $10 \log_{10} (G_e/T_e)$ and (regrettably) assigned units dB K^{-1}.

8.7.2 A satellite link

Let us put approximate values into (8.19). Equation (8.13) allows us to estimate G_s: putting α, the half beamwidth, as 10° gives G_s as 21 dB. This is likely to be an overestimate because of the nature of (8.13), and it is also necessary to maintain field strength at the edges of the earth coverage region, so that the half beamwidth must be larger than the 10° taken. A practical value is 13–14 dB at the beam edges. The carrier power is typically about 10 W, covering a bandwidth of 500 MHz. We can thus substitute in (8.19)

$$P_s = 10 \text{ W}, \qquad G_s = 13 \text{ dB}, \qquad B = 500 \text{ MHz}$$

$$\lambda = 6 \text{ cm}, \qquad r = 36\,000 \text{ km}, \qquad k = 1.38 \times 10^{-23}$$

We find for the signal-to-noise ratio in decibels

$$S/N = 10 \log_{10} (G_e/T_e) - 33$$

Thus the first term has to be of the order of 43 dB to achieve a signal-to-noise ratio of 10 dB. The sky temperature at 5 GHz is of the order of a few degrees Kelvin, and we must allow for degradation in the receiver. It is therefore necessary to have a low-noise receiver: originally cooled masers were used, now superseded by parametric amplifiers, with noise temperatures of about 4 K. Since the line-of-sight from earth station to satellite makes a low angle with the horizon, the antenna has to be designed for low side-lobe level, so that noise from the earth's surface is negligible. All these factors lead to an antenna gain of some 53 dB, necessitating a parabolic mirror some 27 m in diameter. Likewise losses in the antenna feed are exceedingly important in maintaining a low overall noise temperature.

The up-path does not demand such careful attention to noise problems, since the radiated power is not limited by supplies, although it is by technology. The noise temperature of the satellite antenna will be much higher, since the beam is almost covered by the earth. It is not therefore necessary to incorporate a very low-noise amplifier in the satellite.

It may be noted that the net loss from earth and satellite antennae and the path is the same for both up- and down-paths, so that it is only the extra transmitter power, offset by increased noise at the satellite antenna, which improves the situation on the up-link.

8.8 Propagation near the earth

The previous section has considered the simplest case with a transmitting antenna and a receiving antenna, each effectively isolated from all other influences. This simplification is not permissible when the transmission takes place between antennae giving a transmission path close to the surface of the earth. This more complicated problem may be studied in several forms. The simplest is to consider a plane, perfectly conducting earth; a model which can be solved exactly for many cases of interest.

This model can be made more realistic by allowing the plane earth to have finite conductivity, although maintaining homogeneity. We can then progress to a plane, inhomogeneous earth, and finally the same sequence can be considered with a spherical earth. An additional complication is added by the fact that the earth's atmosphere has electrical properties significantly different from those of free space.

The electromagnetic problems encountered when trying to solve for any of the above models become progressively more difficult. Results of practical use can be obtained in the more complicated situations by means of ray theory. In the following sections the problems will be treated in an elementary manner to lead to such practical results.

8.8.1 A plane, perfectly conducting earth

We shall restrict our considerations to a point source above the surface. The source may be for example a short electric or magnetic dipole. The extension to larger arrays is straightforward as long as the current distribution in the antennae is not affected by the presence of the earth. This is usually the case if the antenna is more than a few wavelengths above the earth. Closer proximity will affect both the radiation pattern and the radiation impedance. The situation is shown diagrammatically in fig. 8.11, with a short electric dipole placed vertically at a height h above the surface. This problem can be solved exactly: it is necessary to find a field additional to the free-space field of the dipole such as to make the tangential electric field at the surface zero. This is provided by an image of the transmitting dipole at $z=-h$, as shown. The field at a

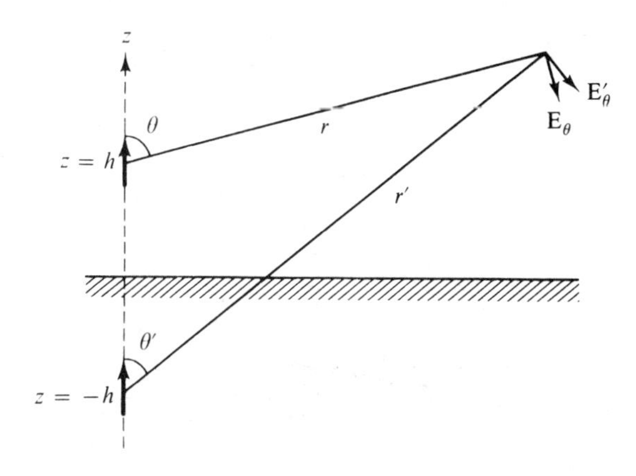

Fig. 8.11. Dipole above a plane earth.

distant point P arising from the transmitting dipole may be written

$$\begin{aligned} \mathrm{E}_\theta &= (A/r)\exp(-\mathrm{j}kr)\sin\theta \\ \mathrm{H}_\phi &= \zeta^{-1}\mathrm{E}_\theta \end{aligned} \tag{8.20}$$

where $$A = \mathrm{j}k\zeta I\,\mathrm{d}l/4\pi$$

Similarly for the image dipole

$$\begin{aligned} \mathrm{E}'_\theta &= (A'/r')\exp(-\mathrm{j}kr')\sin\theta' \\ \mathrm{H}'_\phi &= \zeta^{-1}\mathrm{E}'_\theta \end{aligned} \tag{8.21}$$

Since E_θ and E'_θ are not parallel, the total electric field has to be found by resolving into components. In particular the horizontal component is equal to $\mathrm{E}_\theta \cos\theta + \mathrm{E}'_\theta \cos\theta'$. This must vanish over the plane $z = 0$, on which $\theta' = \pi - \theta$, $r' = r$. Substitution from (8.20) and (8.21) shows that this condition is satisfied provided $A' = A$. The problem is therefore equivalent to that of two radiating sources of equal strength, of the type which has been considered in chapter 4. When P' is very distant we may use the same approximation as used in considering antenna arrays: $\theta' = \theta$, $r' = r + 2h\cos\theta$. The field then becomes

$$\mathrm{E}_\theta \approx (A/r)\exp(-\mathrm{j}kr)\sin\theta[1 + \exp(-2\mathrm{j}hk\cos\theta)] \tag{8.22}$$

The radiation pattern is shown in fig. 8.12. A succession of nulls appear, determined by the condition

$$2hk\cos\theta = (2n+1)\pi \qquad (n = 0, 1, 2, \ldots)$$

or $$\cos\theta = \tfrac{1}{4}\lambda(2n+1)$$

At ground level a gain of two over the free-space field occurs.

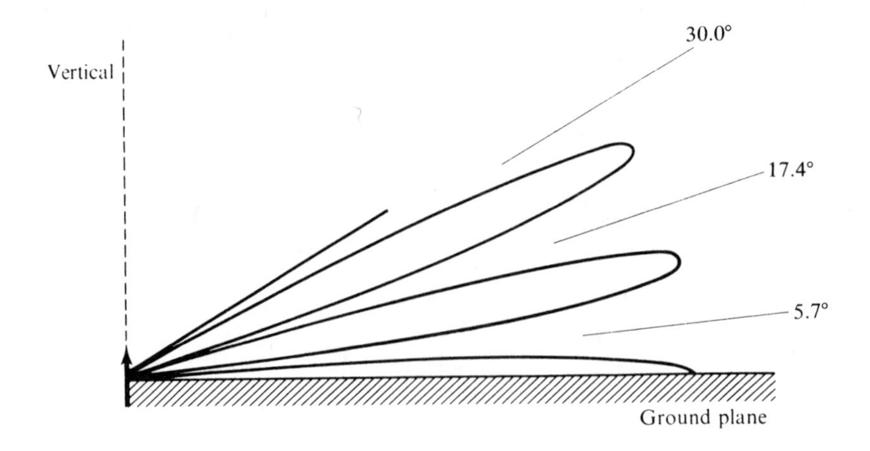

Fig. 8.12. Radiation pattern of dipole at height 2.5λ above a plane conducting earth.

It is also instructive to look at the situation when the dipole is acting as a receiving antenna, as shown in fig. 8.13. The dipole finds itself in the combined field of two waves, the incident wave together with a reflection. The incident wave will be correctly polarised when the electric vector is in a vertical plane, and the results of §1.5.2 will apply. The direction and magnitude of the field components are as shown in fig. 8.13. The reflected wave will be delayed, and reference to fig. 8.13 shows that this delay is $2h \cos \theta$. The total electric field parallel to Oz is

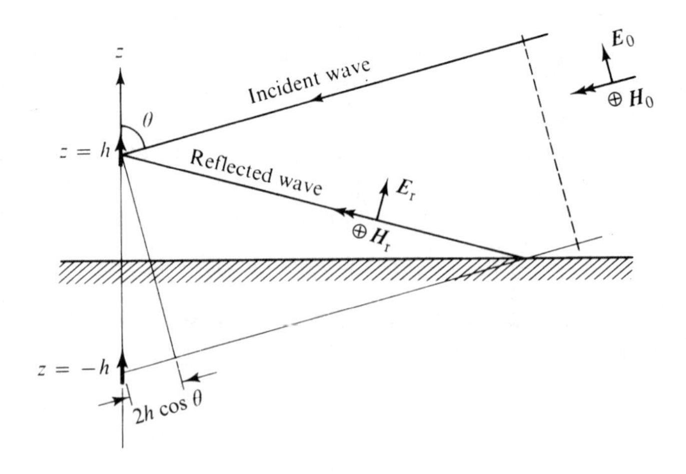

Fig. 8.13. Reception by dipole above a plane earth.

found to be equal to

$$E_0[1+\exp(-j2hk\cos\theta)]\sin\theta.$$

The open-circuit voltage from the dipole will therefore be

$$V = E_0\,dl\sin\theta[1+\exp(-j2hk\cos\theta)] \tag{8.23}$$

It will be seen that this result is compatible with that expressed by (8.22), as it indeed should be. In the present situation both methods are equally simple, but we shall find that in other cases one method may be preferable to the other.

8.8.2 A plane, lossy-dielectric earth

We may extend the analysis of the last section to the case when the earth is taken to be a lossy dielectric. It is simplest to approach the problem, since we are interested in the distant field, by considering the antenna in reception. The situation is the same as shown in fig. 8.13 except that the earth is now not a perfect conductor. The only analytical difference between the two cases is in the nature of the reflected wave. The problem of the reflection of plane waves from plane dielectric boundaries was discussed in chapter 1.

The ratio of the amplitudes of the reflected and the incident waves, taken at the point of incidence, can be described by Fresnel reflection coefficients, $R_v(\theta)$, $R_h(\theta)$, introduced in (1.103) and (1.104). The suffix v refers to the electric field in a vertical plane, the suffix h to the electric field parallel to the earth's surface. For the new situation we may rewrite (8.23) in the form

$$V = E_0\,dl\sin\theta[1+R_v(\theta)\exp(-j2hk\cos\theta)] \tag{8.24}$$

or, if the dipole was horizontal, in the form

$$V = E_0\,dl\sin\theta[1+R_h(\theta)\exp(-j2hk\cos\theta)] \tag{8.25}$$

For point-to-point communication with antennae raised above the earth and separated by a distance far greater than the height, as shown in fig. 8.14, the angle θ will be close to $\frac{1}{2}\pi$. The behaviour of $R_v(\theta)$ and $R_h(\theta)$ was discussed in chapter 1. For earth $\sqrt{\varepsilon_r}>3$, $R_h(\theta)$ never departs greatly from -1, which is the value a perfect conductor would have. The behaviour of $R_v(\theta)$ is more complicated: for high frequencies with θ close to $\frac{1}{2}\pi$ we may take $R_v(\theta)\approx-1$. At low frequencies conduction currents predominate, and this increases the effective value of $|\varepsilon_r|$, making $R_v(\theta)\approx+1$ unless θ is very close indeed to $\frac{1}{2}\pi$. Before using

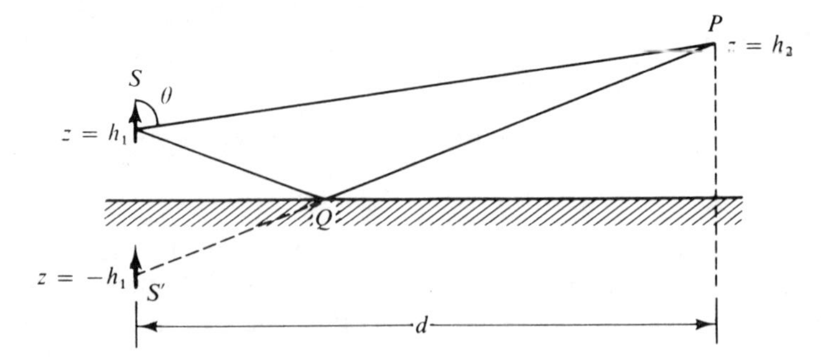

Fig. 8.14. Transmitting and receiving dipole above plane earth.

approximations we will discuss the situation when the earth is taken to be inhomogeneous.

8.8.3 A plane, inhomogeneous earth

To solve the electromagnetic problem of a plane wave reflected from an arbitrarily inhomogeneous surface some simplification is necessary: this may be done by interpreting the situation as an optical problem, in which ray theory is used. In fig. 8.15 is shown the same geometrical configuration as before, but with two rays from a distant source, one direct and one reflected. It is evident that provided we associate an amplitude and polarisation with each ray we get the same result as in §8.8.1 for a perfect reflector. Similarly we get the results of §8.8.2 when

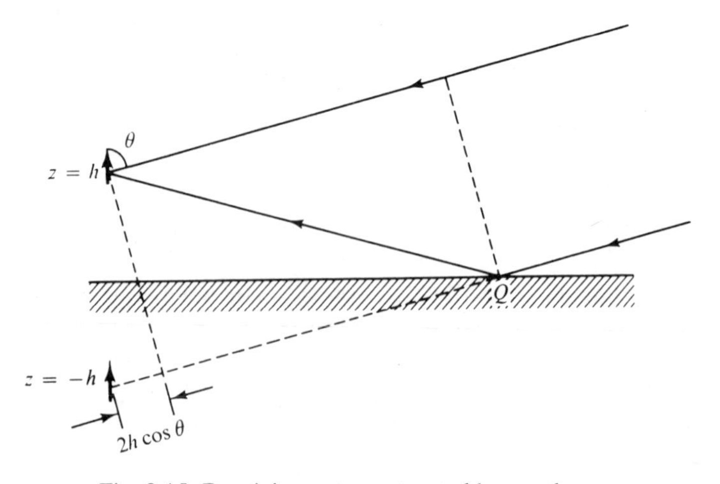

Fig. 8.15. Receiving antenna treated by ray theory.

the reflection coefficient is taken as $R_v(\theta)$ or $R_h(\theta)$. What fig. 8.15 shows is that on a ray basis only one reflected ray arrives at the receiving dipole. More properly we should consider rays near this one as well, since the dipole has a finite receiving area, but the area from which reflection takes place is limited to a region around the point of reflection Q. We would therefore expect that if in (8.24) and (8.25) we use values of $R_v(\theta)$ or $R_h(\theta)$ appropriate to the properties of the earth at the point Q we shall get a close approximation to the correct answer. The only restriction would be that the properties must not change too rapidly, on a wavelength scale. Thus a land–sea boundary could not be treated by this method.

8.8.4 Point-to-point transmission near the surface of the earth

We consider the situation shown in fig. 8.14, in which $d \gg h_1, h_2$. Let us take as unity the strength of the electric field at P by direct transmission from a source at S in the absence of a reflected wave. The strength of the field arising from the reflected wave may then be written in the form $R_v(\theta) \exp[-jk(r_1 - r_2)]$, in which r_1 is the length SP, r_2 the length $SQ + QP$. The total field strength at P is therefore the free-space value multiplied by the factor

$$H = 1 + R_v(\theta) \exp[-jk(r_1 - r_2)] \tag{8.26}$$

We will assume that it is permissible to approximate $R_v(\theta)$ by -1. We then have to express $r_1 - r_2$ in terms of antenna heights and separation. Observing in fig. 8.14 that $QS = QS'$, we have

$$r_1^2 = d^2 + (h_2 - h_1)^2$$

$$r_2^2 = d^2 + (h_2 + h_1)^2$$

The second term in both expressions is much less than the first, so that we may write

$$r_1 \approx d[1 + \tfrac{1}{2}(h_2 - h_1)^2/d^2]$$

$$r_2 \approx d[1 + \tfrac{1}{2}(h_2 + h_1)^2/d^2]$$

whence

$$r_2 - r_1 \approx 2h_1h_2/d$$

Substitution into (8.26) then gives

$$H \approx 1 - \exp(-j2kh_1h_2/d)$$

Hence

$$|H| = 2 \sin(kh_1h_2/d)$$

$$= 2 \sin(2\pi h_1h_2/\lambda d)$$

$|H|$ is often termed the 'height-gain' factor. From the point of view of a receiving station, h_2 may be regarded as a variable. At ground level the signal strength is zero (on account of the phase reversal implied by $R_v(\theta) = -1$). As the antenna is raised a succession of nulls will be encountered, determined by the relation

$$2\pi h_1 h_2/\lambda d = n\pi \qquad (n = 1, 2, 3, \ldots)$$

Example

A half-wave vertical dipole radiating at 100 MHz is situated 50 m above the ground. A similar receiving antenna is situated 20 km away at a height of 20 m above the ground. Estimate the available power from the receiving antenna when the transmitter radiates 100 W.

Solution. (i) First calculate the available power with a free-space link. The configuration given corresponds very closely to the maximum antenna power gain of 2.1 dB. The free-space transmission loss is equal to

$$20 \log_{10} (4\pi \times 2 \times 10^4/3) = 98.5 \text{ dB}$$

The radiated power is 20 dB W, so that the available received power on a direct path is

$$20 - 98.5 + 2 \times 2.1 = -74.3 \text{ dBW}$$

(ii) The effect of the earth will be to multiply the free-space *amplitude* by

$$2 \sin (2\pi h_1 h_2/\lambda d)$$

With the figures given this becomes a ratio of 0.209, or a loss of 13.6 dB. The available power is therefore −87 dBW, or 1.9×10^{-9} W.

8.8.5 Antenna at ground level

It is evident that the results of the last section are inapplicable to antennae at ground level. It is particularly necessary to consider such a situation at low frequencies when it may be impossible to make the antenna dimensions significant in terms of the wavelength and equally impossible to raise it many wavelengths above the ground. A simple example is that of a medium wave broadcast antenna for which the wavelength may be 300 m. It is just about possible to construct a vertical radiator of height $\frac{1}{4}\lambda$ which together with its image in the earth constitutes a half-wave dipole, as shown in fig. 8.16. If the ground were

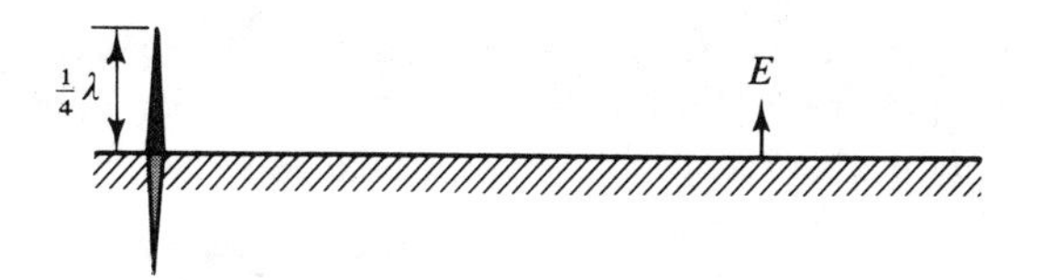

Fig. 8.16. Quarter-wavelength radiator and image.

perfectly conducting, the earth plane would not affect the field configuration since the electric field of a half-wave dipole is everywhere normal to the equatorial plane.

An earth which is not perfectly conducting modifies the situation very considerably: mathematically this is the result of placing $R_v(\theta)$ equal to -1, whereas for a perfect conductor $R_v(\theta)$ equals $+1$. The theory is beyond the scope of this book, and it must suffice to quote the result that when the distance r between antennae on the surface is sufficiently large the appropriate form for $|H|$ is given by

$$|H| \approx \sigma\zeta\lambda^2/4\pi^2 r$$

in which σ is the ground conductivity. In deriving this formula it has been assumed that the frequency is so low that conduction currents in the earth predominate over displacement current. Since $|H|$ has to be multiplied by the free-space value of field intensity, it can be seen that the field amplitude falls off as the inverse square of the distance. This formula refers to a plane earth: it is necessary to modify it for propagation over a spherical earth.

8.8.6 Further factors affecting point-to-point communication

It must be noted in connection with the last four sections that although the model has become successively more refined nevertheless there are further factors which can appreciably modify the results in a practical situation. It has been assumed that the earth's surface is smooth, which is quite untrue. The reflected wave will be determined not only by the electrical properties but by the roughness of the surface, measured on a wavelength scale. At lower wavelengths the reflection coefficient may be only a fraction of that for a smooth surface of the same properties. There is thus a statistical element in all calculations which has not been explicitly brought out.

The medium of propagation has been assumed to be free space. This is also untrue, and at microwave frequencies propagation in the lower 10 km of the earth's atmosphere can be appreciably affected by atmospheric conditions. Considerable and variable attenuation will occur because of precipitation, but the more regular change of properties with height also has effects. These will now be considered.

8.9 Propagation in a stratified troposphere

We will consider only propagation in the lower part of the earth's atmosphere extending up to some 10 km in altitude, the troposphere. This is quite adequate for microwave links. Lower frequencies will propagate via the ionosphere at much higher altitudes, but we shall not consider this situation.

We shall not attempt to use the full treatment of propagation in an inhomogeneous medium. In the relevant practical situations geometrical configurations tend to be on a scale large compared with the wavelength used so that we can consider effects distant from a source as we would for an optical source, by means of rays. We have shown in chapter 1 that the laws of reflection and refraction of plane waves are precisely those obeyed by optical rays. We have seen also that the field at a distance from any source is quasi-plane, so that we expect optical ray theory to be of some use in the situations we are considering. It is only in the region near radiators that we have to worry about finite wavelength and diffraction.

The properties of the atmosphere can be described in terms of a refractive index 'n' equal to $(\varepsilon/\varepsilon_0)^{\frac{1}{2}}$, which will vary in a way depending on the density and composition of the atmosphere, its temperature and its water vapour content. A useful average may be defined, and leads to a refractive index given as a function of altitude z by

$$n = 1 + \kappa z, \qquad \kappa = -4\times10^{-8}\ \text{m}^{-1} \tag{8.27}$$

This is called a 'standard' atmosphere.

We are thus lead as a first approximation to consider ray propagation through a stratified medium of spherical symmetry.

8.9.1 Ray propagation in a horizontally stratified medium

We will firstly look at a medium stratified horizontally over a plane earth, and subsequently extend the treatment to allow for sphericity. Consider a medium divided into horizontal layers, in each of which the refractive index is constant. This is shown in fig. 8.17(a). A ray starting at a point P on the earth's surface making an angle of θ_0 with the vertical will be straight until it meets the first boundary. It will then be refracted

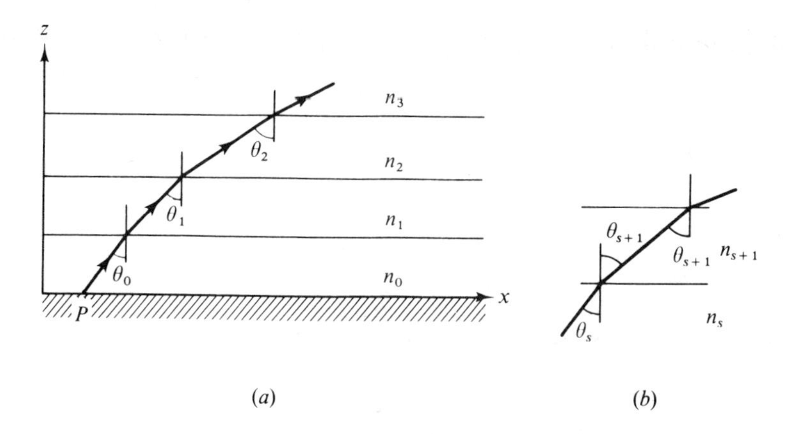

Fig. 8.17. Ray paths in a horizontally stratified medium. (*a*) Ray path. (*b*) Detail.

according to Snell's law and continue straight until the process is repeated at the next boundary. The situation at the boundary between the layers $n = n_s$, $n = n_{s+1}$ is shown in fig. 8.17(*b*). Snell's law relates the angles of incidence and refraction by the equation

$$n_s \sin \theta_s = n_{s+1} \sin \theta_{s+1} \tag{8.28}$$

Applying this successively to all boundaries we have

$$n_{s+1} \sin \theta_{s+1} = n_s \sin \theta_s = \ldots = n_1 \sin \theta_1 = n_0 \sin \theta_0$$

We note that the thickness of the layers does not enter into this equation, so that for a continuously varying refractive index $n(z)$ we have

$$n(z) \sin \theta = n_0 \sin \theta_0 \tag{8.29}$$

The angle θ determines the slope of the ray at any point according to

$$\tan \theta = \mathrm{d}x/\mathrm{d}z$$

or

$$\sin \theta = [1 + (\mathrm{d}z/\mathrm{d}x)^2]^{-\frac{1}{2}}$$

We therefore have for the equation of the ray

$$\left[1 + \left(\frac{\mathrm{d}z}{\mathrm{d}x}\right)^2\right]^{\frac{1}{2}} = \frac{n(z)}{n_0 \sin \theta_0}$$

Rearranging we find

$$\mathrm{d}x/\mathrm{d}z = \{[n(z)/n_0 \sin \theta_0]^2 - 1\}^{-\frac{1}{2}}$$

or

$$x = \int_0^z \{[n(z)/n_0 \sin \theta_0]^2 - 1\}^{-\frac{1}{2}} \mathrm{d}z \tag{8.30}$$

This is an explicit equation for a ray starting at $z = 0$, where $n = n_0$, at an angle θ_0 to the vertical.

Example

A medium has a refractive index given by

$$n^2(z) = n_0^2[1 + \alpha z] \tag{8.31}$$

Determine the equations of the ray paths.

Solution. The integrand in (8.30) takes the form

$$\left(\frac{1+\alpha z}{\sin^2\theta_0} - 1\right)^{-\frac{1}{2}} = \sin\theta_0(\cos^2\theta_0 + \alpha z)^{-\frac{1}{2}}$$

Hence

$$x = \sin\theta_0 \int_0^z \frac{dz}{(\alpha z + \cos^2\theta_0)^{\frac{1}{2}}}$$

$$= \frac{2}{\alpha}\sin\theta_0\left[(\alpha z + \cos^2\theta_0)^{\frac{1}{2}}\right]_0^z$$

$$= \frac{2}{\alpha}\sin\theta_0[(\alpha z + \cos^2\theta_0)^{\frac{1}{2}} - \cos\theta_0] \tag{8.32}$$

Rearrangement gives this equation in the form

$$\alpha z = (\cos\theta_0 + \alpha x/2\sin\theta_0)^2 - \cos^2\theta_0 \tag{8.33}$$

This is the equation of a parabola, vertex at point

$$x = -\alpha^{-1}\sin 2\theta_0, \qquad z = -\alpha^{-1}\cos^2\theta_0 \tag{8.34}$$

When α is positive, (8.31) shows that the refractive index increases with height and the ray is progressively bent towards the normal. If α is negative, a ray will be bent away from the normal. Some examples are shown in fig. 8.18(*a*) and (*b*).

8.9.2 Spherical stratification

In considering ground-to-ground microwave links we are only concerned with distances of the order of 50–100 km, a maximum of about 1% of the earth's radius. This seems to indicate that a plane-earth model should suffice. However, the refractive index of the troposphere, as given by (8.27), is very close to unity so that the slight effect of the earth's curvature is of equal importance. We therefore have to consider a spherically stratified medium, although it will be shown that the effect

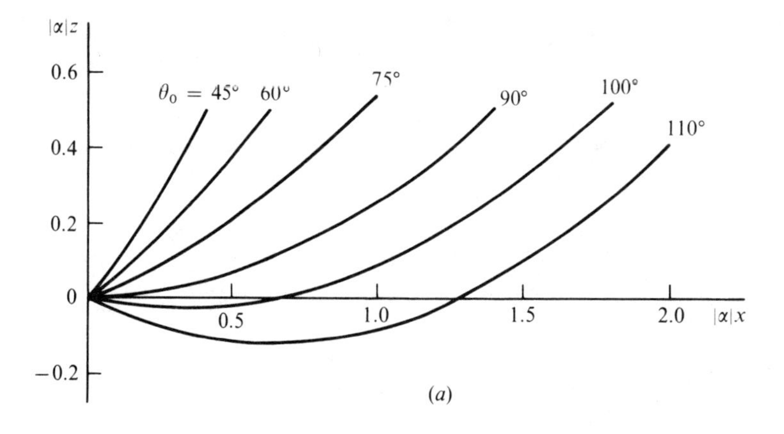

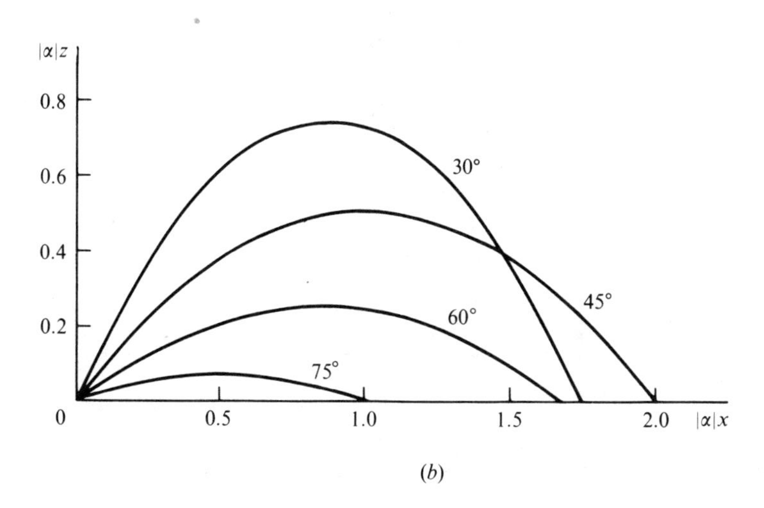

Fig. 8.18. Ray paths for different values of θ_0 in an atmosphere for which $n^2 = n_0^2(1+\alpha z)$. (*a*) $\alpha > 0$. (*b*) $\alpha < 0$.

can be allowed for in the situation under consideration by using a modified refractive index in a planar geometry.

The translation of the situation shown in fig. 8.17 to spherical geometry is shown in fig. 8.19. With the variables in the diagram we have to replace (8.28) by

$$n_s \sin \theta_s = n_{s+1} \sin \theta_s' \tag{8.35}$$

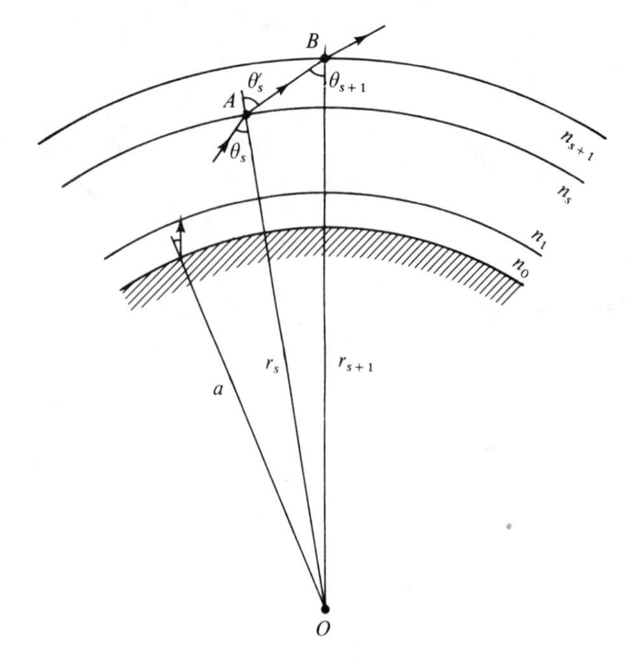

Fig. 8.19. Spherical stratification.

The sine rule for triangles applied to $\triangle AOB$ in the form

$$\frac{\sin O\hat{A}B}{OB} = \frac{\sin O\hat{B}A}{OA}$$

enables us to write

$$\frac{\sin \theta_s'}{r_{s+1}} = \frac{\sin \theta_{s+1}}{r_s}$$

whence (8.35) may be expressed as

$$n_s \sin \theta_s = n_{s+1} \sin \theta_{s+1}(r_{s+1}/r_s)$$

or
$$n_s r_s \sin \theta_s = n_{s+1} r_{s+1} \sin \theta_{s+1}$$

The equivalent to (8.29) is therefore

$$n(r)r \sin \theta = n_0 a \sin \theta_0 \tag{8.36}$$

in which a is the radius from which the ray is launched. This equation can be integrated in a manner similar to that for (8.29). For our purpose however we have already noted that the region of interest is a thin, short layer near the earth's surface. We therefore put in (8.36) $r = a + z$ giving

$$n(z)(1 + z/a) \sin\theta = n_0 \sin\theta_0$$

We then note that this is identical in form with (8.29) if we introduce an effective refractive index $\bar{n}(z)$ given by

$$\bar{n}(z) = n(z)(1 + z/a) \tag{8.37}$$

and write

$$\bar{n}(z) \sin\theta = n_0 \sin\theta_0 \tag{8.38}$$

We remember that $n(z)$ is very close to unity, and write it as $1 + \delta(z)$. Substituting (8.37) and using the fact that both δ and z/a are small we have

$$\bar{n}(z) = 1 + \delta + z/a \tag{8.39}$$

The use of (8.38) enables us to use solutions found for a plane-stratified medium. The curves of fig. 8.18 for example apply if we interpret x as arc distance along the earth's surface, z being altitude as before.

8.9.3 Propagation in a standard atmosphere

The variation of refractive index with height in a standard atmosphere was given by (8.27)

$$n(z) = 1 + \kappa z, \qquad \kappa = -4 \times 10^{-8}\ \text{m}^{-1}$$

We may therefore write the effective refractive index from (8.39) as

$$\bar{n}(z) = 1 + (\kappa + a^{-1})z$$

Using the value 6370 km for a we find

$$\bar{n}(z) = 1 + \kappa_e z, \qquad \kappa_e = 1.2 \times 10^{-7}\ \text{m}^{-1} \tag{8.40}$$

The effective refractive index has therefore a positive gradient in the upward direction even though the actual refractive index has a negative gradient. We can also write (8.40) in the form

$$\bar{n}(z) = 1 + z/a_e, \quad a_e \approx 4a/3$$

The effect of a standard atmosphere is therefore as though propagation were over an earth of radius $4a/3$ which had no atmosphere. It must be pointed out that if an (x, z) ray path is calculated, the points on the ray are then to be laid out as (arc distance, altitude) on the actual earth.

If we take the maximum height of the troposphere as 10 km, the deviation from unity of $\bar{n}(z)$ is only 0.1%, so that from (8.40) we have, to a very close approximation,

$$\bar{n}^2(z) = 1 + 2\kappa_e z$$

This is the form taken in (8.31) for which the ray paths of fig. 8.18 apply, so that we may take the ray paths as parabolic.

Example

Determine the maximum distance at which a ray from an antenna at height h above a (smooth) earth surface and propagating in a standard atmosphere can touch the earth's surface.

Solution. The general form of possible rays is as shown in fig. 8.18(a). It is easiest to take the origin of co-ordinates at the antenna so that the earth's surface is at $z = -h$. The general pattern of rays is shown in fig. 8.20. The ray required is that touching the surface at D, which will have a specific launching angle. This implies that the vertex of the ray is at $x = d$, $z = -h$, so that from (8.34) we have

$$d = -\alpha^{-1} \sin 2\theta_0$$

$$-h = -\alpha^{-1} \cos^2 \theta_0$$

Thus θ_0 is greater than $\frac{1}{2}\pi$ in order to satisfy the first equation. Given numerical values, θ_0 can be found from the second, and hence d. For the standard atmosphere and spherical earth, $\alpha = 2\kappa_e = 2.4 \times 10^{-7}$. αh is

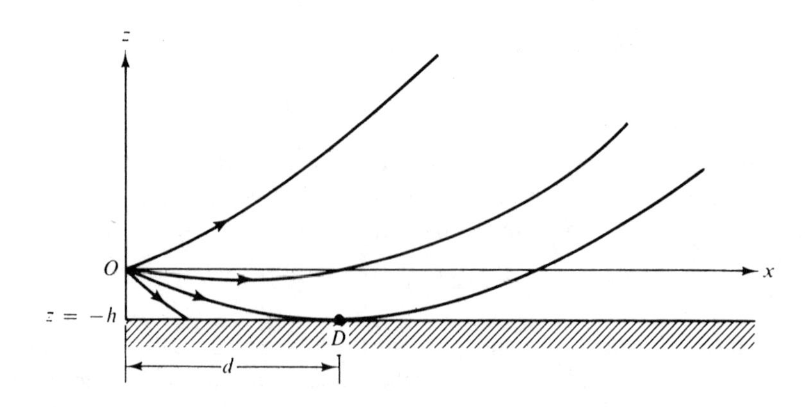

Fig. 8.20. Rays from antenna above the earth's surface.

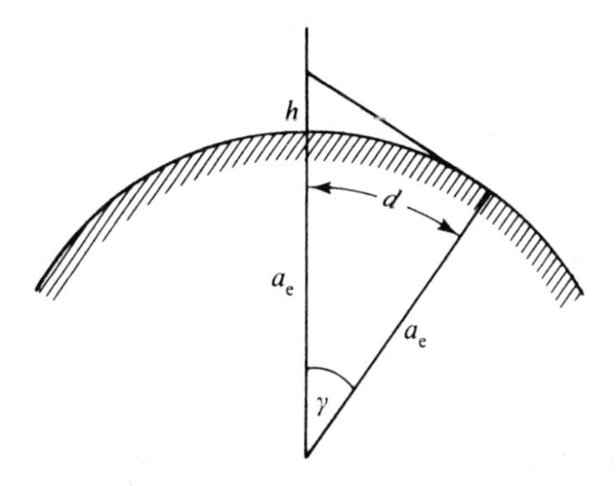

Fig. 8.21. Ray calculation using equivalent earth radius.

therefore small, and θ_0 is near $\frac{1}{2}\pi$. Let us substitute $\theta_0 = \frac{1}{2}\pi + \psi$. The equations then become

$$\alpha d = \sin 2\psi \approx 2\psi$$

$$\alpha h = \sin^2 \psi \approx \psi^2$$

Hence $$\alpha d \approx 2(\alpha h)^{\frac{1}{2}}$$

or $$d \approx (2h/\kappa_e)^{\frac{1}{2}}$$

For $h = 30$ m, this gives $d = 22$ km.

We should be able to work out this problem by assuming an atmosphereless earth, radius $a_e = 1/\kappa_e$. This method is shown in fig. 8.21. The angle γ is given by

$$\cos \gamma = a_e/(a_e + h) \approx 1 - h/a_e$$

γ is small, so that we may replace $\cos \gamma$ by $1 - \frac{1}{2}\gamma^2$, giving $\gamma \approx (2h/a_e)^{\frac{1}{2}}$. The arc distance is given by γa_e. Hence

$$d \approx \gamma a_e = (2ha_e)^{\frac{1}{2}} = (2h/\kappa_e)^{\frac{1}{2}}$$

and is hence in agreement with the previous method.

Example

Two stations of a microwave link are sited 30 km apart at heights of 100 m above a smooth earth surface. Calculate the path of the ray between them in a standard atmosphere. Find the angle of arrival of the ray and the least clearance above the surface.

Solution. As in the last problem we take the origin of co-ordinates at one antenna, and the earth's surface is at $z = -100$. The path is as indicated

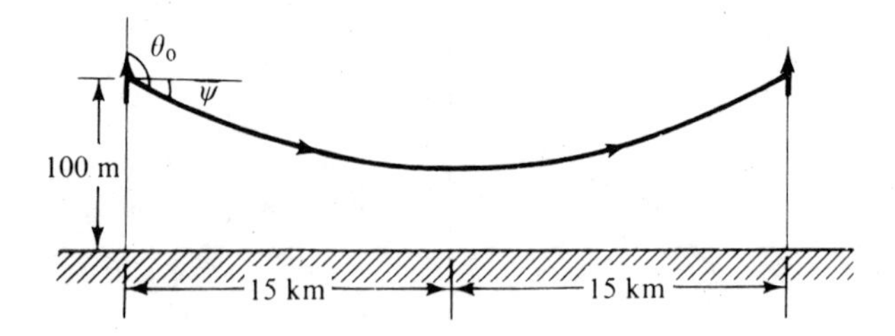

Fig. 8.22. Path between transmitter and receiver.

in fig. 8.22. It is symmetrically placed with the vertex midway between the sites. Equation (8.34) gives, as in the previous example,

$$\alpha d = \sin 2\psi$$

We can now obtain a numerical value for ψ, or as before approximate by

$$\begin{aligned}\psi &\approx \tfrac{1}{2}\alpha d = \kappa_e d \\ &= 1.8 \times 10^{-3} \text{ radians} \\ &= 0.10 \text{ degrees}\end{aligned}$$

The angle of arrival is thus 0.10° below the horizontal.

The co-ordinate height of the vertex is given by

$$\begin{aligned}z = -\alpha^{-1} \sin^2 \psi &= -\alpha^{-1}(\tfrac{1}{2}\alpha d)^2 \\ &= -\tfrac{1}{4}\alpha d^2 \\ &= -\tfrac{1}{2}\kappa_e d^2 \\ &= -13.5 \text{ m}\end{aligned}$$

The clearance above the nominal surface is therefore 86.5 m.

†8.9.4 Estimation of field strength from ray paths

The field strength associated with any ray has to be determined by considering a cone of neighbouring rays as shown in fig. 8.23(a). The energy flowing into the cone from the source is conserved, and the same rate of flow crosses the surfaces $S_1, S_2, S_3, \ldots$. If the rays are straight lines this obviously leads to the inverse square law for power, so that the field decays inversely as distance. This result will be slightly modified when the rays are bent, but since the bending is small the effect is also small.

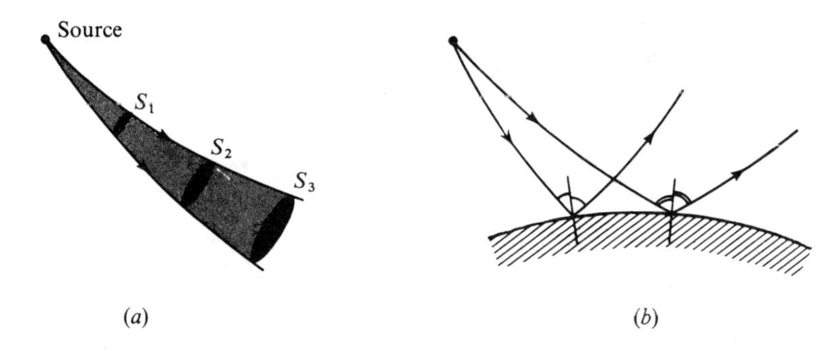

Fig. 8.23. Calculating field strength from ray paths. (*a*) Cone of rays. (*b*) Reflection at a curved surface.

A ray which hits the earth will be reflected. The appropriate Fresnel coefficient must be used in estimating the strength of the reflected ray. It is also necessary to allow for a further reduction in strength because of reflection at a curved surface, as shown in fig. 8.23(*b*).

Results obtained by application of these various factors can be compared with values obtained by electromagnetic calculations, and shown to be satisfactory. We can then apply the ideas to more complicated variations of refractive index with height.

8.9.5 Propagation beyond the horizon

The first example of §8.9.3, to which fig. 8.20 refers, showed that there is a ray which touches the earth's surface. Rays launched nearer to the horizontal will not reach the surface. The ray *OD* of fig. 8.20 therefore defines the optical horizon. The region beyond *D* is in shadow. Since radio waves have a much longer wavelength than light, there will in fact be a signal beyond *D*, arising from diffraction not allowed for by ray theory. The signal will however decrease rapidly as the shadow region is entered. The result of a typical calculation is shown in fig. 8.24.

For some atmospheric conditions, propagation beyond the nominal radio horizon can be much enhanced, due to the formation of 'ducts'.

8.9.6 Ducts

The rate of change of the modified refractive index κ_e is positive for a standard atmosphere. This is because the correction term z/a to allow for a curved earth outweighs the normal decrease of refractive index with height. It is possible to have meteorological conditions for which

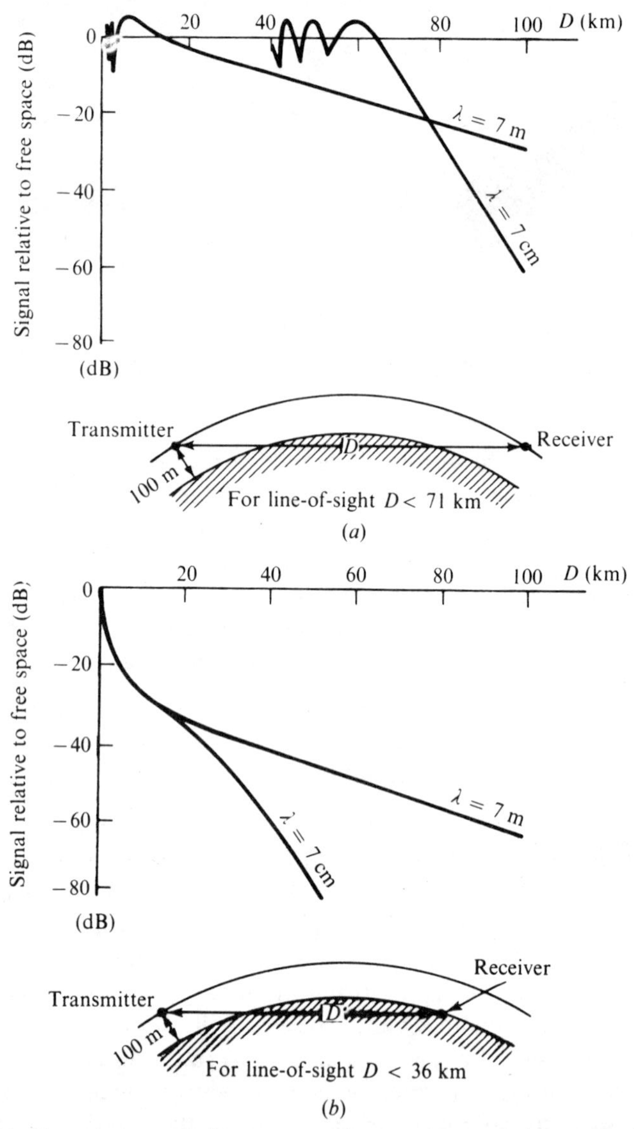

Fig. 8.24. Propagation of electromagnetic waves round the earth. (*a*) Both transmitter and receiver elevated. (*b*) Receiver at ground level. (After van der Pohl & Bremmer, *Phil. Mag. Series* 7, vol. 25, pp. 817–34.)

the rate of decrease of refractive index is sufficiently large to make $\kappa_e < 0$ over some distance. The ray pattern when $\kappa_e < 0$ is shown in fig. 8.25. The rays are reflected back to earth, and from (8.34), the maximum height reached is $(2|\kappa_e|)^{-1} \cos^2 \theta_0$. This situation will obviously

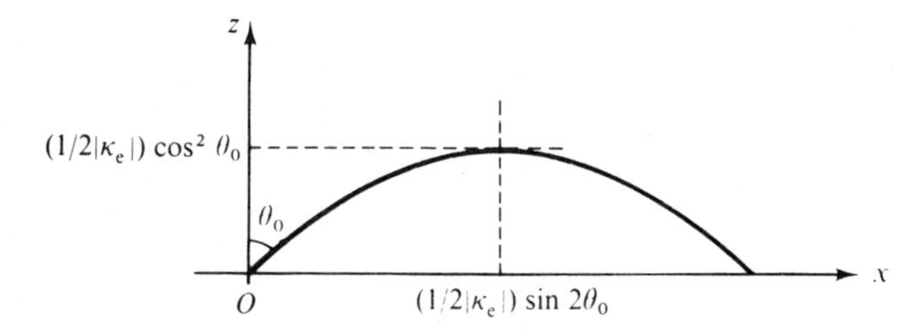

Fig. 8.25. Ray path when the gradient of the modified refractive index is negative.

influence the signal strength along the earth's surface. Calculations of the effect must allow for phase changes at reflections and along the path.

The modified refractive index cannot decrease indefinitely, so that a minimum must be reached, as indicated in fig. 8.26(a). It can be seen from (8.30) that no ray path can exist along which $n(z)$ is less than $n_0 \sin \theta_0$. This implies that the height at which a ray turns over is such that $n^2(z)=n_0^2 \sin^2 \theta_0$, and at this height $\theta=\frac{1}{2}\pi$. If the minimum of n is $n_m(<n_0)$, a ray reaching the minimum must start with $\sin \theta_0=n_m/n_0$. We must therefore have the ray pattern as shown in fig. 8.26(b). We thus have a partial trapping of radiation.

The vertical thickness of ducts is restricted in practice to 10–100 m, with n decreasing by only a few parts in 10^6. Summing rays, as indicated above, leads to the conclusion that for strong reinforcement of the transmitted wave the wavelength must be a small fraction of the duct height. Enhanced propagation because of such conditions is therefore encountered primarily at microwave frequencies.

Conditions may exist whereby n increases or is substantially constant up to some height h after which it decreases to a minimum and then

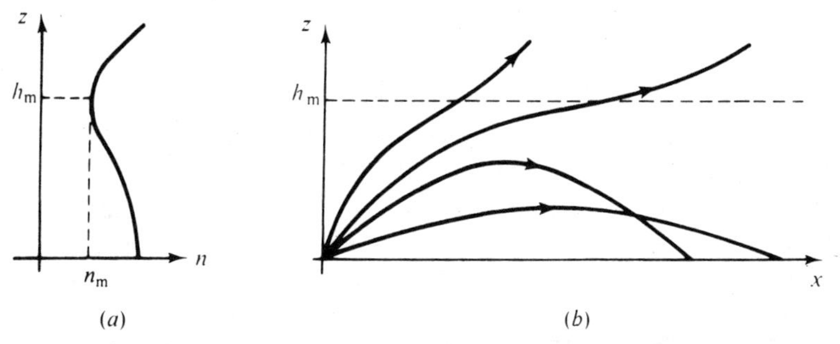

Fig. 8.26. The effect of minimum in refractive index. (a) Variation of refractive index. (b) Ray paths.

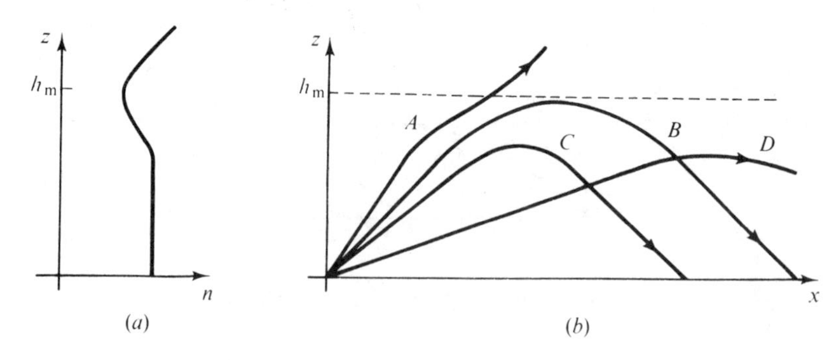

Fig. 8.27. Elevated duct. (*a*) Variation of refractive index. (*b*) Ray paths.

increases again, as shown in fig. 8.27(*a*). The ray pattern is shown in fig. 8.27(*b*). Up to height *h*, rays are straight. Above that height the considerations of the last section apply. Ray *A* escapes; ray *B* is very slowly brought round to the horizontal, since *n* is approaching the minimum, and it then returns to earth. Similarly with rays *C* and *D*, although these are bent round more quickly since the rate of change of *n* with height is greater than for ray *B*. The point of return to earth decreases from ray *B* through *C* and then increases with ray *D*. There is thus a certain distance from the source where no return ray arrives: this is known as the skip distance. This phenomenon will occur for any layer with reflecting properties similar to those discussed, and it is well known as a feature of transmission via the ionosphere.

8.10 Propagation over obstacles

In the discussion of the last sections we have ignored the roughness of the earth's surface. In practice allowance has to be made for undulations in the ground and other obstacles between transmitter and receiver. The only aspect of this problem we shall consider is that arising in point-to-point microwave links.

We have used above the idea of rays. If the direct ray between a transmitter and receiver is interrupted by an obstacle it is evident that signal strength is likely to be reduced. At the same time it is likely that a 'bare miss' of the obstacle by the ray path is likely to give less than the free-space signal strength. In order to assess the least clearance between ray path and obstacle which will give satisfactory performance it is necessary to take into account more properties of radiation than are accounted for by rays. This may be done using Huyghens' principle, without going into the full electromagnetic theory.

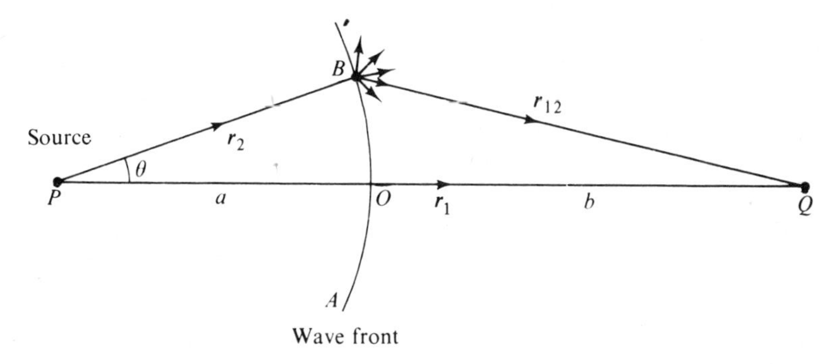

Fig. 8.28. Huyghens' principle.

†8.10.1 Huyghens' principle: Fresnel zones

Huyghens' principle may be illustrated by considering the configuration of fig. 8.28, in which a source P is received at a point Q. The field produced by P will be determined by the factor r^{-1} exp $(-jkr)$. Huyghens argued that the way in which the field 'arrived' at Q was by continual re-radiation from each point on a wave front. The surface OAB is spherical, centred on P, and is therefore an equi-phase surface. At B for example re-radiation takes place in all forward directions with a strength depending on the field at B and a factor depending on the angle of propagation away from B. The field at Q should be obtainable by summation for all points on the wavefront of the radiated fields. This theory is difficult to sustain rigorously but mathematically its expression may indeed be used to calculate the field at P. This being so, the result of obscuring part of the wavefront can be found by summing only over those parts of the wavefront visible from Q. Since the separation between P and Q in our case is large, we need only concern ourselves with directions not far off the normal to the wavefront, and are therefore concerned with expressions for field intensity of the form

$$\mathrm{I}(\boldsymbol{r}_1)=\int r_{12}^{-1}\psi(\boldsymbol{r}_2)\exp(-\mathrm{j}kr_{12})\,\mathrm{d}S \tag{8.41}$$

in which $\psi(\boldsymbol{r}_2)$ is the field strength at the point $\boldsymbol{r}_2$ on the wavefront, and r_{12} is the distance from an element of area dS of the wavefront to Q. The integral may, in principle, be evaluated, but Fresnel observed that under certain practical conditions a simple interpretation could be made. He divided the wavefront into zones, as shown in fig. 8.29. The zone boundaries are defined by the requirement that each point on a boundary is equidistant from Q and for successive boundaries, the distance is

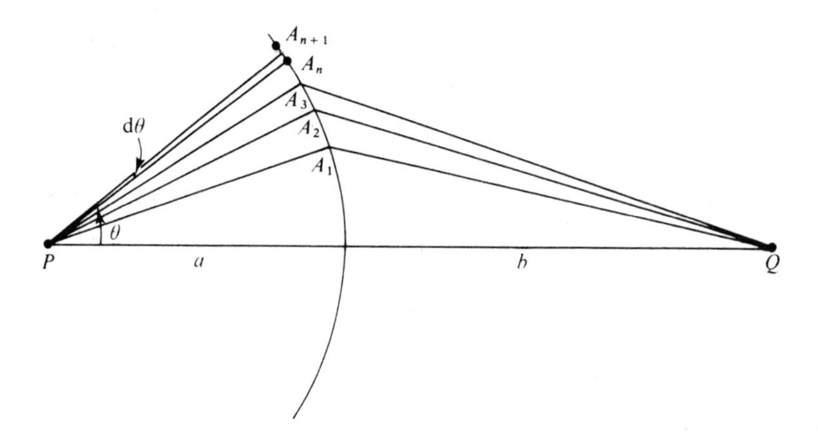

Fig. 8.29. Fresnel zones.

increased by $\frac{1}{2}\lambda$. Thus in fig. 8.29 $QA_1 = b + 0.5\lambda$, $QA_2 = b + \lambda$, $QA_3 = b + 1.5\lambda$, ... These zones are of course bands on the surface of the sphere, formed by revolution about *PQ*.

He next showed that successive bands make contributions to the integral in (8.41) which are opposite in sign. To show this, consider the zone between A_n and A_{n+1}. The area of the zone, from fig. 8.29, is $2\pi a^2 \sin\theta \, d\theta$, so that the contribution to I is equal to

$$2\pi a^2 \sin\theta \, d\theta \, r_{12}^{-1} \, \psi(\mathbf{r}_2) \exp(-jkr_{12}) \tag{8.42}$$

Now from the geometry of fig. 8.29

$$r_{12}^2 = a^2 + (a+b)^2 - 2a(a+b)\cos\theta$$

This equation relates r_{12} and θ, so that the change in r_{12} along the wavefront can be related to the change in angle. Differentiating

$$r_{12} dr_{12} = +a(a+b)\sin\theta \, d\theta$$

The contribution given in (8.42) may therefore be written

$$\frac{2\pi a}{(a+b)} \psi(r_{12}) \exp(-jkr_{12}) \, dr_{12}$$

Integrating over the zone A_nA_{n+1} therefore gives

$$\delta \mathrm{I}_n = \frac{2\pi a}{(a+b)} \int_{b+n\lambda/2}^{b+(n+1)\lambda/2} \psi(r_{12}) \exp(-jkr_{12}) \, dr_{12}$$

The zones will be quite small, so if we assume an average value ψ_n, we have

$$\delta I_n = \frac{2\pi a}{(a+b)}\psi_n \int_{b+n\lambda/2}^{b+(n+1)\lambda/2} \exp(-jkr_{12})\,dr_{12}$$

$$= \frac{4\pi a}{(a+b)}\frac{1}{jk}\psi_n(-1)^n \exp(-jkb)$$

The result for the whole surface can therefore be written in the form

$$I = (\psi_0 - \psi_1 + \psi_2 - \psi_3 + \ldots)\frac{4\pi a}{(a+b)}\frac{1}{jk}$$

$$= \{\tfrac{1}{2}\psi_0 + [\tfrac{1}{2}(\psi_0+\psi_2) - \psi_1] + [\tfrac{1}{2}(\psi_2+\psi_4) - \psi_3] + \ldots\}\frac{4\pi a}{(a+b)}\frac{1}{jk}$$

The first term in square brackets is the difference between the average amplitude over zones 0 and 2 and the amplitude over zone 1, and is therefore likely to be small. Similar arguments apply to the remaining brackets. Fresnel therefore deduced that

$$I \approx \psi_0 \frac{2\pi a}{(a+b)}\frac{1}{jk}$$

From this it appears that it is necessary only to have contributions at Q from the first zone, so that obstruction to the wavefront which does not exclude this first zone should not have much effect on the field strength at Q. This is the basis on which clearances are estimated.

8.10.2 Clearance of obstacles

In fig. 8.30 is shown a transmitter–receiver link PQ distance d in length. We can draw wavefronts for radiation from P, giving successive surfaces S_1, S_2, . . . On each of these we may construct the boundary defining the first Fresnel zone. From fig. 8.29 this is defined by the fact that the length $PA_1 + A_1Q$ exceeds PQ by $\frac{1}{2}\lambda$. In the variables of fig. 8.30 this gives

$$(\rho^2 + z^2)^{\frac{1}{2}} + [\rho^2 + (d-z)^2]^{\frac{1}{2}} = d + \tfrac{1}{2}\lambda$$

ρ will be much less than z (since λ is small) except at the ends, so we may approximate to the two square roots with use of the binomial theorem, giving

$$z\left(1 + \frac{1}{2}\frac{\rho^2}{z^2}\right) + (d-z)\left[1 + \frac{1}{2}\frac{\rho^2}{(d-z)^2}\right] = d + \tfrac{1}{2}\lambda$$

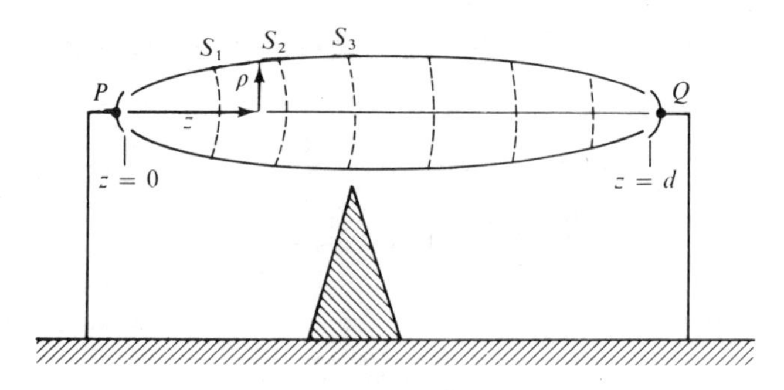

Fig. 8.30. Estimating clearance over obstacles.

Simplifying we find

$$\rho^2 = \lambda z(d - z)/d \tag{8.43}$$

If we shift the origin of the z-co-ordinate to the mid-point by writing $z = \bar{z} + \frac{1}{2}d$, (8.43) becomes

$$\rho^2 = \frac{\lambda}{d}\left(\frac{1}{2}d + \bar{z}\right)\left(\frac{1}{2}d - \bar{z}\right)$$

or

$$\frac{4}{\lambda d}\rho^2 + \frac{4}{d^2}\bar{z}^2 = 1 \tag{8.44}$$

This is the equation of an ellipse of major axis $\frac{1}{2}d$ and minor axis $\frac{1}{2}\surd(\lambda d)$. Thus the maximum distance between the boundary of the first Fresnel zone and the direct ray path is $\frac{1}{2}\surd(\lambda d)$ occurring at the mid-point of the link. If, for example, $d = 30$ km, $\lambda = 6$ cm, this distance is 21 m. A clearance of this order of magnitude is therefore necessary at mid-path.

8.11 Microwave relay systems

Many of the principles discussed in previous sections find application in the design of microwave communication systems. The maximum distance that can be covered in a single hop is limited by several factors: the transmitter power available, propagation phenomena in the troposphere, curvature of the earth, available sites for transmitter and receiver, and others. Depending on terrain and other factors, distances of 30–70 km are possible. Longer distances can be covered by using relay stations in which the incoming microwave signal is demodulated, amplified and re-transmitted. Such relay stations can operate unattended for long periods. The way in which telephone and television channels are combined is very similar to the methods used in line communication, discussed in chapters 6 and 7. The only major difference is that because of the nature of the microwave generators available it is found best to transmit the combined channels as a frequency modulation of the microwave carrier.

In the following sections features relating to propagation rather than to communications will be briefly discussed, using as an example the practice of the British Post Office as exemplified in the microwave network in the United Kingdom.

8.11.1 General considerations

The range of frequencies suitable for use is determined by many considerations. It is desirable to be able to use highly directed beams, which puts a lower limit on frequency, and at higher frequencies attenuation in the propagation path increases. The British Post Office has links operating around 2, 4, 6, 7 and 11 GHz, with a 19 GHz system in development. The 4 and 6 GHz frequencies are also shared by satellite services, with which microwave links have much in common.

A major constraint on design is set by the power available for transmission. Much existing equipment uses travelling-wave tube amplifiers, producing up to 10 W. Solid-state devices are now available, with somewhat smaller output powers. Typically, the bandwidth in the receive or transmit channel may be 250 MHz. Capacity is increased by the use of two carriers with mutually perpendicular polarisations. These can be separated at the receiving antenna with satisfactory discrimination.

8.11.2 Antennae

As discussed in §8.2.5, the gain and receiving area of a microwave antenna is set by its physical size. The simplest antenna is a parabolic dish, which at 4 GHz and above is equipped with waveguide feed. The effective area and radiation pattern are then controlled by the distribution of field over the aperture, which is determined by the feed. This simple antenna might take the form of fig. 8.31(*a*). An alternative to the paraboloid is a horn, as shown in fig. 8.31(*b*). The size of such devices is limited by constructional tolerances, by wind forces, and by expense. In practice it is found that 3–4 m is acceptable. The performance of these antennae can be improved by paying attention to the feed system, tailoring the field across the aperture to obtain high gain and also to control the side lobes. An alternative to feeding with a horn

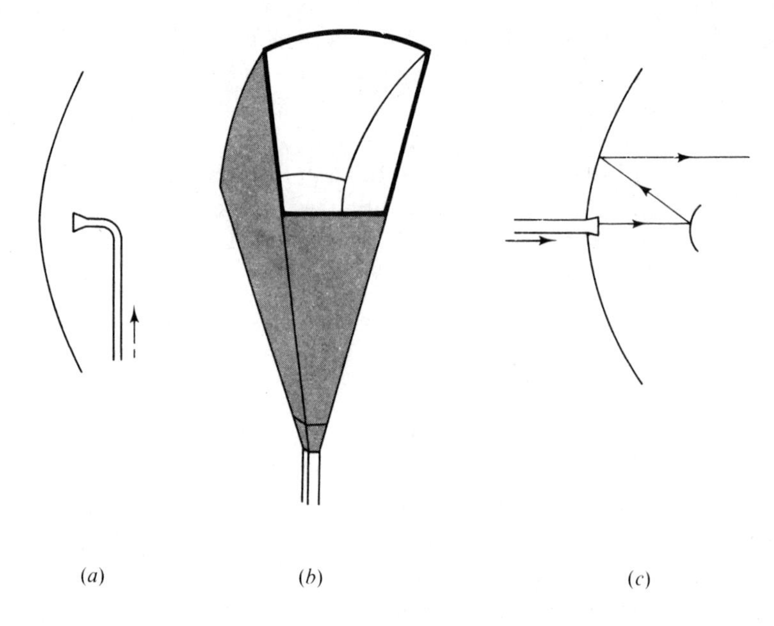

Fig. 8.31. Microwave antennae. (*a*) Parabolic mirror. (*b*) Horn. (*c*) Cassegrain feed.

is to use a sub-reflector, as indicated in fig. 8.31(*c*); after the telescope using this principle, such an antenna is referred to as a Cassegrain antenna. Since the sub-reflector is small its shape does not have to conform to a simple contour, and further the losses are smaller than with a waveguide feed: this is important in earth stations of satellite links when low noise temperatures are required.

Table 8.2 gives some details of antennae used by the British Post Office.

8.11.3 Path planning

From the last section we see antenna gains of around 50 dB are obtainable in 4–6 GHz region. The direct path loss over a 50 km distance at 5 GHz is, from (8.16), about 140 dB. With transmit and receive antennae each of 50 dB gain, the net loss is about 40 dB. The actual path loss over a route will vary considerably from this figure. It will, for example, be modified by reflection from the ground, by the nature of the terrain and in particular by the clearance of obstacles, by variations in propagation as atmospheric conditions change. At microwave frequencies the ground is usually to be regarded as rough, so that the reflections may be

Table 8.2 *Typical microwave antennae used by the BPO*

Type of antenna	Polarisation	Year of introduction	Typical power gain* (dB)	Typical aperture efficiency (%)	Remarks
3 m diameter paraboloid	Monopolar	Pre-1960	39.4 (4 GHz)	55 (4 GHz)	Heavy cast reflector, poor side lobe performance
Large horn (9 m high)	Bipolar	1961	45 (6 GHz)	52 (6 GHz)	Wideband (4–11 GHz), good electrical performance, mechanical construction creates maintenance problems
Early Cassegrain (3.7 m diameter)	Bipolar	1963	43 (6 GHz)	35 (6 GHz)	Single frequency band, good side lobe performance, low efficiency
Focal plane paraboloid (3.7 m diameter)	Bipolar	1965	44.5 (6 GHz)	45 (6 GHz)	Single frequency band, light spun-aluminium reflector, inexpensive to produce
Latest Cassegrain	Bipolar	1975	45.8 (6 GHz)	68 (6 GHz)	Single frequency band, good electrical performance, high aperture efficiency obtained by slotted taper feed and optimisation of secondary reflector
Future	Bipolar	1979	Design objectives yet to be determined		Dual upper 6 GHz and 11 GHz frequency band, good electrical performance with high efficiency

* Measured relative to the gain of an isotropic antenna
From Martin-Royle, Dudley & Ferin, POEEJ **70**, 45–54.

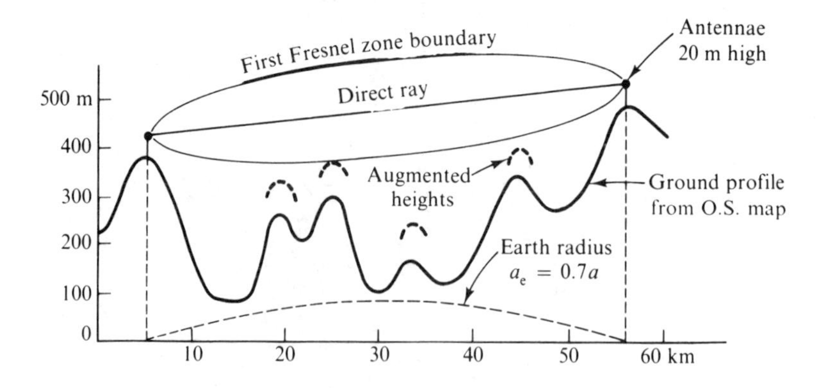

Fig. 8.32. Route planning for microwave link.

small. This would not be the case, for example, over a sea path. Each case will have to be considered on its merits.

The clearance required of obstacles has been discussed in §8.10 and the effect of tropospheric propagation in §8.9. These two effects have to be combined in choosing a route: the clearance between an obstacle and the ray path must be a minimum of 0.6 of the first Fresnel zone, and the ray path will in general be curved. The device of using an equivalent earth radius was discussed in §8.9.3, where for a standard atmosphere it was shown that this equivalent radius was the actual radius multiplied by $\frac{4}{3}$. For conditions when the refractive index increases with height this factor will be reduced to less than unity, and it is found in the UK that a factor of 0.7 covers most propagation conditions. The way in which this information is used in route planning is shown in fig. 8.32. The ground profile is plotted on a horizontal base line from an Ordnance Survey map. The deviation due to the equivalent earth radius is plotted and used to raise the height of each intervening obstacle. The ray path is a straight line, and the boundary of the first Fresnel zone is plotted from (8.43) or (8.44). The clearance can then be checked.

8.11.4 Fading

The propagation path will be changing, not merely from day to day, but continuously. Air currents moving regions of hot or cold air, for example, will cause a ray path to fluctuate. Such fading can only be assessed on an experimental basis for a particular path. Where a strong reflected ray arrives at the receiving antenna as well as a direct ray, the inter-

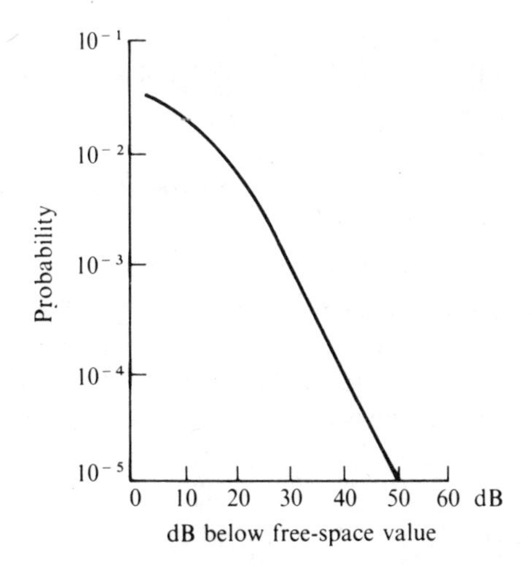

Fig. 8.33. Probability curve for estimating fading. The ordinate indicates the probability that, at any given instant, the signal level is below that given by the abscissa; e.g. the signal will be more than 30 dB below the free-space value for 0.1% of the time.

ference between them, which determines the resultant signal, is critically dependent on the path length of each ray. These are likely to alter, and give strong fading. Where this is the case it is possible to reduce the effect by using two antennae at different heights, when either one or the other will be satisfactory. The direction of a ray path is also likely to fluctuate, and this may have an effect when using a highly directional antenna, with a beamwidth of less than 0.5°.

The magnitude of fading can only be expressed on a probabilistic basis. The probability can be expressed in the form of percentage of time for which the signal will be less than the free-space value. A typical result is shown in fig. 8.33, but the fading will be worse on some paths than others and at some time of the year.

8.11.5 Noise

The effective noise temperature of the antennae in a microwave link is usually near ground temperature, since the directional antennae are operating near grazing incidence. Side lobes in addition help to bring this about. It is not therefore necessary to use very low noise receivers, as it is in satellite earth stations.

The existence of fading influences noise properties somewhat, since it is necessary to equip receivers with automatic gain control if the output is to be constant. This increases the noise at times of fade. This is of course the same thing as saying that the signal-to-noise ratio gets worse, but in this application the signal is kept constant. We can get some idea of the signal-to-noise requirements by considering a particular case.

A 50 km path has a free-space loss of 140 dB. The received power for a transmitted output of 10 W with antenna gains of 50 dB is then $(10-140+2\times 50)$ or 30 dB below one watt. If we allow worst-case fading as 40 dB we finally have 70 dB below one watt. The noise on a 250 MHz bandwidth at room temperature is kTB, or 10^{-12} W. The receiver noise factor will increase this, by about 6 dB, say, giving a noise power of 114 dB below one watt. We thus have in hand 44 dB. We have allowed nothing for intermodulation noise, so that a final signal-to-noise ratio of between 30 and 40 dB is likely. This is of the order of magnitude of the figure required for satisfactory performance on a public service link. We may conclude, for example, that any decrease in transmitter power would shorten the distance between relay stations.

8.12 The radar equation

Another example of a microwave system is provided by radar. A radar transmitter emits a pulse of radiation which falls on a target. This target will scatter some of the incident radiation back to the radar, which from the delay between transmission and return calculates the target distance. The parameters entering into a calculation must be much the same as for the microwave link: transmitter power P_t, antenna gain G (for both transmission and reception), distance d, and the effective scattering area of the target, σ. This latter is defined as the ratio of total power scattered by the target to the power density of the incident wave.

At the target the power density is given by

$$P = P_t G/4\pi d^2 \text{ W m}^{-2}$$

The scattered power is σP, and assuming this is scattered isotropically the power density on return to the radar is given by

$$\sigma P/4\pi d^2$$

Hence the received power P_r is given by

$$P_r = \frac{\lambda^2 G}{4\pi} \cdot \frac{\sigma}{4\pi d^2} \cdot \frac{P_t G}{4\pi d^2}$$

$$= \frac{\sigma}{4\pi} P_t \left(\frac{\lambda G}{4\pi d^2}\right)^2$$

In terms of antenna aperture we may write

$$P_r = \frac{\sigma}{4\pi} P_t \left(\frac{A}{\lambda d^2}\right)^2 \tag{8.45}$$

In a radar system the aim is usually to extend the distance from which the minimum detectable P_r is returned. If A is fixed (by dimension) it would appear profitable to decrease λ. However P_t is limited by available technology, and decreases as λ decreases. Similarly the least detectable power gets greater as λ decreases (it also depends on the pulse length of transmission).

As an example let us take

$$\lambda = 25 \text{ cm}$$
$$A = 15 \text{ m}^2$$
$$\sigma = 1 \text{ m}^2$$
$$P_t = 1 \text{ mW}$$
$$P_r = 5 \times 10^{-14} \text{ W}$$

We have from (8.45)

$$d^4 = \frac{\sigma P_t A^2}{4\pi P_r \lambda^2}$$

whence $d = 330$ km

It should be observed that this calculation is based on 100% certainty of detecting the signal at the power level given (a factor of about 5 greater than noise). In fact this would represent a probability of detection.

8.13 Worked example

A radio navigation system operates on a frequency of 200 kHz. A transmitting antenna may be considered to be a vertical conducting mast 40 m in height insulated from the earth, assumed to be plane and perfectly conducting. The antenna is fed between its base and earth. The current delivered to the base of the mast is 100 A peak.

Estimate the peak electric field strength at the surface of the earth, at a distance of 150 km from the antenna. The formula for the magnetic field from a Hertzian dipole may be used as the starting point of any derivation. Estimate also the radiation resistance of the antenna. Any assumptions made should be carefully stated. The field strength in a practical situation would be less than predicted by this model. To what causes might this be attributed?

Antennae such as those described are often equipped with insulated horizontal wires connected to the top of the mast. What might this be expected to achieve?

(The magnetic -far-field, H_ϕ, radiated by a short current element $I\,dl$ is given by

$$H_\phi = \frac{I dl}{4\pi}\frac{jk}{r} e^{-jkr} \sin\theta$$

where $k = 2\pi/\lambda$, $\boldsymbol{r}$ is the distance and θ is the angle between $d\boldsymbol{l}$ and $\boldsymbol{r}$.)

Solution. The wavelength at 200 kHz is 1.5×10^3 m. Hence the antenna is electrically short. It is necessary to make some assumption about the distribution of current up the mast, although it is not critical. We further have to allow for the conducting earth. This however is equivalent to an image making the mast into a dipole of twice its length in free space. Assuming the current to fall linearly from I_0 at the base to zero at the top, and taking l as the height of the mast we have for the distant field

$$H_\phi = \tfrac{1}{2}I_0 \cdot 2l \cdot \frac{jk}{4\pi r} e^{-jkr} \sin\theta$$

The shortness of the mast means that we do not have to worry about relative phase delays. At the earth's surface $\theta = \frac{1}{2}\pi$, and hence

$$|E| = \sqrt{\left(\frac{\mu_0}{\varepsilon_0}\right)}\,|H_\phi| = \sqrt{\left(\frac{\mu_0}{\varepsilon_0}\right)}\,\frac{I_0 lk}{4\pi r} = \frac{\mu_0 I_0 l f}{2r}$$

With the figures given, $|E| = 3.4$ mV m^{-1}. The radiated power can be determined using Poynting's theorem.

$$P = \tfrac{1}{2}R_a I_0^2 = \int_0^{\pi/2} \tfrac{1}{2}EH_\phi^* 2\pi r^2 \sin\theta\, d\theta$$

$$= \sqrt{\left(\frac{\mu_0}{\varepsilon_0}\right)}\left(\frac{I_0 lk}{4\pi}\right)^2 \int_0^{\pi/2} \pi \sin^3\theta\, d\theta$$

$$= \sqrt{\left(\frac{\mu_0}{\varepsilon_0}\right)}\left(\frac{I_0 l}{2\lambda}\right)^2 \pi \cdot \tfrac{2}{3}$$

Integration is only over the top of the earth's surface. Hence

$$R_a = \sqrt{\left(\frac{\mu_0}{\varepsilon_0}\right)}\left(\frac{l}{\lambda}\right)^2 \frac{\pi}{3} = 0.28\ \Omega$$

The actual figure will be different because the earth is not a perfect conductor, and attenuation will take place. The inhomogeneous nature of the earth's surface will also affect the value.

The 'capacity-hat' serves to make the current more uniform, since it need not then vanish at the top.

8.14 Summary

This chapter has been concerned with the propagation of radio waves through space as a means of communication. The parameters needed to describe transmitting and receiving antennae are given, and applied to a simple free-space link.

The complications introduced by the presence of the earth and its troposphere have been considered, and solutions derived by the use of ray theory. The effect of the earth is to introduce a reflected ray as well as the direct ray, with resultant interference between the two.

The effect of the refractive index of the troposphere is comparable with the effect of the earth's curvature, and in standard conditions can be allowed for by using an equivalent radius for an earth without atmosphere.

A rule has been established for determining the necessary clearance of obstacles in a microwave link.

Formulae

Antennae

$$G = 4\pi \times (\text{Power per unit solid angle/Total power})$$

$$A = \lambda^2 G/4\pi$$

Propagation 'loss' in free-space link

$$20 \log_{10}(4\pi r/\lambda)\ \text{dB}$$

Ray equation for horizontal stratification

$$x = \int_0^z \{[n(z)/n_0 \sin\theta_0]^2 - 1\}^{-\frac{1}{2}}\, \mathrm{d}z$$

If $n^2 = n_0^2(1 + \alpha z)$,

$$\alpha z = [\alpha x/2 \sin\theta_0 + \cos\theta_0]^2 - \cos^2\theta_0$$

Standard atmosphere

$$n = 1 - 4 \times 10^8 z$$

8.15 Problems

1. A microwave receiving antenna has a physical aperture of 1 m^2. At $\lambda =$ 3 cm its effective receiving area is 50% of the physical aperture when receiving in the optimum direction. Calculate the power gain in that direction.

Two such antennae are 1 km apart, adjusted for optimum transmission. What is the maximum received power when 1 kW is radiated from one of them? Assume free-space propagation.

2. Write brief notes on the propagation of radio waves over the range 10–50 MHz, in a terrestrial environment.

A transmitter operating at 45 MHz radiates 200 kW effectively omnidirectionally. The antenna is 350 m high. Estimate the signal strength at a receiving antenna 10 m above the earth, distance 30 km from the transmitter, neglecting the curvature of the earth, and stating any assumptions made.

(Bristol University, 1975.)

3. A certain system employs an antenna with a noise temperature of 29 K which is connected to a parametric amplifier through a transmission line with 1.2 dB loss. The parametric amplifier has a gain of 16 dB and a noise temperature of 80 K. The amplified output is taken to a travelling-wave amplifier with a gain of 20 dB and noise figure of 6 dB. The output of the travelling-wave amplifier is taken to a mixer whose noise figure is 13 dB. Calculate the effective noise temperature, (i) at the input terminals of the parametric amplifier, (ii) at the antenna terminals with the antenna disconnected. (The standard temperature is 290 K.)

(Cambridge University; part question.)

4. A transmitting antenna has a radiation resistance of 70 Ω and a power gain of 20 dB in the direction of a receiver 60 km distant. If the antenna is excited by a current of 1.2 A, r.m.s., determine the wave intensity and the electric field strength at the receiver. If the receiving antenna has an effective aperture of 10^{-3} m^2, determine the maximum power available to the receiver and the overall transmission loss in dB.

(London University, 1972; part question.)

5. Discuss the term 'effective aperture' of an antenna. What is the relationship between the gain of an antenna and its effective aperture?

A ground station is to be designed to receive domestic television signals from a satellite which radiates 100 kW effective power from a directional antenna at 12 GHz.

The receiver has the following characteristics:

(*a*) Bandwidth = 5 MHz.

(*b*) Input noise figure above 300 K = 10 dB.

(*c*) Input signal/noise ratio = 20 dB.

If the satellite is 40 000 km from the ground station, find the gain of the antenna for the ground station.

(Southampton University.)

6. Derive an expression for the effective attenuation between two isotropic antennae in space. The gain and aperture of an antenna are related by the expression

$$G_0 = 4\pi A/\lambda^2$$

where G_0 is the gain with respect to an isotropic radiator.

A passive satellite has an isotropic scattering aperture of 10 m^2, and an alternative active satellite has an antenna gain of 6dB for reception and transmission with a transmitter power of 10 W at 1 GHz. Assuming that the ground-station receiver requires 1 pW of signal, compare the maximum range for communication via the passive and active satellites. The ground station has an antenna gain of 30 dB and a transmitter power of 10 kW.

7. The parabolic dish of a 3 cm wavelength radar has a diameter of 1 m. It is supplied with 1 MW pulses and is scanned to detect a target, which scatters the radiation isotropically and which has an effective cross-sectional area of 4 m^2, at a range of 100 km. The same antenna is used to receive the scattered signal.

Show that the signal strength received is about 10^{-12} W.

What factors limit the range of radar?

(London University, 1974; part question.)

8. A radio link at a frequency of 4 GHz is set up over flat terrain between two antennae 20 km apart on towers 40 m high. The nature of the terrain is such that the ratio of the amplitudes of reflected to transmitted rays is 0.3.

Show that the strength of the received signal is about 1.7 dB above that of free-space transmission.

Show also that as the receiving antenna is raised and lowered variations of about 5 dB would be expected.

9

Optical transmission

9.1 Introduction

The guiding of electromagnetic waves by metallic structures and propagation in space has been studied in chapters 3 and 4. It has been seen in chapter 7 that techniques associated with optics are of use at microwave frequencies. The use of optical frequency carriers for the transmission of information has made the study of wave guidance at these frequencies of great importance, and it is the purpose of the present chapter to outline the application of Maxwell's equations to propagation and transmission of waves of optical frequency, both in dielectric waveguides and in free space.

The waveguides studied in chapter 4 were metallic structures, and it was shown that their behaviour was governed largely by the size of the structure in terms of the wavelength. Dielectric structures, such as rods, can also be shown to guide electromagnetic waves, but at microwave frequencies the use of such devices has been as antennas rather than as waveguides. At optical frequencies the dimensions are extremely small, and the losses in metals extremely high, so that conventional metal waveguides are of no interest. Dielectric waveguides, however, do not necessarily suffer the same high loss, and can be used at optical frequencies. The fields associated with propagation on a dielectric waveguide extend to the region outside the waveguide so that the permittivity ε cannot be taken as constant through space as has previously been assumed. It will be seen that this greatly complicates the analysis. Before considering the approximate methods which have been developed, a case which in principle can be solved by the techniques of chapter 4 will be considered: the step-index fibre.

9.2 The step-index fibre

Geometrically this is to be considered as a cylindrical dielectric rod immersed in a space of lower dielectric constant, as indicated in fig. 9.1. As mentioned above, the electromagnetic field will exist both in the rod and in the outer region, referred to as the cladding, and in general

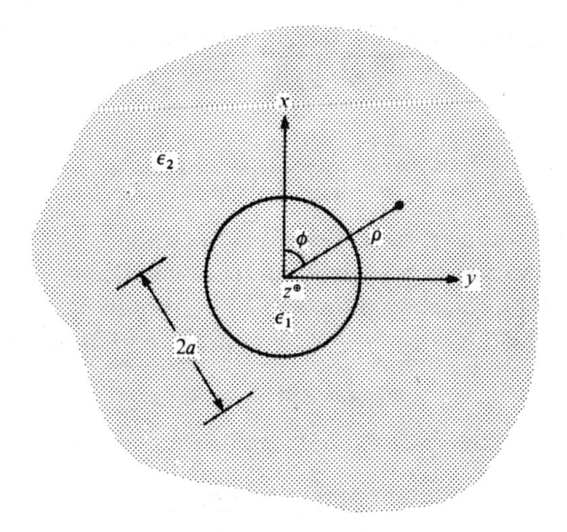

Fig. 9.1. Cross-section of step-index fibre.

radiation to infinity will take place. Guidance is associated with power transmission along the rod, and in this case the field in the cladding must decrease radially at a rate faster than ρ^{-1}. Since in either region the dielectric is homogeneous, all the theory of §3.3 applies in each, although different values of ε will appertain to the two regions. At the boundary, continuity of tangential components of $\boldsymbol{E}$ and $\boldsymbol{H}$ is necessary. It will be seen that, in general, neither E_z nor H_z can be zero, so that modes cannot be classified as *TE* and *TM*. As in §3.8 E_z and H_z satisfy the equation

$$\frac{1}{\rho}\frac{\partial}{\partial \rho}\left(\rho\frac{\partial \psi}{\partial \rho}\right)+\frac{1}{\rho^2}\frac{\partial^2 \psi}{\partial \phi^2}+\kappa^2\psi=0 \tag{9.1}$$

where

$$\kappa^2=\gamma^2+\omega^2\mu\varepsilon$$

One solution suitable for use inside the dielectric rod is given by (3.61),

$$\psi=J_m(\kappa_1\rho)\cos m\phi\exp(-\gamma z) \tag{9.2}$$

where

$$\kappa_1^2=\gamma^2+\omega^2\mu_0\varepsilon_1$$

The theory of Bessel functions shows that $J_m(u)$ is proportional to $u^{-1/2}$ as m becomes very large, so that (9.2) does not give a solution of the type we look for in the cladding. To obtain more rapid decrease in the radial direction the constant κ^2 in (9.1) must be negative, so that for $\rho > a$

$$\kappa^2 = \gamma^2 + \omega^2\mu_0\varepsilon_2 = -\kappa_2^2 \tag{9.3}$$

The required solution is then written

$$\psi = K_m(\kappa_2\rho)\cos m\phi \tag{9.4}$$

where $K_m(w)$ is the modified Bessel function which tends exponentially to zero as u becomes large. Graphical representation of $J_m(u)$ was given in fig. 3.12; a representation for $K_m(w)$ is given in fig. 9.2.

For a propagating wave $\gamma = j\beta$, so that, from (9.2) and (9.3),

$$\omega^2\mu_0\varepsilon_1 > \beta^2 > \omega^2\mu_0\varepsilon_2 \tag{9.5}$$

For optical media $\varepsilon = n^2\varepsilon_0$, so that this equation can be written

$$\frac{\omega n_1}{c} > \beta > \frac{\omega n_2}{c} \tag{9.6}$$

in which c is the velocity of light in free space.

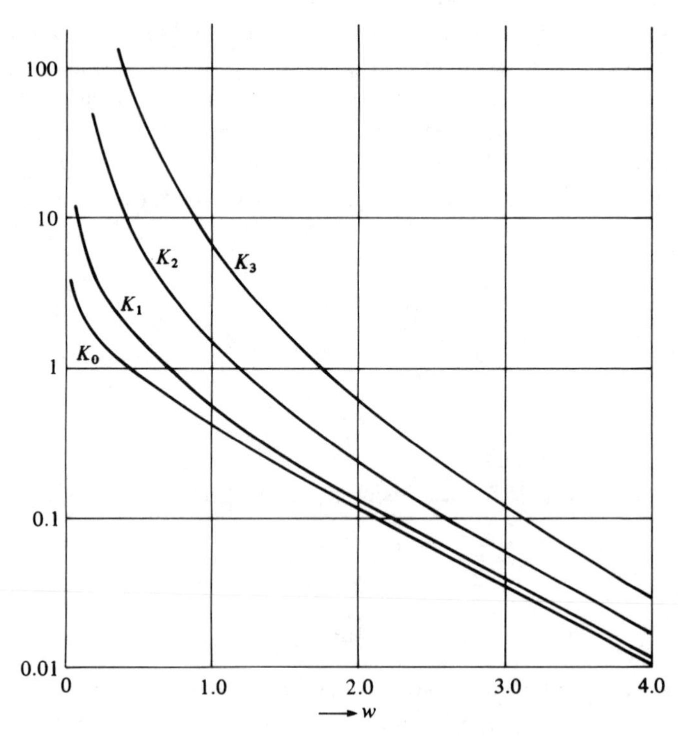

Fig. 9.2. The modified Bessel functions $K_m(w)$.

The characteristic equation
Suitable forms for H_z and E_z are given by

$$H_z = A\frac{J_m(\kappa_1\rho)}{J_m(\kappa_1 a)}\cos m\phi \exp(-\gamma z) \quad (0<\rho<a) \tag{9.7}$$

$$= A\frac{K_m(\kappa_2\rho)}{K_m(\kappa_2 a)}\cos m\phi \exp(-\gamma z) \quad (\rho>a) \tag{9.8}$$

$$E_z = B\frac{J_m(\kappa_1\rho)}{J_m(\kappa_1 a)}\sin m\phi \exp(-\gamma z) \quad (0<\rho<a) \tag{9.9}$$

$$= B\frac{K_m(\kappa_2\rho)}{K_m(\kappa_2 a)}\sin m\phi \exp(-\gamma z) \quad (\rho>a) \tag{9.10}$$

These are chosen so that, as well as satisfying (9.1), both E_z and H_z are continuous at $\rho=a$. An arbitrary sum of cos $m\phi$ and sin $m\phi$ might have been taken, but a choice of origin for ϕ can make H_z proportional to cos $m\phi$, as given. It will be seen later that other continuity conditions then require the form sin $m\phi$ for E_z. From (3.48), (3.49) and the equivalents in § 3.3.2, using the cylindrical polar expression in §10.7 it is found that the azimuthal components of $\boldsymbol{E}$ and $\boldsymbol{H}$ are, for $0<\rho<a$, given by

$$\begin{aligned} E_\phi &= \frac{j\omega\mu}{\kappa_1^2}\frac{\partial H_z}{\partial\rho} - \frac{j\beta}{\kappa_1^2}\frac{1}{\rho}\frac{\partial E_z}{\partial\phi} \\ H_\phi &= -\frac{j\beta}{\kappa_1^2}\frac{1}{\rho}\frac{\partial H_z}{\partial\phi} - \frac{j\omega\varepsilon_1}{\kappa_1^2}\frac{\partial E_z}{\partial\rho} \end{aligned} \tag{9.11}$$

in which E_z, H_z are given by equations (9.9) and (9.7). For $\rho>a$, we use (9.8) and (9.10) with $\kappa^2=-\kappa_2^2$:

$$\begin{aligned} E_\phi &= -\frac{j\omega\mu}{\kappa_2^2}\frac{\partial H_z}{\partial\rho} + \frac{j\beta}{\kappa_2^2}\frac{1}{\rho}\frac{\partial E_z}{\partial\phi} \\ H_\phi &= \frac{j\beta}{\kappa_2^2}\frac{1}{\rho}\frac{\partial H_z}{\partial\phi} + \frac{j\omega\varepsilon_2}{\kappa_2^2}\frac{\partial E_z}{\partial\rho} \end{aligned} \tag{9.12}$$

Both E_ϕ and H_ϕ must be continuous at $\rho=a$. Because of the choice of sin $m\phi$ in the expressions for E_z each term in equations (9.11) and (9.12) has the same azimuthal behaviour. The formal equation (the *characteristic equation*) for determining β can now be set up. It is useful to adopt the notation

$$\left[\frac{\mathrm{d}}{\mathrm{d}\rho}\frac{J_m(\kappa_1\rho)}{J_m(\kappa_1 a)}\right]_{\rho=a}=\kappa_1\frac{J'_m(\kappa_1 a)}{J_m(\kappa_1 a)}=\kappa_1 f_m(\kappa_1 a) \tag{9.13}$$

$$\left[\frac{\mathrm{d}}{\mathrm{d}\rho}\frac{K_m(\kappa_2\rho)}{K_m(\kappa_2 a)}\right]_{\rho=a}=\kappa_2\frac{K'_m(\kappa_2 a)}{K_m(\kappa_2 a)}=\kappa_2 g_m(\kappa_2 a) \tag{9.14}$$

The equation of continuity for E_ϕ can then be written

$$A\frac{\omega\mu_0}{u}f_m(u)-B\frac{m\beta}{u^2}=-A\frac{\omega\mu_0}{w}g_m(w)+B\frac{m\beta}{w^2} \tag{9.15}$$

where $u=\kappa_1 a$, $w=\kappa_2 a$. With the same notation the equation for H_ϕ is found to be

$$A\frac{m\beta}{u^2}-B\frac{\omega\varepsilon_2}{u}f_m(u)=-A\frac{m\beta}{w^2}+B\frac{\omega\varepsilon_2}{w}g_m(w) \tag{9.16}$$

Each of these equations gives a value for A/B, and only for certain values of β will these be identical. Thus the equation for β becomes

$$\omega^2\mu_0\left(\frac{f_m}{u}+\frac{g_m}{w}\right)\left(\varepsilon_1\frac{f_m}{u}+\varepsilon_2\frac{g_m}{w}\right)=(m\beta)^2\left(\frac{1}{u^2}+\frac{1}{w^2}\right)^2 \tag{9.17}$$

In general this equation requires numerical techniques to solve for β, and roots correspond to modes for which neither E_z nor H_z is zero. However, *TE* and *TM* modes occur in the special case of axial symmetry, $m=0$. Thus in this case equation (9.16) is satisfied by $B=0$, leaving equation (9.15) as

$$\frac{1}{u}f_m(u)=-\frac{1}{w}g_m(w) \tag{9.18}$$

By definition of u, w,

$$u^2+w^2=\omega^2\mu_0(\varepsilon_1-\varepsilon_2)a^2=v^2 \tag{9.19}$$

so that for a given ω, u and w can be determined. These roots correspond to *TE* modes. Similarly *TM* modes are given by $A=0$, satisfying equation (9.15) together with

$$\frac{\varepsilon_1}{u}f_m(u)=-\frac{\varepsilon_2}{w}g_m(w) \tag{9.20}$$

The recurrence relations for Bessel functions given in §10.6 enable equations (9.18) and (9.20) to be written in the form

$$\frac{1}{u}\frac{J_1(u)}{J_0(u)} = -\frac{1}{w}\frac{K_1(w)}{K_0(w)} \tag{9.21}$$

$$\frac{\varepsilon_1}{u}\frac{J_1(u)}{J_0(u)} = -\frac{\varepsilon_2}{w}\frac{K_1(w)}{K_0(w)} \tag{9.22}$$

These equations will be considered further in §9.2.1 below.

Further progress towards classifying modes may be made by considering the *weak-guidance* case, when ε_1 and ε_2 differ by only a small amount, typically less than 1%. This is in fact the situation of practical interest in optical fibres.

9.2.1 The weak-guidance approximation

Although we wish to consider the case $\varepsilon_2 \to \varepsilon_1$, it is clear that to obtain non-trivial results we must keep u, w as different variables: this implies from equation (9.19) that

$$v = a\omega[\mu_0(\varepsilon_1 - \varepsilon_2)]^{1/2}$$

is considered finite even though $\varepsilon_1 - \varepsilon_2$ is small. The reformulation of equation (9.17) is considered in more detail in appendix 6, §10.6, but the simplest approximation corresponds to putting $\varepsilon_1 = \varepsilon_2$, $\beta^2 = \omega^2\mu_0\varepsilon_1$ in that equation.

It is shown in §10.6 that two sets of roots derive from this approximation, summarised in tables 9.1 and 9.2.

Cut-off frequencies

The characteristic equations in tables 9.1 and 9.2 have to be solved to specification of v in equation (9.19). As with metal waveguides, real solutions for u and w will only exist for certain ranges of v. The characteristic of a mode on a dielectric waveguide is that the field is confined to the core and neighbouring cladding: cut-off may be defined as the frequency at which the field of a particular mode extends throughout space; mathematically this corresponds to $w \to 0$. Thus, as detailed in §10.6, the cut-off for *EH* modes of azimuthal number m is found to be given by $J_m(v_c) = 0$; for *HE* modes $m \geqq 2$, the equation becomes $J_{m-2}(v_c) = 0$ excluding $v_c = 0$. For $m = 1$ one mode is possible for which $v_c = 0$, HE_{11}. Apart from this, the values in table 3.3 will show that the lowest value of v_c is the first zero of $J_0(v_c)$, 2.405. Thus HE_{11} is a true dominant mode in the range $0 < v < 2.405$.

Table 9.1 *EH modes*

$\frac{1}{u}\frac{J_{m+1}(u)}{J_m(u)} = -\frac{1}{w}\frac{K_{m+1}(w)}{K_m(w)}$; $A = B(\varepsilon_1/\mu_0)^{1/2}$		
E_ϕ	$0<\rho<a$	$-Aj\omega\mu_0 a\frac{J_{m+1}(u\rho/a)}{uJ_m(u)}\cos m\phi$
	$\rho>a$	$Aj\omega\mu_0 a\frac{K_{m+1}(w\rho/a)}{wK_m(w)}\cos m\phi$
E_ρ	$0<\rho<a$	$Aj\omega\mu_0 a\frac{J_{m+1}(u\rho/a)}{uJ_m(u)}\sin m\phi$
	$\rho>a$	$-A\omega\mu_0 a\frac{K_{m+1}(w\rho/a)}{wK_m(w)}\sin m\phi$

Table 9.2. *HE modes*

$\frac{1}{u}\frac{J_{m-1}(u)}{J_m(u)} = \frac{1}{w}\frac{K_{m-1}(w)}{K_m(w)}$; $A = -B(\varepsilon_1/\mu_0)^{1/2}$		
E_ϕ	$0<\rho<a$	$Aj\omega\mu_0 a\frac{J_{m-1}(u\rho/a)}{uJ_m(u)}\cos m\phi$
	$\rho>a$	$Aj\omega\theta_0 a\frac{K_{m-1}(w\rho/a)}{wK_m(w)}\cos m\phi$
E_ρ	$0<\rho<a$	$Aj\omega\mu_0 a\frac{J_{m-1}(u\rho/a)}{uJ_m(u)}\sin m\phi$
	$\rho>a$	$Aj\omega\mu_0 a\frac{K_{m-1}(w\rho/a)}{wK_m(w)}\sin m\phi$

Example
Two particular step-index fibres have refractive indices for core and cladding of 1.465, 1.460 respectively. One has a core diameter of 10 μm, the other 50 μm. For each fibre calculate the optical free space wavelength corresponding to cut-off of the four lowest order modes.

Solution. From equation (9.19) we can write

$$v = \left(\frac{2\pi a}{\lambda}\right)(n_1^2 - n_2^2)^{1/2}$$

thus

$$\lambda_c = \left(\frac{2\pi a}{v_c}\right)(n_1^2 - n_2^2)^{1/2}$$

The numerical aperture $NA = (n_1^2 - n_2^2)^{1/2} = 0.121$ and in this problem $2a = 50$ μm or 10 μm. (The larger diameter is typical of that used for multimode fibres.)

The HE_{11} (LP_{01}) mode does not cut off. Behaviour of the higher modes is summarised in the following table, with the values of v_c obtained from table 3.3

	Cut-off wavelength λ_c $2a = 50$ μm	$2a = 10$ μm
TE_{01}, TM_{01} and HE_{21} (approx.) (LP_{11}) cut-off at $v_c = 2.405$	7.90 μm	1.58 μm
HE_{12}, HE_{31} and HE_{11} have $v_c = 3.832$	4.96 μm	0.99 μm
HE_{41} and EH_{21} have $v_c = 5.136$	3.70 μm	0.74 μm
TE_{02}, TM_{02} and HE_{22} (approx.) have $v_c = 5.52$	3.44 μm	0.69 μm

The dominant mode; HE_{11}

Putting $m = 1$ in the expression in table 9.2 will show that the radial and azimuthal components of the transverse electric field correspond to a field directed in the y-direction and in strength proportional to $J_0(u\rho/a)/[uJ_1(u)]$. In fig. 9.3 are shown curves of u, w for this mode as functions of v in the dominant range. In fig. 9.4 is shown the variation of field strength with radius for $v = 1$, 1.5, 2.0. It will be seen that as cut-off is approached the field extends further into the cladding. This puts a limit on the lowest frequency that can be used in a practical fibre of finite cladding radius.

It is of interest to evaluate a practical case: typical values for the refractive indices could be $n_1 = 1.46$, $(n_1 - n_2)/n_1 = 0.002$. We may write from equation (9.19)

$$v = \frac{2\pi a}{\lambda}(n_1^2 - n_2^2)^{1/2} \cong \frac{2\sqrt{(2\pi)}an_1}{\lambda}\left(\frac{n_1 - n_2}{n_1}\right)^{1/2}$$

Thus for $v = 2.405$ we find $(a/\lambda) = 4.15$. For a free-space wavelength of 0.85 μm, the fibre core must be less than 3.5 μm in order to propagate only the dominant mode.

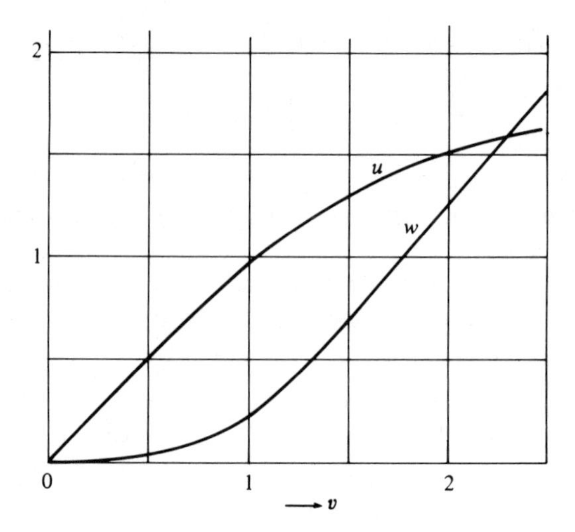

Fig. 9.3. Showing the parameters u, w for the HE_{11} mode as functions of v.

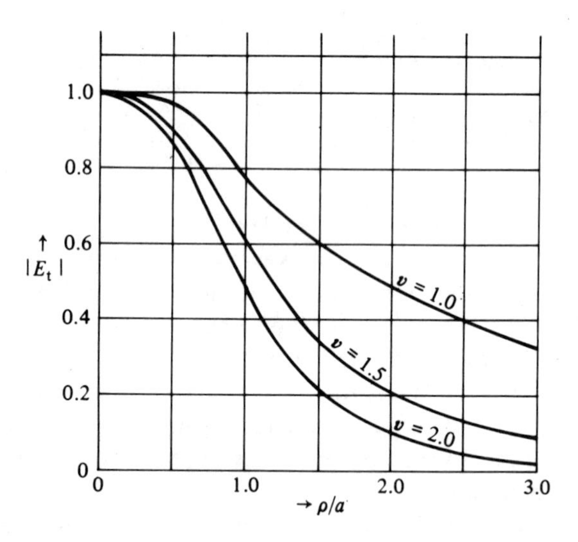

Fig. 9.4. Showing the variation of transverse electric field for HE_{11} as a function of radius for three values of the parameter v.

Degeneracy

The application of the recurrence relation for Bessel functions given in §10.6 will show that the characteristic equations for HE_{m+2} and EH_m modes are the same, and the modes are therefore degenerate. The theory given above concerns the approximation $\varepsilon_1 \to \varepsilon_2$, and the first order correction to this introduces multipliers n_1, n_2 in the numerators of these equations. The equations then differ, and the modes will have slightly different propagation constants.

Propagation constant

Although in deriving the simplified equations the propagation constant β was replaced by $\omega^2\mu_0\varepsilon_1$, the approximate results may be used to improve this estimate.

A quantity b is defined by

$$b=\frac{w^2}{v^2}=\frac{\beta^2-\omega^2\mu_0\varepsilon_2}{\omega^2\mu_0(\varepsilon_1-\varepsilon_2)} \tag{9.23}$$

from which

$$\beta^2=\omega^2\mu_0\varepsilon_1\left[1-\frac{\varepsilon_1-\varepsilon_2}{\varepsilon_1}(1-b)\right]$$

Thus in the range of frequency for which the mode propagates

$$0<b<1$$

The form of $b(v)$ for the HE_{11} mode can be derived from fig. 9.3.

9.3 The wave equation in a radially inhomogeneous medium

Because of the structure of the step-index fibre the rigorous analysis of the previous section is possible. Not only is it complicated but no manufactured fibre of micron dimensions is likely to have an abrupt change of refractive index. It is therefore necessary to investigate fibres with a continuously changing refractive index. The previous derivation of the wave equation in chapter 3 assumed homogeneous dielectric so that a new approach is required.

Maxwell's equations are given in terms of Cartesian components in equations (3.1)–(3.6). We make explicit use of the relation

$$\operatorname{div}\boldsymbol{D}=\operatorname{div}(\varepsilon\boldsymbol{E})=0$$

Expanding this equation gives

$$\boldsymbol{E}\cdot\nabla\varepsilon+\varepsilon\operatorname{div}\boldsymbol{E}=0$$

In the situation we wish to investigate the permittivity varies across the fibre, but the profile is independent of length. Thus we have

$$\varepsilon(x,y)\operatorname{div}\boldsymbol{E}+\mathrm{E}_x\frac{\partial\varepsilon}{\partial x}+\mathrm{E}_y\frac{\partial\varepsilon}{\partial y}=0.$$

or

$$\text{div}\,\boldsymbol{E} = -\boldsymbol{E}_t \cdot \nabla_t\,(\ln \varepsilon) \tag{9.24}$$

in which the suffix 't' denotes transverse components, as in §3.3. Straightforward substitution of equations (3.2), (3.3) into equation (3.4) yields the equation

$$\left(\frac{\partial^2}{\partial x^2}+\frac{\partial^2}{\partial y^2}+\frac{\partial^2}{\partial z^2}\right)\mathrm{E}_x + \omega^2\mu\varepsilon\,\mathrm{E}_x = \frac{\partial}{\partial x}\,(\text{div}\,\boldsymbol{E})$$

Finally, therefore, allowing for a propagation term $\exp(-\mathrm{j}\beta z)$, we find

$$\nabla_t^2\mathrm{E}_x + (\omega^2\mu\varepsilon - \beta^2)\mathrm{E}_x = -\frac{\partial}{\partial x}[\boldsymbol{E}_t \cdot \nabla_t(\ln \varepsilon)] \tag{9.25}$$

A precisely similar analysis applies to the component E_y to give

$$\nabla_t^2\mathrm{E}_y + [\omega^2\mu\varepsilon(x, y) - \beta^2]\mathrm{E}_y = -\frac{\partial}{\partial y}[\boldsymbol{E}_t \cdot \nabla_t(\ln \varepsilon)] \tag{9.26}$$

Equations (9.25) and (9.26) form a simultaneous pair for the solution of E_x, E_y. Provided that it is remembered that $\boldsymbol{E}_t$ represents the transverse electric field in Cartesian components, these equations may be combined in the form

$$\nabla_t^2\boldsymbol{E}_t + (\omega^2\mu\varepsilon - \beta^2)\boldsymbol{E}_t = -\nabla_t[\boldsymbol{E}_t \cdot \nabla_t(\ln \varepsilon)] \tag{9.27}$$

The operation ∇_t^2 takes place on scalar quantities and we may use any equivalent expression in other co-ordinate systems. If equation (9.27) can be solved for E_x, E_y then E_z can be determined from equation (9.24), followed by the other field components, using equations (3.1)–(3.3).

It is clear that the magnitude of the right-hand side of equation (9.27) depends on the spatial derivative of the permittivity: in most cases of interest in optical fibre work, the change of ε in a distance of one-wavelength is so small that the term may to a good approximation be taken as zero. With this approximation both E_x and E_y independently satisfy the scalar wave equation, albeit with non-uniform permittivity. In the absence of discontinuities both E_x and E_y must be continuous, everywhere finite and, further, must die away with distance from the core for bound modes. Thus we are interested in solutions of

$$\nabla_t^2\psi + (\omega^2\mu\varepsilon - \beta^2)\psi = 0 \tag{9.28}$$

which are continuous and tend to zero at infinity. This equation and the conditions for its solution are of the same form as Schrödinger's equation in quantum mechanics, from which it is known that the above conditions suffice to determine possible values of β and the corresponding solution. As an example of this process we will consider the case of a parabolic variation of refractive index.

9.3.1 Parabolic profile

In this case we write

$$\varepsilon = \varepsilon_0 n_0^2 \left[1 - 2\Delta\left(\frac{\rho}{a}\right)^2\right] \tag{9.29}$$

in which Δ is a dimensionless constant, usually small. This profile is not physically realisable since ε is negative for large enough values of ρ. However we shall show that solutions exist which are closely confined about the axis $\rho = 0$ and which have very small fields in regions where ε is non-physical.

In cylindrical polar co-ordinates equation (9.28) becomes

$$\frac{\partial^2 \psi}{\partial \rho^2} + \frac{1}{\rho}\frac{\partial \psi}{\partial \rho} + \frac{1}{\rho^2}\frac{\partial^2 \psi}{\partial \phi^2} + \left[k^2 n_0^2 - \beta^2 - 2k^2 n_0^2 \Delta\left(\frac{\rho}{a}\right)^2\right]\psi = 0 \tag{9.30}$$

in which $k^2 = \omega^2 \mu_0 \varepsilon_0 = (2\pi/\lambda)^2$. We restrict consideration to purely symmetrical solutions, for which $\partial\psi/\partial\phi \equiv 0$. It may be shown by direct substitution that the expression

$$\psi = \exp(-\rho^2/w^2) \tag{9.31}$$

is a solution provided that

$$w = \left[\sqrt{\left(\frac{2}{\Delta}\right)}\frac{a}{kn_0}\right]^{1/2} = \left[\frac{\lambda a}{\pi n_0}\sqrt{\left(\frac{1}{2\Delta}\right)}\right]^{1/2}$$

$$\beta^2 = k^2 n_0^2 (1 - \lambda\sqrt{(2\Delta)}/\pi n_1 a)$$

Typical values might be $a = 25\ \mu\text{m}$, $n_0 = 1.5$, $\Delta = 0.01$, $\lambda = 1\ \mu\text{m}$. In fig. 9.5 is shown both the profile of refractive index and the distribution of ψ. The constant w has the value 6.1 μm, so that at $\rho = a$, ψ is about 2×10^{-8} of its magnitude at $\rho = 0$. This situation will therefore be closely maintained if the profile corresponding to equation (9.29) is replaced by a constant for $\rho > a$, as indicated by the pecked line in fig. 9.5. This is an entirely realisable profile. This result indicates that guidance is achieved with only very small changes in refractive index.

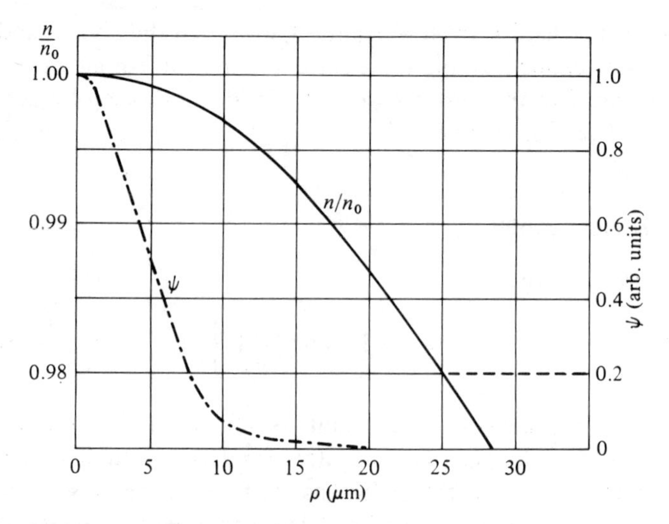

Fig. 9.5. Showing transverse field and refractive index for a parabolic profile: $a = 25\ \mu\text{m}$, $n_0 = 1.5$, $\Delta = 0.01$, $\lambda = 1\ \mu\text{m}$.

9.3.2 Approximate treatment of the step-index fibre

In the approximation represented by equation (9.28) a possible transverse field configuration is

$$0 < \rho < a\mathrm{E}_y = A\frac{J_l(\kappa_1\rho)}{J_l(\kappa_1 a)}\cos l\phi \tag{9.32}$$

$$\mathrm{E}_x = 0$$

$$\rho > a\mathrm{E}_y = A\frac{K_l(\kappa_2\rho)}{K_l(\kappa_2)}\cos l\phi \tag{9.33}$$

$$\mathrm{E}_x = 0$$

in which symbols have the same significance as in §9.2.

These forms satisfy continuity of E_y at $\rho = a$. As discussed earlier, for bound modes

$$\omega^2\mu_0\varepsilon_1 > \beta^2 > \omega^2\mu_0\varepsilon_2$$

so that

$$\kappa_1^2, \kappa_2^2 < \omega^2\mu_0(\varepsilon_1 - \varepsilon_2)$$

The ratios κ_1/β, κ_2/β, therefore satisfy the inequality

$$\frac{\kappa_1^2}{\beta^2}, \frac{\kappa_2^2}{\beta^2} < \frac{\varepsilon_1 - \varepsilon_2}{\varepsilon_1} = 2\Delta \cong 2\frac{n_1 - n_2}{n_1}, \quad \Delta \ll 1 \tag{9.34}$$

Since in each region separately ε is constant, we have, from equation (9.24),

$$\operatorname{div} \boldsymbol{E} = 0$$

whence in this case

$$\mathrm{j}\beta \mathrm{E}_z = \frac{\partial \mathrm{E}_y}{\partial y}$$

Further, from equation (3.3),

$$\mathrm{j}\omega\mu_0 \mathrm{H}_z = -\frac{\partial E_y}{\partial x}$$

Substitution of equations (9.32) and (9.33) with use of the recurrence relations given in §10.6 yields the results

$$0 < \rho < a,\ \mathrm{E}_z = \frac{A\kappa_1}{2\mathrm{j}\beta J_l(\kappa_1 a)} \times [-J_{l-1}(\kappa_1\rho)\sin(l-1)\phi - J_{l+1}(\kappa_1\rho)\sin(l+1)\phi] \tag{9.35}$$

$$\mathrm{H}_z = \frac{A\kappa_1}{2\mathrm{j}\omega\mu_0 J_l(\kappa_1 a)} \times [-J_{l-1}(\kappa_1\rho)\cos(l-1)\phi + J_{l+1}(\kappa_1\rho)\cos(l+1)\phi] \tag{9.36}$$

$$\rho > a,\ \mathrm{E}_z = \frac{A\kappa_2}{2\mathrm{j}\beta K_l(\kappa_2 a)} \times [K_{l-1}(\kappa_2\rho)\sin(l-1)\phi - K_{l+1}(\kappa_2\rho)\sin(l+1)\phi] \tag{9.37}$$

$$\mathrm{H}_z = \frac{A\kappa_2}{2\mathrm{j}\omega\mu_0 K_l(\kappa_2 a)} \times [K_{l-1}(\kappa_2\rho)\cos(l-1)\phi + K_{l+1}(\kappa_2\rho)\cos(l+1)\phi] \tag{9.38}$$

We also note

$$\mathrm{H}_x = -\frac{\beta}{\omega\mu_0}\mathrm{E}_y - \frac{1}{\mathrm{j}\omega\mu_0}\frac{\partial \mathrm{E}_z}{\partial y} \tag{9.39}$$

$$\mathrm{H}_y = \frac{1}{\mathrm{j}\omega\mu_0}\frac{\partial \mathrm{E}_z}{\partial x} \tag{9.40}$$

It has been shown that $\kappa_1/\beta \ll 1$, so that $E_z \ll E_y$. Carrying out the differentiations in equations (9.39) and (9.40) will provide another factor of κ_1 or κ_2 in the numerator, so that the first term in equation (9.39) is much greater than the second and also than H_y. As might be expected the assumption of only a very small difference in permittivities results in a wave which is locally plane.

We have to impose continuity of E_z, H_z, E_ϕ, H_ϕ at $\rho = a$. By choice of equations (9.32) and (9.33) E_ϕ is continuous. Continuity of E_y ensures continuity of H_ϕ if we neglect the small order terms. For E_z, H_z we require equality in the two separate terms in $\cos(l \pm 1)\phi$. This requires

$$\kappa_1 \frac{J_{l-1}(\kappa_1 a)}{J_l(\kappa_1 a)} = -\kappa_2 \frac{K_{l-1}(\kappa_2 a)}{K_l(\kappa_2 a)} \tag{9.41}$$

$$\kappa_1 \frac{J_{l+1}(\kappa_1 a)}{J_l(\kappa_1 a)} = \kappa_2 \frac{K_{l+1}(\kappa_2 a)}{K_l(\kappa_2 a)} \tag{9.42}$$

The same equations assure continuity of both E_z and H_z. Further, the recurrence relations of §10.6 show that either of these equations can be reduced to the other.

Comparison of equations (9.35)–(9.38) with tables 9.1 and 9.2 show that the assumed field corresponds to the sum of EH_{l-1} modes with HE_{l+1} modes. This is confirmed by comparison of equations (9.41) and (9.42) with the characteristic equations in those tables.

Equations (9.32) and (9.33) are said to define a *linear polarised mode*, designated LP_{ln}. It is apparent that LP mode is the sum of an EH_{l-1} and an HE_{l+1} mode, and a more accurate analysis will give slightly different propagation constants instead of the identical ones in this simplest approximation. The LP modes are not therefore true modes in which the transverse field propagates unchanged. However, for many purposes they may be so regarded. The case $l = 0$ corresponds to HE modes, and LP_{01} refers to HE_{11}. Variants on equations (9.32) and (9.33) are

$$E_y = A \frac{J_l(\kappa_1 \rho)}{J_l(\kappa_1 a)} \sin l\phi, \quad E_x = 0;$$

$$E_x = A \frac{J_l(\kappa_1 \rho)}{J_l(\kappa_1 a)} \cos l\phi, \quad E_y = 0;$$

$$E_x = A \frac{J_l(\kappa_1 \rho)}{J_l(\kappa_1 a)} \sin l\phi, \quad E_y = 0.$$

Analysis shows that these too can be regarded as similar sums of EH_{l-1} and EH_{l+1} modes. (NB. in an *EH* mode H_z leads E_z in the angular variable ϕ, in *HE* modes H_z lags on E_z.) Figure 9.6 indicates the possibilities for LP_{11} modes.

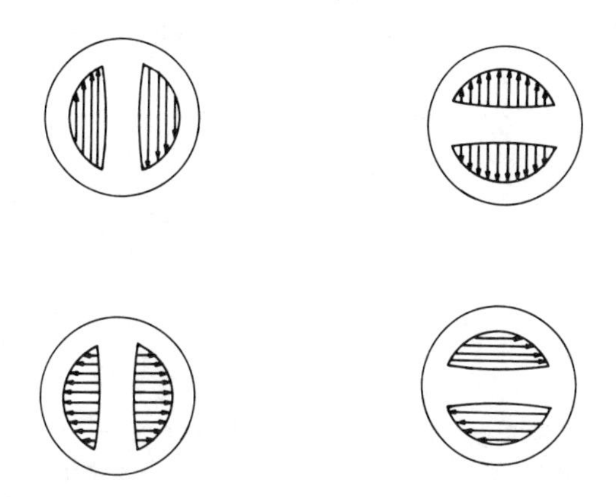

Fig. 9.6. Schematic presentation of the four degenerate LP_{11} modes.

9.3.3 Other approximation methods

As mentioned in §9.3, equation (9.28) is of a type well known in quantum mechanics, and various methods of finding approximate solutions have been devised which have their place in the study of optical fibres. The variational method for determining β, and the W.K.B. method may be instanced. These are mathematically more complicated and are covered in references cited at the end of the chapter.

9.4 Multi-mode fibres: ray optics

The previous sections have considered an optical fibre as a dielectric waveguide. This is necessary when considering mono-mode fibres, which are of interest because the field pattern is maintained over very long distances. The penalty for this is the need to make a core on the micron scale. Thick fibres, with a core of the order of 50 μm diameter and 150 μm overall, are much easier to make but will support many bound modes. Analysis can be carried out on the lines of earlier sections, but it is simpler to use ray optics, as was done in §8.9.

9.4.1 Rays in a graded fibre

The passage of rays through plane and spherically stratified media was analysed in §8.9. In such terms a fibre may be described as cylindrically stratified. A ray which starts from the axis will subsequently lie always in a plane containing the axis, and apart from co-ordinate notation, will be as portrayed in fig. 8.17. the analysis leading up to equation (8.30) gives, in the notation of fig. 9.7,

$$\left(\frac{\mathrm{d}\rho}{\mathrm{d}z}\right)^2=\left\{\frac{n(\rho)}{n_0\cos\gamma_0}\right\}^2-1 \tag{9.43}$$

and

$$z=\int_0^\rho\{[n(\rho)/n_0\cos\gamma_0]^2-1\}^{-1/2}\mathrm{d}\rho \tag{9.44}$$

A ray will be contained within the fibre if at some value of ρ within the core $\mathrm{d}\rho/\mathrm{d}z$ vanishes, indicating a maximum. This will occur when

$$n(\rho)=n_0\cos\gamma_0$$

Alternatively, if the refractive index of the core decreases monotonically to the cladding index n_c, rays for which $\gamma_0<\gamma_m$ will be bound within the core, where

$$\cos\gamma_m=n_c/n_0 \tag{9.45}$$

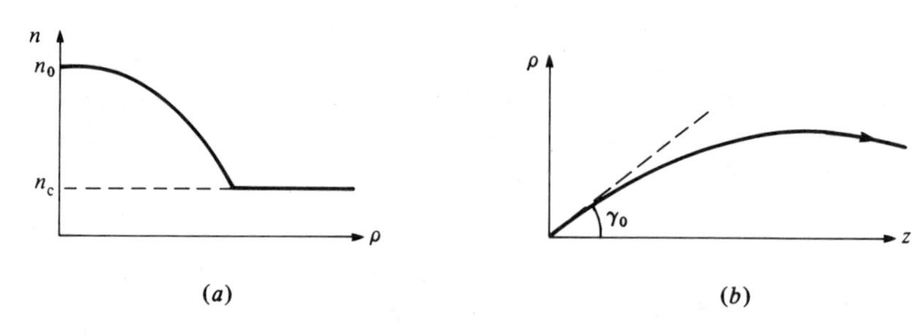

Fig. 9.7. Graded index fibre: (*a*) refractive index; (*b*) meridional ray.

Skew rays

In addition to the meridional rays just discussed, it is possible to have rays which do not pass through the axis. These are skew rays, and may be analysed by an extension of the above analysis. It has been possible for meridional rays to restrict attention to a single plane of the co-ordinate system. For skew rays this is not so. We will first derive a concise expression of Snell's law. This law contains two statements: the incident and refracted rays and the normal to the interface are coplanar, and the angles of incidence and refraction are related by $n_1 \sin \theta_1 = n_2 \sin \theta_2$. In fig. 9.8($a$) the incident ray is described by the unit vector $\boldsymbol{t}_1$, the refracted ray by $\boldsymbol{t}_2$ and the normal to the interface by $\boldsymbol{\nu}$. It can be seen that both statements are included in the vector equation

$$n_1 \boldsymbol{\nu} \times \boldsymbol{t}_1 = n_2 \boldsymbol{\nu} \times \boldsymbol{t}_2 \tag{9.46}$$

The situation with a cylindrically stratified medium is shown in fig. 9.8(b).

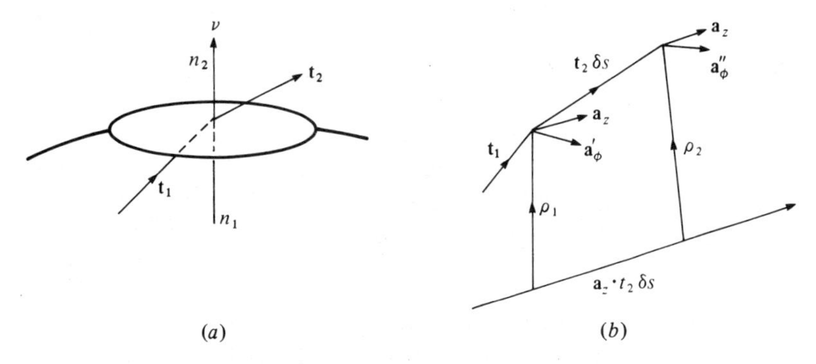

Fig. 9.8. The geometry of Snell's law: (a) showing a ray at the interface between two media; (b) illustrating the progress of a skew ray in a graded-index fibre.

Let the vector $\boldsymbol{t}_1$ make angles α_1, β_1, γ_1 with the axis vectors $\boldsymbol{a}'_\rho$, $\boldsymbol{a}'_\phi$, $\boldsymbol{a}_z$ at P', and the vector $\boldsymbol{t}_2$ angles α'_2, β'_2, γ'_2 with the same local axes. With respect to the local axes at $P''\boldsymbol{t}_2$ makes angles α_2, β_2, γ_2. Equation (9.46) then gives

$$n_1 \cos \gamma_1 = n_2 \cos \gamma'_2$$

$$n_1 \cos \beta_1 = n_2 \cos \beta'_2$$

By use of the equations

$$\cos \beta'_2 = \boldsymbol{t}_2 \cdot \boldsymbol{a}_z \times \boldsymbol{a}'_\rho$$

$$\cos \beta_2 = \boldsymbol{t}_2 \cdot \boldsymbol{a}_z \times \boldsymbol{a}''_\rho$$

it can be shown for the geometry of fig. 9.8(b) that

$$\rho_1 \cos \beta'_2 = \rho_2 \cos \beta_2$$

Since $\boldsymbol{a}_z$ is constant in space $\gamma'_2 = \gamma_2$, so that we have the two relations

$$\left.\begin{aligned} n_1 \cos \gamma_1 &= n_2 \cos \gamma_2 \\ n_1 \rho_1 \cos \beta_1 &= n_2 \rho_2 \cos \beta_2 \end{aligned}\right\} \tag{9.47}$$

The vector $\boldsymbol{t}$ is expressed in terms of co-ordinates (ρ, ϕ, z) of a point on the ray by the equation

$$\boldsymbol{t} = \dot{\rho}\boldsymbol{a}_\rho + \rho\dot{\phi}\boldsymbol{a}_\phi + \dot{z}\boldsymbol{a}_z$$

where the dot signifies differentiation with respect to the arc length of the curve, s. Thus we may identify

$$\cos\beta = \rho\dot{\phi}, \qquad \cos\gamma = \dot{z}$$

and note that

$$\dot{\rho}^2 + \rho^2\dot{\phi}^2 + \dot{z} = 1 \tag{9.48}$$

Finally therefore in the limit of continuous grading, equations (9.47) yield the two continuity equations

$$n(\rho)\frac{\mathrm{d}z}{\mathrm{d}s} = E \tag{9.49}$$

$$n(\rho)\rho^2\frac{\mathrm{d}\phi}{\mathrm{d}s} = El \tag{9.50}$$

in which E, l are constants determined by initial conditions.

Equation (9.49) can be used to change the independent variable from s to z: equation (9.50) then becomes

$$\rho^2\frac{\mathrm{d}\phi}{\mathrm{d}z} = l \tag{9.51}$$

and equation (9.48) yields

$$\left(\frac{\mathrm{d}\rho}{\mathrm{d}z}\right)^2 = g(\rho) = \left[\frac{n(\rho)}{E}\right]^2 - \frac{l^2}{\rho^2} - 1 \tag{9.52}$$

If the constant l is made zero, equations (9.43) and (9.52) become identical, since equation (9.49) gives $E = n_0 \cos\gamma_0$. In general, if the ray is launched at (ρ_0, ϕ_0, z_0), making angles $(\alpha_0, \beta_0, \gamma_0)$ with *local* cylindrical polar axes, the constants are given by

$$\left.\begin{aligned} E &= n(\rho_0)\cos\gamma_0 \\ l &= \rho_0\cos\beta_0\sec\gamma_0 \end{aligned}\right\} \tag{9.53}$$

The angle α made between $\boldsymbol{t}$ and $\boldsymbol{a}_\rho$ may be regarded as the angle of refraction at a layer, so that the condition $\dot{\rho} = 0$ corresponds to the critical case in refraction from a dense to a less dense medium.

Classification of rays
A ray is specified by the values of E and l, together with a point on the ray. Equation (9.52) can be used to obtain z as a function of ρ, and ϕ may then be found by integrating equation (9.51). Evidently this process is possible only if the expression on the right-hand side of equation (9.52), $g(\rho)$, is positive. We will consider the case when $n(\rho)$ decreases monotonically from a value n_0 on the axis to n_c at the core-cladding interface, equal to the uniform cladding refractive index. When $l=0$ we have the condition expressed by equation (9.45) to ensure that a ray remains entirely within the core. In this case the function $g(\rho)$ is as indicated in fig. 9.9(a); for the value of E illustrated, the turning point

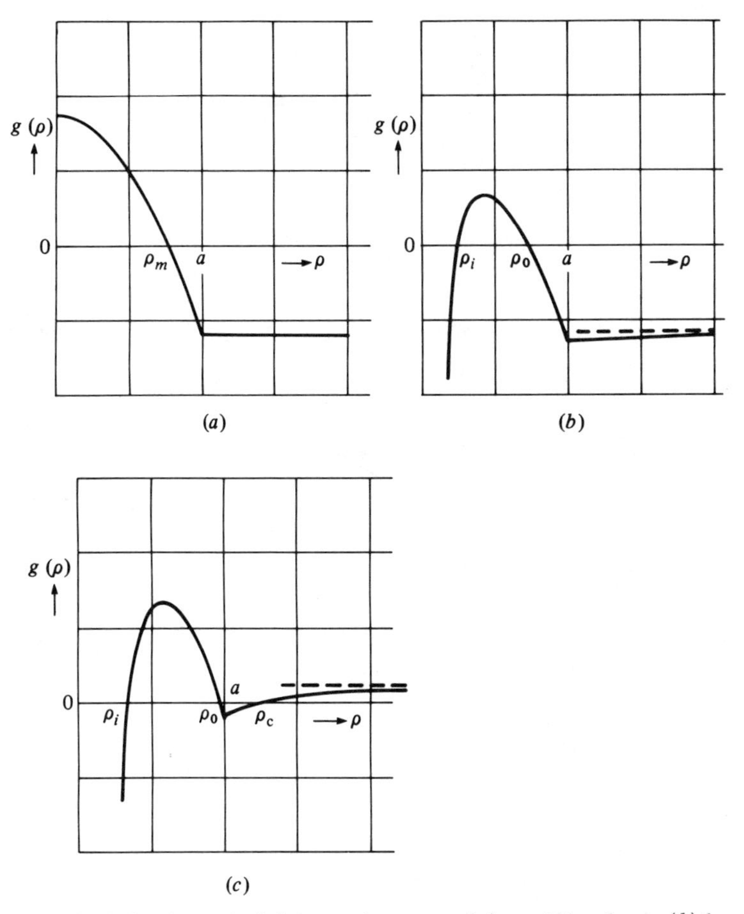

Fig. 9.9. The behaviour of $g(\rho)$ for various rays: (a) meridional ray; (b) bound skew ray; (c) tunnelling ray.

of the ray occurs when $\dot{\rho}=0$, at radius $\rho_m < a$. The form of $g(\rho)$ for the same value of E but with a non-zero l is shown in fig. 9.9(b): the skew ray is restricted to the annulus $\rho_i < \rho < \rho_0 < \rho_m$. In both cases if $n_0 > E > n_c$, $g(\rho) > 0$ only for a region within the core, and thus represents a bound ray. A lesser value of E can apparently be permitted when $l > 0$ whilst still retaining the skew ray to be within the core, as shown in fig. 9.9(c). This diagram indicates that a ray is possible within the cladding outside the radius ρ_c. Investigation using wave theory shows that this ray is excited by the ray within the core, so that energy is no longer confined within the core. Such a ray is termed a '*tunnelling ray*'. Thus we may summarise

$$n_0 > E > n_c \qquad \text{bound rays}$$

$$\left.\begin{array}{l} E < n_c \\ l^2 > a^2\left(\dfrac{n_c^2}{E^2} - 1\right) \end{array}\right\} \qquad \text{tunnelling rays}$$

$$\left.\begin{array}{l} E < n_c \\ l^2 < a^2\left(\dfrac{n_c^2}{E^2} - 1\right) \end{array}\right\} \qquad \text{refracted rays}$$

The last named are so termed because the incident ray will be refracted into the cladding and its energy lost from the core.

9.4.2 Rays in a step-index fibre: transit time

The analysis of the last section can be used when the core refractive index is uniformly equal to n_0. The condition $n_0 > E > n_c$ then corresponds to total internal reflection at the interface. Ray paths are straight lines. A meridional ray for $\gamma_0 < \gamma_m$ is shown in fig. 9.10. Figure 9.11 depicts a skew ray. The meridional ray is characterised by the angle γ_0, the skew ray by γ_0 and one other angle.

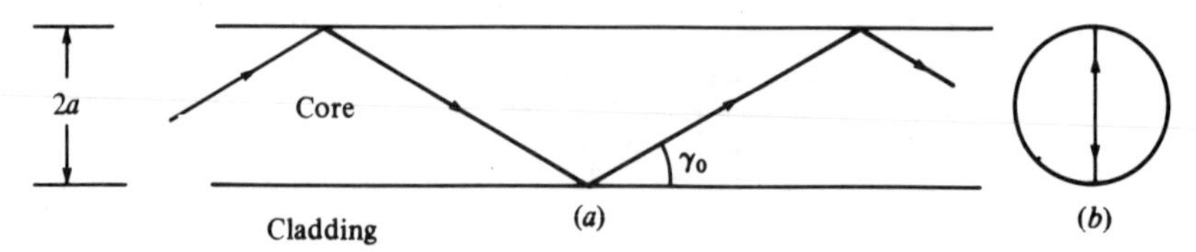

Fig. 9.10. Meridional ray in step-index fibre: (a) longitudinal section; (b) transverse section.

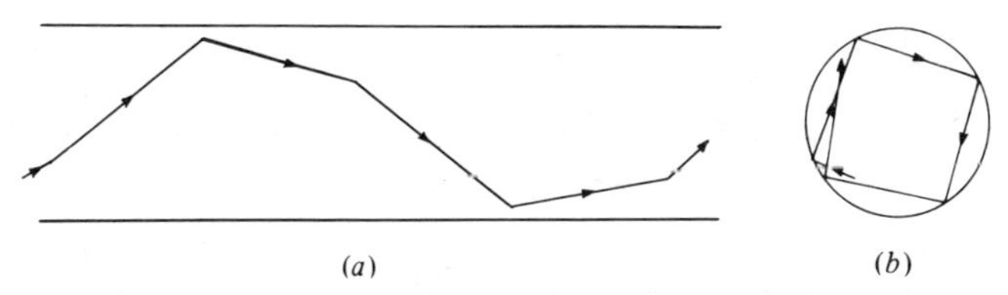

Fig. 9.11. Skew ray in step-index fibre: (a) side-view; (b) transverse section. section.

For a meridional ray, the path length between successive reflections, z_{p}, is from the geometry of fig. 9.10 given by

$$z_{\text{p}} = 2a \cot \gamma_0$$

The path length between the same two points is equal to $2a$ cosec γ_0, corresponding to a transit time of $(2an_0/c)$ cosec γ_0. Over an axial length L containing many reflections the transit time is given to a close approximation by the expression

$$\begin{aligned} \tau &= \frac{L}{z_{\text{p}}} \frac{2an_0}{c} \operatorname{cosec} \gamma_0 \\ &= \frac{Ln_0}{c} \sec \gamma_0 \end{aligned} \tag{9.54}$$

The minimum delay occurs for the axial ray $\gamma_0 = 0$, the maximum for that with $\gamma_0 = \gamma_m = \cos^{-1}(n_{\text{c}}/n_0)$. The difference between these two extremes, τ_{d}, is thus given by

$$\tau_{\text{d}} \cong \frac{Ln_0}{c} \left(\frac{n_0}{n_{\text{c}}} - 1 \right)$$

A similar calculation can be made for skew rays, and it is found that equation (9.54) is correct in that case also.

It is customary to define

$$\Delta = \frac{1}{2} \frac{n_0^2 - n_{\text{c}}^2}{n_0^2} \cong \frac{n_0 - n_{\text{c}}}{n_0}, \qquad \Delta \ll 1 \tag{9.55}$$

With this notation,

$$\tau_{\text{d}} \cong \frac{Ln_0}{c} \Delta \tag{9.56}$$

9.4.3 Parabolic profile

In general, with a graded profile a skew ray has the appearance indicated in fig. 9.12, being restricted within the annulus $\rho_0 > \rho > \rho_i$, as

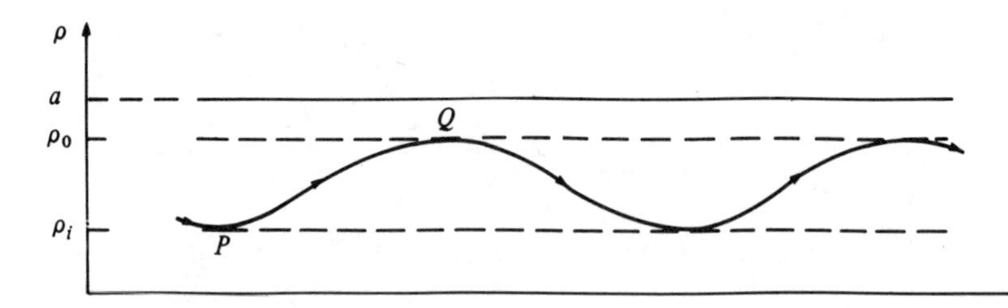

Fig. 9.12. Showing variation of radial co-ordinate of a skew ray as function of axial distance.

discussed above. The quantity z_p is calculated from the equation

$$z_p = \int_P^Q dz = \int_{\rho_i}^{\rho_0} g^{-1/2}(\rho)\, d\rho \tag{9.57}$$

The transit time between the points P and Q is equal to

$$\frac{1}{c}\int_P^Q n(\rho)\, ds$$

Thus over a length L the transit time is given to a close approximation by

$$\tau = \frac{L}{z_p c}\int_{\rho_i}^{\rho_0} n^2(\rho) g^{-1/2}(\rho)\, d\rho \tag{9.58}$$

The constant of the ray, E, in equation (9.52) lies between the limits n_0 and n_c for a bound ray. Hence we may always define an angle γ_r by the equation

$$E = n_0 \cos \gamma_r \tag{9.59}$$

If the ray is meridional γ_r is the angle between ray and z-axis. For a skew ray it does not have a direct significance.

For the parabolic profile, retaining the notation of equation (9.55) with n_0 referring to the refractive index on the axis, we write

$$n^2(\rho) = n_0^2\{1 - 2\Delta(\rho/a)^2\} \tag{9.60}$$

Substitution of this expression and of equation (9.59) into equation (9.52) yields the form

$$g(\rho) = \left[\frac{\tan \gamma_r}{\rho \rho_m}\right]^2 (\rho_0^2 - \rho^2)(\rho^2 - \rho_i^2) \tag{9.61}$$

where ρ_m, ρ_i, ρ_0 have the significance given in fig. 9.9(a, b). Further relations between these quantities are

$$\rho_m = [a/\sqrt{(2\Delta)}] \sin \gamma_r$$

$$\rho_m^2 = \rho_0^2 + \rho_i^2$$

$$\rho_0 \rho_i = l\rho_m \cot \gamma_r$$

Thus from equation (9.57)

$$z_p = \rho_m \cot \gamma_r \int_{\rho_i}^{\rho_0} \rho[(\rho_0^2 - \rho^2)(\rho^2 - \rho_i^2)]^{-1/2} \, d\rho$$

The integration may be carried out to give

$$z_p = \frac{\pi}{2} \rho_m \cot \gamma_r \tag{9.62}$$

Substitution into equation (9.58) and re-arrangement gives for the transit time

$$\tau = \frac{Ln_0}{cz_p} \left[z_p \sec \gamma_r - (\sqrt{(2\Delta)}/a) \int_{\rho_i}^{\rho_0} \rho^3 [(\rho_0^2 - \rho^2)(\rho^2 - \rho_i^2)]^{-1/2} \, d\rho \right]$$

The integral may be evaluated and is equal to $\pi(\rho_0^2 + \rho_i^2)/4 = \pi\rho_m^2/4$. Finally, we find

$$\tau = \frac{Ln_0}{2c} (\cos \gamma_r + \sec \gamma_r) \tag{9.63}$$

Since this result does not depend on the parameter l the transit time and the axial distance between adjacent maxima and minima are independent of the skewness of the ray, as for the step-index fibre. This, however, is not generally true. For the maximum value of γ_r, γ_m, we have

$$\tau = \frac{n_0 z_p}{2c} \left(\frac{n_c}{n_0} + \frac{n_0}{n_c} \right)$$

$$\cong \frac{n_0 z_p}{c} (1 + \tfrac{1}{2}\Delta^2)$$

Hence the differential between the extreme and axial rays over a distance L is given by

$$\tau_d \cong \frac{Ln_0}{2c} \Delta^2 \tag{9.64}$$

Comparison with the result for the step-index fibre, equation (9.56), shows that a much smaller variation occurs with the parabolic profile.

Typical values of $n_0 = 1.5$, $\Delta = 0.01$ gives a figure of 50 ns km^{-1} for the step-index fibre, as opposed to 0.25 ns km^{-1} for the parabolic profile.

9.4.4 Clad-power-law profiles

A more general form of refractive index variation, containing those already studied, has the form

$$n^2(\rho) = n_0^2[1 - 2\Delta(\rho/a)^\alpha] \quad (\rho < a) \tag{9.65}$$

$$= n_c^2 \quad (\rho > a)$$

Analytic expressions may be found for most of the ray parameters. It can be shown that transit time does not depend on skewness and, further, that a suitable choice of α can be made to minimise the spread of transit times.

9.4.5 The ray equation

Ray paths have been derived above by direct application of Snell's law in the form of equation (9.46) to a cylindrically stratified medium. A general differential equation for ray paths will now be deduced. If we consider the application of equation (9.46) to a graded medium, over a small distance we may identify the variables in that equation as follows:

$$n_1 = n(\boldsymbol{r})$$

$$n_2 = n(\boldsymbol{r}) + \mathrm{d}s\, \boldsymbol{t} \cdot \nabla n$$

$$\boldsymbol{\nu} = \boldsymbol{\nabla} \boldsymbol{n} / |\boldsymbol{\nabla} \boldsymbol{n}|$$

$$\boldsymbol{t}_1 = \boldsymbol{t}$$

$$\boldsymbol{t}_2 = \boldsymbol{t} + \mathrm{d}\boldsymbol{t}$$

$$\boldsymbol{t} \cdot \mathrm{d}\boldsymbol{t} = 0$$

In these expressions a point on the ray path is identified by the arc length s, and $\boldsymbol{r}$ is to be considered as a function of s, so that $\boldsymbol{t} \equiv \mathrm{d}\boldsymbol{r}/\mathrm{d}s$. The final relation results from the fact that $\boldsymbol{t}$ is of unit magnitude. Substituting in equation (9.46) and neglecting second order terms we find

$$n\, \boldsymbol{\nabla} \boldsymbol{n} \times \mathrm{d}\boldsymbol{t} = -\boldsymbol{t} \cdot \boldsymbol{\nabla} \boldsymbol{n}\, \mathrm{d}s\, \boldsymbol{\nabla} \boldsymbol{n} \times \boldsymbol{t}$$

Forming the vector product of both sides with $\boldsymbol{t}$ yields

$$n\, \mathrm{d}\boldsymbol{t} + \mathrm{d}s\, \boldsymbol{t}\, \boldsymbol{t} \cdot \boldsymbol{\nabla} \boldsymbol{n} = \mathrm{d}s\, \boldsymbol{\nabla} \boldsymbol{n}$$

or

$$\frac{\mathrm{d}}{\mathrm{d}s}(n\boldsymbol{t}) = \nabla n$$

The equation for $\boldsymbol{r}(s)$ on the ray path is therefore

$$\frac{\mathrm{d}}{\mathrm{d}s}(n\mathrm{d}\boldsymbol{r}/\mathrm{d}s) = \nabla n$$

Applied to a graded-index fibre this vector equation yields in cylindrical polar co-ordinates the three equations for a point (ρ, ϕ, z) on a ray

$$\frac{\mathrm{d}}{\mathrm{d}s}\left(n\frac{\mathrm{d}\rho}{\mathrm{d}s}\right) - n\rho\left(\frac{\mathrm{d}\phi}{\mathrm{d}s}\right)^2 = \frac{\mathrm{d}n}{\mathrm{d}\rho}$$

$$n\frac{\mathrm{d}}{\mathrm{d}s}\frac{\mathrm{d}\phi}{\mathrm{d}s} + \frac{\mathrm{d}}{\mathrm{d}s}\left(n\rho\frac{\mathrm{d}\phi}{\mathrm{d}s}\right) = 0$$

$$\frac{\mathrm{d}}{\mathrm{d}s}\left(n\frac{\mathrm{d}z}{\mathrm{d}s}\right) = 0$$

For paraxial rays, lying in a meridional plane, we may replace to a good approximation $\mathrm{d}/\mathrm{d}s$ by $\mathrm{d}/\mathrm{d}z$. the equation for such a ray becomes

$$\frac{\mathrm{d}^2\rho}{\mathrm{d}z^2} = \frac{1}{n}\frac{\mathrm{d}n}{\mathrm{d}\rho}$$

9.4.6 Illumination

Light in a fibre has to be derived from an external source. In the case of multi-mode fibres this is often a diffuse source such as a light-emitting diode. The radiation from a diffuse source occurs in all directions from each elemental area of its radiating surface. The intensity of radiation may be described by the equation

$$\mathrm{d}P = I(\theta)\,\mathrm{d}A\,\mathrm{d}\Omega \tag{9.66}$$

in which $\mathrm{d}A$ is an element of area and $\mathrm{d}\Omega$ an element of solid angle in the direction of the polar angle θ. These variables are shown in fig. 9.13. For a perfectly diffuse, or Lambertian, source the intensity function $I(\theta)$ is given by

$$I_0 \cos\theta \quad (0 < \theta < \pi/2) \tag{9.67}$$

We are concerned with transmission across an end face, as indicated in fig. 9.14(a). Because of the discontinuity between the source region,

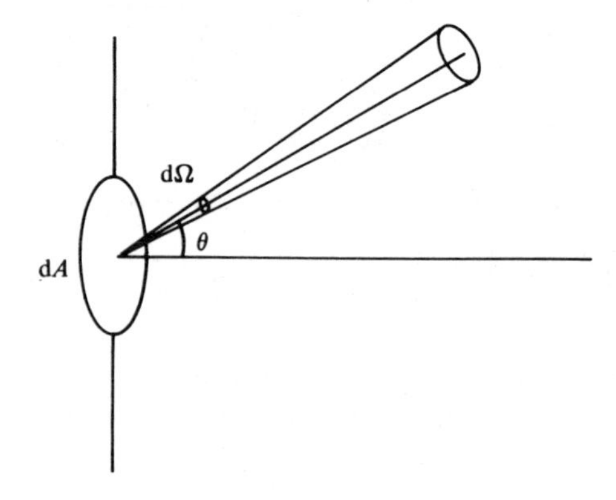

Fig. 9.13. Illustrating variables involved in radiation from a diffuse source.

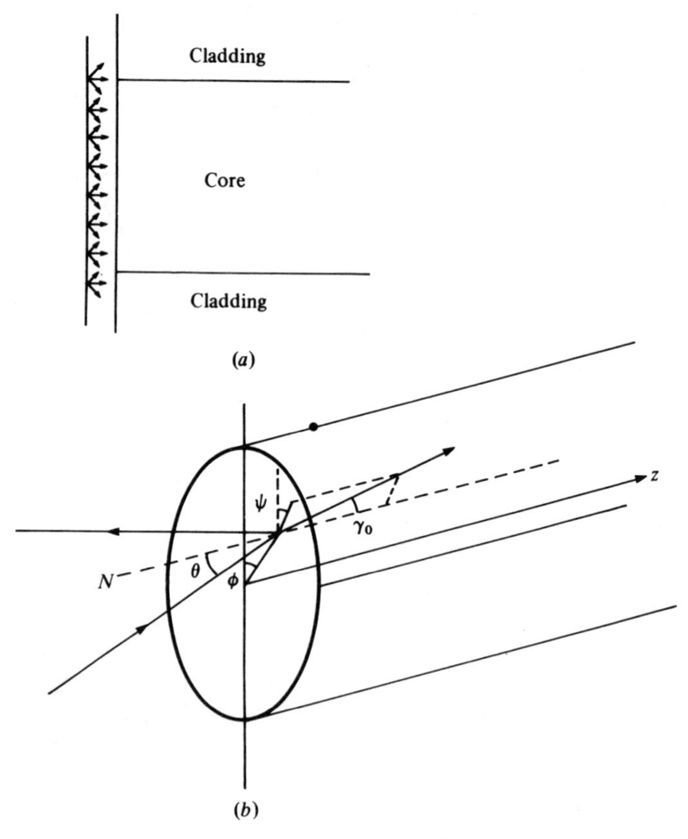

Fig. 9.14. Illumination of a fibre: (a) extended diffuse source; (b) ray behaviour at end-face.

usually air, and the end of the fibre, some power will be reflected. This can be estimated using the Fresnel formulae (§1.5). This power loss for normal values of refractive index is only a few percent, and will not be considered here. We wish to calculate the power accepted from the source as bound rays, i.e. rays for which just inside the fibre

$$E = n(\rho)\cos\gamma(\rho) < n_c \tag{9.68}$$

Figure 9.14(b) shows a ray impinging on the end-face at an angle θ with the axis. If n_e is the refractive index of the external medium, by Snell's law

$$n_e \sin\theta = n(\rho)\sin\gamma(\rho)$$

There is therefore a maximum value for θ, $\theta_m(\rho)$, given by

$$n_e \sin\theta_m = [n^2(\rho) - n_c^2]^{1/2} \tag{9.69}$$

For a step-index fibre, $n(\rho) = n_0$ so that θ_m is characteristic of the fibre. The expression in equation (9.69) is then defined to be the *numerical aperture* of the fibre. In the case of a graded profile, the expression is referred to as the 'local numerical aperture'.

If we assume that the source is placed very close to the end-face and covers the core area, the total power falling on the core face is given by

$$P_t = \int_0^a d\rho \int_0^{2\pi} \rho\, d\phi \int_0^{\pi/2} d\theta \int_0^{2\pi} I(\theta)\sin\theta\, d\theta\, d\psi$$

Using equation (9.67) we find

$$P_t = \pi^2 a^2 I_0 \tag{9.70}$$

The accepted power is given by

$$P_a \int_0^a d\rho \int_0^{2\pi} \rho d\phi \int_0^{\theta_m(\rho)} d\theta \int_0^{2\pi} I_0 \cos\theta \sin d\theta\, d\theta\, d\psi$$

$$= \frac{2\pi^2}{n_e^2} I_0 \int_0^a \rho[n^2(\rho) - n_c^2]\, d\rho$$

Thus for a step-index fibre,

$$P_a = \pi^2 a^2 I_0 (n_0/n_e)^2 \sin^2\gamma_m$$

$$= \pi^2 a^2 I_0 \sin^2\theta_m \tag{9.71}$$

whilst for a parabolic profile,

$$P_a = \tfrac{1}{2}\pi^2 a^2 (n_0/n_e)^2 \sin^2\gamma_m \tag{9.72}$$

where

$$\cos \gamma_m = n_c / n_0$$

Comparing equations (9.70), (9.71) and (9.72) we see that only a very small part of the total power is channelled into bound rays, and further that the parabolic profile channels only one-half the power of the step-index profile.

It is evident that a diffuse source would be unsuitable for exciting a single mode fibre. To do this efficiently it is necessary to create the required in-phase field distribution across the end-face; semi-conductor lasers have been developed for this purpose.

Example

A step-index fibre has a core diameter of 60 μm with refractive index 1.512. The cladding is of refractive index 1.485. Determine the numerical aperture, the maximum angle a bound ray can make with the axis and the distance between successive reflections for such a ray. Estimate the maximum time-differential between rays over a distance of 5 km.

Solution. The numerical aperture $= (n_1^2 - n_2^2)^{1/2} = \underline{0.28} = NA$

$$\Delta = \frac{n_1 - n_2}{n_1} = 0.0179$$

$$\sin \gamma_m = 1 - n_2^2/n_1^2 = (NA)^2/n_1^2$$

$$\cos \gamma_m = 1.485/1.512 \quad \therefore \underline{\gamma_m = 10.8°}$$

The axial path length

$$z_p = \frac{60}{\tan \gamma_m} = 310\ \mu\text{m}.$$

The diagonal path length

$$d = z_p \sec \gamma_m.$$

Hence

$$d - z_p = z_p(\sec \gamma_m - 1),$$

and the time differential is given by

$$\tau = \frac{n_1}{c}(d - z_p)\frac{L}{z_p}, \text{ where } L = 5.10^3 \text{ m}.$$

$$= \frac{n_1 L}{c}\left(\frac{n_1}{n_2} - 1\right) = \underline{0.458\ \mu\text{s}}.$$

For the case of a parabolic profile,

$$\tau = \frac{Ln}{2c} \cdot \Delta^2 = 4.02 \times 10^{-9}\ \text{s} = \underline{0.402\ \text{ps}}.$$

9.5 Gaussian beams

It has been seen that a medium with a parabolic profile supports a beam of Gaussian cross-section of intensity (§9.3.1). It is of interest to enquire what happens when such a beam passes out of the guiding medium into free space. It has been seen also (§4.6.4) that a beam with an initial Gaussian cross-section maintains that section, albeit expanding, as it propagates in free space. We are concerned with radial rates of change which are very small over one wavelength, so that approximations can be made in the wave equation. It has been shown in §9.3 that in a homogeneous medium each Cartesian component of transverse electric field obeys the scalar wave equation in the form

$$\nabla^2 \psi + k^2 \psi = 0 \tag{9.73}$$

where $k^2 = \omega^2 \mu \varepsilon$. Since the cross-section changes only slowly, the beam will approximate locally to a plane wave so that we can look for solutions of the form

$$\psi = f(x, y, z) \exp(-\mathrm{j}kz) \tag{9.74}$$

in which f changes only slowly with respect to its variables. Substituting in equation (9.73) gives

$$\frac{\partial^2 f}{\partial x^2} + \frac{\partial^2 f}{\partial y^2} + \frac{\partial^2 f}{\partial z^2} = 2\mathrm{j}k \frac{\partial f}{\partial z}$$

The assumption that f changes little in one-wavelength implies that

$$\left| \frac{\partial^2 f}{\partial z^2} \right| \ll 2k \left| \frac{\partial f}{\partial z} \right|$$

so that f obeys the approximate equation

$$\nabla_{\mathrm{t}}^2 f = 2\mathrm{j}k \frac{\partial f}{\partial z} \tag{9.75}$$

in which ∇_{t}^2 has the same significance as in earlier sections. The form of equation (4.59) suggests a symmetrical solution of the form

$$f = \exp[-a(z) - \rho^2 b(z)]$$

in which a, b may be complex. In cylindrical polar co-ordinates equation (9.75) becomes

$$\frac{1}{\partial}\frac{\partial}{\partial\rho}\left(\rho\frac{\partial f}{\partial\rho}\right)+\frac{1}{\rho^2}\frac{\partial^2 f}{\partial\phi^2}=2\mathrm{j}k\frac{\partial f}{\partial z} \tag{9.76}$$

Substitution of the trial solution shows that it is necessary to have

$$2\mathrm{j}k\frac{\mathrm{d}a}{\mathrm{d}z}=4b;\quad 2\mathrm{j}k\frac{\mathrm{d}b}{\mathrm{d}z}=-4b^2$$

Integration gives

$$b=\frac{\mathrm{j}k}{2}(z+\mathrm{j}\kappa)^{-1} \tag{9.77}$$

$$a=\ln(z+\mathrm{j}\kappa) \tag{9.78}$$

It may be noted that the constant of integration has to be imaginary for a non-trivial solution: a real part corresponds to a change of origin. Thus, using equation (9.74),

$$\psi(\rho, z)=(z+\mathrm{j}\kappa)^{-1}\exp\left[-\mathrm{j}\frac{k\rho^2}{2}(z+\mathrm{j}\kappa)^{-1}\right]\exp(-\mathrm{j}kz) \tag{9.79}$$

Rationalising the reciprocals we find

$$|\psi(\rho, z)|=(z^2+\kappa^2)^{-1/2}\exp[-k\kappa\rho^2/2(z^2+\kappa^2)] \tag{9.80}$$

$$\arg\psi(\rho, z)=-kz-\tan^{-1}\kappa/z-k\rho^2 z/2(z^2+\kappa^2) \tag{9.81}$$

The first of these equations provides a significance for the constant κ: at $z=0$ the beam section has the form $\exp(-k\rho^2/2\kappa)$, so that we may write

$$\kappa=kw_0^2/2$$

where w_0 is the beam radius to $1/e$. For other values of z the beam radius is given by

$$w(z)=w_0[1+(z/\kappa)^2]^{1/2} \tag{9.81}$$

It may be noted that for $\lambda=1\ \mu$m, $w_0=1$ mm, $\kappa=\pi$ m. The equation for $\arg\psi$ concerns curvature of the equi-phase surfaces: if we denote

$$\arg\psi=A(z)+\rho^2 B(z)$$

the equation of the equi-phase surface passing through $(0, z)$ is

$$A(z+\zeta)+\rho^2 B(z+\zeta)=A(z)$$

in which ζ denotes the off-axis change in axial co-ordinate. For small changes this reduces to

$$\rho^2 = -2R(z)\zeta \tag{9.82}$$

where $R(z) = A'(z)/2B(z)$. The form of equation (9.82) shows $R(z)$ to be the radius of curvature of the surface on axis. Hence we find

$$R(z) = z + \frac{\kappa^2}{z} \tag{9.82}$$

neglecting a term w_0^2/z in comparison with κ^2/z. The beam is portrayed schematically in fig. 9.15.

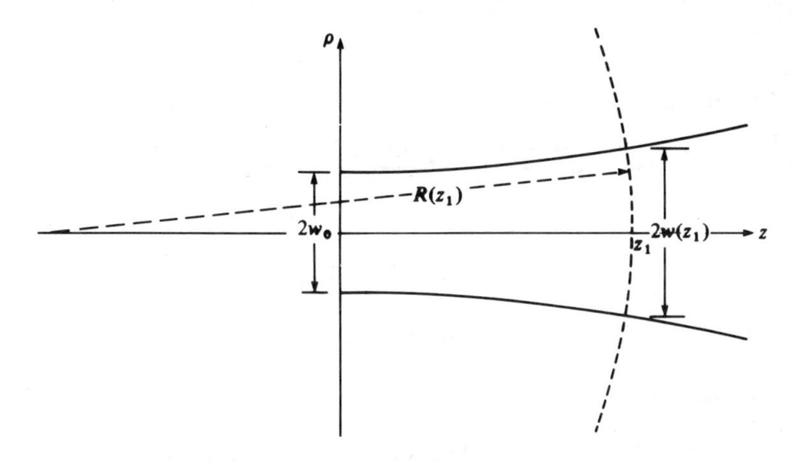

Fig. 9.15. Illustrating a symmetrical Gaussian beam.

The function $z + j\kappa$ is customarily denoted by the symbol $q(z)$ and is regarded as the parameter of a Gaussian beam. It may be expressed in terms of radius of curvature and spot size by the equation

$$\frac{1}{q(z)} = \frac{1}{R(z)} - j\frac{2}{kw^2(z)} \tag{9.83}$$

Example

A typical helium–neon laser of wavelength 0.633 nm has a beam of radius 0.5 mm. Assuming this is the waist of Gaussian beam with $w_0 = 0.5$ mm, calculate the value of w at a distance of 10 km.

Repeat the exercise for a laser of initial spot size 5 mm and wavelength 1.06 μm, at a distance of 3.84×10^5 km (distance of earth to moon).

Show that for a given target distance there is a value of w_0 which will produce a minimum spot size at the target. Calculate this value in the two cases above.

Solution. (*a*) From equation (9.81) we have

$$w(z) = w_0[1+(z/\kappa)^2]^{1/2}, \quad \text{where } \kappa^2 = \frac{k^2 w_0^4}{4}, \quad k = \frac{2\pi}{\lambda}$$

thus

$$w(z) = w_0\left[1+\left(\frac{\lambda z}{\pi w_0^2}\right)^2\right]^{1/2}$$

Given $w_0 = 5\times10^{-4}$, $\lambda = 0.633\times10^{-6}$, $z = 10^4$, substitution gives

$$\underline{w(z) = 4.03 \text{ m.}}$$

(*b*) In the second part,

$$\frac{\lambda z}{\pi w_0^2} = \frac{1.06\times10^{-6}\times3.84\times10^8}{\pi(5\times10^{-3})^2} = 5.18\times10^6$$

and

$$\underline{w(z) = 25.9 \text{ km.}}$$

(*c*)
$$w^2 = w_0^2 + \left(\frac{\lambda z}{\pi}\right)^2 w_0^{-2}$$

For minimum

$$\frac{d(w^2)}{d(w_0^2)} = 1 - \left(\frac{\lambda z}{\pi}\right)^2 w_0^{-4} = 0, \quad \text{whence } w_0^2 = \frac{\lambda z}{\pi}$$

thus $\underline{w^2 = 2w_0^2}$.
In case (*a*) above

$$w_0^2 = \frac{0.633\times10^{-6}\times10^4}{\pi}, \quad \underline{w_0 = 0.0449 \text{ m}}$$

and (*b*)

$$w_0^2 = \frac{1.06\times10^{-6}\times3.84\times10^8}{\pi}, \quad \underline{w_0 = 11.4 \text{ m}}$$

9.5.1 Gaussian beams in optical systems

The type of beam considered in the last section is frequently the form produced by lasers, and the subsequent passage through optical systems is of interest. It transpires that propagation of Gaussian beams through lenses and other axially symmetric optical components can be simply related to parameters determined from geometrical optics. The reason behind this is that geometrical optics deals with spherical wavefronts, and in the paraxial approximation the Gaussian beam theory relates curvature of the wavefront with amplitude distribution through the complex parameter q.

This may be seen by comparing equation (9.79) with a similar equation for a spherical wave emanating from a point source, as shown in fig. 9.16. For this situation

$$\psi = \frac{1}{r} \exp(-jkr)$$

where

$$r = (\rho^2 + z^2)^{1/2} \cong z + \rho^2/2z$$

Substituting

$$\psi = \frac{1}{z} \exp\left(-jkz - j\frac{k\rho^2}{2z}\right) \tag{9.84}$$

With the origin chosen, $z \equiv R$, the radius of curvature of the wavefront.

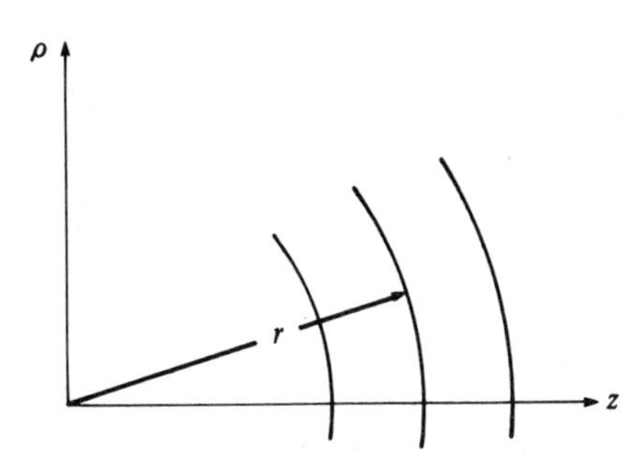

Fig. 9.16. Waves from point source.

The comparison shows that z in equation (9.84) is replaced by $q(z)$ in equation (9.79). Since the formulae of geometrical optics are essentially relations between curvature of wavefronts derived by continuity of ψ across boundaries, they may be used to give relations between the parameters q in the various regions. As a simple example consider the case of a Gaussian beam normally incident on the plane surface of discontinuity between two regions of differing refractive index, as shown

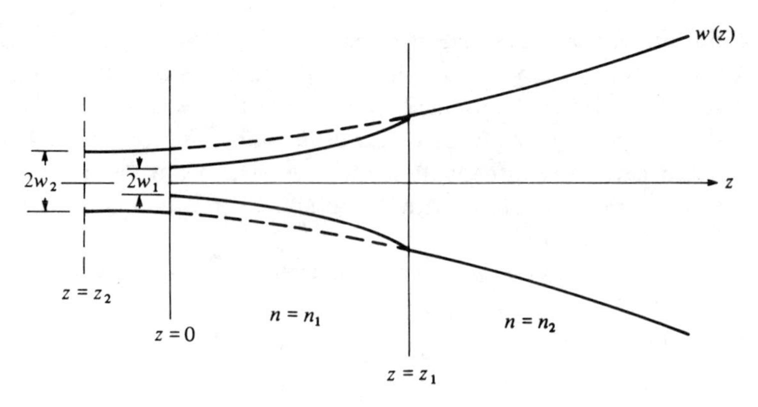

Fig. 9.17. Gaussian beams at a discontinuity in refractive index.

in fig. 9.17. The origin of z is chosen at the waist of the Gaussian beam of free space wavelength λ launched where $n=n_1$, and the beam will be described by

$$\psi=\frac{1}{q'}\exp[-jk_1 z-jk_1\rho^2/2q']$$

where

$$q'=z+j\kappa_1, \qquad \kappa_1=\pi n_1 w_1^2/\lambda, \qquad k_1=2\pi n_1/\lambda.$$

A Gaussian beam in the region $z>z_1$, $n=n_2$ with its waist located at $z=z_2$ will be described by

$$\psi''=\frac{A}{q''}\exp[-jk_2 z-jk_2\rho^2/2q'']$$

where

$$q''=z-z_2+j\kappa_2, \qquad \kappa_2=\pi n_2 w_2^2/\lambda, \qquad k_2=2\pi n_2/\lambda.$$

This contains arbitrary parameters of z_2 as the position of the waist and w_2 as the waist radius. On the plane $z=z_1$, ignoring small reflections, for all ρ we must satisfy

$$\psi'(\rho, z_1)=\psi''(\rho, z_1) \qquad (9.85)$$

Equating coefficients of ρ^2 yields

$$q_2=\frac{n_2}{n_1}q_1$$

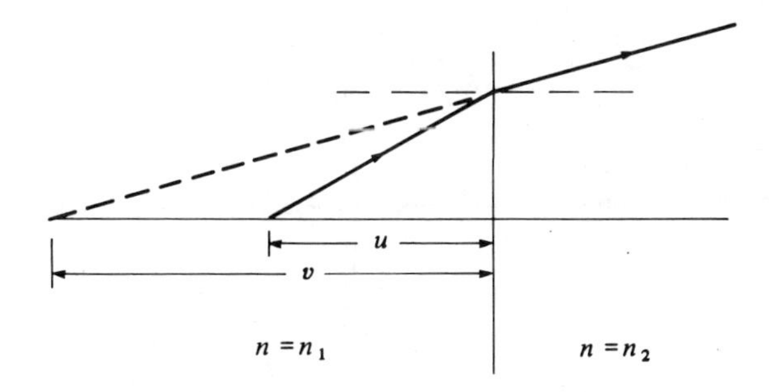

Fig. 9.18. Geometrical optics of discontinuity in refractive index.

implying

$$z_1 - z_2 = \frac{n_2}{n_1} z_1 \tag{9.86}$$

and

$$\kappa_2 = \frac{n_2}{n_1} \kappa_1$$

The geometrical optical situation is shown in fig. 9.18. Application of Snell's law, assuming the angles are small, gives

$$v = \frac{n_2}{n_1} u \tag{9.87}$$

In terms of z_1, z_2,

$$u = z_1, \qquad v = z_1 - z_2$$

giving an identical relation to equation (9.86). The beam diameter for $z > z_1$ is given through the waist diameter w_2, related to w_1 by

$$w_2 = w_1 \sqrt{[(n_1/n_2)]}$$

In addition we must have, for equation (9.85) to be satisfied,

$$q_1^{-1} \exp(-jk_1 z_1) = A q_2^{-1} \exp(-jk_2 z_1)$$

Hence the constant A is given by

$$A = \frac{n_2}{n_1} \exp[j(k_2 - k_1) z_1]$$

As a further example we may take the example of a thin lens, shown in fig. 9.19. Taking an origin for the z-co-ordinate at the object point

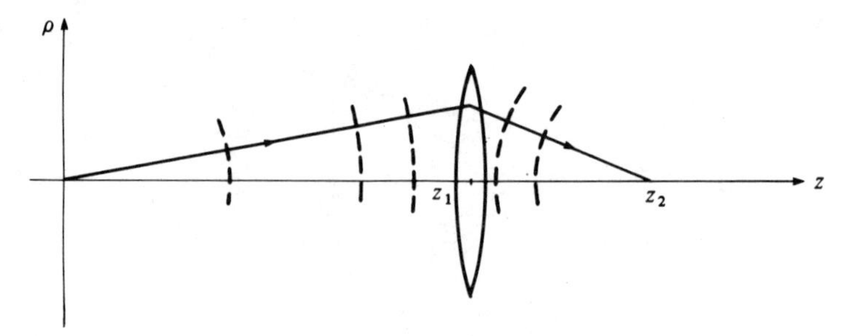

Fig. 9.19. Image formation by thin lens.

and with the notation of the diagram, standard thin lens formulae give

$$\frac{1}{z_1}+\frac{1}{z_2-z_1}=\frac{1}{f} \tag{9.88}$$

An incident Gaussian beam with waist at the origin will be characterised by a parameter $q'=z+\mathrm{j}\kappa_1$, and in the image space an emergent beam by the parameter $q''=z-z_2+\mathrm{j}\kappa_2$. Translating equation (9.88) we have

$$\frac{1}{q_1}-\frac{1}{q_2}=\frac{1}{f} \tag{9.89}$$

where

$$q_1=z_1+\mathrm{j}\kappa_1, \qquad q_2=z_1-z_2+\mathrm{j}\kappa_2$$

Hence z_2 and κ_2 may be found.

9.5.2 Ray transmission matrices

The passage of a ray through an optical system is determined once its initial position and slope are given. For paraxial rays linear relations exist between the initial co-ordinate and slope and those at any other point. A typical ray is indicated in fig. 9.20. The linear relations may be written

$$\rho_2=A\rho_1+Bm_1$$
$$m_2=C\rho_1+Dm_1$$

the matrix

$$\begin{bmatrix} A & B \\ C & D \end{bmatrix}$$

is the *ray transmission matrix.*

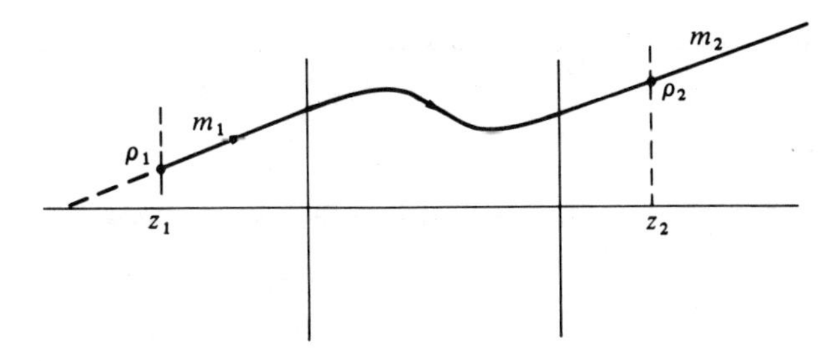

Fig. 9.20. Illustrating an arbitrary ray obeying geometrical optics.

In the paraxial case the radius of curvature of the wave front at (ρ_1, z_1) is

$$R_1 \cong \rho_1/m_1$$

At (ρ_2, z_2) it is

$$R_2 \cong \rho_2/m_2$$

Thus an alternative way of relating the q-parameters of input and output Gaussian beams is given by

$$q_2 = \frac{Aq_1 + B}{Cq_1 + D} \tag{9.90}$$

The two examples considered in §9.5.1 are represented in fig. 9.21 (a) and (b). From fig. 9.21(a) we have

$$\rho_1 = \rho_2; \qquad m_2 = m_1(n_1/n_2)$$

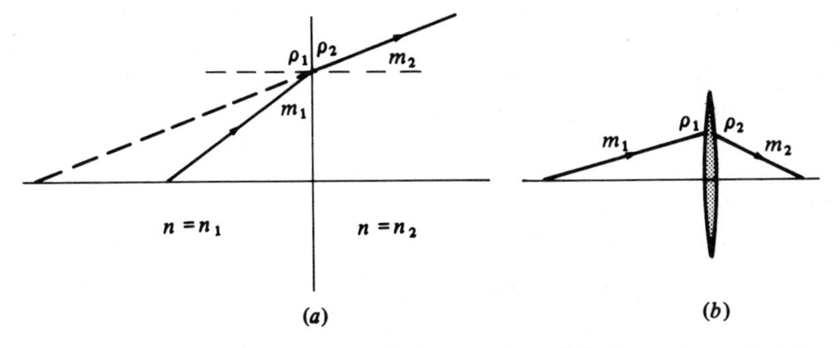

Fig. 9.21. Parameters for ray transmission matrices: (a) discontinuity in refractive index; (b) thin lens.

whence the ray transmission matrix is

$$\begin{bmatrix} 1 & 0 \\ 0 & n_2/n_1 \end{bmatrix}$$

For fig. 9.21(b), $\rho_2 = \rho_1$, and equation (9.88) gives, since $m_2 < 0$,

$$\frac{m_1}{\rho_1} - \frac{m_2}{\rho_2} = \frac{1}{f}$$

whence the ray transmission matrix is

$$\begin{bmatrix} 1 & 0 \\ -f^{-1} & 1 \end{bmatrix}$$

The use of ray transmission matrices provides a systematic way of following the passage of a Gaussian beam through a series of surfaces of discontinuity.

9.6 Summary

Techniques for applying electromagnetic theory to optical transmissions have been presented. The rigorous theory for the step-index fibre has been outlined, together with the weak-guidance approximation. It has been shown that in an inhomogeneous medium the scalar wave equation applies approximately to the Cartesian components of electric field, and particular cases have been investigated.

Propagation in thick fibres has been investigated by ray optics, and applied to step-index and graded-profile fibres. Ray optics has been applied to consideration of illumination of thick fibres by a diffuse source.

The scalar wave equation applied to the propagation of light beams in space has been shown to lead to beams of Gaussian cross-section. It has been shown how the propagation of such beams in an optical system may be calculated using the parameters of paraxial ray optics.

Formulae

Step-index fibre – approximate equations for cut-off parameters:

EH modes	$J_m(v_c) = 0,\ v_c \neq 0$
HE_{11}	$v_c = 0$
$HE_m,\ m \geqq 2$	$J_{m-2}(v_c) = 0,\ v_c \neq 0$

Radially inhomogeneous medium:

Approximate scalar wave equations: $\nabla_t^2\psi+[\omega^2\mu_0\varepsilon(\rho)-\beta^2]\psi=0$

$$\psi=\mathrm{E}_x \text{ or } \mathrm{E}_y$$

Parabolic profile: $\varepsilon=n_0^2\varepsilon_0[1-2\Delta(\rho/a)^2]$

$$\psi=\exp(-\rho^2/w^2)$$

$$w^2=\frac{\lambda a}{\pi n_0}\sqrt{\left(\frac{1}{2\Delta}\right)}$$

Rays in graded profile:

$$n(\rho)\frac{\mathrm{d}z}{\mathrm{d}s}=E;\quad n(\rho)\rho^2\frac{\mathrm{d}\phi}{\mathrm{d}s}=El$$

E, l constants

$$\left(\frac{\mathrm{d}\rho}{\mathrm{d}z}\right)^2=\left[\frac{n(\rho)}{E}\right]^2-\frac{l^2}{\rho^2}-1$$

Transit time differential:

$$\text{Step-index: }\frac{Ln_0}{c}\Delta$$

$$\text{Parabolic profile: }\frac{Ln_0}{2c}\Delta^2$$

Ray equation:

$$\frac{\mathrm{d}}{\mathrm{d}s}(n\,\mathrm{d}\boldsymbol{r}/\mathrm{d}s)=\boldsymbol{\nabla}\boldsymbol{n}$$

Numerical aperture (step-index fibre): $(n_0^2-n_c^2)^{1/2}$
Gaussian beams:

Approximate solution:

$$\psi=f(\boldsymbol{r})\exp(-\mathrm{j}kz)$$

$$\nabla_t^2 f=2\mathrm{j}k\frac{\partial f}{\partial z}$$

Symmetrical solution:

$$f=q^{-1}\exp[-\mathrm{j}k\rho^2/2q],\quad q=z+\mathrm{j}\kappa$$

$$\frac{1}{q}=\frac{1}{R(z)}-\mathrm{j}\frac{2}{kw^2(z)}$$

Ray transmission matrix:

$$\begin{pmatrix} \rho_2 \\ m_2 \end{pmatrix} = \begin{bmatrix} A & B \\ C & D \end{bmatrix} \begin{pmatrix} \rho_1 \\ m_1 \end{pmatrix}$$

9.7 *Problems*

(All wavelengths are '*free-space*' values.)

1. A step-index fibre is made with a cladding material of refractive index 1.444 and a core such that $\Delta = 0.015$. The fibre is required to propagate a single mode when excited by a laser source of wavelength 1.55 μm. Estimate the maximum permissible core diameter.

2. A step-index fibre is excited in the HE_{11} mode. Calculate the core diameter which would make $v = 2.0$ for a wavelength of:

(i) 0.85μm, (ii) 1.27 μm, (iii) 1.35 μm, (iv) 1.55 μm.

Calculate for each case the wavelength corresponding to cut-off of the next higher mode.

3. An approximate solution of the wave equation in a fibre with a parabolic profile is given by $E_x = \psi$, $E_y = 0$ with ψ given by equation (9.31). For these conditions show that to the first power of Δ

$$\frac{\partial}{\partial x}[\boldsymbol{E}_t \cdot \boldsymbol{\nabla}_t \ln \varepsilon] \cong -\frac{4\Delta}{a^2}\left(1 - \frac{2\rho^2}{w^2}\cos^2\phi\right)\exp(-\rho^2/w^2)$$

Compare this with the term $k^2 n_0^2 \psi$ and show that neglect of the right-hand side of equation (9.27) is equivalent to neglecting $(\lambda/an_0)^2\Delta$ in comparison with unity. Evaluate this quantity for the example given in §9.3.1.

4. The approximate wave equation (9.28) can be written in the case of the parabolic profile given by equation (9.29) in the form:

$$\frac{\partial^2\psi}{\partial x^2} + \frac{\partial^2\psi}{\partial y^2} + \left[k^2 n_0^2 - \beta^2 - 2k^2 n_0^2 \Delta \frac{x^2 + y^2}{a^2}\right]\psi = 0.$$

It is known that the equation $d^2u/d\xi^2 + (A - \xi^2)u = 0$ has solutions which are everywhere finite only if $A = 2n + 1$, being an integer. The corresponding solution is given by $u = H_n(\xi)\exp(-\xi^2/2)$ in which $H_n(\xi)$ are Hermite polynomials. Expressions for the first few are given in §10.9. Show that a suitable expression for ψ is given by:

$$\psi = H_m(x\sqrt{2}/w)H_n(y\sqrt{2}/w)\exp[-(x^2 + y^2)/w^2]$$

with

$$\beta^2 = k^2 n_0^2 - (4/w^2)(m + n + 1)$$

where w has the value given in equation (9.31).

5. Investigate the intensity patterns represented by the solution of question 4 for small values of m, n.

6. Show that for the problem of question 4

$$\beta \cong kn_0 - \frac{\sqrt{(2\Delta)}}{a}(m+n+1)$$

and hence that the group velocity $d\omega/d\beta$ is approximately the same for all modes.

7. Show that for a step-index fibre

$$\beta \cong \frac{\omega n_1}{c} - \frac{u^2\sqrt{(2\Delta)}}{2av}$$

and hence that

$$\frac{1}{v_g} = \frac{d\beta}{d\omega} = \frac{n_1}{c}\left[1+\Delta\left(\frac{u^2}{v^2} - \frac{2u}{v}\frac{du}{dv}\right)\right]$$

Using the results of fig. 9.3 estimate the term in Δ for the HE_{11} mode for $v=1.0$, 2.0. Hence show that when $n_1 = 1.5$, $\Delta = 0.01$ the group delay over 1 km length is approximately 40 ps less at $v=1$ than $v=2$.

8. Given that a ray propagating in a material of non-uniform refractive index, n, will follow a trajectory given by

$$\frac{d}{ds}(n\, d\boldsymbol{r}/ds) = \boldsymbol{\nabla n}$$

where ds is an element of the trajectory at a vector distance $\boldsymbol{r}$ from the origin, show that a parabolic index profile

$$\begin{aligned} n(\rho) &= n(0)[1-2\Delta(\rho/a)^2] \quad 0<\rho<a \\ &= n(0)[1-2\Delta] \qquad\qquad \rho>a \end{aligned}$$

will cause paraxial, meridional rays to be brought periodically to focus.

In a particular fibre $n(0)=1.5$ and $2\Delta = 0.005$. Calculate the maximum angle θ_m that a ray entering the fibre on the axis may make with the axis at entry if it is to remain guided by the fibre core.

(University of Bristol.)

9. Contrast and explain the basic properties of single-mode and highly multi-mode fibres for optical communications purposes.

A light ray travelling in a medium of refractive index n obeys the '*ray equation*'

$$\frac{d}{ds}[n\, d\boldsymbol{r}/ds] = \boldsymbol{\nabla n}.$$

where s is the distance measured along the ray. Starting from this equation, find the possible trajectories of rays trapped in a graded-index fibre with core radius

a within which the refractive index is given by

$$\frac{n_1 - n(\rho)}{n_1} = k\left(\frac{\rho}{a}\right)^2$$

where k is a constant and ρ is the radial distance from the core axis.

(Cambridge University; Electrical Sciences Tripos part question.)

10. The ray equation for an optical medium with continuously varying refractive index n is

$$\frac{\mathrm{d}}{\mathrm{d}s}[n\,\mathrm{d}\boldsymbol{r}/\mathrm{d}s] = \boldsymbol{\nabla} \boldsymbol{n}$$

where s is the distance along a ray path and $\boldsymbol{r}$ is a position vector. Starting from the above equation and using cylindrical co-ordinates (ρ, ϕ, z) show that there are two ray invariants, $n \cos\theta = \text{constant}$ and $\rho^2\,\mathrm{d}\phi/\mathrm{d}z = \text{constant}$, associated with an optical fibre which has circular symmetry and constant properties in the axial direction, where θ is the angle made by the ray and the fibre axis.

Discuss qualitatively the use of graded refractive index fibres with reference to optical source/fibre coupling and pulse spreading.

(Cambridge University; Electrical Sciences Tripos part question.)

11. A small plane diffuse source emits a total power P_0. It illuminates a step-index fibre from an axial point close to the plane end face of the fibre. Show that the power accepted as bound rays by the fibre is given by

$$P = P_0(n_1^2 - n_2^2)/n_e^2$$

in which n_1, n_2, n_e are the refractive indices of core, cladding and source regions respectively.

12. A source illuminates a fibre whose core refractive index is 1.5. Assuming that reflection losses are those which occur for normal incidence, estimate the fraction of incident power reflected from the end of the fibre when the source is in a medium of refractive index (i) unity, (ii) 3.7. The Fresnel formulae are given in §1.5.

13. Justify the use of a ray model to describe optical propagation in multi-mode fibres and indicate the limitations of the model.

A step-index polymer-clad silica fibre has a core refractive index of 1.46 and a cladding refractive index of 1.40. A small diffuse pulsed optical source is attached to the end of the fibre at right-angles to the axis and aligned with the axis. Assuming that the source diameter is much less than the fibre core diameter, estimate the fraction of the total emitted optical power which will be collected and propagated by the fibre. What increase in pulse width can be expected over a distance of 1 km?

(University of Bristol.)

14. It is proposed that an optical fibre local data network should use a 200 μm core diameter coated fibre. The refractive indices of core and cladding are 1.46 and 1.40 respectively. Calculate:

(*a*) the numerical aperture of the fibre;
(*b*) the fraction of light generated in a source of refractive index 3.7 that may be coupled into the fibre. (Assume that the light is generated at a small, diffuse, plane source and that absorption and Fresnel reflection can be made negligible.);
(*c*) the overall time dispersion.

15. The resonator of a helium–neon laser ($\lambda = 0.633\ \mu$m) is 0.4 m long. It has a plane mirror at one end and a mirror of radius of curvature 1.0 m at the other. Assuming a Gaussian beam, determine the spot size at each mirror.

16. A CO_2 laser ($\lambda = 10.6\ \mu$m) has mirrors 1.5 m apart, the one of radius of curvature 10 m, the other 20 m. Determine the position of the waist and the value of w_0.

17. A helium–neon laser ($\lambda = 0.633\ \mu$m) has a beam waist diameter of 0.5 mm. It is directed axially onto a thin lens of focal length 10 mm placed 10 cm from the laser. Determine the characteristics of the emergent beam and the diameter of the beam 1 m from the lens.

18. Determine the ray transmission matrix for the following cases:
(i) a concave spherical mirror, radius R, with common reference planes at the centre of curvature;
(ii) a distance d of free space;
(iii) two thin lenses focal lengths f_1, f_2, separated by distance d. Input and output reference planes at the planes of the two lenses:
(iv) show how the result of (iii) can be calculated from (ii) and the thin lenses individually.

10

Appendices

10.1 Power flow in circular waveguide

The purpose of this section is to derive expressions for the coefficients A_{mn}, B_{mn} of (3.72), (3.73). This is most conveniently done by proving a more general result, that the power flow may be expressed as the integral of either $|H_z|^2$ or $|E_z|^2$ over the cross-section. We first use a result derivable from Gauss' theorem.

Consider a real function ψ which satisfies the equation

$$\nabla_t^2\psi + \kappa^2\psi = 0$$

and which satisfies either $\psi = 0$ or $\partial\psi/\partial n = 0$ on a closed boundary C. ψ is then a suitable function to give H_z or E_z in a waveguide of cross-section C.

We apply Gauss's theorem

$$\oint \boldsymbol{F} \cdot \mathrm{d}\boldsymbol{S} = \int \operatorname{div} \boldsymbol{F}\, \mathrm{d}v$$

to a volume consisting of a cylinder defined by the contour C of unit length as shown in fig. 10.1. Take for $\boldsymbol{F}$ the vector $\psi\boldsymbol{\nabla}_t\psi$. ψ is not a function of the axial co-ordinate, so that $\boldsymbol{F}$ is purely transverse. The end surfaces therefore do not contribute to the left-hand side of this equation. The remaining part of the surface integral also vanishes because $\boldsymbol{F}\cdot \mathrm{d}\boldsymbol{S} = \psi\boldsymbol{\nabla}_t\psi \cdot \boldsymbol{n}\, \mathrm{d}S$, and either ψ or $\partial\psi/\partial n$ vanishes on the bounding contour by supposition. Therefore

$$\int \operatorname{div}\,(\psi\boldsymbol{\nabla}_t\psi)\, \mathrm{d}v = 0$$

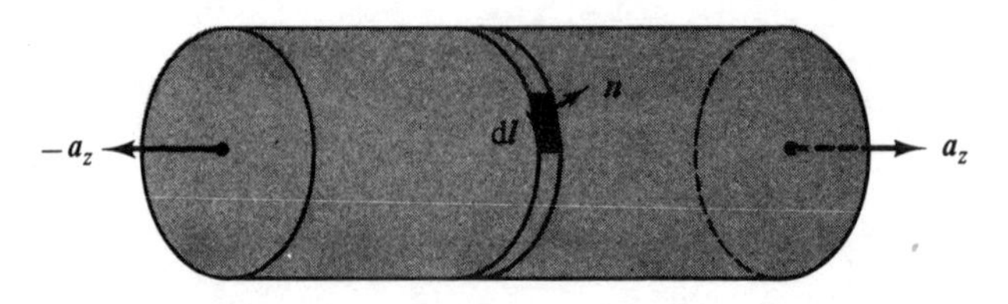

Fig. 10.1. Surface of integration in §10.1.

Since ψ does not depend on z this reduces to

$$\int \nabla_t \cdot (\psi \nabla_t \psi)\, dS = 0$$

in which the integral is taken over the cross-section of the guide. Finally therefore we have

$$\int \nabla_t \psi \cdot \nabla_t \psi \, dS = -\int \psi \nabla_t^2 \psi \, dS$$
$$= \kappa^2 \int \psi^2 \, dS \qquad (10.1)$$

The field variables for a propagating *TE* mode may be expressed in terms of ψ by the equations

$$H_z = \psi \exp(-j\beta z)$$

$$\mathbf{H}_t = -\frac{j\beta}{\kappa^2} \nabla_t \psi \exp(-j\beta z)$$

$$\mathbf{E}_t = \frac{\omega\mu}{\beta} \mathbf{H}_t \times \boldsymbol{a}_z$$

The longitudinal component of the Poynting vector is therefore given by

$$S = \frac{\omega\mu}{\beta} \mathbf{H}_t \cdot \mathbf{H}_t^* = \frac{\omega\mu}{\beta} \frac{\beta^2}{\kappa^4} \nabla_t \psi \cdot \nabla_t \psi$$

and the total power is therefore

$$P = \frac{\omega\mu\beta}{2\kappa^4} \int \nabla_t \psi \cdot \nabla_t \psi \, dS$$

where the integral is over the cross-section. The result of (10.1) enables us to write this in the form

$$P = \frac{\omega\mu\beta}{2\kappa^2} \int \psi^2 \, dS \qquad (10.2)$$

For a *TM* mode when we identify

$$E_z = \psi \exp(-j\beta z)$$

$$P = \frac{\beta^3}{2\omega\varepsilon\kappa^2} \int \psi^2 \, dS \qquad (10.3)$$

For the circular waveguide we have the form of the function ψ given by (3.61), with κa given by either s_{mn} or t_{mn} from (3.62) and (3.63).

Equation (10.2) therefore gives

$$P=\frac{\omega\mu\beta a^4}{2\kappa^2 a^4}\int_0^a\int_0^{2\pi}[J_m(\kappa\rho)\cos m\phi]^2\rho\,d\rho\,d\phi$$

$$=\frac{\omega\mu\beta a^4}{2(\kappa a)^4}\int_0^{\kappa a}\int_0^{2\pi}[J_m(u)\cos m\phi]^2 u\,du\,d\phi$$

We thus identify from (3.72), since $\kappa a=s_{mn}$ for a *TE* mode,

$$A_{mn}=\frac{1}{2s_{mn}^4}\int_0^{s_{mn}}\int_0^{2\pi}[J_m(u)\cos m\phi]^2 u\,du\,d\phi$$

A similar calculation from (10.3) for a *TM* mode, for which $\kappa a=t_{mn}$, gives the result

$$B_{mn}=\frac{1}{2t_{mn}^4}\int_0^{t_{mn}}\int_0^{2\pi}[J_m(u)\cos m\phi]^2 u\,du\,d\phi$$

Thus the coefficients are given by the same integral with different limits. The integration with respect to ϕ may be performed immediately, giving

$$\frac{\pi\varepsilon_m}{2\alpha^4}\int_0^{\alpha}[J_m(u)]^2 u\,du\,\cdot$$

where $\varepsilon_0=2$, $\varepsilon_m=1$, $m\neq 0$, and α takes on the value s_{mn} for A_{mn} and t_{mn} for B_{mn}. The integral is given in E. Jahnke & F. Emde, *Tables of functions* (Dover, 1943), p. 146, yielding

$$\frac{\pi\varepsilon_m}{2\alpha^4}\cdot\frac{\alpha^2}{2}\{[J_m(\alpha)]^2-J_{m-1}(\alpha)J_{m+1}(\alpha)\}$$

Bessel functions satisfy the recurrence relation

$$\frac{dJ_m}{du}=J'_m(u)=-\frac{m}{u}J_m(u)+J_{m-1}(u)=\frac{m}{u}J_m(u)-J_{m+1}(u)$$

enabling the integral to be expressed in the form

$$\frac{\pi\varepsilon_m}{2\alpha^4}\cdot\frac{\alpha^2}{2}\left\{\left(1-\frac{m^2}{\alpha^2}\right)[J_m(\alpha)]^2+[J'_m(\alpha)]^2\right\}$$

Hence, since $J'_m(s_m)=0$, $J_m(t_m)=0$,

$$A_{mn}=\frac{\pi\varepsilon_m}{4}\frac{1}{s_{mn}^2}\left(1-\frac{m^2}{s_{mn}^2}\right)[J_m(s_{mn})]^2$$

$$B_{mn}=\frac{\pi\varepsilon_m}{4}\frac{1}{t_{mn}^2}[J'_m(t_{mn})]^2$$

The numerical values are given in Abramowitz & Segun, *Handbook of mathematical functions* (Dover, 1968), table 9.5, from whence the values in tables 3.5 and 3.6 arc derived.

10.2 Evaluation of integral of §4.2.1

It is required to evaluate the integral

$$I=\int_0^{\pi}\frac{\cos(\frac{1}{2}\pi\cos\theta)}{\sin\theta}\,d\theta$$

Put $u=\cos\theta$. Then

$$\begin{aligned}I&=\int_{-1}^{+1}\frac{\cos^2(\pi u/2)}{1-u^2}\,du\\&=\int_{-1}^{+1}\tfrac{1}{2}\cos^2(\pi u/2)\left[\frac{1}{1-u}+\frac{1}{1+u}\right]du\\&=\int_{-1}^{+1}\cos^2(\pi u/2)\frac{du}{1-u}\end{aligned}$$

Now put $v=1-u$

$$\begin{aligned}I&=\int_0^2\cos^2[\pi(1-v)/2]\frac{dv}{v}\\&=\int_0^2\sin^2(\pi v/2)\frac{dv}{v}\\&=\int_0^2\tfrac{1}{2}(1-\cos\pi v)\frac{dv}{v}\\&=\tfrac{1}{2}\int_0^{2\pi}(1-\cos t)\frac{dt}{t}\end{aligned}$$

The cosine integral $\mathrm{Ci}(z)$ is defined by the equation

$$\mathrm{Ci}(z)=\gamma+\ln z+\int_0^z\frac{\cos t-1}{t}\,dt$$

in which $\gamma=0.5772$. Hence

$$I=\tfrac{1}{2}[\gamma+\ln(2\pi)-\mathrm{Ci}(2\pi)]$$

From tables we find $\mathrm{Ci}(2\pi)=-0.0228$, so that the numerical value of I is 1.219.

10.3 Evaluation of integral of §4.2.4

The integral part of (4.58) can be expressed as the product of two integrals

$$\int_{-\infty}^{\infty} \exp\left[-\xi^2\left(\frac{1}{\rho_0^2}+\frac{jk}{2r}\right)+\frac{jkx\xi}{r}\right] d\xi$$

$$\int_{-\infty}^{\infty} \exp\left[-\eta^2\left(\frac{1}{\rho_0^2}+\frac{jk}{2r}\right)+\frac{jky\eta}{r}\right] d\eta$$

It is known that

$$\int_{-\infty}^{\infty} \exp(-\beta\xi^2+j\omega\xi)\, d\xi = \sqrt{\left(\frac{\pi}{\beta}\right)} \exp\left(-\frac{\omega^2}{4\beta}\right)$$

for Re $(\beta)>0$. Hence the product of the integrals become

$$\frac{\pi}{\beta}\exp\left[-\frac{k^2}{r^2}(x^2+y^2)\frac{1}{4\beta}\right]$$

where $\beta=(1/\rho_0^2)+(jk/2r)$. Substitution in (4.58) gives

$$E_x=\frac{jk}{4\pi}\frac{2E_0}{r}e^{-jkr}\frac{\pi}{\beta}\exp\left(-\frac{k^2\rho^2}{4\beta r^2}\right)$$

10.4 Lumped loading of audio cables

The conditions for distortionless transmission on a transmission line with primary constants R, L, C and G was seen to be such that $L/R=C/G$. Consider a typical twisted pair audio cable, with primary constants $R=27\ \Omega\ \mathrm{km}^{-1}$, $G=0.6\ \mu\mathrm{S\ km}^{-1}$, $L=0.6\ \mathrm{mH\ km}^{-1}$, $C=0.04\ \mu\mathrm{F\ km}^{-1}$.

If we keep R, G and C fixed, the inductance necessary to satisfy the distortionless requirement is $L=RC/G=1.8\ \mathrm{H\ km}^{-1}$. It is clearly impracticable to consider increasing the inductance by a factor of 3000, but a compromise solution can be achieved by introducing lumped inductance (or loading coils) at discrete intervals, such that the effective inductance per unit loop length is significantly increased. If we assume, under these conditions, that $\omega L \gg R$ over the operational frequency range, and since, for a typical twisted pair, $\omega C \gg G$, we can approximate the propagation coefficient $\gamma=[(R+j\omega L)(G+j\omega C)]^{\frac{1}{2}}$ as

$$\gamma=\alpha+j\beta=\tfrac{1}{2}R\sqrt{(C/L)}+\tfrac{1}{2}G\sqrt{(L/C)}+j\omega\sqrt{(LC)}$$

As previously noted, this has the required characteristics for a distortionless propagation, since the attenuation is independent of

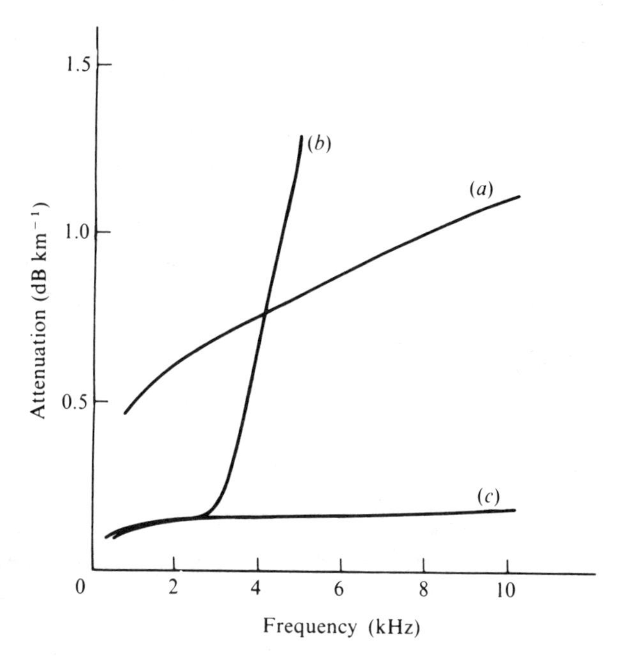

Fig. 10.2. Gain-frequency response for loaded line. (*a*) Unloaded. (*b*) Lump-loaded. (*c*) Uniformly loaded.

frequency, whilst the phase shift is proportional to frequency. The loading coils used in practice are precision, balanced, toroidal coils, typically adding 88 mH per 1850 m of line.

A truly distributed inductance would reduce the distortion over a wide band of frequencies. Figure 10.2 shows the attenuation v. frequency characteristic for the unloaded, lump-loaded and uniformly loaded cable. We can see that the lump loading effectively produces a low-pass filter characteristic. By neglecting R and G, and treating a section of line with loading as a simple T or Π reactive network, we can show that the cut-off frequency of the equivalent low-pass filter is given by

$$f_c = \frac{1}{\pi[(L_c + lL)Cl]^{\frac{1}{2}}}$$

where l is the spacing between loading coils, L_c is the added inductance, and L and C are the original line parameters. We see that f_c is inversely proportional to l, or the greater the spacing of the loading coils for a given additional loading per unit length, the lower the cut-off frequency. Using the values quoted above we find $f_c = 3.92$ kHz. This is clearly adequate for a normal telephone speech channel.

As a point of practical interest, the spacing of 1850 m is utilised in the multiplexed pulse code modulation system to house the regenerative repeaters.

10.5 The properties of ferrites at radio frequencies

It is the purpose of this section to put into context the property of ferrites used in chapter 8, that a ferrite medium propagates waves circularly polarised in different senses with different velocities.

Magnetic relationships

It is necessary to recall certain results appertaining to magnetic materials. Firstly, it is possible to define within a material certain average magnetic fields, which are related by the equation

$$\boldsymbol{H} = \mu_0^{-1}\boldsymbol{B} - \boldsymbol{M} \qquad (10.4)$$

$\boldsymbol{M}$ is the magnetisation vector, which is equal to the magnetic dipole moment per unit volume. The magnetisation arises because the magnetic fields act on atoms constituting the medium. An individual atom finds itself in a magnetic field of local flux density $\boldsymbol{B}_{\text{loc}}$ defined as the flux density which would exist at the site of the atom were it removed, all else remaining constant. This local flux density is related to the flux density in (10.4) by

$$\boldsymbol{B}_{\text{loc}} = \boldsymbol{B} + \alpha\boldsymbol{M} \qquad (10.5)$$

where α is a constant which can be estimated for a given atomic configuration.

The magnetic dipoles may be thought of as current loops, so that the torque $\boldsymbol{T}$ exerted on a dipole $\boldsymbol{m}$ in a field $\boldsymbol{B}$ is given by

$$\boldsymbol{T} = \boldsymbol{m} \times \boldsymbol{B} \qquad (10.6)$$

The magnetisation vector is related to an individual dipole moment $\boldsymbol{m}$ through the equation

$$\boldsymbol{M} = N\boldsymbol{m} \qquad (10.7)$$

in which N is the number of dipoles per unit volume. For a real material there is evidently a statistical element to be taken into account, but the model provided by the above equations is adequate to our purpose.

Ferromagnetic materials

In ferromagnetic materials the individual dipoles are provided by unpaired electrons, each of which has a magnetic dipole moment. This magnetic dipole moment $\boldsymbol{m}$ is related to the angular momentum $\boldsymbol{\Omega}$ of the electron by the relation

$$\boldsymbol{\Omega} = -\gamma^{-1}\boldsymbol{m} \tag{10.8}$$

γ is a constant for the electron.

In metallic ferromagnetics the conductivity is too high to allow use at high frequencies. Ferrites however have ferromagnetic properties combined with very low conductivity, enabling them to be used at microwave frequencies.

Gyro-magnetic precession

Consider a single electron in a (free-space) magnetic field of flux density $\boldsymbol{B}$. It will experience a torque given by (10.7) and the consequent rate of change of angular momentum will be given by

$$\mathrm{d}\boldsymbol{\Omega}/\mathrm{d}t = \boldsymbol{m} \times \boldsymbol{B}$$

or, using (10.8),

$$\mathrm{d}\boldsymbol{m}/\mathrm{d}t = -\gamma\boldsymbol{m} \times \boldsymbol{B} \tag{10.9}$$

If the magnetic field $\boldsymbol{B}$ is steady, then in equilibrium we expect $\mathrm{d}\boldsymbol{m}/\mathrm{d}t$ to be zero, and the magnetic dipole will be aligned with magnetic field. Let us consider the result of disturbing the alignment.

From (10.9) $\mathrm{d}\boldsymbol{m}/\mathrm{d}t$ is perpendicular to $\boldsymbol{m}$, and hence $\boldsymbol{m}$ cannot change in magnitude but only in direction. If we decompose $\boldsymbol{m}$ into one component $\boldsymbol{m}_{\parallel}$ parallel to $\boldsymbol{B}$ and another $\boldsymbol{m}_{\perp}$ perpendicular to $\boldsymbol{B}$ we have

$$\mathrm{d}\boldsymbol{m}_{\perp}/\mathrm{d}t = -\boldsymbol{m}_{\perp} \times \boldsymbol{B}$$

The significance of this equation is shown diagrammatically in fig. 10.3. From the geometry we see that $\boldsymbol{m}_{\perp}$ rotates clockwise looking along $\boldsymbol{B}$

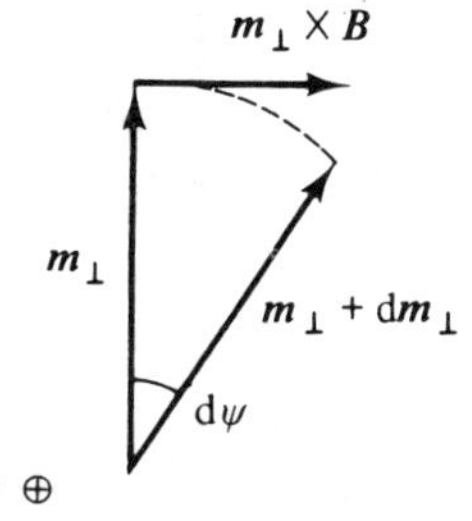

Fig. 10.3. Rate of change of magnetic dipole moment.

with an angular velocity given by

$$\mathrm{d}\psi = \gamma B \,\mathrm{d}t$$

or

$$\frac{\mathrm{d}\psi}{\mathrm{d}t} = \gamma B$$

Since we are considering a single electron in free space, $\boldsymbol{B}$ is related to $\boldsymbol{H}$ by the free-space constant μ_0, so that this characteristic angular frequency ω_0 may be expressed as

$$\omega_0 = 2\pi f_0 = \gamma \mu_0 H \tag{10.10}$$

The constant $\gamma\mu_0/2\pi$ is found to have the numerical value 0.035 MHz per A m^{-1}. This rotation of $\boldsymbol{m}$ is indicated in fig. 10.4, and is known as *precession.* In a real substance, interaction with other atoms will introduce loss, and after some time the dipoles will come into alignment with the magnetic field.

Behaviour in a radio-frequency field

We now consider the situation when a static magnetic field aligns the dipole moment in one direction, and an alternating magnetic field is superposed.

Let us write

$$\boldsymbol{B} = \boldsymbol{B}_0 + \boldsymbol{B}_1(t)$$

$$\boldsymbol{m} = \boldsymbol{m}_0 + \boldsymbol{m}_1(t)$$

where $\boldsymbol{B}_1$ and $\boldsymbol{m}_1$ depend on time, and are small compared with $\boldsymbol{B}_0$, $\boldsymbol{m}_0$. We have

$$\begin{aligned}\mathrm{d}\boldsymbol{m}_1/\mathrm{d}t &= -\gamma[(\boldsymbol{m}_0 + \boldsymbol{m}_1) \times (\boldsymbol{B}_0 + \boldsymbol{B}_1)] \\ &= -\gamma(\boldsymbol{m}_0 \times \boldsymbol{B}_0 + \boldsymbol{m}_1 \times \boldsymbol{B}_0 + \boldsymbol{m}_0 \times \boldsymbol{B}_1 + \boldsymbol{m}_1 \times \boldsymbol{B}_1)\end{aligned}$$

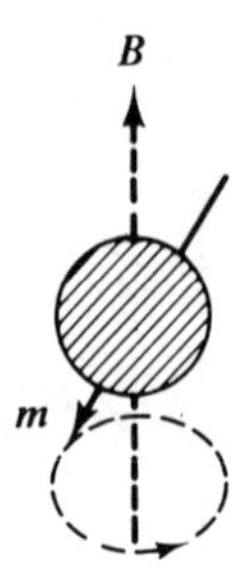

Fig. 10.4. Precession of spinning electron.

The first term is zero since $\boldsymbol{m}_0$ and $\boldsymbol{B}_0$ will be parallel, the last is the product of two small terms and will be neglected. Hence

$$\mathrm{d}\boldsymbol{m}_1/\mathrm{d}t = -\gamma\{\boldsymbol{m}_1 \times \boldsymbol{B}_0 + \boldsymbol{m}_0 \times \boldsymbol{B}_1\}$$

This equation is linear in the quantities $\boldsymbol{m}_1$, $\boldsymbol{B}_1$ and we may represent them by phasor quantities. Assume this to be done, enabling us to write

$$\mathrm{j}\omega\mathbf{m}_1 = -\gamma\{\mathbf{m}_1 \times \boldsymbol{B}_0 + \boldsymbol{m}_0 \times \mathbf{B}_1\} \tag{10.11}$$

Let us express this equation in Cartesian components, taking Oz as parallel to $\boldsymbol{B}_0$ and $\boldsymbol{m}_0$. We find

$$\mathrm{j}\omega\mathrm{m}_{1x} = -\gamma(\mathrm{m}_{1y}B_0 - m_0\mathrm{B}_{1y}) = -\omega_0\left(\mathrm{m}_{1y} - \frac{m_0}{B_0}\mathrm{B}_{1y}\right)$$

$$\mathrm{j}\omega\mathrm{m}_{1y} = -\gamma(-\mathrm{m}_{1x}B_0 + m_0\mathrm{B}_{1x}) = \omega_0\left(\mathrm{m}_{1x} - \frac{m_0}{B_0}\mathrm{B}_{1x}\right)$$

in which ω_0 is the gyromagnetic frequency given by (10.10).

Solving these equations for m_{1x} and m_{1y} we have

$$\left.\begin{aligned} \mathrm{m}_{1x} &= \left(1 - \frac{\omega^2}{\omega_0^2}\right)^{-1} \frac{m_0}{B_0}\left(\mathrm{B}_{1x} + \frac{\mathrm{j}\omega}{\omega_0}\mathrm{B}_{1y}\right) \\ \mathrm{m}_{1y} &= \left(1 - \frac{\omega^2}{\omega_0^2}\right)^{-1} \frac{m_0}{B_0}\left(-\frac{\mathrm{j}\omega}{\omega_0}\mathrm{B}_{1x} + \mathrm{B}_{1y}\right) \end{aligned}\right\} \tag{10.12}$$

Equations (10.12) tell us that if $\mathbf{B}_1$ has a fixed direction Ox (say), then a magnetic dipole moment is produced which not only has a component m_{1x} parallel to B_{1x} and in phase with it, but also a component m_{1y} lagging m_{1x} by $\pi/2$ in time. If $\mathbf{B}_1$ is aligned along Oy, a similar result is produced with m_{1y} still lagging m_{1x}. This is because of the prescribed direction of precession.

We may use (10.11) to derive the form of relationship to be expected between $\mathbf{B}_1$ and the alternating magnetic field $\mathbf{H}_1$ in a ferrite medium.

The field occurring in (10.9) must, in a medium, be interpreted as $\boldsymbol{B}_{\mathrm{loc}}$ of (10.5), giving

$$\mathrm{d}\boldsymbol{m}/\mathrm{d}t = -\gamma\boldsymbol{m} \times \boldsymbol{B}_{\mathrm{loc}}$$

Summing over a unit volume, using (10.7), we have

$$\mathrm{d}\boldsymbol{M}/\mathrm{d}t = -\gamma\boldsymbol{M} \times \boldsymbol{B}_{\mathrm{loc}}$$

Substituting from (10.5) and (10.4) we have

$$\mathrm{d}\boldsymbol{M}/\mathrm{d}t = -\gamma\mu_0\boldsymbol{M} \times \boldsymbol{H}$$

Assuming alternating components $\mathbf{M}_1$ and $\mathbf{H}_1$ as before we find

$$j\omega \mathbf{M}_1 = -\gamma\mu_0(\mathbf{M}_1 \times \boldsymbol{H}_0 + \boldsymbol{M}_0 \times \mathbf{H}_1)$$

Comparison with (10.11) and (10.12) gives

$$\left.\begin{aligned} M_{1x} &= \left(1-\frac{\omega^2}{\omega_0^2}\right)^{-1}\frac{M_0}{H_0}\left(H_{1x}+\frac{j\omega}{\omega_0}H_{1y}\right) \\ M_{1y} &= \left(1-\frac{\omega^2}{\omega_0^2}\right)^{-1}\frac{M_0}{H_0}\left(-\frac{j\omega}{\omega_0}H_{1x}+H_{1y}\right) \end{aligned}\right\} \qquad (10.13)$$

Equation (10.4) enables us to write

$$\left.\begin{aligned} B_{1x} &= \mu_0(H_{1x}+M_{1x}) \\ B_{1y} &= \mu_0(H_{1y}+M_{1y}) \end{aligned}\right\} \qquad (10.14)$$

$\mathbf{B}_1$ has been assumed to be perpendicular to $\boldsymbol{B}_0$. Provided $\boldsymbol{B}_0$ is large enough to saturate the medium, an increase in flux density is related only to the increase in the free-space magnetic field intensity. We therefore can add

$$B_{1z} = \mu_0 H_{1z}$$

Substituting (10.13) in (10.14) and collecting terms we finally have

$$\left.\begin{aligned} B_{1x} &= \mu H_{1x} + j\kappa H_{1y} \\ B_{1y} &= -j\kappa H_{1x} + \mu H_{1y} \\ B_{1z} &= \mu_0 H_{1z} \end{aligned}\right\} \qquad (10.15)$$

where

$$\mu = \mu_0\left[1+\frac{M_0/H_0}{1-\omega^2/\omega_0^2}\right]$$

$$\kappa = \frac{\omega}{\omega_0}\frac{\mu_0 M_0/H_0}{1-\omega^2/\omega_0^2}$$

$$\omega_0 = \gamma\mu_0 H_0$$

The significance of this **B–H** relationship may be seen as follows. Consider the situation in which the **H** vector rotates in a clockwise direction when looking against the direction of the magnetic field (i.e. along the negative z direction). This corresponds, for example, to the situation in a TEM wave of left-handed circular polarisation as described in chapter 1, equation (1.50). We may write, taking $H_{1z}=0$,

$$\mathbf{H}_1 = H_1(\boldsymbol{a}_x + j\boldsymbol{a}_y)$$

or

$$H_{1x} = H_1, \qquad H_{1y} = jH_1$$

Thus $$B_{1x} = H_1(\mu - \kappa)$$

$$B_{1y} = H_1(-j\kappa + j\mu) = jH_1(\mu - \kappa)$$

whence $$\mathbf{B}_1 = (\mu - \kappa)H_1(\boldsymbol{a}_x + j\boldsymbol{a}_y)$$

Hence in this case **B** is proportional to **H** with an effective permeability $(\mu - \kappa)$. Similar consideration for the opposite rotation gives

$$\mathbf{H}_1 = H_1(\boldsymbol{a}_x - j\boldsymbol{a}_y)$$

$$\mathbf{B}_1 = (\mu + \kappa)H_1(\boldsymbol{a}_x - j\boldsymbol{a}_y).$$

If therefore we work in terms of circularly polarised waves in the direction of magnetisation we may describe the magnetic properties at radio frequencies by means of scalar permeabilities. It is to be noted that the directions of rotation are referred uniquely to the magnetisation vector of the ferrite. The results are shown diagrammatically in fig. 10.5.

By substitution from (10.15) it may be shown that

$$\left.\begin{aligned} \mu_+ = \mu + \kappa = \mu_0\left(1 + \frac{M_0/H_0}{1 - \omega/\omega_0}\right) \\ \mu_- = \mu - \kappa = \mu_0\left(1 + \frac{M_0/H_0}{1 + \omega/\omega_0}\right) \end{aligned}\right\} \quad (10.16)$$

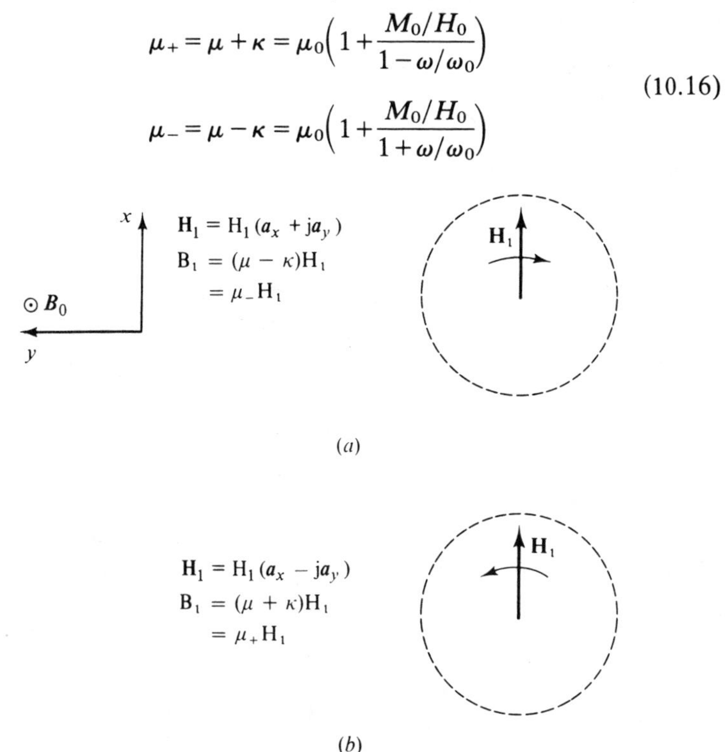

Fig. 10.5. Relation between B_1 and H_1. Rotation looking into $\boldsymbol{B}_0$(*a*) clockwise, (*b*) anticlockwise.

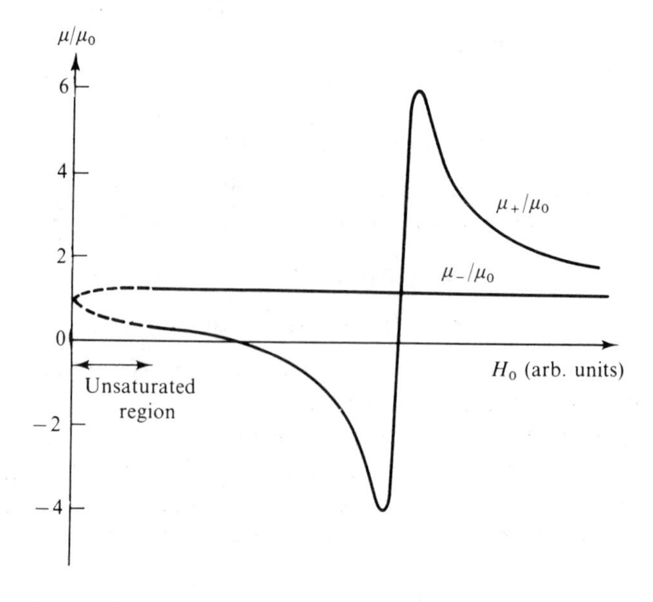

Fig. 10.6. Schematic dependence of μ_+ and μ_- on H_0, for fixed radio frequency.

Further substitution for H_0 reduces these to

$$\mu_+ = \mu_0\left[1+\frac{\gamma\mu_0 M_0}{\omega_0-\omega}\right]$$

$$\mu_- = \mu_0\left[1+\frac{\gamma\mu_0 M_0}{\omega_0+\omega}\right]$$

Once the ferrite is magnetically saturated M_0 will be independent of H_0, so the variation of H_0 is equivalent to variation of ω_0. The form of these expressions as functions of H_0 is shown in fig. 10.6.

TEM waves in an unbounded medium

Provided that we restrict ourselves to waves propagating parallel or anti-parallel to the direction of magnetisation and which are also circularly polarised we may take over the results of chapter 1. The most general *TEM* wave propagating in the direction of the magnetisation vector is of the form

$$\mathbf{E} = \mathrm{E}_0'(\boldsymbol{a}_x - \mathrm{j}\boldsymbol{a}_y)\exp(-\mathrm{j}\beta_+ z) + \mathrm{E}_0''(\boldsymbol{a}_x + \mathrm{j}\boldsymbol{a}_y)\exp(-\mathrm{j}\beta_- z) \qquad (10.17)$$

where

$$\beta_+ = \omega[\varepsilon(\mu+\kappa)]^{\frac{1}{2}}$$

$$\beta_- = \omega[\varepsilon(\mu-\kappa)]^{\frac{1}{2}}$$

For waves in the reverse direction we change the sign of the exponents. It is to be noted that whereas the first term is associated with left-handed circular polarisation, in the reverse direction this term will be associated with right-handed polarisation. Similar remarks apply to the second term.

Faraday rotation

Consider a forward wave as given by (10.17) in which $E_0' = E_0'' = \frac{1}{2}E_0$. Collecting terms

$$\mathbf{E} = E_0 \exp\left[-\tfrac{1}{2}j(\beta_+ + \beta_-)z\right]\left[\boldsymbol{a}_x \cos\tfrac{1}{2}(\beta_+ - \beta_-)z - \boldsymbol{a}_y \sin\tfrac{1}{2}(\beta_+ - \beta_-)z\right]$$

This represents a wave for which, at any given value of z, the **E** vector lies in a fixed plane which rotates as z increases. Looking into the oncoming waves, the **E** vector rotates in a clockwise direction through an angle $\frac{1}{2}(\beta_+ - \beta_-)l$ in going from $z = 0$ to $z = l$, as shown in fig. 10.7. For

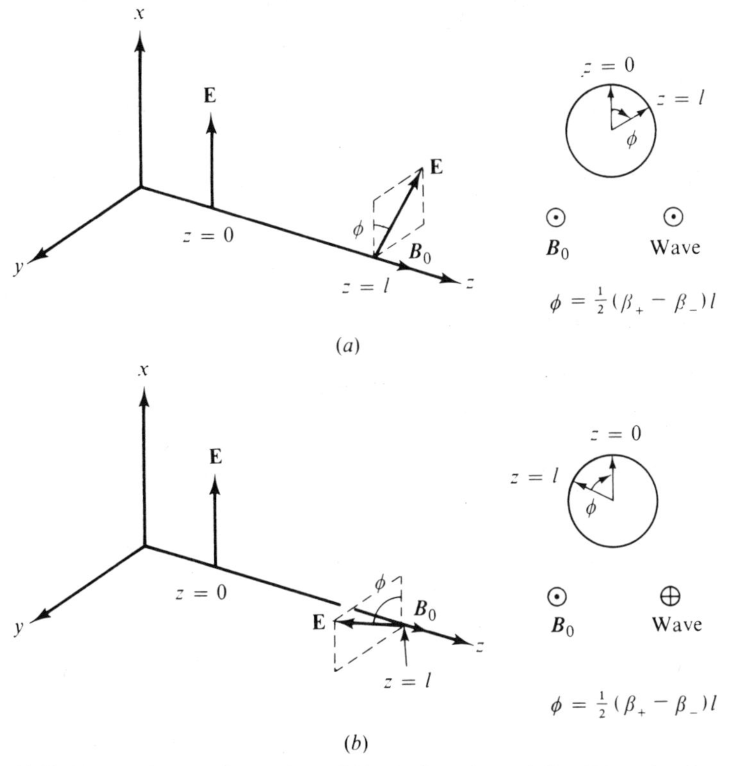

Fig. 10.7. Dependence of rotation of **E** on direction of $\boldsymbol{B}_0$. Wave in direction (a) parallel to $\boldsymbol{B}_0$, (b) antiparallel to $\boldsymbol{B}_0$.

a wave in the reverse direction we find

$$\mathbf{E} = E_0 \exp[\tfrac{1}{2}\mathrm{j}(\beta_+ + \beta_-)z][\boldsymbol{a}_x \cos\tfrac{1}{2}(\beta_+ - \beta_-)z + \boldsymbol{a}_y \sin\tfrac{1}{2}(\beta_+ - \beta_-)z]$$

Still maintaining the same viewpoint as before (now looking along the retreating wave), the **E** vector will rotate through the same angle in the same direction as before in going from $z = l$ to $z = 0$. If viewed looking into the oncoming wave this rotation is progressively anti-clockwise, opposite in sense to that for the forward wave. The medium is thus non-reciprocal.

We can use this rotation effect if we can launch a plane-polarised wave into the ferrite medium and extract a similar wave after passage through a distance of ferrite. The effect was first noticed by Faraday in experiments on transmission of light through a glass rod immersed in an axial magnetic field. In this case ω will be very large, and (10.16) enables us to write

$$\mu \pm \kappa \approx \mu_0\left[1 \mp \frac{\omega_0 M_0}{\omega H_0}\right]$$

whence $$\beta_+ - \beta_- \approx -\omega\sqrt{(\mu_0\varepsilon)}\frac{\omega_0 M_0}{\omega H_0} = -\sqrt{(\mu_0\varepsilon)}\gamma\mu_0 M_0$$

The angle of rotation per unit length is therefore $\tfrac{1}{2}\gamma\mu_0\sqrt{(\varepsilon_r)}M_0/c$ anti-clockwise looking into the oncoming light.

10.6 The step-index fibre

This section indicates the derivation of the results given in §§9.2 and 9.3.2.

Recurrence relations

In the analysis repeated use is made of the following relations between Bessel functions:

$$\frac{2m}{u}J_m(u) = J_{m-1}(u) + J_{m+1}(u)$$

$$J'_m(u) = \tfrac{1}{2}[J_{m-1}(u) - J_{m+1}(u)]$$

$$= J_{m-1}(u) - \frac{m}{u}J_m(u)$$

$$= -J_{m+1}(u) + \frac{m}{u}J_m(u)$$

$$\frac{2m}{w}K_m(w)=K_{m+1}(w)-K_{m-1}(w)$$

$$K'_m(w)=-\tfrac{1}{2}[K_{m+1}(w)+K_{m-1}(w)]$$

$$=-K_{m-1}(w)-\frac{m}{w}K_m(w)$$

$$=-K_{m+1}(w)+\frac{m}{w}K_m(w)$$

The characteristic equation
Using the notation

$$2\Delta=(\varepsilon_1-\varepsilon_2)/\varepsilon_1\cong 2(n_1-n_2)/n_1$$

we have

$$\varepsilon_2=\varepsilon_1(1-2\Delta)$$

$$\beta^2=\omega^2\mu_0\varepsilon_1(1-2\Delta u^2/v^2)$$

Equation (9.17) can then be written in the form

$$W_m\left[W_m-2\Delta\frac{g_m(w)}{w}\right]=m^2\left(\frac{1}{u^2}+\frac{1}{v^2}\right)^2\left(1-2\Delta\frac{u^2}{v^2}\right) \tag{10.18}$$

where

$$W_m=\frac{1}{u}\frac{J'_m(u)}{J_m(u)}+\frac{1}{w}\frac{K'_m(w)}{K_m(w)}=\frac{f_m(u)}{u}+\frac{g_m(w)}{w} \tag{10.19}$$

The first approximation is given by putting $\Delta=0$, whence

$$W_m=\pm m\left(\frac{1}{u^2}+\frac{1}{v^2}\right)$$

The use of the recurrence relations then yields in the case of the upper sign (*EH* modes)

$$\frac{1}{u}\frac{J_{m+1}(u)}{J_m(u)}=-\frac{1}{w}\frac{K_{m+1}(w)}{K_m(w)}$$

whilst the lower sign (*HE* modes) gives

$$\frac{1}{u}\frac{J_{m-1}(u)}{J_m(u)}=\frac{1}{w}\frac{K_{m-1}(w)}{K_m(w)}$$

These equations are solved for u, w subject to

$$u^2+w^2=v^2$$

Second approximation

To obtain a better approximation we write

$$W_m = \pm m\left(\frac{1}{u^2}+\frac{1}{v^2}\right)+\delta_\pm \tag{10.20}$$

Substituting in (10.18) and neglecting second order terms we find

$$\delta_\pm = \Delta\left[\frac{g_m(w)}{w} \mp m\left(\frac{1}{u^2}+\frac{1}{w^2}\right)\frac{u^2}{v^2}\right]$$

$$= \Delta\left[\frac{g_m(w)}{w} \mp \frac{m}{w^2}\right]$$

Substituting in (10.20) and using (10.19) we find

$$\frac{f_m}{u} \mp \frac{m}{u^2} + \frac{n_2}{n_1}\left(\frac{g_m}{w} \mp \frac{m}{w^2}\right) = 0$$

whence for *EH* modes (upper sign)

$$\frac{n_1}{u}\frac{J_{m+1}(u)}{J_m(u)} = -\frac{n_2}{w}\frac{K_{m+1}(w)}{K_m(w)}$$

and for *HE* modes (lower sign)

$$\frac{n_1}{u}\frac{J_{m-1}(u)}{J_m(u)} = \frac{n_2}{w}\frac{K_{m-1}(w)}{K_m(w)}$$

Cut-off parameters

For the functions $K_m(w)$ as $w \to 0$

$$K_m(w) \cong \tfrac{1}{2}(m-1)!\left(\frac{2}{w}\right)^m \quad m > 1$$

$$K_0(w) \cong -\ln w$$

Thus

$$\frac{K_{m+1}(w)}{wK_m(w)} \to \infty \quad w \to 0 \quad m \geqq 0$$

Hence the cut-off for *EH* modes corresponds to

$$J_m(u) = 0 \quad u \neq 0$$

The value $u = 0$ has to be excluded since

$$\lim_{u \to 0} [J_{m+1}(u)/uJ_m(u)] = \tfrac{1}{2}$$

For HE modes, the case $m=0$ is as for EH modes since $J_{-1}=-J_1$, $K_{-1}=K_1$. For $m=1$, $K_0/wK_1\to\infty$, so that cut-off corresponds to zeros of $uJ_1(u)$ including $u=0$ since $J_0(0)=1$. For $m\geqq 2$, $K_{m-1}/wK_m\to 1/2(m-1)$, giving

$$\frac{n_1}{u}\frac{J_{m-1}(u)}{J_m(u)}=\frac{n_2}{2(m-1)}$$

Use of the recurrence relations reduces this equation to

$$J_{m-2}(u)=-\Delta J_m(u).$$

10.7 Vector formulae

The angle θ between two vectors $\boldsymbol{A}$, $\boldsymbol{B}$ is defined as that turned through in going from $\boldsymbol{A}$ to $\boldsymbol{B}$ in a right-handed corkscrew motion.

The scalar product $\boldsymbol{A}\cdot\boldsymbol{B}$ equals $AB\cos\theta$.

The vector product $\boldsymbol{A}\times\boldsymbol{B}$ is the direction of the right-handed corkscrew motion going from $\boldsymbol{A}$ to $\boldsymbol{B}$, and of magnitude $AB\sin\theta$.

The triple scalar product is

$$\boldsymbol{A}\cdot\boldsymbol{B}\times\boldsymbol{C}=\boldsymbol{B}\cdot\boldsymbol{C}\times\boldsymbol{A}=\boldsymbol{C}\cdot\boldsymbol{A}\times\boldsymbol{B}.$$

The symbols $\cdot$ and $\times$ may be interchanged. If the order of vectors is changed once the sign is changed.

Triple vector product

$$\boldsymbol{A}\times(\boldsymbol{B}\times\boldsymbol{C})=(\boldsymbol{A}\cdot\boldsymbol{C})\boldsymbol{B}-(\boldsymbol{A}\cdot\boldsymbol{B})\boldsymbol{C}$$

$$\begin{aligned}\boldsymbol{A}\times\boldsymbol{B}\cdot\boldsymbol{C}\times\boldsymbol{D}&=(\boldsymbol{A}\times\boldsymbol{B})\times\boldsymbol{C}\cdot\boldsymbol{D}\\&=[(\boldsymbol{A}\cdot\boldsymbol{C})\boldsymbol{B}-(\boldsymbol{B}\cdot\boldsymbol{C})\boldsymbol{A}]\cdot\boldsymbol{D}\\&=(\boldsymbol{A}\cdot\boldsymbol{C})(\boldsymbol{B}\cdot\boldsymbol{D})-(\boldsymbol{B}\cdot\boldsymbol{C})(\boldsymbol{A}\cdot\boldsymbol{D})\end{aligned}$$

Gauss theorem $\quad \oint\boldsymbol{F}\cdot d\boldsymbol{S}=\int \operatorname{div}\boldsymbol{F}\,dv$

Stoke's theorem $\quad \oint\boldsymbol{F}\cdot d\boldsymbol{l}=\int \operatorname{curl}\mathbf{F}\cdot d\boldsymbol{S}$

$$\operatorname{div}(\boldsymbol{E}\times\boldsymbol{H})=\boldsymbol{H}\cdot\operatorname{curl}\boldsymbol{E}-\boldsymbol{E}\cdot\operatorname{curl}\boldsymbol{H}$$

Cartesian forms: unit vectors $\boldsymbol{a}_x$, $\boldsymbol{a}_y$, $\boldsymbol{a}_z$

$$\operatorname{curl}\boldsymbol{F}=\left\{\frac{\partial F_z}{\partial y}-\frac{\partial F_y}{\partial z}\right\}\boldsymbol{a}_x+\left\{\frac{\partial F_x}{\partial z}-\frac{\partial F_z}{\partial x}\right\}\boldsymbol{a}_y+\left\{\frac{\partial F_y}{\partial x}-\frac{\partial F_x}{\partial y}\right\}\boldsymbol{a}_z$$

$$\operatorname{div}\boldsymbol{F}=\frac{\partial F_x}{\partial x}+\frac{\partial F_y}{\partial y}+\frac{\partial F_z}{\partial z}$$

$$\nabla V = \frac{\partial V}{\partial x}\boldsymbol{a}_x + \frac{\partial V}{\partial y}\boldsymbol{a}_y + \frac{\partial V}{\partial z}\boldsymbol{a}_z$$

$$\nabla^2 V = \frac{\partial^2 V}{\partial x^2} + \frac{\partial^2 V}{\partial y^2} + \frac{\partial^2 V}{\partial z^2}$$

Spherical polar forms (fig. 2.4)

$$\text{curl}\,\boldsymbol{F} = \frac{1}{r\sin\theta}\left[\frac{\partial}{\partial\theta}(\sin\theta F_\phi) - \frac{\partial F_\theta}{\partial\phi}\right]\boldsymbol{a}_r + \frac{1}{r}\left[\frac{1}{\sin\theta}\frac{\partial F_r}{\partial\phi} - \frac{\partial}{\partial r}(rF_\phi)\right]\boldsymbol{a}_\theta + \frac{1}{r}\left[\frac{\partial}{\partial r}(rF_\theta) - \frac{\partial F_r}{\partial\theta}\right]\boldsymbol{a}_\phi$$

$$\text{div}\,\boldsymbol{F} = \frac{1}{r^2}\frac{\partial}{\partial r}(r^2 F_r) + \frac{1}{r\sin\theta}\frac{\partial}{\partial\theta}[\sin\theta F_\theta] + \frac{1}{r\sin\theta}\frac{\partial F_\phi}{\partial\phi}$$

$$\nabla V = \frac{\partial V}{\partial r}\boldsymbol{a}_r + \frac{1}{r}\frac{\partial V}{\partial\theta}\boldsymbol{a}_\theta + \frac{1}{r\sin\theta}\frac{\partial V}{\partial\phi}\boldsymbol{a}_\phi$$

$$\nabla^2 V = \frac{1}{r^2}\frac{\partial}{\partial r}\left(r^2\frac{\partial V}{\partial r}\right) + \frac{1}{r^2\sin\theta}\frac{\partial}{\partial\theta}\left(\sin\theta\frac{\partial V}{\partial\theta}\right) + \frac{1}{r^2\sin^2\theta}\frac{\partial^2 V}{\partial\phi^2}$$

Cylindrical polar forms (fig. 3.2)

$$\text{curl}\,\boldsymbol{F} = \left(\frac{1}{\rho}\frac{\partial F_z}{\partial\phi} - \frac{\partial F_\phi}{\partial z}\right)\boldsymbol{a}_\rho + \left(\frac{\partial F_\rho}{\partial z} - \frac{\partial F_z}{\partial\rho}\right)\boldsymbol{a}_\phi + \frac{1}{\rho}\left[\frac{\partial}{\partial\rho}(\rho F_\phi) - \frac{\partial F_\rho}{\partial\phi}\right]\boldsymbol{a}_z$$

$$\text{div}\,\boldsymbol{F} = \frac{1}{\rho}\frac{\partial}{\partial\rho}(\rho F_\rho) + \frac{1}{\rho}\frac{\partial F_\phi}{\partial\phi} + \frac{\partial F_z}{\partial z}$$

$$\nabla V = \frac{\partial V}{\partial\rho}\boldsymbol{a}_\rho + \frac{1}{\rho}\frac{\partial V}{\partial\phi}\boldsymbol{a}_\phi + \frac{\partial V}{\partial z}\boldsymbol{a}_z$$

$$\nabla^2 V = \frac{1}{\rho}\frac{\partial}{\partial\rho}\left(\rho\frac{\partial V}{\partial\rho}\right) + \frac{1}{\rho^2}\frac{\partial^2 V}{\partial\phi^2} + \frac{\partial^2 V}{\partial z^2}$$

10.8 Physical constants

Boltzmann's constant	$k = 1.38 \times 10^{-23}\ \text{J K}^{-1}$
Planck's constant	$h = 6.63 \times 10^{-34}\ \text{J s}$
Velocity of light	$c = 3 \times 10^{8}\ \text{m s}^{-1}$
Primary magnetic constant	$\mu_0 = 4\pi \times 10^{-7}\ \text{H m}^{-1}$

Primary electric constant	$\varepsilon_0 = 8.854\times10^{-12}$ F m^{-1}
Wave impedance of free space	$\zeta_0 = \sqrt{(\mu_0/\varepsilon_0)} = 377\ \Omega$
At 290 K	$kT = 4\times10^{-21}$ WHz^{-1}

Conductors

	μ/μ_0	$\varepsilon/\varepsilon_0$	$\sigma(\mathrm{Sm}^{-1})$	R_s (mΩ)	l (μm)
Copper	1	1	5.9×10^7	$8.1\sqrt{f_G}$	$2.1f_G^{-\frac{1}{2}}$
Aluminium	1	1	3.7×10^7	$10\sqrt{f_G}$	$2.6f_G^{-\frac{1}{2}}$

f_G frequency in GHz.

Dielectrics

	$\varepsilon/\varepsilon_0$	$\tan\delta$
Polyethylene	2.3	2×10^{-4}
Alumina	8.5	2×10^{-3}
Mica	7	3×10^{-4}
Water	80	0.23

10.9 Hermite polynomials

These may be defined by the generating function

$$\exp(-s^2+2s\xi) = \sum_{n=0}^{\infty} H_n(\xi)\frac{s^n}{n!}$$

$$H_0(\xi)=1;\quad H_1(\xi)=2\xi;\quad H_2(\xi)=4\xi^2-2$$

Acknowledgments and further reading

The material of chapters 1, 2, 3 and 4 is very much standard electromagnetic theory, and can be amplified with the aid of one of the large number of more advanced books on the subject: for example, J. A. Stratton, *Electromagnetic theory* (McGraw-Hill, 1941), or S. Ramo, J. R. Whinnery & T. Van Duzer, *Fields and waves in communication electronics* (Wiley, 1965). Part of the treatment in chapter 4, particularly that on reciprocity, is based on E. D. Monteath's *Applications of the electromagnetic reciprocity principle* (Pergamon, 1973). Attention may also be drawn to: E. C. Jordan and K. G. Balmain, *Electromagnetic Waves and Radiating Systems*; R. H. Clarke and J. Brown *Diffraction Theory and Antennas*; R. F. Harrington, *Time-Harmonic Electromagnetic Fields.* The transmission-line theory of chapter 5 is paralleled by the treatment in many texts on transmission lines. Chapters 6, 7 and 8 are centred around particular practical systems, information on which is scattered throughout the literature. Material on signals and noise has been included in chapter 6 in order to place in perspective some of the effects arising from the propagation along lines. This material forms the subject of a very large literature. One book is to be published in this series. The information on coaxial lines is heavily dependent on papers published by British Post Office workers in the *Post Office Electrical Engineers Journal* (*POEEJ*), **66**, October 1973, pp. 131–205. In particular, figs. 6.26, 6.27, 6.28, 6.29 and 6.30 are based on the paper by L. H. Still, W. J. B. Stephens & R. C. H. Bundy, '60 MHz FDM transmission system', pp. 174–8. For further information on aspects of digital transmission, reference may be made to *Digital transmission systems* by D. G. W. Ingram & P. W. J. Bylanski (Peter Peregrenus, 1976).

With regard to chapter 7, reference must be made to *Principles of microwave circuits* (McGraw-Hill, 1948; new ed. Dover, 1965) by C. G. Montgomery, R. H. Dicke & E. M. Purcell, *Waveguide handbook* (McGraw-Hill, 1951) by N. Marcuvitz, and *Radar System engineering* (McGraw-Hill, 1947) by L. N. Ridenour. Figure 7.24 is taken from *Principles of microwave circuits.* Microwave measurements have been treated only briefly. Further reference may be made to H. M. Barlow

& A. L. Cullen's *Microwave measurements* (Constable, 1966). The original work on the long-haul waveguide is available in *Bell Systems Technical Journal* **33**, November 1954, pp. 1209–66: 'Waveguide as a communications medium', by S. E. Miller. For a compendium on microwave systems, see A. F. Harvey's *Microwave engineering* (Academic Press, 1963).

The description of the long-haul waveguide trials conducted by the British Post Office is again dependent on the *Post Office Electrical Engineering Journal*: 'The millimetric waveguide system: the design, production and installation of the waveguide', W. K. Ritchie & C. E. Rowlands, *POEEJ* **69**, 1976, pp. 79–86, and 'The millimetric waveguide system: terminal and repeater equipment', I. A. Ravenscroft, *POEEJ* **69**, 1977, pp. 250–7.

Chapter 8 is more of a summary of practical systems than either of chapters 6 or 7. This arises from the much greater mathematical problems arising in propagation in space rather than on a line. The curves of fig. 8.24 are after B. van der Pohl & M. Bremmer, *Phil. Mag. Series 7*, **25**, 1938, pp. 817–34. The description of microwave relay systems relies heavily on *POEEJ* **69**, 1976–7, pp. 225–34 and **70**, 1977, pp. 45–54; parts 1, 2 and 3 of 'A review of the British Post Office microwave relay network', by R. D. Martin-Royle, L. W. Dudley & R. J. Ferin. The book by Monteath contains much of interest relating to the theory of antennae from an applications standpoint.

Chapter 9 aims to give an introduction to a wide-ranging development of modern importance. Attention has been mainly restricted to aspects of optical propagation rather than to optical communication systems. Optical waveguide theory is developed in the following texts: D. Marcuse, *Light Transmission Optics* (Van Nostrand Reinhold, 1972); *Theory of Dielectric Waveguides* (Academic Press, 1974); *Principles of Optical Fibre Measurements* (Academic Press, 1981); A. W. Snyder & J. D. Love, *Optical Waveguide Theory* (Chapman and Hall, 1983). System aspects are developed in J. Gowar, *Optical Communication Systems* (Prentice-Hall, 1984).

Answers to problems

Chapter 1

1. (i) 2.07×10^8 m s^{-1}, 1.32×10^{-4} dB m^{-1}, (ii) 1.5×10^8 m s^{-1}, 5.5×10^{-2} dB m^{-1}. 3·86, 12.2, 38.6 dB m^{-1}.

2. $\boldsymbol{E}(z, t) = \mathrm{Re}\,[(E_1\boldsymbol{a}_x + \mathrm{j}E_1\boldsymbol{a}_y)\exp \mathrm{j}(\omega t - \beta_1 z)]$
$= \boldsymbol{a}_x E_1 \cos(\omega t - \beta_1 z) - \boldsymbol{a}_y E_1 \sin(\omega t - \beta_1 z)$

(a) $\boldsymbol{H} = E_1\sqrt{(\varepsilon/\mu)}\,(-\mathrm{j}\boldsymbol{a}_x + \boldsymbol{a}_y)\exp[\mathrm{j}(\omega t - \beta_1 z)]$.

(b) $(E_1\boldsymbol{a}_x - \mathrm{j}E_1\boldsymbol{a}_y)\exp[\mathrm{j}(\omega t - \beta_1 z)]$

If field is $E\boldsymbol{a}_x$ at $z=0$, then at any other point $\boldsymbol{E} = E[\boldsymbol{a}_x \cos(\beta_1-\beta_2)z/2 + \boldsymbol{a}_y \sin(\beta_1-\beta_2)z/2]\exp[-\mathrm{j}(\beta_1+\beta_2)z/2]$.

3. Use (1.101) and (1.102) in conjunction with (1.98), taking $\mu_1 = \mu_2 = \mu_0$. The former can only vanish if $\theta_i = \phi$, which is contrary to supposition. The latter requires $\theta_i = \frac{1}{2}\pi - \phi$, leading to result from which θ_i can always be found.

4. Using the result of the worked example; aluminium 2.2×10^7; silver 1.1×10^8; steel 1.4×10^6 S m^{-1}.

5. 16.9×10^5 nepers m^{-1}, 16.9×10^5 rad m^{-1}, 8.9×10^4 m s^{-1}, 3.7 μm. Skin depth = 5.9×10^{-7} m, relaxation time = 2.9×10^{-19} s.

Chapter 2

1. Power = $\frac{1}{2}V_0 I_0$.

2. Evaluate $\frac{1}{2}\,\mathrm{Re}\,(\mathrm{E}_x\mathrm{H}_y^*)$.

3. Application of (1.63) and (1.64) show that at the surface $\mathrm{E}_x/H_y = \sqrt{(\mathrm{j}\omega\mu/\sigma)}$. The power is then given by $\frac{1}{2}\,\mathrm{Re}\,(\mathrm{E}_x\mathrm{H}_y^*)$ W m^{-2}.

4. The analysis is in § 1.4.2. The current sheet is $2\mathrm{E}_i/\zeta$ A m^{-1}. Ignoring the field produced by this sheet, we find the force between the incident wave, $\mathrm{H} = \mathrm{E}_i/\zeta$ and the current. We find the pressure is $\varepsilon_0\mathrm{E}_i^2$. In terms of incident power density, $P_i = \mathrm{E}_i^2/2\zeta$, pressure is $2P_i/c$. Check against § 2.3.1.

5. The pressure for a number of plane waves perfectly reflected will be the sum of the individual pressures $(4/3)\times10^{-5}$ N m^{-2}.

Chapter 3

1. 10.6 mm diameter; 26.8 kV m^{-1}.

2. 1.9×10^8 m s^{-1}, 1.2 Ω.

3. Cavity implies short-circuit at both ends, and is therefore $\frac{1}{2}\lambda$ long. $f = 1.5$ GHz. With dielectric $f = 0.61$ GHz. Use method of §§ 3.10.3 and 3.11.

4. Bookwork leading to TE_{m0} modes, for which $f_{m0} = mc/2a$. Dominant mode $f_{10} = c/2a$. With $a = 2$ cm, $f_{10} = 7.5$ GHz. Taking $a = 1$ cm, $f_{10} = 15$ GHz. Maximum E is $|\omega\mu Aa/\pi|$ Vm^{-1}. Use § 3.7 to calculate $|A|$ from P, $|A| =$ 51.5 Am^{-1}. $E_m = 31$ KVm^{-1}.

5. (i) $a = 2b = 3.75$ cm. (ii) Use table 3.4, diameter 4.4 cm. Taper circular into rectangular, observing similarity of TE_{10} rectangular to TE_{11} circular.

6. At resonance, length is $\lambda_g/2$. Hence $\lambda_g = 5$ cm. From $2\pi/\lambda_g = \beta =$ $(2\pi/c)(f^2 - f_{01}^2)^{\frac{1}{2}}$ and $f_{01} = 15$ GHz, we find $f = 16.2$ GHz. Mode pattern is that of dominant mode turned through a right angle.

7. Table 3.4 shows TE_{11} and TM_{01} are propagated. From (3.71) and fig. 3.12, TM_{01} has zero electric field at $\rho = 0$. Using fig. 3.12 to show $J_1'(0) =$ $\lim_{\rho\to 0} J_1(\kappa\rho)/\kappa\rho = \frac{1}{2}$, (3.68) shows that **H** on the axis is in direction $\phi = 0$ and of value $jA\beta/(2\kappa)$, and hence on axis $|E| = |0.5\,\omega\mu A/\kappa| = 0.5|A|(f/f_{11})\sqrt{(\mu/\varepsilon)}$. To find $|A|$ use (3.72) with $A_{11} = 55.3 \times 10^{-3}$ from table 3.5. For $P = 2$ MW, $|A| =$ 3.74×10^3 A m^{-1} and E = 0.96 MV m^{-1}.

8. Use $\alpha = -(1/2P)(dP/dz)$. 16 dB km^{-1}.

Chapter 4

1. Electrically the dipole is short, length $\lambda/20$. $R_a = 2.0\ \Omega$. $E_m = 3.0$ mV m^{-1}.

2. Still a short dipole and contribution from individual elements add in phase. $R_a = 0.5\ \Omega$. $E_m = 3.0$ mV m^{-1}.

3. Using $R_a = 73\ \Omega$, I = 0.17 A and E = 9.9 mV m^{-1}.

4. Proceed from (4.18) using proper limits and $I(\zeta) = I_0\, e^{-jk\zeta}$

$$E = (\zeta_0/2\pi r) \cot(\theta/2) \sin[kl \sin^2(\theta/2)]$$

5. Analysis of § 4.4.

6. 10 mV; 0.17 μW.

7. Use geometry of fig. 4.14. Contribution to the distant electric field from an element $d\xi\, d\eta$ is

$$(j\omega\mu_0 k/4\pi r)\, d\xi\, d\eta\, \exp[-jkr_1 + jk(\xi \sin\alpha + \eta \sin\beta)] \cos\beta$$

(Note θ in question is $\frac{1}{2}\pi - \beta$ of fig. 4.14.) Radiation pattern is

$$\frac{\sin(ka \sin\alpha)}{ka \sin\alpha} \cdot \frac{\sin(kb \sin\beta)}{kb \sin\beta} \cos\beta$$

Chapter 5

1. (i) 500 MHz. (ii) $Z_T = 160\ \Omega$. (iii) $\frac{8}{9}$.
If the minimum occurs 0.3 m from the end, $Z_T = 40\ \Omega$.
2. Load impedance = $100 + j116\ \Omega$.
First voltage maximum 25 cm from the load.
VSWR = 3.
Matching stub 75 cm from load, 33.9 cm long.

3. $42.16\ \Omega$.
4. $\alpha = 0.087\ \text{dB m}^{-1}$.
5. Stub located 0.247 m from load, 1.353 m long.

For a VSWR of 1.4, the total admittance at the stub can be anywhere on a circle through $S = 1.4$. For any point on this circle, if we subtract a susceptance equal to that of the short-circuited stub, we can produce a new locus on the Smith chart. The maximum and minimum points on this locus, with respect to the origin, indicate the range of normalised resistance allowable at the load.

$R_{L\max} = 103.4\ \Omega$, $R_{L\min} = 54.5\ \Omega$.

6. First pulse = 3.95 kV, second pulse = 2.39 kV.

Chapter 6

1. Signal-to-noise power ratio = 38 dB.
2. See 6.4.3 and 9.4.

Maximum spacing = 1.258 miles (2.01 km).

Using (6.9) and (6.10) with $G = 0$, unloaded attenuation = $0.58\ \text{dB mile}^{-1}$, after-loading attenuation = $0.14\ \text{dB mile}^{-1}$.

3. Maximum number of channels = 25.
4. Spacing less than 5.49 km.
5. Operating level = $22 - 25 = -3$ dB m. The noise level must be below -58 dB m. Gain per stage = $1000/N$. Substitution gives $N = 16$.
6. A plot of input impedance against frequency gives a mean frequency between maxima of 16.25 kHz. If the fault occurs distance l from the sending end, the phase shift at frequency f_1 for a go and return path distance $2l$ will be $2\beta_1 l + \phi_1$, where $\beta_1 = 2\pi f_1/c$ and ϕ_1 is the phase shift associated with the fault. Thus maxima will occur when $(2\beta_2 l + \phi_2) - (2\beta_1 l + \phi_1) = 2\pi$. Assuming $\phi_1 \simeq \phi_2$, this gives $l = c/2(f_2 - f_1) = 9.23$ km.

Since the input impedance tends to zero as the frequency decreases, we can presume a short-circuit at the fault.

7. Total resistance $R = 2\alpha Z_0 = 1.5\ \Omega\ \text{m}^{-1}$. Since $R_{\text{inner}} = R_s/2\pi a = 0.92\ \Omega\ \text{m}^{-1}$, $R_{\text{outer}} = R - R_{\text{inner}} = 0.58\ \Omega\ \text{m}^{-1}$.

Assuming a solid conductor, $R_{\text{outer}} = R_s/2\pi b = 0.28\ \Omega\ \text{m}^{-1}$.

Chapter 7

1. 462.4 kW.
2. For $a = 25$ mm, $f_{10} = 6$ GHz; $a = 12$ mm, $f_{10} = 12.5$ GHz. In both cases there is zero power flow: in the narrow guide because power cannot be transmitted in an evanescent mode, and in the larger because of complete reflection with $|k| = 1$. Attenuation of evanescent mode = $157\ \text{nepers m}^{-1}$.
3. 152.7 W.
5. $y_{11} = 59.8\angle -62.3°\ \text{mS}$, $y_{12} = -1.23\angle 12.5°\ \text{mS}$
$y_{21} = -106\angle -17.5°\ \text{mS}$, $y_{22} = 34.8\angle 10.4\ \text{mS}$

Chapter 8

1. Power gain 38.5 dB. Maximum received power 0.28 W.
2. 17.9 mV m^{-1}. Assumptions: (*a*) plane earth; (*b*) phase reversal at reflection; (*c*) equal amplitude direct and reflected waves.
3. (i) 195.2 K. (ii) 228.4 K.
4. Wave intensity 2.23×10^{-7} W m^{-2}. Electric field strength at the receiver 13 mV m^{-1}. Maximum power available to the receiver 2.23×10^{-10} W. Overall transmission loss 116.6 dB.
5. 48.77 dB.
6. Passive satellite 46 km. Active satellite 4773 km.
7. Power received $= 2.18\times10^{-12}$ W.

Chapter 9

1 4.8 μm.
2. 2.16, 3.23, 4.64, 3.95; 1.022, 1.53, 1.62, 1.86 μm.
3. 7×10^{-6}.
4. Put $\xi = x\sqrt{2}/w$, $\eta = y\sqrt{2}/w$, when $\partial^2\psi/\partial x^2 = (2/w^2)\partial^2\psi/\partial\xi^2$, etc.
6. $d\beta/d\omega = n_0/c$. Material dispersion has not been included.
7. $d(u^2v^{-1})/du \cong 0.71, -0.11$.
8. 5.7°
12. 4, 18%.
13. 0.08, 0.2 μs.
14. 0.414, 0.0063, 209 ns km^{-1}.
15. Make phase front of beam conform to mirror curvature. 0.31, 0.42 mm.
16. Waist 1.03 m from 10 m curvature mirror, 3.2 mm.
17. Waist 10.1 mm from lens, $w_0 = 7.8$ μm; 0.45 mm.
18. $\begin{bmatrix} -1 & 0 \\ 2/R & 1 \end{bmatrix}$; $\begin{bmatrix} 1 & d \\ 0 & 1 \end{bmatrix}$; $\begin{bmatrix} 1-d/f_1 & d \\ f_1^{-1}+f_2^{-1}-d/f_1f_2 & 1-d/f_2 \end{bmatrix}$

Index